Sustainable Hybrid Energy Systems

Sustainable Hybrid Energy Systems

Carbon Neutral Approaches, Modeling, and Case Studies

Jiuping Xu and Fengjuan Wang

WILEY-VCH

Authors

Prof. Jiuping Xu
Sichuan University
Institute of New Energy and
Low-Carbon Technology
No. 24 South Section 1
Yihuan Road
Chengdu, 610064
China

Dr. Fengjuan Wang
Sichuan University
Institute of New Energy and
Low-Carbon Technology
No. 24 South Section 1
Yihuan Road
Chengdu, 610064
China

Cover image: © The img/Shutterstock

Library of Congress Card No.: applied for

British Library Cataloguing-in-Publication Data
A catalogue record for this book is available from the British Library.

Bibliographic information published by the Deutsche Nationalbibliothek The Deutsche Nationalbibliothek lists this publication in the Deutsche Nationalbibliografie; detailed bibliographic data are available on the Internet at <http://dnb.d-nb.de>.

Print ISBN: 978-3-527-35243-2
ePDF ISBN: 978-3-527-84325-1
ePub ISBN: 978-3-527-84326-8
oBook ISBN: 978-3-527-84327-5

Typesetting Straive, Chennai, India
Printing and Binding CPI Group (UK) Ltd, Croydon, CR0 4YY

C9783527352432_230124

Contents

List of Figures

List of Tables

Preface

Humans cannot rely only on a single energy source but on a mixture of multiple energy sources. Hybridization is the eternal theme of human energy utilization. For example, in the agricultural era, humans used natural power such as animal power, wind power, and hydro power; in the industrial age, firewood, coal, and electric energy were used at the same time; in the information age, oil, natural gas, and coal were the main energy sources; now and in the future, nuclear energy, wind energy, and solar energy and other low-carbon energy sources will gradually become mainstream. High-quality hybridization of different energy resources is closely related to energy security and stability, and requires systematic research. Since 2020, the carbon-neutral mission has put forward new requirements for energy hybridization, and how to build a sustainable hybrid energy system under this goal has become a key issue of global concern.

Fully considering the global mission of achieving carbon neutrality and the global passion for promoting energy transition, this book discusses the comprehensive approaches to building sustainable hybrid energy systems. The hybrid energy system considered in this book is multi-input (wind, solar, coal, etc.) and single-output (electricity). We clearly distinguish it into three kinds, the co-located hybrid energy system, the co-combusted hybrid energy system, and the co-operated hybrid energy system. We also explore four groups of comprehensive approaches to exploring sustainable hybrids, including deployment optimization, emission quota allocation, scheduling coordination, and optimal policy implementation. This book contains six parts and including 14 chapters.

The first part introduces background in Chapter 1. The global mission of achieving carbon neutrality, the global passion for promoting energy transition, and the global status of developing hybrid energy systems are first described; then the general definition of hybrid energy systems is presented and the detailed characteristics of co-located hybrid energy systems, co-combusted hybrid energy systems, and co-operated hybrid energy systems are identified separately; advantages of energy system hybridization are also introduced. A logical relationship diagram of all the chapters of the book is also presented.

The second part presents the deployment optimization of co-operated hybrid energy systems, including Chapters 2–4. Co-operated hybrid energy systems deployed next to customers are significantly important nowadays as they can avoid long-distance transmission loss and provide flexible choices. In these three chapters, the optimal deployment problem of distributed level co-operated hybrid energy systems is discussed from three aspects, the industrial decarbonization-oriented deployment, the sustainable operation-oriented deployment, and the disaster resilience-oriented deployment. The general mathematical optimization models are proposed, and three case studies are conducted to demonstrate the practicability and effectiveness of the proposed models.

The third part provides emission quota allocation of co-combusted hybrid energy systems, as seen in Chapters 5–7. Due to the fossil fuel crisis, generation fuel hybridization has played and will continue to play an important role in generating electricity from thermal power plants. In these three chapters, the optimal emission quota allocation problem of co-combusted hybrid energy systems is discussed, and three kinds of different generation sources co-combustion scenarios, including coal-biomass co-combustion, coal-municipal solid waste co-combustion, and coal-sewage co-combustion, are explored. These studies are conducted from concentrated power plant level, three bi-level emission quota allocation optimization models are proposed, and related case studies are conducted.

The fourth part grants the scheduling coordination of co-operated hybrid energy systems, as seen in Chapters 8–10. Due to the inherent uncertainties and the unstorable features, the coordinated scheduling problem of conventional and new energy sources is very complex. In these three chapters, short-term scheduling problems of hybrid solar–hydro, solar–hydro–wind, and solar–hydro–gas systems under reliable, economical, and social equilibrium requirements are explored. These studies are conducted from concentrated multi-power plants level, the problems are described in detail and the corresponding multi-objective optimization model are proposed and solved, and related case studies are also conducted.

The fifth part covers policy mechanism toward hybrid energy systems, as seen in Chapters 11–13. Policy mechanism is one of the most efficient tools to integrate new energy into traditional power systems, thus forming the hybrid power systems. Nowadays, the renewable portfolio standards policy has been widely applied. In these three chapters, three multi-objective renewable portfolio standards implementation strategies, the strategy considering both the power suppliers and users, considering the equity and economy equilibrium, and considering the carbon trade and green certificate trade are separately studied. These studies are conducted from the regional grid level, the corresponding mathematical optimization model are proposed, and related case studies are conducted.

The last part gives emerging hotspot of developing hybrid energy storage systems in Chapter 14. The urgent need for energy storage in future energy systems characterized by a high percentage of renewable energy sources is presented; existing electrical, electrochemical, chemical, mechanical, and thermal energy storage technologies and their properties in terms of power and energy are described; and the advantages and future challenges of hybrid energy storage systems are presented.

Many people contributed directly or indirectly to this book; we wish to give special thanks to the Wiley publishing team (Lifen Yang, Monica Chandra Sekar, Merlyn Hema Daniel, and Shiji Sreejish) for their professional work. The authors would like to thank the anonymous referees for their insightful comments and suggestions to improve this book, and the authors listed in the references for their excellent work. The authors would also like to thank the Major Programs of the National Social Science Foundation of China (22&ZD142), the Youth Programs of the National Natural Science Foundation of China (No. 72301185), and the Postdoctoral Science Foundation of China (2023M732420) for providing us foundation support. We would also like to thank the members of our research group for their efforts in studying sustainable hybrid energy system–related research, which provide important reference and solid support for this book, including Chengwei Lv, Qian Huang, Chuangdang Zhao, Mengyuan Zhu, Liying Liu, Yalou Tian, Hongyan Tao, and Guocan Yang.

15th July 2023
Chengyi Building

Jiuping Xu and Fengjuan Wang

1 Introduction

To undertake rapid reductions thereafter in accordance with best available science, so as to achieve a balance between anthropogenic emissions by sources and removals by sinks of greenhouse gases in the second half of this century.

Article 4 of the Paris Agreement Apr. 2016

As mankind cannot rely only on a single energy source but on a mixture of multiple energy sources, hybridization is the eternal theme of human energy utilization, and the restructuring of the energy system is related to global security and stability, which requires systematic research. This chapter introduces the background of constructing hybrid energy systems (HESs), provides a detailed definition and classification of HESs, describes the motivation for hybridization of energy systems, and gives a logical relationship diagram of the whole book chapters.

1.1 Background

This section introduces the global mission of achieving carbon neutrality, the global passion of promoting energy transition, and the global status of developing HESs.

1.1.1 Global Mission of Achieving Carbon Neutrality

In the past few decades, global population has continued to grow, social economy has developed rapidly, living standards have improved significantly, electricity supply and demand have increased dramatically, and a large amount of fossil energy has been continuously exploited and utilized (Chen et al. 2016a). According to the *BP Statistical Review of World Energy 2021*, global power generation reached 2.68 billion kilowatt hour in 2020, of which coal, natural gas, and oil power generation accounted for 35.12%, 23.37%, and 2.82%, respectively. However, if these fossil fuels are still mined at 2020 levels, the world's remaining reserves of coal, natural gas, and oil will only last 139, 48.8, and 50 years (BP Global 2021). Not only that, the excessive

Sustainable Hybrid Energy Systems: Carbon Neutral Approaches, Modeling, and Case Studies, First Edition. Jiuping Xu and Fengjuan Wang.

burning of fossil fuels has resulted in massive emissions of sulfur dioxide, carbon dioxide, nitrogen oxides, and other gases, causing and exacerbating environmental problems such as global acid rain, greenhouse effect, and photochemical pollution (Liu et al. 2017). Statistics released by the International Energy Agency (IEA) in 2021 show that in 2019, the world's carbon emissions from fuel combustion reached 34.234 billion tonnes, of which 14.068 billion tonnes were directly attributable to heat and power production (International Energy Agency 2021a). Therefore, climate change caused by carbon dioxide and other greenhouse gas emissions has become the biggest non-traditional security risk in the twenty-first century, and reducing greenhouse gas emissions from energy use has become a common global issue.

In recent decades, the international community has made a series of efforts to address the climate crisis. In 1992, more than 150 countries and the European Economic Community signed the United Nations Framework Convention on Climate Change (UNFCC) in Rio de Janeiro, Brazil, and agreed to control the concentration of greenhouse gases in the atmosphere at a stable level. In 1997, the third Conference of the Parties to the Convention (COP3) was held in Kyoto, Japan, and formulated the first document in human history, the Kyoto Protocol, to limit greenhouse gas emissions in the form of regulations. The protocol sets greenhouse gas emission reduction targets for developed countries and countries with economies in transition but does not impose greenhouse gas emission reduction obligations for developing countries (The Ministry of Foreign Affairs of the People's Republic of China 2008). In 2009, at the 15th Meeting of the Parties to the Convention, relevant authorities from 193 countries and regions signed the Copenhagen Accord as a follow-up plan after the expiration of the Kyoto Protocol. The Copenhagen Accord, although not legally binding, lays the foundation for the first truly global agreement to limit and reduce greenhouse gas emissions in the future, the Paris Agreement (United Nations 2009). In 2015, the Paris Agreement, the first global emission reduction agreement covering nearly 200 countries and regions in human history, was finally reached, becoming the second legally binding agreement after the Kyoto Protocol. The Paris Agreement made a unified arrangement for the global response to climate change after 2020, and put forward the goal of controlling the global temperature rise within the range of 2 °C and working toward 1.5 °C.

According to the 2018 report of the United Nations Intergovernmental Panel on Climate Change (IPCC), to achieve the goal of no more than 2 °C, the world needs to reach carbon neutrality around 2070. To achieve the 1.5 °C target, carbon neutrality needs to be brought forward to around 2050. Therefore, since 2018, many parties have made carbon-neutral commitments. According to data from the World Resources Institute (WRI), at least 83 countries around the world have made carbon-neutrality commitments through legislation, promulgation of policies, and submission of statements of intent to the United Nations Framework Convention on Climate Change (Climate Watch 2022). China is the world's largest energy producer, the world's largest energy consumer, and the world's largest carbon emitter (International Energy Agency 2021a). At the federation conference in September 2020, China also made a solemn commitment to achieve carbon peaking by 2030 and carbon neutrality by 2060. According to data from the WRI, the common transition period in developed countries is 40–60 years from carbon peaking to carbon neutrality. For example, the United Kingdom and France achieved carbon peaks in 1991 and committed to carbon neutrality by 2050, with a transition period of 59 years. The U.S. reached its carbon peak in 2007 and pledged to achieve carbon neutrality by 2050, with a 43-year transition period. Japan achieved carbon peak in 2012, pledge to achieve carbon neutrality by 2050, with a transition period of 38 years.

The global mission of achieving carbon neutrality is pressuring energy transition from fossil fuel-dominated energy systems toward renewable energy-dominated energy systems; details are described in the following part.

1.1.2 Global Passion for Promoting Energy Transition

Due to the rapid increase in consumption of fossil energy and the increasingly severe impact of climate change, renewable energy has ushered in an unprecedented development opportunity; the third round of the energy revolution characterized by large-scale utilization of renewable energy is booming around the world. According to the IEA, the main sources of global carbon dioxide are the electricity and heat sector (power generation, 44%), the transportation sector (land, shipping, air transport, 26%), the industrial sector (metal smelting and chemical manufacturing, 20%), and the construction sector (building construction and home life, 9%) (International Energy Agency 2021b). Since electricity and heat production is the industry with the highest carbon emission in the world, if power supply cannot be very clean, the emission reduction target will be difficult to achieve (Energy Foundation 2021). Converting the dependence of electricity production on traditional fossil energy to clean energy and ensuring a diversified, stable, efficient, and clean power supply are becoming increasingly crucial to achieving the net-zero emission goal and promoting the transformation of the power structure.

Table 1.1 shows the power generation structure of several countries in 2020. Overall, the proportion of fossil fuel power generation in developing countries like China and India is very high, 60.75% and 70.56% of which comes from less clean coal, far exceeding the world's average coal power generation ratio of 33.79%. Although in terms of proportion, the proportion of fossil energy power generation in Japan is also close to 70%, more than 30% is from gas power with a higher degree of cleanliness. The same situation applies to the United States, although the proportion of fossil energy power generation in the United States also exceeds 60%, but more than 40% is from gas power. From the perspective of renewable energy power, European Union (EU) has a large number of wind and hydropower stations, and the proportion of renewable energy power generation (38.16%) is the highest around the world.

Based on the existing power supply structure, many organizations have forecast global power supply structure in 2050, as shown in Table 1.2. Among them, the World Energy Scenarios (2013) publication *World Energy Scenario: 2050 Energy Future* was released before the Paris Agreement was signed in 2013, the International Renewable Energy Agency publication *Global Energy Transition: 2050 Roadmap* was released after the agreement was signed in 2018, and the International Energy Agency's (2020) publication *Net Zero Emissions to 2050: A Roadmap*

Table 1.1 The power generation structure of different countries in 2020.

	Non-renewable power				Renewable power				
Country	Coal (%)	Natural gas (%)	Oil (%)	Nuclear (%)	Hydro (%)	Wind (%)	Solar (%)	Others (%)	Total (%)
World	33.79	22.8	4.36	10.12	16.85	6.15	3.27	2.71	28.98
EU 27	13.16	19.94	3.97	24.77	12.6	14.34	5.24	5.98	38.16
Japan	29.09	31.28	8.77	4.57	9.02	1.13	8.97	7.17	26.29
USA	19.11	40.23	0.71	19.5	7.06	8.31	3.27	1.8	20.44
India	70.56	3.88	0.02	3.32	12.19	4.5	4.38	1.05	22.22
China	60.75	3.32	2.1	4.8	17.78	6.12	3.42	1.7	29.02

Table 1.2 Proportion of global power generation by various power generation methods in 2050 under different scenarios.

	Non-renewable power			Renewable power				
Scenarios	Coal (%)	Natural gas (%)	Nuclear (%)	Hydro (%)	Biomass (%)	Wind (%)	Solar (%)	Geothermal (%)
WEC-Symphony	17.7	19.9	14.5	16.1	5.7	8.4	16.2	1.4
IREA-REmap	1	10	4	12	4	36	26	3
IEA-Netzero	0.9	1.3	7.7	11.9	4.6	34.8	34.9	1.2

for the Global Energy Sector comes at a time when parties to the agreement have made carbon-neutral commitments in 2021. The three documents are staggered in time and can be used to compare changes due to international policy. As can be seen from Table 1.2, in the World Energy Council's Energy Sustainability Forecast Scenario, the power generation ratio of non-renewable energy power is 52% (coal power 17.7%, gas power 19.9%, nuclear power 14.5%), the proportion of renewable energy power generation is 48%, of which photovoltaic (PV) and hydropower have the highest proportions, both exceeding 16.2%. The International Renewable Energy Agency's Renewable Energy Pathway Scenario (IREA-REmap) is based on the premise that the global temperature rise is kept below 2 °C. The proportion of wind power and PV power generation will rise to 36% and 26%, respectively. The IEA's net-zero emissions scenario (IEA-Netzero) is based on the premise of a temperature control of 1.5 °C. Under this premise, net-zero emissions need to be achieved in 2050, and the proportion of renewable energy in the power structure exceeds 87.4%. As can be seen from the data, renewable energy will become the main force of power supply in 2050.

Table 1.3 shows the proportion of power generation by different power generation methods in the world in 2020 and 2050. It can be seen that from 2020 to 2050, the global power structure will require major adjustments. The most obvious change is that the proportion of coal-fired power generation will be significantly reduced,

Table 1.3 Proportion of different types of power generation in the world in 2020 and 2050.

		World	
Category	**Generation**	**2020 (%)**	**2050 (%)**
Non-renewable	Coal	33.79	0.9–17.7
	Natural gas	22.8	1.3–19.9
	Nuclear	10.12	4.0–14.5
Renewable	Hydro	16.85	11.9–16.1
	Wind	6.15	8.4–36.0
	Solar	3.27	16.2–34.9

from 33.79% to 0.9–17.7% globally. The proportion of gas-electric power generation in the world will decrease from 22.8% to 1.3–17.7%. On the whole, the proportion of hydropower generation will be slightly reduced. This is because the development of hydropower is relatively mature and will be close to saturation. The increased power demand in the future will be mainly provided by new energy sources such as solar and wind power. It can be seen from the table that the proportion of wind power and PV power generation in the world will increase to 35–36%. In general, in order to achieve the goal of net zero emissions as much as possible, the electricity supply will show a trend of high cleanliness by 2050. The proportion of coal power supply would have dropped sharply, and clean energy such as solar, wind power, and hydropower will take over 60–90% of the power supply tasks in the world.

It can be seen that converting the dependence of electricity production on traditional fossil energy to clean energy has become an irreversible trend. As an effective solution to mitigate this issue, establishing HESs becomes the most feasible option and has drawn wide attention around the world, as shown in the following part.

1.1.3 Global Status of Developing Hybrid Energy Systems

Here we use the Web of Science (WOS) database to search HES-related researches to find out the global development status.

The search codes in WOS are "hybrid energy system*" (Topic) or "hybrid renewable energy system" (Topic) or "multi-energy system*," and the article type includes "Article" or "Review Article" or "Proceeding Paper" or "Early Access," and totally 2585 papers written in English were extracted from Science Citation Index Expanded (SCI-Expanded) and Social Sciences Citation Index (SSCI) databases. The authors conducted the search on 30 June 2023.

Ranking the retrieved literature by their publication year, the results can be seen in Figure 1.1. It can be seen that the first literature appeared in 1984, and since then a small amount of literature has focused on HESs every year, but until 2008, the number of papers per year was still less than 10; since 2008, the number of papers

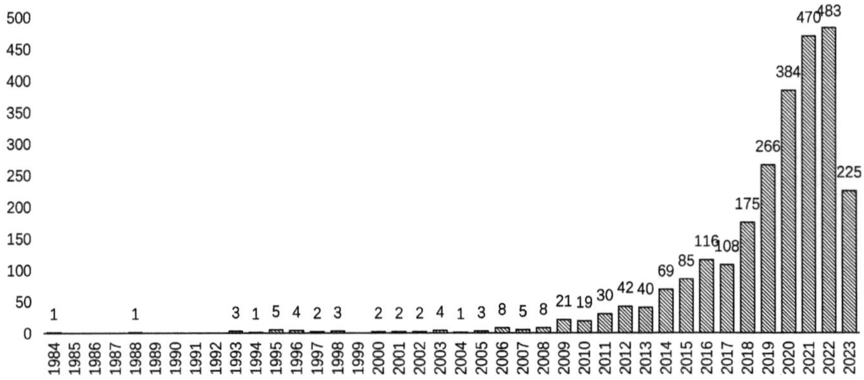

Figure 1.1 Research papers on hybrid energy systems (in order of year).

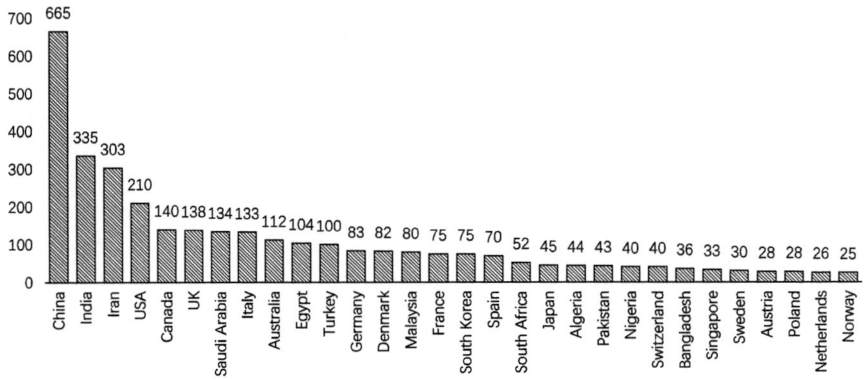

Figure 1.2 Research papers on hybrid energy systems (in order of country).

has increased, but still less than 100 papers per year; since 2016, the number of papers has increased rapidly, and in 2022 the number of papers in just one year reached 483.

Ranking the retrieved literature by their publication country, the results can be seen in Figure 1.2. As can be seen from the figure, developing countries are the main force in HESs research, with China, India, and Iran authors accounting for the top three places among all countries, and China leading the way with 665 papers accounting for 26% of the total literature. Developed countries USA, Canada, and UK also published 130–210 relevant papers, respectively, ranking 4th to 6th.

We exported the retrieved literature and imported it into the literature analysis software Citespace to visualize research hotspots. In Citespace, we set the Time Slicing from 1993 to 2023 and one year per slice, Set the Nodetype as Reference, and chose the Selection Criteria as g-index (g = 25); the pruning method is pathfinder, the sliced networks and the merged network are pruned at the same time to simplify the keyword network and highlight the structures of these researches. Then the knowledge map of references' clusters is obtained, as shown in Figure 1.3. It can be seen that current research mainly focuses on four categories.

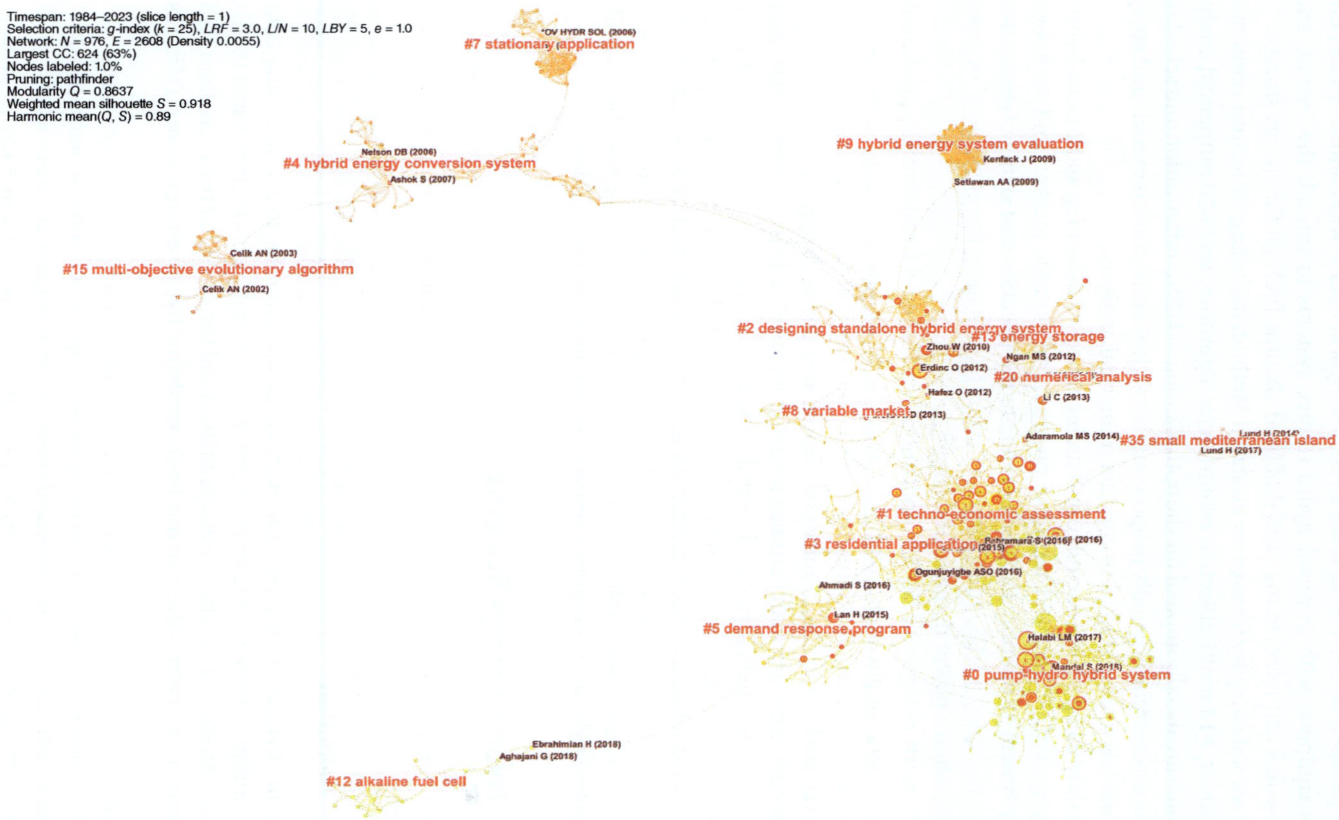

Figure 1.3 Cluster of references.

The first category focuses on Research Objects, including #0 pumped-hydro hybrid system, #4 hybrid energy conversion system, and #12 alkaline fuel cell, where pumped-hydro hybrid system refers to hybrid hydro–wind and/or PV system that is equipped with pumped hydro storage systems to smooth the output, and alkaline fuel cell refers to the HES that used alkaline fuel cell for energy storage.

The second category focuses on Research Method, including #1 techno-economic assessment, #15 multi-objective evolutionary algorithm, and #20 numerical analysis, which are usually applied for the optimization of HES configuration and scheduling; these researchers usually propose mathematical optimization models and apply these models to case studies to conduct numerical analysis.

The third category is Research Content, including #2 designing standalone hybrid energy system, #5 demand response program, #8 variable market, and #9 hybrid energy system evaluation. Cluster #2 refers to the selection and sizing of generation sources to deploy a HES, and Cluster #5 refers to managing the customers' demand so as to balance power demand and supply.

The fourth category is HES Applications, including #3 residential application, #7 stationary application, and #35 small Mediterranean island; most of the above-mentioned applications are focused on distributed power plant level, like industrial parks, and community and remote area power supply.

Although there are many studies focusing on specific hybrid systems, few studies have defined and categorized hybrid system's at a macro level. To the best of the authors' knowledge, only Arent et al. (2021) and U.S. Department of Energy (2021) have defined HESs at a macro level, but the former is a perspective paper and the latter focuses on research opportunities. To fill this gap, this book clearly distinguished it into three kinds and explored four groups of comprehensive approaches to build sustainable hybrids through mathematical modeling and case studies.

1.2 Hybrid Energy Systems

This section presents the definition, classification, and motivation for establishing HESs.

1.2.1 Definition

The term "hybrid" is employed in many disciplines to refer to a system that comprises multiple distinct constituent parts that are combined and integrated to take advantage of each one's unique characteristics and the synergies between them (U.S. Department of Energy 2021). In this book, we define hybrid energy system (HES) as:

> *Multiple energy generation technologies, generation fuels, and storage technologies that are integrated-through co-location, co-operation, and co-combustion-to achieve cost savings, efficiency, and environmental performances etc. when producing energy products (electricity, heat, liquid fuels etc.) and non-energy products (freshwater, chemicals, materials etc.).*

The generation technologies that can be integrated to form a hybrid system are very wide, including fossil fuel generation technology (coal, natural gas, diesel), renewable generation technology (wind, solar, hydro, biomass, geothermal), fuel cell generation, or nuclear. The hybridization pathways include co-location, co-combustion, and co-operation. And the system products include energy products (electricity, heat, hydrogen, and liquid fuel) and non-energy products (freshwater, chemicals, and materials)

The universe of possibilities for HESs that result from this definition is summarized in Figure 1.4. Multiple forms of HES have been established and used across the energy sector today, and they can be classified as Multi-input Single-output Hybrid Energy Systems, Single-input Multi-output Hybrid Energy Systems, and Multi-input Multi-output Hybrid Energy Systems (Arent et al. 2021). Here we refer to those integrating at least two generation resources or generation fuels or storage technologies as inputs to generate electricity as multi-input single-output HESs. An example is coal-biomass co-combustion thermal power plants, which integrates two generation fuels in the generation process for enhanced environmental performance. Another example is co-operation of wind, solar, and thermal power plants within an energy base, which optimizes power outputs to increase power supply quality and reliability. Those using a single-generation source to produce two or more energy or non-energy products are single-input multi-output HESs, the cogeneration, combined heat and power plant (CHP) that uses one fuel to produce electricity and heat as a vivid example. Those integrating multiple generation technologies and producing multi-products are called multi-input multi-output HESs, the distributed wind–solar-storage system that generates both electricity and hydrogen is an example.

1.2.2 Classification

This book focuses on multi-input single-output HES (as shown in the gray square in Figure 1.4) that generates electricity (Dykes et al. 2020). Fuels, generation technologies, and storage technologies that may potentially be integrated in the HESs include coal, biomass (agroforestry waste, solid waste, sewage, etc.); wind turbines, photovoltaics, concentrated solar power, hydro, geothermal, natural gas, nuclear; chemical storage; mechanical storage; and thermal storage. Based on the hybridization pathway, the multi-input single-output HES can be further categorized as co-located HES, co-combusted HES, and co-operated HES. Details are as follows.

Co-located HES refers to the distribution-level hybrid system that sites multiple generation technologies at the consumer side (as shown in Figure 1.5), for example, the hybrid wind–solar–natural gas system deployed in industrial parks, hybrid wind–solar-storage diesel system deployed in remote areas, and hybrid solar-storage system deployed on buildings. The design, configuration, and operation of the constituent technologies are fully integrated, and the system is operated under a single control scheme so that the entire system can operate more efficiently and be treated as a single resource.

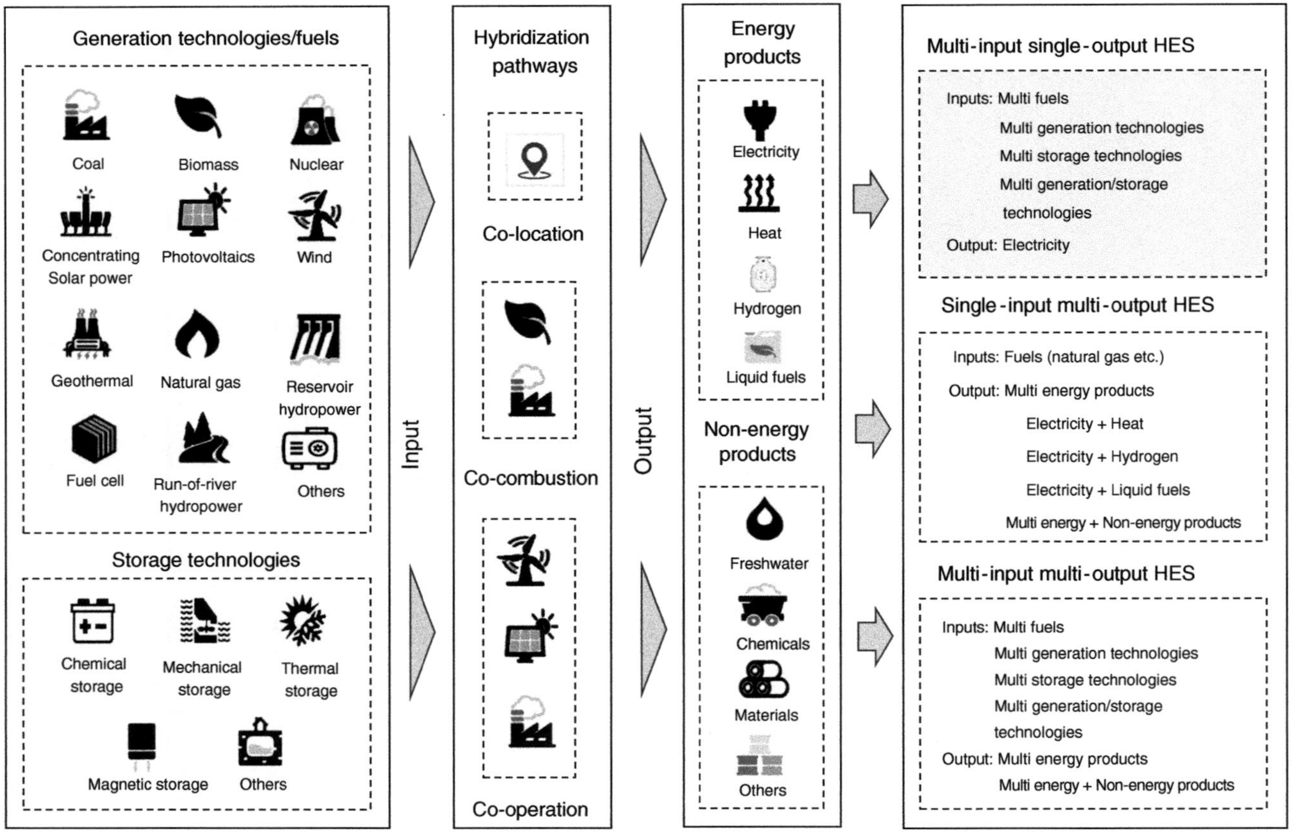

Figure 1.4 Definition of hybrid energy systems (this book focus on multi-input single-output hybrid energy systems, shown in gray).

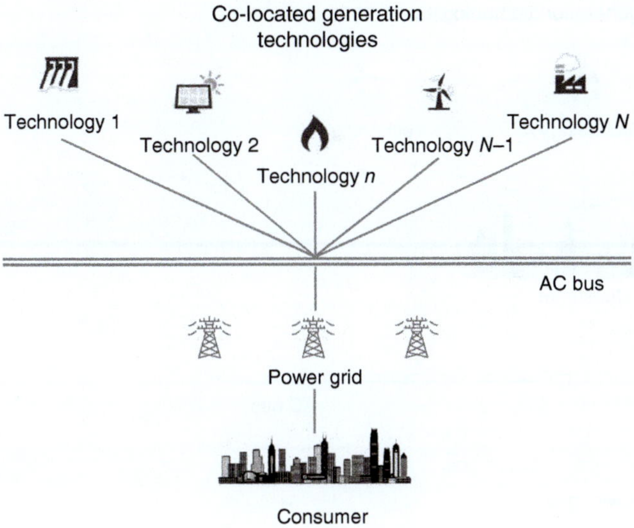

Figure 1.5 Co-located hybrid energy system.

Co-combusted HES involves two or more generation fuels that are burned together within a thermal power plant (as shown in Figure 1.6). For example, a coal-sewage co-combustion power plant, a coal-solid waste co-combustion power plant, or an agroforestry waste-municipal solid waste co-combustion power plant. Co-combustion of different fuels provides opportunities for carbon emission reductions caused by coal combustion and is treated as an important transformation path of traditional coal-fired power plants.

Co-operated HESs involve multiple generation technologies that are linked locationally, for example, they are located within an energy base and share the

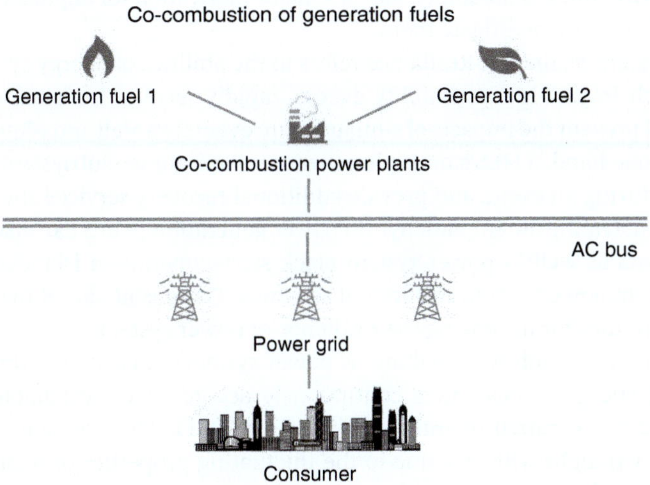

Figure 1.6 Co-combusted hybrid energy system.

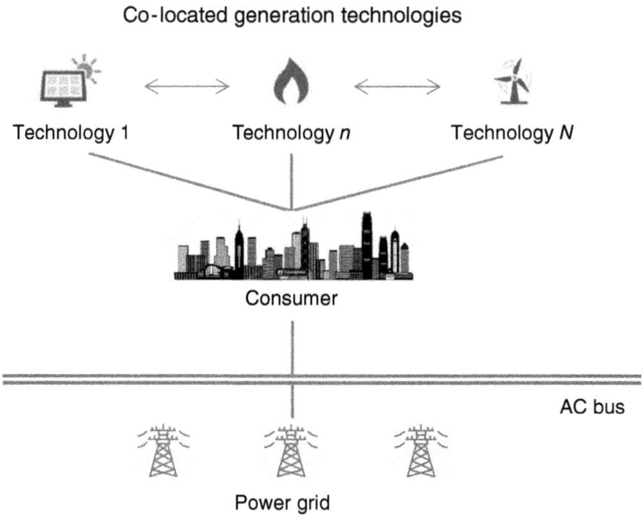

Figure 1.7 Co-operated hybrid energy system.

same interconnection for power sending (as shown in Figure 1.7). They are usually deployed far from the consumer and operated at the concentrated multi-power plant level by unit commitment. The power generation technologies (wind, solar, coal, hydro) are usually owned by different enterprises, and the operation of these power plants in co-operated HES is independent.

1.2.3 Advantages

The hybridization of multi-generator power sources possesses several advantages and offers opportunities for building energy systems with greater resilience, reliability, affordability, security, and sustainability, and is a major motivation for engineers and researchers worldwide to investigate them.

Improved power system resilience. Resilience refers to the abilities of energy system to withstand high impact-low probability events, rapidly recover from these disruptive events, and prevent the impacts of similar future events (Panteli and Mancarella 2015). On the one hand, a HES's multiple generation and storage subsystems can maintain power during an event, and provide additional recovery services after an event has happened. On the other hand, hybrid power generation plants can also provide self-black starts as well as power system black starts, operate in islanded mode, and participate in power system restoration schemes. There is no doubt that HESs offer new possibilities for improving the resilience of power systems.

Increased power supply reliability. Reliability is power system's capacity to offer power and electrical energy to customers continuously according to acceptable quality standards and the required quantity (ZMSkvcable 2021). The production of renewable energy is fraught with risk due to the fluctuating properties of wind and solar resources and how much they depend on their environment. A HES combines variable renewable energy sources with reliable fossil fuel engines and energy storage facilities, offering an excellent way to reduce variation (Tian and Seifi 2014).

Enhanced power system flexibility. Flexibility denotes energy system's capacity to adjust promptly to changes in electricity supply and demand (Babatunde et al. 2020). A HES can be created to offer great flexibility and a wide range of grid services to an increasingly high-renewables grid by integrating renewable energy, storage, controls, and/or flexible loads (ESIG 2022). For example, hydropower has the advantage of being flexible to start and stop, and can be used as a backup station for PV power generation. When PV power output is affected by weather conditions and day/night turnover, hydropower mitigates the intermittency of PV by quickly adjusting output. Compared with just using PV power supply, it obviously improves the flexible adjustment ability of the power supply side.

Improved energy security. An important element of energy security is a reliable and assured energy supply, so it is crucial that HESs make a positive contribution to securing an adequate and stable electricity supply. In fact, HESs are instrumental in the development of new energy sources, and more power plants offer the possibility of securing an adequate electricity supply; moreover, because renewable energy sources such as wind and photovoltaics are highly volatile, HESs use more stable conventional electricity to smooth out these fluctuations, thus seemingly allowing the system to output more stable, higher quality electricity. At the same time, diversification is an important component of energy security. HESs are developed in a variety of modes and play an important role in ensuring the security of electricity supply from diversification.

Increased energy economics. Linking various technologies could improve the use of facilities and delay the need for additional infrastructure investment by enabling the sharing of electrical and physical infrastructure. Meanwhile, the high fuel costs of thermal units can be replaced by energy arbitrage through hybridization with renewable energy systems. In addition, the hybridization of conventional baseload thermal with other energy sources enables resource operational optimization to prevent off-design operations that raise emissions and unit operating costs.

Improved environmental sustainability. Replacing coal, oil, and natural gas with electricity at the end consumption side is considered a more sustainable way to consume energy, and customer-side HESs provide a new way to do so. On the other hand, the large-scale development of new energy sources such as wind power and PV will squeeze the power generation capacity of coal power, force the flexibility of traditional coal power to be modified, and reduce the carbon emission of the power generation process. Therefore, HESs play an important role in environmental sustainability.

1.3 Chapter Organization

This book focuses on four key challenges of HESs. Namely, the deployment optimization of co-located HES, the emission quota allocation of co-combusted HES, the scheduling coordination of co-operated HES, and the optimal implementation

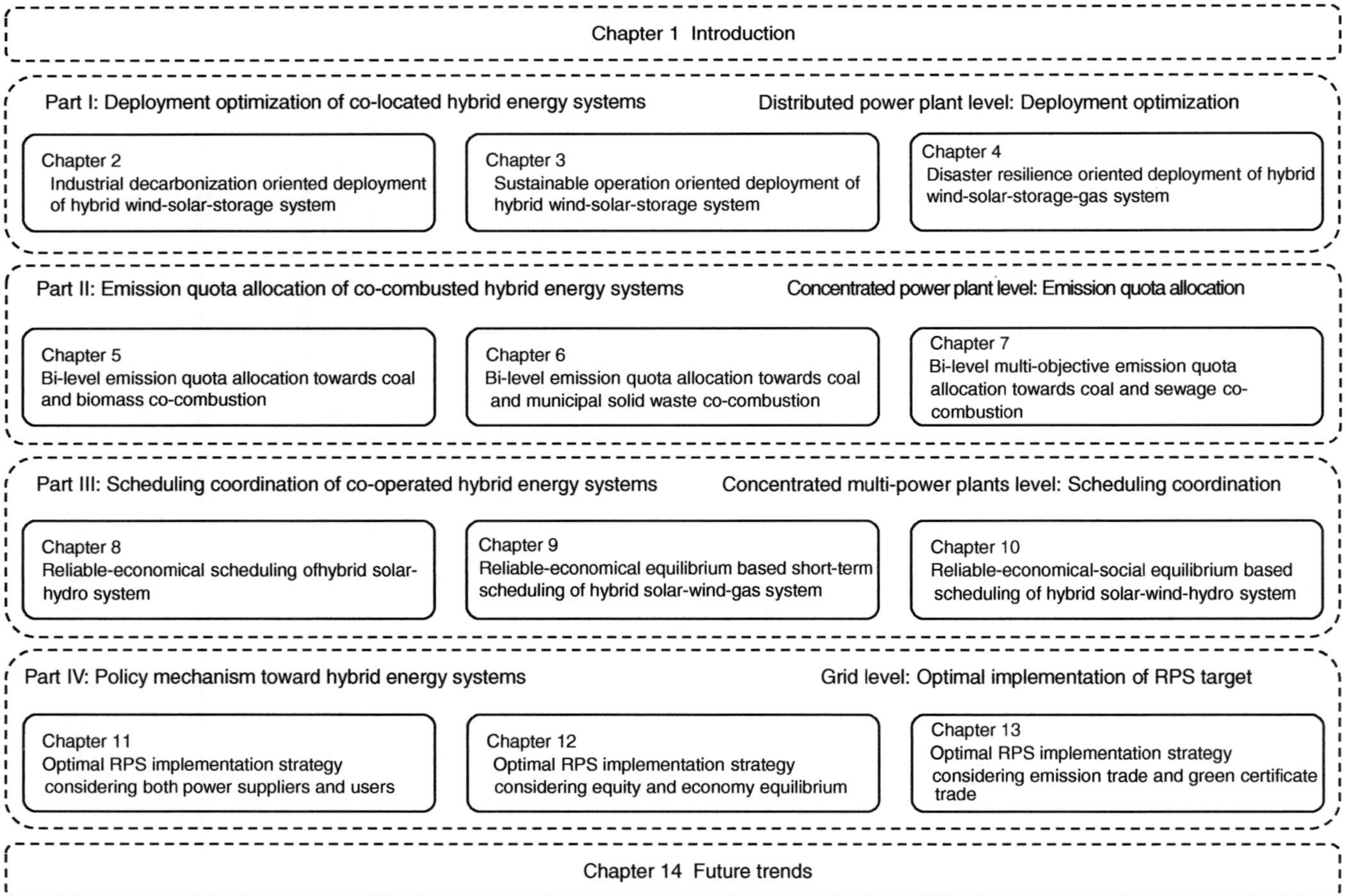

Figure 1.8 Structure of book chapters.

of renewable portfolio standard (RPS) policy toward HESs. Structure and detail contents of book chapters can be seen in Figure 1.8.

The deployment optimization of co-located HESs is discussed in Chapters 2–4. Distributed power plant level co-located HES can be deployed in residential communities, office buildings, commercial areas, campuses, data centers, hospitals, and airports, but the performance must first be guaranteed through configuration optimization (Zhang and Wei 2022). During the system deployment process, various generation technologies and storage technologies are integrated, the type and capacity of each source will be determined, and the operation strategy will be proposed to improve the system's affordability, energy effectiveness, and environmental friendliness (Hakimi et al. 2022). Chapters 2–4 focus on industrial decarbonization-oriented HES deployment optimization, sustainable operation-oriented HES deployment optimization, and disaster resilience-oriented HES deployment optimization separately.

The emission quota allocation of co-combusted HESs is discussed in Chapters 5–7. Reasonable and effective apportion of carbon emission responsibility is an important foundation for controlling total carbon emissions (Wang et al. 2019a). According to the "top-down" approach of allocating carbon emission quotas, the quotas are first distributed to each province in accordance with the amount of energy used in each region, and then they are distributed to industries in accordance with the amount of energy consumed or the nature of the production process (Zhang and Fan 2022). This book focuses on the more micro-level emission quota allocation – allocation among different power plants in the generation industry. In Chapters 5–7, the multi-objective emission quota allocation toward coal-biomass HES, bi-level emission quota allocation toward coal-biomass HES, and bi-level multi-objective emission quota allocation toward coal-biomass HES will be discussed in detail.

The scheduling coordination of co-operated HES is discussed in Chapters 8–10. Power generation scheduling analyzes the operation of the power system network and the economic dispatch of each power plant to optimize the overall energy delivery under given constraints, such as power supply reliability, economy, and environmental friendliness (Gartner Glossary 2022). This was originally a topic widely studied traditionally by researchers and engineers, but the large-scale access of uncertain new energy sources makes the traditional scheduling scheme no longer applicable, and how to achieve coordinated scheduling by taking into account various attributes from a system perspective becomes a new difficulty (Wang et al. 2019b). In Chapters 8–10, the reliable-economical scheduling of hybrid solar–hydro system, reliable-economical scheduling of hybrid solar–wind–gas system, and reliable-economical-social equilibrium based scheduling of hybrid solar–wind–hydro system will be studied.

The optimal implementation of RPS target to build HESs is discussed in Chapters 11–13. The RPS has been adopted worldwide to encourage the use of renewables (Yu et al. 2023; Rouhani et al. 2016). It is a policy instrument that mandates a minimum market share for clean energy technologies in order to stimulate the development of those technologies (Shayegh and Sanchez 2021). Setting reasonable standards for each region and enforcing them rigorously can serve the purpose of stimulating investment and use of renewable electricity, thus creating a virtuous cycle. Therefore, in Chapters 11–13, the optimal RPS implementation strategy considering both power supplier and users, equity and economy equilibrium, and emission trade and green certificate trade will be studied.

The deployment optimization of co-located HES, the emission quota allocation of co-combusted HESs, the scheduling coordination of co-operated HES, and the optimal implementation of RPS target provide carbon-neutral approaches for energy systems, and are organized from the distributed power plant level, the concentrated power plant level, the concentrated multi-power plant level, and the grid level. Each of the above-mentioned chapters will be developed along the lines of background introduction – key problem statement – mathematical modeling and model solving – case study. The main methods used to build mathematical model are multi-objective planning and bi-level planning, which have been widely applied to characterize objective conflicts and achieve equilibrium among decision makers.

References

Adaramola, M.S., Paul, S.S., and Oyewola, O.M. (2014). Assessment of decentralized hybrid PV solar-diesel power system for applications in Northern part of Nigeria. *Energy for Sustainable Development* 19: 72–82. https://doi.org/10.1016/j.esd.2013.12.007.

Aghajani, G. and Ghadimi, N. (2018). Multi-objective energy management in a micro-grid. *Energy Reports* 4: 218–225. https://doi.org/10.1016/j.egyr.2017.10.002.

Ahmadi, S. and Abdi, S. (2016). Application of the Hybrid Big Bang–Big Crunch algorithm for optimal sizing of a stand-alone hybrid PV/wind/battery system. *Solar Energy* 134: 366–374. https://doi.org/10.1016/j.solener.2016.05.019.

Arent, D.J., Bragg-Sitton, S.M., Miller, D.C. et al. (2021). Multi-input, multi-output hybrid energy systems. *Joule* 5 (1): 47–58.

Ashok, S. (2007). Optimised model for community-based hybrid energy system. *Renewable Energy* 32 (7): 1155–1164. https://doi.org/10.1016/j.renene.2006.04.008.

Babatunde, O.M., Munda, J.L., and Hamam, Y.J.E.R. (2020). Power system flexibility: a review. *Energy Reports* 6: 101–106.

Bahramara, S., Moghaddam, M.P., and Haghifam, M.R. (2016). Optimal planning of hybrid renewable energy systems using HOMER: a review. *Renewable and Sustainable Energy Reviews* 62: 609–620. https://doi.org/10.1016/j.rser.2016.05.039.

BP Global (2021). Statistical Review of World Energy. https://www.bp.com/content/dam/bp/business-sites/en/global/corporate/pdfs/energy-economics/statistical-review/bp-stats-review-2021-full-report.pdf (accessed 09 October 2023).

Celik, A.N. (2002). Optimisation and techno-economic analysis of autonomous photovoltaic–wind hybrid energy systems in comparison to single photovoltaic and wind systems. *Energy Conversion and Management* 43 (18): 2453–2468. https://doi.org/10.1016/S0196-8904(01)00198-4.

Celik, A.N. (2003). Techno-economic analysis of autonomous PV-wind hybrid energy systems using different sizing methods. *Energy Conversion and Management* 44 (12): 1951–1968. https://doi.org/10.1016/S0196-8904(02)00223-6.

Chen, Y., He, L., Li, J. et al. (2016a). An inexact bi-level simulation-optimization model for conjunctive regional renewable energy planning and air pollution control for electric power generation systems. *Applied Energy* 183: 969–983.

Chen, J., Huang, Y., Zhang, N. et al. (2016b). Micro-cable structured textile for simultaneously harvesting solar and mechanical energy. *Nature Energy* 1: 16138. https://doi.org/10.1038/nenergy.2016.138.

Climate Watch (2022). Net-Zero Tracker. https://www.climatewatchdata.org/net-zero-tracker?showEUCountries=true (accessed 09 October 2023).

Dykes, K., King, J., DiOrio, N. et al. (2020). *Opportunities for Research and Development of Hybrid Power Plants*. Golden, CO: National Renewable Energy Laboratory. NREL/TP-5000-75026, https://www.nrel.gov/docs/fy20osti/75026.pdf.

Ebrahimian, H., Barmayoon, S., Mohammadi, M., and Ghadimi, N. (2018). The price prediction for the energy market based on a new method. *Economic Research-Ekonomska Istraživanja* 31 (1): 313–337. https://doi.org/10.1080/1331677X.2018.1429291.

Energy Foundation (2021). Five strategies to achieve China's 2060 carbon neutrality goal. https://cgs.umd.edu/sites/default/files/2020-09/final_5%20Strategies_China%202060_Chinese%20%281%29.pdf (accessed 09 October 2023).

Erdinc, O. and Uzunoglu, M. (2010). Recent trends in PEM fuel cell-powered hybrid systems: investigation of application areas, design architectures and energy management approaches. *Renewable and Sustainable Energy Reviews* 14 (9): 2874–2884. https://doi.org/10.1016/j.rser.2010.07.060.

ESIG (2022). Unlocking the flexibility of hybrid resources: a report of the energy systems integration group hybrid resources task force.

Fathima, A.H. and Palanisamy, K. (2015). Optimization in microgrids with hybrid energy systems–a review. *Renewable and Sustainable Energy Reviews* 45: 431–446. https://doi.org/10.1016/j.rser.2015.01.059.

Gartner Glossary (2022). Generation scheduling. https://www.gartner.com/en/information-technology/glossary/generation-scheduling (accessed 09 October 2023).

Hafez, O. and Bhattacharya, K. (2012). Optimal planning and design of a renewable energy based supply system for microgrids. *Renewable Energy* 45: 7–15. https://doi.org/10.1016/j.renene.2012.01.087.

Hakimi, S.M., Hasankhani, A., Shafie-khah, M. et al. (2022). Optimal sizing of renewable energy systems in a microgrid considering electricity market interaction and reliability analysis. *Electric Power Systems Research* 203: 107678.

Halabi, L.M., Mekhilef, S., Olatomiwa, L., and Hazelton, J. (2017). Performance analysis of hybrid PV/diesel/battery system using HOMER: a case study Sabah, Malaysia. *Energy Conversion and Management* 144: 322–339. https://doi.org/10.1016/j.enconman.2017.04.070.

International Energy Agency's (2020). Net Zero by 2050: A Road Map for the Global Energy Sector. https://www.iea.org/reports/net-zero-by-2050 (accessed 09 October 2023).

International Energy Agency (2021a). Key world energy statistics. https://www.iea.org/reports/key-world-energy-statistics-2021 (accessed 09 October 2023).

International Energy Agency (2021b). Global CO_2 Emissions by Sector, 2019. https://www.iea.org/data-and-statistics/charts/global-co2-emissions-by-sector-2019 (accessed 09 October 2023).

Kenfack, J., Neirac, F.P., Tatietse, T.T. et al. (2009). Microhydro-PV-hybrid system: sizing a small hydro-PV-hybrid system for rural electrification in developing countries. *Renewable Energy* 34 (10): 2259–2263. https://doi.org/10.1016/j.renene.2008.12.038.

Lan, H., Wen, S., Hong, Y.Y. et al. (2015). Optimal sizing of hybrid PV/diesel/battery in ship power system. *Applied Energy* 158: 26–34. https://doi.org/10.1016/j.apenergy.2015.08.031.

Li, C., Ge, X., Zheng, Y. et al. (2013). Techno-economic feasibility study of autonomous hybrid wind/PV/battery power system for a household in Urumqi, China. *Energy* 55: 263–272. https://doi.org/10.1016/j.energy.2013.03.084.

Liu, Y., He, L., and Shen, J. (2017). Optimization-based provincial hybrid renewable and non-renewable energy planning - a case study of Shanxi, China. *Energy* 128: 839–856.

Lund, H., Werner, S., Wiltshire, R. et al. (2014). 4th Generation District Heating (4GDH): integrating smart thermal grids into future sustainable energy systems. *Energy* 68: 1–11. https://doi.org/10.1016/j.energy.2014.02.089.

Lund, H., Østergaard, P.A., Connolly, D., and Mathiesen, B.V. (2017). Smart energy and smart energy systems. *Energy* 137: 556–565. https://doi.org/10.1016/j.energy.2017.05.123.

Mandal, S., Das, B.K., and Hoque, N. (2018). Optimum sizing of a stand-alone hybrid energy system for rural electrification in Bangladesh. *Journal of Cleaner Production* 200: 12–27. https://doi.org/10.1016/j.jclepro.2018.07.257.

Nelson, D.B., Nehrir, M.H., and Wang, C. (2006). Unit sizing and cost analysis of stand-alone hybrid wind/PV/fuel cell power generation systems. *Renewable Energy* 31: 1641–1656. https://doi.org/10.1016/j.renene.2005.08.031.

Ngan, M.S. and Tan, C.W. (2012). Assessment of economic viability for PV/wind/diesel hybrid energy system in southern Peninsular Malaysia. *Renewable and Sustainable Energy Reviews* 16 (1): 634–647. https://doi.org/10.1016/j.rser.2011.08.028.

Ogunjuyigbe, A.S.O., Ayodele, T.R., and Akinola, O.A. (2016). Optimal allocation and sizing of PV/Wind/Split-diesel/Battery hybrid energy system for minimizing life cycle cost, carbon emission and dump energy of remote residential building. *Applied Energy* 171: 153–171. https://doi.org/10.1016/j.apenergy.2016.03.051.

*Ovonic Hydrogen Solutions (2006). WEB2006, Ovonic Hydrogen Solution. http://www.ovonic-hydrogen.com/products/portable.Html (accessed 19 October 2023).

Panteli, M. and Mancarella, P. (2015). Influence of extreme weather and climate change on the resilience of power systems: impacts and possible mitigation strategies. *Electric Power Systems Research* 127: 259–270.

Perera, A.T.D. (2013). A hybrid tool to combine multi-objective optimization and multi-criterion decision making in designing standalone hybrid energy systems. *Applied Energy* 107: 412–425. https://doi.org/10.1016/j.apenergy.2013.02.049.

Rouhani, O.M., Niemeier, D., Gao, H.O., and Bel, G. (2016). Cost-benefit analysis of various California renewable portfolio standard targets: is a 33% RPS optimal? *Renewable and Sustainable Energy Reviews* 62: 1122–1132.

Setiawan, A.A., Zhao, Y., and Nayar, C.V. (2009). Design, economic analysis and environmental considerations of mini-grid hybrid power system with reverse osmosis desalination plant for remote areas. *Renewable Energy* 34 (2): 374–383. https://doi.org/10.1016/j.renene.2008.05.014.

Shayegh, S. and Sanchez, D.L. (2021). Impact of market design on cost-effectiveness of renewable portfolio standards. *Renewable and Sustainable Energy Reviews* 136: 110397.

The Ministry of Foreign Affairs of the People's Republic of China (2008). United Nations Framework Convention on Climate Change and its Kyoto Protocol. https://www.fmprc.gov.cn/chn/pds/ziliao/tytj/tyfg/t23942.htm (accessed 09 October 2023).

Tian, Z. and Seifi, A.A. (2014). Reliability analysis of hybrid energy system. *International Journal of Reliability, Quality and Safety Engineering* 21 (03): 1450011.

U.S. Department of Energy (2021). Hybrid energy systems: opportunities for coordinated research. https://www.nrel.gov/docs/fy21osti/77503.pdf (accessed 10 October 2023).

United Nations (2009). Copenhagen climate change conference ends with non-legally binding deal. https://news.un.org/zh/story/2009/12/123802 (accessed 09 October 2023).

Wang, B.J., Zhao, J.L., and Wei, Y.X. (2019a). Carbon emission quota allocating on coal and electric power enterprises under carbon trading pilot in China: mathematical formulation and solution technique. *Journal of Cleaner Production* 239: 118104.

Wang, F., Xie, Y., and Xu, J. (2019b). Reliable-economical equilibrium based short-term scheduling towards hybrid hydro-photovoltaic generation systems: case study from China. *Applied Energy* 253: 113559.

World Energy Scenarios (2013). Composing energy futures to 2050. https://www.worldenergy.org/assets/downloads/World-Energy-Scenarios_Composing-energy-futures-to-2050_Executive-summary.pdf (accessed 10 October 2023).

Yu, S., Zhou, S., Dai, Y., and Zhang, J. (2023). Impacts of rps and fit on inter-regional power transmission line layout in China: considerations of high renewable energy penetration. *Energy Policy* 178: 113615.

Zhang, X. and Fan, D. (2022). Carbon emission quota allocation of high energy consumption industries in undeveloped areas - a case study of Inner Mongolia Autonomous Region. *Heliyon* 8 (11): e11241.

Zhang, J. and Wei, H. (2022). A review on configuration optimization of hybrid energy system based on renewable energy. *Frontiers in Energy Research* 10: 977925. https://doi.org/10.3389/fenrg.2022.977925.

Zhou, W., Lou, C., Li, Z. et al. (2010). Current status of research on optimum sizing of stand-alone hybrid solar–wind power generation systems. *Applied Energy* 87 (2): 380–389. https://doi.org/10.1016/j.apenergy.2009.08.012.

ZMSkvcable (2021). What is power system reliability?

2

Industrial Decarbonization-Oriented Deployment of Hybrid Wind–Solar-Storage System

Leveraging advancements in low-carbon electricity from both grid and onsite renewable generation sources will be critical to decarbonization efforts.

Industrial Decarbonization Roadmap
The U.S. Department of Energy, Sep. 2022

The deployment of hybrid wind–solar-storage systems is becoming an increasingly popular strategy for achieving industrial decarbonization. Combining renewable energy sources such as wind and solar with energy storage technologies can provide reliable, sustainable energy that can be used to power industrial processes and reduce dependence on fossil fuels. One area where the deployment of hybrid systems is particularly promising is the wastewater treatment sector. By using the floor area of existing wastewater treatment plants to deploy distributed renewable energy (DRE), such as wind and solar, it is possible to promote a clean and efficient energy transition while also achieving industrial decarbonization goals. However, determining the ideal capacity for such systems has been a challenge, as few studies have focused on this issue. To address this challenge, this chapter proposes a grid-connected wind–solar-storage system to retrofit existing wastewater treatment plants and explore their regional potential. A multi-objective model that takes into account economic, technological, and environmental considerations is built and the ε-constraint method is proposed for model solving, which can be used to identify the optimal potential, treatment scale, and layout area for hybrid wind–solar-storage systems of wastewater treatment plant. A case study in the Yangtze River Delta (YRD) region in China has demonstrated the effectiveness of this approach. Overall, the deployment of hybrid wind–solar-storage systems represents a promising approach to achieving industrial decarbonization goals. By optimizing the regional potential for such systems, governments and industries can improve space utilization and energy efficiency while reducing dependence on fossil fuels and promoting a clean and sustainable energy future.

Sustainable Hybrid Energy Systems: Carbon Neutral Approaches, Modeling, and Case Studies,
First Edition. Jiuping Xu and Fengjuan Wang.

2.1 Background Review

The industry sector includes a wide range of activities, such as manufacturing, mining, construction, and transportation, and it is responsible for approximately 30% of global greenhouse gas emissions. To achieve a net-zero emissions future, it is important to reduce the carbon footprint of this sector by adopting more sustainable and cleaner technologies (Our World in Data 2020). As industrial parks need to serve for decades, upgrading the facilities of industrial parks to reduce carbon emissions is inevitable (De Pee et al. 2018). Installing renewable energy systems such as solar panels or wind turbines to produce energy on-site has been proven to be a relatively economical and accessible approach to decrease the energy utilization–related emissions for industry parks (McIlwaine et al. 2021; Schallenberg-Rodríguez and Montesdeoca 2018). By using on-site renewables, industrial parks can reduce their reliance on the traditional grid and minimize their exposure to volatile electricity prices. It also provides a reliable and sustainable source of energy, ensuring continuous electricity supply even during power outages (O'Meara 2020). With the rapid industrialization of developing countries, it is important to explore and implement renewable energy applications in the industrial sector.

The wastewater treatment industry, which requires significant amounts of energy, is responsible for roughly 1.6% of the world's carbon dioxide emissions and consumes around 3% of the world's electricity (Lu et al. 2018). The complex infrastructure and high operating costs associated with wastewater treatment pose significant obstacles for many developing countries, resulting in environmental degradation and threatening public health (Grant et al. 2012). To address the significant energy demands and land usage requirements of wastewater treatment facilities, an optimal solution would be to implement on-site hybrid renewable generation systems (Van Loosdrecht and Brdjanovic 2014). As urbanization continues to grow at an unprecedented rate, there is a corresponding increase in wastewater production and energy consumption. Therefore, the aim of this chapter is to optimize hybrid generation systems for use in wastewater treatment facilities.

To address the issue of balancing energy consumption with environmental protection in the wastewater treatment industry, three main solutions have been proposed. Firstly, biomass and heat recovery has been proven to be a feasible solution in technical plant operations. Secondly, energy efficiencycan be improved through measures such as conducting energy input–output audits and utilizing ameliorative treatment technologies. For instance, a study by Huang et al. (2021) utilized a data envelopment analysis model to assess the energy efficiency of 270 wastewater treatment parks, revealing that 253 were inefficient due to input excesses or output shortfalls. This provides valuable information for optimizing the management of wastewater treatment parks. Additionally, Skouteris et al. (2020) discovered that the use of pure oxygen in aerobic wastewater treatment can enhance efficiency. Thirdly, the development of DRE sources such as solar and wind power can provide additional energy to wastewater treatment plants (WWTPs). For example, a study conducted on WWTPs in California found that on-site solar photovoltaic (PV) systems could fulfill 30%–100% of their energy demands (Strazzabosco et al. 2019). Previous research has contributed significantly to the understanding of the potential for energy transition in WWTPs, particularly with regard to energy recovery and efficiency, and these findings have been applied in industrial settings. However, the practical implementation of DRE in WWTPs, particularly in developing countries, remains limited. Therefore, additional academic and policy research is necessary to explore DRE application approaches in the wastewater sector.

Deploying DRE in the floor area of the wastewater sector can be an effective approach to promoting a clean and efficient energy transition (Griggs et al. 2013). Various DRE technologies such as solar, wind–solar, and wind–solar-heat have been demonstrated to be feasible (Nakkasunchi et al. 2021; Saleh and Al-Nimr 2020). For instance, Helal et al. (2013) utilized PV, wind turbines, and micro-turbines to simulate a system for a WWTP in Egypt. Bey et al. (2021) investigated the energy and economic feasibility of grid-connected PV systems in Algeria. Wang et al. (2021) proposed a wind–solar-storage system for a drinking water treatment plant in China to examine the feasibility of using a complementary daily wind–solar system to supply power. Nawarkar and Salkar (2019) assessed the optimal operating parameters of a continuous electrocoagulation technique for wastewater treatment using solar power and batteries in India. Similarly, Fernández-Marchante et al. (2021) calculated the carbon emissions of wastewater treatment using wind and solar energy, and demonstrated that environmental risks can be reduced. However, while most previous studies have focused on the micro-technical feasibility or benefits of utilizing renewable power, the regional potential of such applications in the wastewater treatment industry has yet to be fully examined.

DRE applications have the potential to reduce carbon emissions associated with coal-fired power plants, but they typically require significant initial capital investment. Previous studies have explored the environmental and economic trade-offs associated with DRE integration. For instance, Campana et al. (2021) determined the trade-off solutions between cost and the DRE share of PV, wind, battery, and hydrogen storage systems for WWTPs. Nguyen et al. (2020) optimized the DRE configuration of a WWTP in South Korea and examined reliability, economic, and environmental performances. However, most previous studies are based on single cases and did not consider the influence of the scale effect. Since energy configuration conditions and demands differ depending on the WWTP scale, single cases cannot provide horizontal WWTP comparisons. To achieve a carbon-neutral environment, it is crucial to understand the economic, environmental, and technical configuration schemes for different WWTP scales to reduce energy costs and meet industrial decarbonization goals.

This chapter proposes a scheme for upgrading existing wastewater treatment plants by deploying a wind–solar-storage system and optimizing its regional potential capacity. Regional datasets are established by analyzing natural resource data, available areas, and energy consumption data. The study divides regional wastewater treatment plants into seven categories, taking into account the impact of treatment scale on floor area and electricity consumption, and estimates the available area and unit electricity consumption of each category using fuzzy numbers. A multi-objective model is then built to optimize the regional application capacity, investment benefits, and decarbonization issues. The proposed method is verified through a case study of the YRD region with 1662 wastewater treatment plants, and the optimal types of wind turbines and PV panels for 24 plants with a typical treatment scale are analyzed. The surplus renewable energy could be exported to the main grid or discarded due to limited transmission capacity. The research paradigm proposed in this study could assist governments in evaluating the potential of DRE in the wastewater treatment industry and benefit other industries or regions aiming to improve space utilization and achieve energy decarbonization.

2.2 Main Issue Description

In this section, the schematic of the proposed hybrid wind–solar-storage system integrated with wastewater treatment plant is introduced, the decarbonization datasets for regional emission reduction potential evaluation is presented, and the optimization scheme of the hybrid energy system deployment is displayed.

2.2.1 System Schematic

To design an energy system–retrofitting scheme that is compliant with the plants' traditional layouts is significantly important in estimating the industrial park's renewable energy integration potential; therefore, it is the first key problem to be solved in this chapter. In a typical wastewater treatment park, settling tanks, treatment tanks, ancillary buildings, and greening areas are fundamental components (Kumar 2011; Patziger 2021). For the space over treatment facility tanks, the flexible PV panels, which require fewer support structures, can be installed there so that the normal operation will not be disrupted, and the space can be efficiently used (Wu et al. 2019). This scheme has been verified to be feasible in a canal in California (McKuin et al. 2021). For the buildings in the park, roof PV panels is perfect choice (Ali et al. 2018). As the design and construction of wastewater treatment park should also meet green requirements, these areas can be utilized to install the wind turbines (Dewi et al. 2019). And as wind and solar resources are complementary in nature, the integration of these two resources can positively improve system's efficiency and resilience (Xu et al. 2018a; Campos et al. 2020). The storage system, on the other hand, is installed in the original electricity distribution rooms to save construction costs. The scheme is illustrated in Figure 2.1.

2.2.2 Decarbonization Datasets

The scale, design, and applied technology of each wastewater treatment plant varies a lot, so it is extremely difficult to accurately obtain the exact floor area of all plants, especially when there are large number of projects. However, with the construction standards in hand, the land use index of different scaled wastewater treatment project can be obtained, and the floor area of each project can be estimated by multiplying the scale and corresponding land use index (Ministry of Construction of

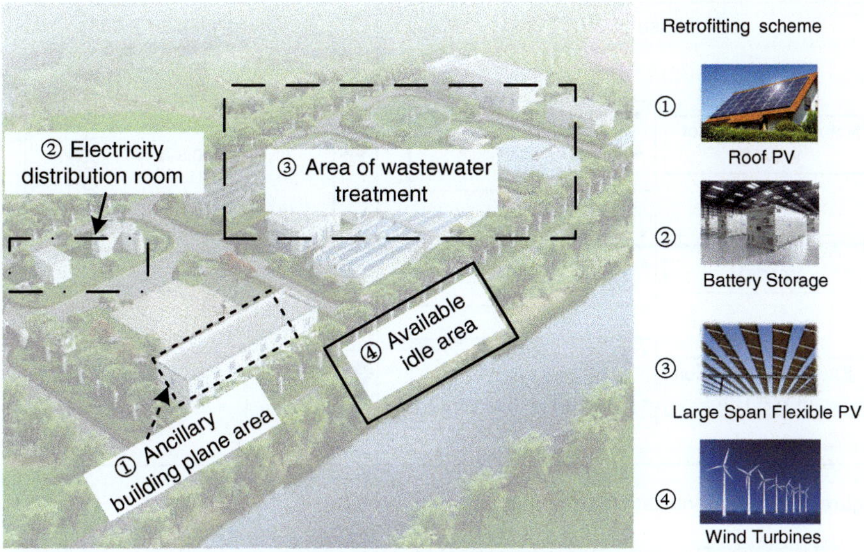

Figure 2.1 The hybrid energy system retrofitting scheme of a wastewater treatment park. (Roof PV) Source: Smileus/Adobe Stock; (Battery Storage) Source: Phonlamaiphoto/Adobe Stock; (Wind Turbines) Source: TebNad/Adobe Stock.

Table 2.1 The index of wastewater treatment plants.

		Category						
Parameter	**Unit**	**I**	**II**	**III**	**IV**	**V**	**VI**	**VII**
Daily Treatment Scale	1×10^4 m^3/day	[0.1,0.5)	[0.5,2)	[2,5)	[5,10)	[10,20)	[20,50)	Above 50
Index of planned land use	$m^2 \cdot day/m^3$	2	2~1.5	1.5~1.2	1.2~0.8	0.8~0.6	0.6~0.5	0.5

the People's Republic of China 2001; Ministry of Ecology and Environment of the People's Republic of China 2021). In this chapter, wastewater treatment park are classified into seven groups based on the scale, and the corresponding land use index is collected, as shown in Table 2.1.

The decarbonization database requires three sets of data: The natural resources data set, the available areas data set, and the energy consumption data set. The natural resources data set includes annual full load hours of PV panels and wind turbines. The annual full load hours are converted by local government units or industry associations from meteorological and energy generation data, and this parameter is adopted to reflect the local wind and solar energy generation situation. These parameters indicate the working hours of wind turbines or PV panels under rated capacity (Wei et al. 2019) and reflect the quality of local renewable energy resources (Sun et al. 2017). The available areas data set include available ancillary building area, facility tank area, and green area, and they can be expressed as a ratio of the

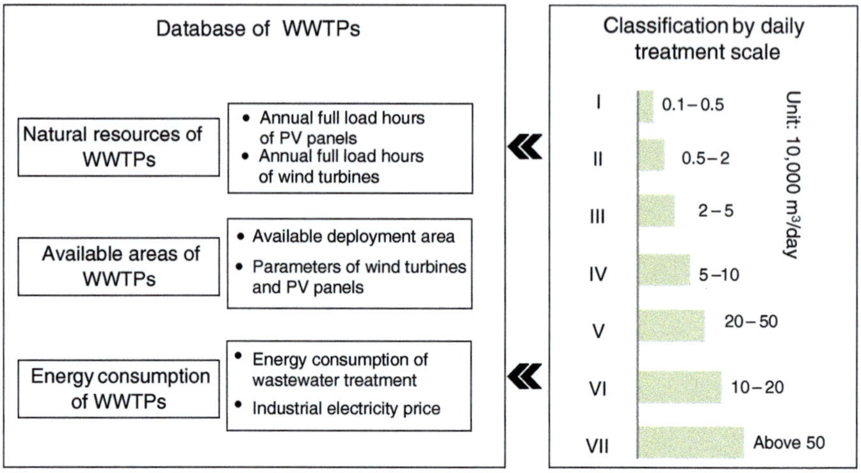

Figure 2.2 Decarbonization potential dataset of WWTPs.

total floor area. However, it is impractical to determine these coefficients as certain parameters, for example, existing research on wastewater treatment parks indicates an ancillary building's plane area to be 5% to 30% of the total area; therefore, the coefficients can be estimated using linear regression (Jarvis and Stuart 2001), that is, 30% and 5% for a wastewater treatment park with a treatment scale of 0.1 and 100, respectively, with the middle interval being estimated using interpolation. Lastly, the energy consumption data and the industrial electricity price data will also be collected. As energy consumption generally decreases with an increase in wastewater treatment park scale and because treatment technology and equipment aging degree influence electricity consumption (Niu et al. 2019), the trapezoidal fuzzy numbers are employed to determine the expected values of the uncertain parameters(Xu and Zhou 2011). The scheme of the data sets is shown in Figure 2.2.

2.2.3 Optimization Scheme

Although installing renewable generation units can reduce carbon emissions, it also leads to high expenditures for the investor (Abbas et al. 2021). Therefore, to simultaneously consider economic costs, environmental benefits and technical conditions are significantly important in the retrofitting process. Here in this chapter, a mixed-integer multi-objective optimization model is developed to deal with the trade-offs. Several critical factors of the decision problem, including the annual green energy generation and wastewater power consumption, are examined to deal with the energy supply and demand. The area that can be utilized at each wastewater treatment park is also estimated with the method introduced above. With these objectives and constraints, the decision maker needs to make a decision on the type and capacity on the wind, PV and battery units, and determine the main grid imported or exported quantity of electricity. It's worth noting that the life cycle assessment is incorporated into the model, so that the initial investment, maintenance costs, and service life of energy transmission are simultaneously considered. The framework is shown in Figure 2.3.

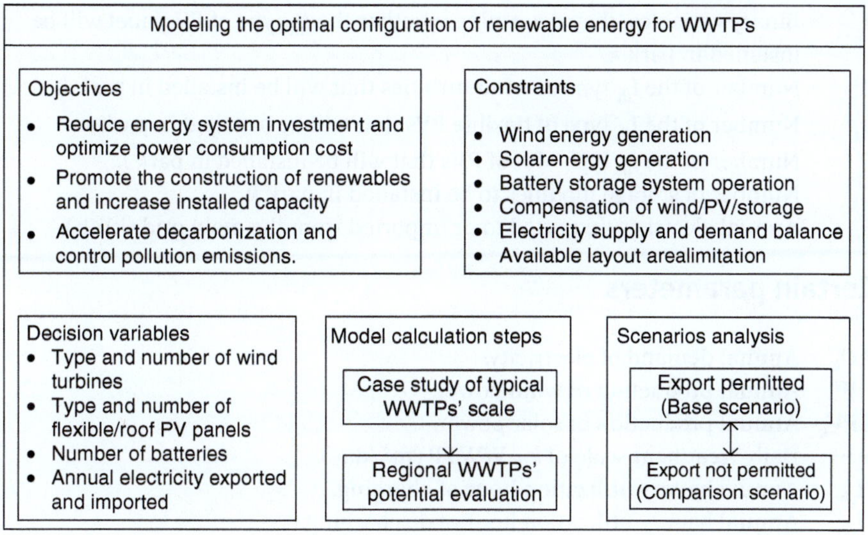

Figure 2.3 Modeling of the optimized potential application of renewable energy.

2.3 Mathematical Modeling

This section presents the mathematical model for the description given above. In order to formulate the multi-objective optimization model, two basic assumptions are made: firstly, throughout the renewable energy system's life cycle, it is assumed that the natural resources remain constant; secondly, the energy storage battery power is reset to zero at the end of every year.

2.3.1 Notations

Notations used in this chapter are listed in this subsection.

Indices

k Index of industry parks (this chapter refers to wastewater treatment park), $k = 1...K$.

n Index of years, $n = 1...N$.

i Index of wind turbine type that can be selected to be installed in the industry park, $i = 1...I$.

j Index of PV panel types that can be selected to be installed in the industry park, $j = 1...J$.

Decision variables

x_k^i Binary variables, they determine whether the i_{th} type of wind turbine will be installed in park k.

y_k^j Binary variables, they determine whether the j_{th} type of PV panel will be installed in park k.
A_k^i Number of the i_{th} type of wind turbines that will be installed in park k.
$B_k^{f,j}$ Number of the j_{th} type of flexible PVs that will be installed in park k.
$B_k^{r,j}$ Number of the j_{th} type of roof PVs that will be installed in park k.
C_k Number of storage modules to be installed in park k.
E_k Annual electricity exported to or imported from the main grid (kWh).

Certain parameters

AD_k Annual demand of electricity.
AW_k Annual production of wind power.
APV_k Annual production of solar power.
S_k Daily treatment scale of k_{th} WWTP (m^3/day).
H_{cha} Annual battery utilization hour of charging.
H_{dis} Annual battery utilization hour of discharging.
U_w Unit cost of wind turbine (per kW).
U_r Unit cost roof PV (per kW).
U_f Unit cost of large-span flexible PV (per kW).
U_b Unit cost of battery module (per kW).
r Discount rate
T_n Total days per year.
TS_k Total area of the k_{th} wastewater treatment plant.
P_w Rated capacity of a wind turbine (kW)
P_{pv} Rated capacity of a unit PV panel (kW)
P_{cha} Rated battery charging capacity (kW).
P_{dis} Rated battery discharging capacity (kW).
P_B Rated capacity of battery system (kW).
Q_k^w Annual full load hours of wind power (Hour/year).
Q_k^s Annual full load hours of solar power (Hour/year).
pr Unit price of industrial electricity from the main grid (CNY/kWh).
e_o Carbon emission coefficient per kWh of thermal electricity (kg/kWh).
e_w Carbon emission coefficient per kWh of wind electricity (kg/kWh).
e_{pv} Carbon emission coefficient per kWh of solar PV electricity (kg/kWh).
In_k Index of planned land use at the k_{th} WWTP.
s_w Floor space needed to install per unit of wind turbine (m^2).
s_r Floor space needed to install per unit of proof PV (m^2).
s_f Floor space needed to install per unit of flexible PV (m^2).
α Decision maker's attitude toward installed renewable generation capacity.
β Decision maker's attitude toward electricity utilization associated carbon emissions.

Uncertain parameters

$\tilde{P}_k$ Electricity consumption for treating each m^3 of wastewater at the k_{th} WWTP (kWh/m^3).

$\tilde{\lambda}_k^w$ Ratio coefficient of the available idle area to the k_{th} total plant area.
$\tilde{\lambda}_k^r$ Ratio coefficient of the ancillary building area to the k_{th} total plant area.
$\tilde{\lambda}_k^f$ Ratio coefficient of the water treatment area to the k_{th} total plant area.

2.3.2 Decarbonized Deployment

This section provides detailed objectives and constraints of the industrial decarbonization-oriented deployment problem of the hybrid wind-solar-storage system.

2.3.2.1 To Reduce the Total Electricity Utilization Costs

Economic feasibility is significantly important for every new project. Up to now, many indicators have been proposed to evaluate the economic situation of a hybrid energy system, for example, the total investment costs (Seker and Kahraman 2021), the annual revenue (Mansouri Kouhestani et al. 2020), and the levelized cost of electricity (LCOE) (Babaelahi et al. 2019) Among all these indicators, LCOE holds the advantage of being able to be used to horizontally compare the investment benefits of two different projects, no matter how different the projects are in sizes or facilities (Jahangiri et al. 2020); therefore, it is a valuable indicator for the administration to assess the economic benefits of the hybrid energy system, and it is thereby chosen here to represent the administration's economic objective, as shown in Eq. (2.1).

$$Minf_1 = \frac{\sum_k \left(U_w \cdot A_k + U_f \cdot B_k^f + U_r \cdot B_k^r + U_b \cdot C_k + \sum_n \frac{E_k \cdot pr}{(1+r)^n} \right)}{\sum_k \sum_n S_k \cdot \tilde{P}_k \cdot T_n} \tag{2.1}$$

The numerator of Eq. (2.1) represents the life cycle electricity utilization expenditure of the industrial park, and the denominator represents its life cycle electricity consumption. U_w, U_f, U_r, and U_b represent total comprehensive costs of wind turbines, PV panels (roof-top and large span flexible), and battery modules, which comprise the initial investment costs (I), annual operation and management cost(OM_n), and equipment replacement cost (R_n). The life cycle comprehensive of different components can be represented as $U = I + \sum_n \frac{(OM_n + R_n)}{(1+r)^n}$. As we discussed in the key problem statement, there are different types of wind turbines and PV panels to be selected for the industry park when designing the hybrid system, and these costs parameters are one-dimensional matrices. A_k, B_k^f, B_k^r, and C_k in Eq. (2.1) are decision variables, which represent the number of wind turbines, PV panels (roof-top and large span flexible), and battery modules. E_k means the annual electricity exported to or imported from the main grid, pr refers to the electricity price, and r shows discount rate. S_k is the daily treatment scale of the kth wastewater treatment plant, $\tilde{P}_k$ represents unit electricity consumption of the wastewater treatment plant to treat each cubic meter of wasted water, and T_n means the total number of days of a certain year.

2.3.2.2 To Promote the Installed Capacity of Wind and Solar Power

The growing concern for global warming stimulates the world to adopt renewable energy installation and utilization targets; for example, India, China, and the USA have listed remarkable threshold percentages according to renewable portfolio standards (Spillias et al. 2020). Consequently, the installed capacity of new energy has become a direct reference data for the central government to guarantee renewable energy utilization (Wu et al. 2021). Meanwhile, a higher installed capacity can provide higher ratio of renewable energy generation, leading to higher self-sufficiency ratio (SSR) for the industry park. Therefore, maximizing the installed capacity of wind and solar power is also an important objective for the administration to plan the installation of hybrid energy systems, as shown in Eq. (2.2).

$$Maxf_2 = \sum_k (P_w \cdot A_k + P_{pv} \cdot B_k) \tag{2.2}$$

where P_w, P_{pv}, NW_k, NPV_k are row vectors of $1 \times i, 1 \times j, i \times 1$, and $j \times 1$, respectively; here, i and j indicate the optional kinds of wind turbines and PV panels, respectively.

2.3.2.3 To Accelerate Decarbonization and Control Pollution Emissions

Given the importance of achieving carbon peaking and carbon neutralization, wind and solar energy production are being encouraged because of their clean production process and low carbon footprint (Lin et al. 2021). The latest World Energy Balance data show that 90.28% of the global energy consumption is supplied by coal, oil, natural gas, biofuels, and only 2.21% of the power generation worldwide is currently generated by solar and wind energy (International Energy Agency 2020). This ratio is even lower in developing countries, such as Brazil (2.18%), India (1.22%), and Egypt (0.46%) (International Energy Agency 2020). Given this situation, innovative efforts are needed to mitigate global climate change. An example is the maximum reduction of the annual carbon emissions in the industry sector; further example of wastewater industry is shown in Eq. (2.3).

$$Maxf_3 = \sum_k \left(AD_k \cdot e_o - E_k \cdot e_o - AW_k \cdot e_w - APV_k \cdot e_{pv}\right) \tag{2.3}$$

where AD_k is the annual demand of electricity; AW_k and APV_k are the annual production of wind and solar electricity, respectively; and E_k is the main grid electricity supply.

2.3.2.4 Wind Power Output

In a short-term schedule, the real-time electricity output of wind farms is characterized by the actual wind speeds and the hub height (Wang et al. 2019). But in the long term, such as monthly scale, annual scale, or the whole life cycle, full load hours can better reflect the characteristics of wind farms' output (Shawon et al. 2013). The full load hour of wind turbine(Q_k^w) is calculated as a turbine's average annual electricity output divided by its rated capacity; it belongs to second-hand data and is converted by the local administration or industry associations (Ng and Ran 2016). Take data in Denmark as an example, in 2021, the median annual full load hours of wind turbines were 3061 hours, then the capacity factor can be calculated as $3061/(365 \times 24) = 34.9\%$ (National Renewable Energy Laboratory of the US 2015). Therefore, if the rated power of a wind turbine is 3 MW, the annual energy production can be measured by $3\times$ $34.9\%\times8760$. Therefore, by introducing the annual full load hours of wind turbines in a certain location, the annual electricity generation of a given installed capacity of wind farms can be estimated. Here in this chapter, the rated capacities of different kinds of wind turbines vary from 0.1 kW to 15 MW, so the binary variable x_k^i is introduced to select the proper type of wind turbine, (see Eq. (2.4)) where i indicates different kinds of turbines.

$$\sum_i x_k^i \leqslant 1, \quad A_k^i \leqslant x_k^i \cdot M \tag{2.4}$$

Here, M refers to a large number associated with artificial variables (Cococcioni and Fiaschi 2021), limiting the selection of turbine type to only one for each industry park. Let Q_k^w be multiplied by the rated capacity (P_w^i) and the total number of wind turbines A_k^i, from which the annual wind energy generation AW_k (Eq. (2.5)) is obtained.

$$AW_k = Q_k^w \cdot \sum_i \left(x_k^i \cdot P_w^i \cdot A_k^i\right) \tag{2.5}$$

2.3.2.5 Solar Power Output

The same as the situation for wind energy generation, the annual solar power output can be represented by the product of PV module rated capacity and the full load hours (Kassem et al. 2018). Typically, PV modules are rated between 0.25 and 0.4 kW; therefore, the binary variable y_k^j is used here to show the decision maker's choice of a different kind of PV module. Let $B_k^{f,j}$ and $B_k^{r,j}$ be the numbers of roof PV and flexible PV panels; then, the constraints are as Eq. (2.6).

$$\sum_j y_k^j \leqslant 1, \quad B_k^{f,j} + B_k^{r,j} = B_k^j, \quad B_k^j \leqslant y_k^j \cdot M \tag{2.6}$$

where B_k^j represents the sum of the i_{th} roof PV and flexible PV. The annual total solar energy generation APV_k is shown in Eq. (2.7).

$$APV_k = B_k^{pv} \cdot \sum_j \left(y_k^j \cdot P_{pv}^j \cdot B_k^j\right) \tag{2.7}$$

2.3.2.6 Operation of Battery Storage System

It is common sense that the battery system's service life will be significantly affected if it were frequently charged and discharged (Zhang et al. 2020). Here we let H_{cha} and H_{dis} be the annual utilization hour bounds of charging and discharging the battery system, respectively, which will be used to restrict the total operating times and durations of the battery system (Kazemi and Zareipour 2017). For example, if the number of full charge and discharge cycles of a battery does not exceed 2 times per day, and the time of full charge and full discharge is calculated as 2.5 hours, the utilization hours per year will not exceed $2 \times 2.5 \times 365 = 1825$ hours. Let P_{cha} and P_{dis} be the rated battery charging and discharging capacity, respectively; and let η_{cha}, η_{dis} be the efficiency coefficients; then, the annualized battery energy EB_k can be written as Eq. (2.8).

$$EB_k = P_{cha} \cdot C_k \cdot H_{cha} \cdot \eta_{cha} = \frac{P_{dis} \cdot C_k \cdot H_{dis}}{\eta_{dis}} \tag{2.8}$$

2.3.2.7 Compensation of Wind, Solar, and Storage Resources

It is widely believed that solar and wind energy can be used as good fit as auxiliary and supplemental power sources for industry parks (Spellman 2013). However, due to the variable and intermittent features of wind and solar resources, their power output is volatile and unstable, thus requiring buffering and regulation using energy storage technology to enable the system to be flexibly dispatched (Xu et al. 2018b). Several documents have suggested the capacity ratio of storage to wind and PV, for example, the *design standard for wind-solar storage combined power stations in China (GB/T51437-2021)* states that the rated power of a storage system should not be less than 10% of the total installed rated power of wind turbines and PV panels to ensure system safety and stability (Ministry of Housing and Urban Rural Development of the People's Republic of China 2021); therefore, the battery installed capacity constraint is as Eq. (2.9).

$$\left(P_w \cdot A_k + P_{pv} \cdot B_k\right) \cdot 10\% \leq P_b \cdot C_k \tag{2.9}$$

2.3.2.8 Electricity Supply and Demand Balance

Industry parks, for example, data center parks, wastewater treatment parks, are often operated 24 hours per day, indicating that the electricity supply needs to be continuous. At the same time, the power demands of the industry parks at a certain time cannot be accurately estimated. To fully ensure the availability of the required electricity supply, this study calculates the full load operations at the industry parks while considering the electricity consumption as a triangular fuzzy parameter. For each industry park, let $\tilde{P}_k$ be the electricity required for treating each cubic meter of wastewater; here, $\tilde{P}_k = (p_{k1}, p_{k2}, p_{k3}, p_{k4})$; p_{k1}, p_{k4}, with p_{k1}, p_{k4}, respectively, being the lower and upper limits and with $[p_{k2}, p_{k3}]$ being the most likely range (Xu and Zhou 2011). Therefore, the required electricity per unit of product is estimated using trapezoidal fuzzy numbers, with $0 \leq \tau \leq 1$ being the confidence level. Then, $\tilde{P}_k$ can be estimated as Eq. (2.10):

$$\tilde{P}_k \rightarrow E\left[\tilde{P}_k\right] = \frac{1-\tau}{2}\left(p_{k1} + p_{k2}\right) + \frac{\tau}{2}\left(p_{k3} + p_{k4}\right) \tag{2.10}$$

The park's yearly electricity requirement (Eq. (2.11)) can be written as the product of the electricity required for treating each unit of product, the production scale S_k, and the total days per year T_n.

$$AD_k = E\left[\tilde{P}_k\right] \cdot S_k \cdot T_n \tag{2.11}$$

The renewable system is connected to the main network to ensure a continuous supply of power when there is insufficient renewable energy and to channel any excess renewable power to the main network. E_k is the electricity exported to or imported from the main grid at the kth industry park every year. Three possible situations can be considered as the follows.

$$\begin{cases} E_k < 0, \text{renewable energy is exported to the main grid} \\ E_k = 0, \text{no energy is exported or imported} \\ E_k > 0, \text{renewable energy is imported from the main grid} \end{cases}$$

The balance of power demand and supply can be shown in Eq. (2.12) On the basis of the above analysis, the electricity supply and demand balance equation is constructed. The left side is the annual energy supply from the wind and solar PV generation sources, to which a connection to the main grid is added. The right side is the energy demand based on a full load at the treatment plant throughout the year.

$$EW_k + EPV_k + E_k = ED_k \tag{2.12}$$

2.3.2.9 Available Area for New Energy Installation

Industry park usually occupy large areas and are thus ideal for laying wind–solar-storage energy systems. Take wastewater treatment parks considered in this chapter as an example, previous studies have proved that wastewater treatment plants of different scales have corresponding land use indices, and the total floor area of each wastewater treatment plant can be calculated as the product of the treatment scale and the index (Liu et al. 2010). As the range of these indexes is determined (as shown in Table 2.1), the land use index corresponding to each scale between each category can be estimated using the interpolation method. Let In_k be the planned land use index. Then the total area of the k_{th} wastewater treatment plant can be written as Eq. (2.13).

$$TS_k = S_k \cdot In_k \tag{2.13}$$

According to the proposed scheme shown in Figure 2.1, it is also necessary to evaluate the floor area of a single equipment and the maximum available area. The number of installed wind turbines and PV panels is limited by the available area at the industry park. Here we let $\tilde{\lambda}_k^w$, $\tilde{\lambda}_k^r$, $\tilde{\lambda}_k^f$ be the ratio coefficients of the available idle area, ancillary building plane area, and wastewater treatment area at the k_{th} industry park; these coefficients can be determined by interpolation and linear regression. Let (S_1, λ_1), (S_2, λ_2) be two possible area coefficients corresponding to two different treatment scales; then, $\tilde{\lambda}_k$ can be estimated as Eq. (2.14).

$$\tilde{\lambda}_k \to E\left[\tilde{\lambda}_k\right] = (S_k - S_1)\frac{\lambda_2 - \lambda_1}{S_2 - S_1} + \lambda_1 \tag{2.14}$$

Let s_w, s_r, and s_f be the area needed to install each wind turbine, roof PV panel, and flexible PV panel. The layout area constraints are calculated as Eqs. (2.15)–(2.17); in addition, the deployment of wind turbines needs the layout interval to be considered, and thus, a maximum number needs to be set.

$$A_k^w \cdot s_w \leq S_k \cdot E\left[\tilde{\lambda}_k^w\right] \tag{2.15}$$

$$B_k^r \cdot s_r \leq S_k \cdot E\left[\tilde{\lambda}_k^r\right] \tag{2.16}$$

$$B_k^f \cdot s_f \leq S_k \cdot E\left[\tilde{\lambda}_k^f\right] \tag{2.17}$$

2.3.3 Global Model

The wastewater treatment industry is among the most energy-intensive sectors worldwide, and integrating hybrid wind–solar-storage systems has been shown to be an effective means for users to generate on-site electricity. Consequently, incorporating renewable wind and solar generation technology into energy systems for wastewater treatment facilities would undoubtedly enhance their energy resilience and optimize space utilization. However, as the primary decision maker, the government must ensure that the regular operations of the energy system are maintained to safeguard social welfare. With this in mind, a mixed-integer multi-objective optimization model has been developed to assess the potential for long-term integration of renewable energy in the wastewater treatment sector, as discussed above. One advantage of this proposed model is that it can be applied not only to individual wastewater treatment plants but also to all such facilities in the surrounding region. The proposed model has three main objectives. The first is economic objective, with the aim of minimizing the average levelized cost of electricity. This provides an overall measure of the economic benefits of renewable energy implementation. The second objective is technological objective, with the aim of maximizing the total installed capacity of wind and solar generation technology, which ensures optimal electricity output under ideal conditions. Finally, the model has an environmental objective, which is to reduce the carbon emissions associated with electricity utilization, primarily from thermal power imported through the main grid. The proposed model focuses on the annual energy balance at wastewater treatment plants rather than on daily operations. As a result, it can serve as an effective tool for guiding investments in wind and solar generation facilities over the long term. The decision variables are, therefore, the types and numbers of wind turbines, roof PV panels, flexible PV panels, and batteries; and the electricity exported to or imported from the main grid for each wastewater treatment plant. By integrating the constraints and objectives, the global model is built as Eq. (2.18).

$$
\begin{aligned}
&\text{Min } f_1 = \frac{\sum_k (U_w \cdot A_k + U_f \cdot B_k^f + U_r \cdot B_k^r + U_b \cdot C_k + \sum_n \frac{E_k \cdot pr}{(1+r)^n})}{\sum_k \sum_{\mathrm{n}} S_k \cdot \tilde{P}_k \cdot T_n} \\
&\text{Max } f_2 = \textstyle\sum_k (P_w \cdot A_k + P_{pv} \cdot B_k) \\
&\text{Max } f_3 = \textstyle\sum_k (AD_k \cdot e_o - E_k \cdot e_o - AW_k \cdot e_w - APV_k \cdot e_{pv}) \\
&s.t. \begin{cases}
\sum_i x_k^i \leq 1, A_k^i \leq x_k^i \cdot M \\
AW_k = Q_k^w \cdot \sum_i (x_k^i \cdot P_w^i \cdot A_k^i) \\
\sum_j y_k^j \leq 1, B_k^{f\,j} + B_k^{r,j} = B_k^j, B_k^j \leq y_k^j \cdot M \\
APV_k = B_k^{pv} \cdot \sum_j (y_k^j \cdot P_{pv}^j \cdot B_k^j) \\
AB_k = P_{cha} \cdot C_k \cdot H_{cha} \cdot \eta_{cha} = \frac{P_{dis} \cdot C_k \cdot H_{dis}}{\eta_{dis}} \\
\left(P_w \cdot A_k + P_{pv} \cdot B_k\right) \cdot 10\% \leq P_b \cdot C_k \\
AD_k = E\left[\tilde{P}_k\right] \cdot S_k \cdot D_t \\
AW_k + APV_k + E^k = AD_k \\
TS_k = S_k \cdot In_k \\
A_k^w \cdot s_w \leq S_k \cdot E\left[\tilde{\lambda}_k^w\right] \\
B_k^r \cdot s_r \leq S_k \cdot E\left[\tilde{\lambda}_k^r\right] \\
B_k^f \cdot s_f \leq S_k \cdot E\left[\tilde{\lambda}_k^f\right] \\
A_k, B_k^r, B_k^f, C_k \in N \\
k = 1,2, \ldots K
\end{cases}
\end{aligned}
\tag{2.18}
$$

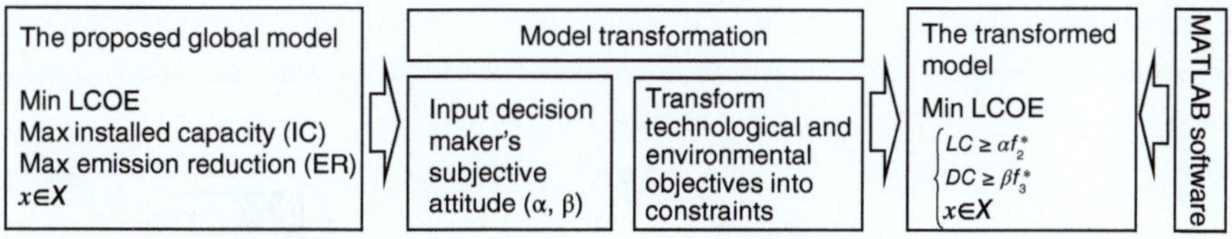

Figure 2.4 Model-solving flowchart.

To improve the model's applicability, this chapter classifies the daily treatment scales of wastewater treatment plants into seven categories. By adjusting the available area or energy consumption parameters, the model can be applied to a single wastewater treatment plant to achieve optimal equipment selection. Furthermore, by integrating the database of wastewater treatment plants in a region, the model can explore optimal regional economic performance, installed capacity, and decarbonization potential for wastewater treatment plants. In addition to wind–solar-storage systems, the proposed model can also be applied to wind-storage and solar-storage systems by adjusting the corresponding parameters. Furthermore, this model has potential for use in other industrial sectors, such as drinking water treatment plants.

2.3.4 Model Solving

It is a well-known fact that the government must balance a variety of objectives when making decisions. However, finding an optimized solution for a multi-objective problem can be challenging. Therefore, the ε-constraint method is introduced to transform and solve the proposed global model (Eq. (2.18)). The ε-constraint method is an effective approach for solving both convex and non-convex multi-objective models by transforming the objectives into constraints. This method has been widely applied to solve multi-objective optimal energy problems (Alcaraz et al. 2020). For example, Shojaei et al. (2020) transformed a probabilistic multi-objective reactive power planning model with objectives including total investment, voltage stability, and loadability into a single-objective model using the ε-constraint method. Similarly, Batista and Batista (2018) solved a multi-objective model of the energy demand-side management problem in smart grids by using the ε-constraint approach. In this study, economic performance is the most critical indicator to motivate stakeholders. Therefore, installed capacity and carbon reduction objectives are transformed into constraints (As show in Figure 2.4). The installed capacity is bounded by parameter f_2^*, and the coefficient α is assigned a value between 0 and 1 according to the decision-maker's attitude. Similarly, the coefficient β and parameter f_3^* are used to transform the carbon reduction objective.

2.4 Case Study

As the largest global carbon dioxide emitter, China has pledged to achieve carbon neutrality by 2060 and is thus driven to adopt wind and solar energy solutions

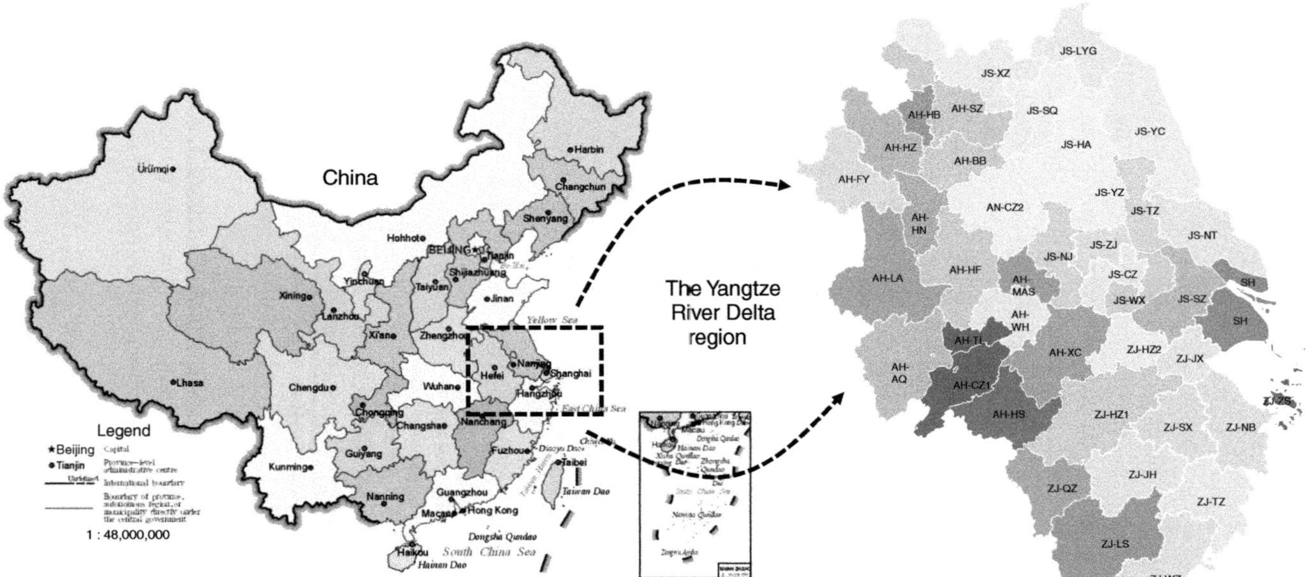

Source of map of China: Map Technology Review Center of the Ministry of Natural Resources of China
Web: http://bzdt.ch.mnr.gov.cn/browse.html?picId=%224o28b0625501ad13015501ad2bfc0292%22

Treatment Scale (TS) Unit: 10, 000 m^3/day

TS	811.25	475.71	329.00	303.62	302.57	275.83	228.73	226.30	196.07	173.74	172.30	170.63	142.88	141.98	138.65	120.17	107.79	97.65	86.75
City	SH	JS-SZ	AH-HF	JS-WX	ZJ-HZ1	JS-NJ	ZJ-SX	ZJ-NB	ZJ-WZ	ZJ-JX	ZJ-JH	JS-NT	JS-CZ	ZJ-TZ	JS-XZ	JS-YC	ZJ-HZ2	JS-SQ	JS-YZ

TS	82.85	80.85	78.14	77.67	72.75	71.00	69.65	65.00	63.95	59.70	55.32	51.40	48.38	47.30	46.50	44.15	41.51	40.25	26.30	22.33	19.70	18.25
City	JS-HA	AH-CZ2	AH-WH	JS-LYG	JS-ZJ	AH-FY	JS-TZ	AH-BB	AH-SZ	AH-AQ	AH-HZ	AH-HN	ZJ-QZ	AH-XC	AH-LA	AH-MAS	ZJ-LS	AH-HB	AH-HS	ZJ-ZS	AH-CZ1	AH-TL

Figure 2.5 Location and wastewater treatment capacity of cities in the Yangtze River Delta region.

(Mallapaty 2020). The YRD is a crucial region in China, responsible for generating 11,287.1 million m^3 of wastewater in 2019, accounting for 21.46% of the nation's total wastewater (National Bureau of Statistics of China 2020a). Therefore, the YRD region could play a crucial role in China's industrial decarbonization efforts. This section aims to study the case of the YRD to demonstrate the efficiency and practicality of the proposed method.

2.4.1 Case Description

The YRD region includes the Shanghai municipality and the Jiangsu, Zhejiang, and Anhui provinces. It has been exploring decarbonization efforts in various industries, including the wastewater treatment sector. With a growing population and increasing industrialization, wastewater treatment has become an important issue in the region. However, traditional wastewater treatment processes rely heavily on fossil fuels, which contribute to greenhouse gas emissions (National Bureau of Statistics of China 2020b). The YRD region is blessed with abundant wind and solar resources, which have the potential to provide a substantial amount of clean and renewable energy. However, despite this potential, the current level of wind and solar development in the region is not yet sufficient to fully harness these resources (Dong et al. 2019; World Resources Institute 2020). Therefore, the region has been looking to renewable energy sources such as wind and solar power to decarbonize the wastewater treatment industry. By integrating these clean energy sources into wastewater treatment processes, the industry can reduce its carbon footprint and move toward a more sustainable future. The statistical data of the 1662 wastewater treatment facilities in the YRD were adopted from the Ministry of Ecology and Environment of the People's Republic of China (2021), with the minimum scale being $0.1 \times 10^4\,\mathrm{m}^3/\mathrm{day}$ and with the maximum treatment scale being $280 \times 10^4\,\mathrm{m}^3/\mathrm{day}$. The statistics of the subregions are summarized in Figure 2.5.

2.4.2 Data Collection

Before executing the case study, necessary data were collected. The annual full load hours of wind turbines and PV panels were collected from each prefecture-level city and were derived from the Chinese solar PV industry survey association [http://www.21spv.com/news/show.php?itemid=96099] and National Energy Adimistration (2020). Ten types of wind turbines with rated power ranging from 10 kW to 1.5 MW (Table 2.2), and two types of PV panels with a rated power of 0.25–0.4 kW (Table 2.3) are considered in this chapter.

In addition, the construction and installation cost information was derived from market research. The operation and maintenance cost of wind turbines was 1.5% while that of PV panels was 1.0% of the construction and installation cost (Ragheb 2015). The discount rate was 5%. The annual battery utilization hours is 1750 hours. Battery parameters are shown in Table 2.4.

Table 2.2 Wind turbines parameters.

Type	Rated power (kW)	NPV factor	IC[a] (10^3 CNY)	O&M[b] (10^3 CNY)	CC[c] (10^3 CNY)	Roter Diameter (m)	Floor area (m^2)	Swept area (m^2)	Cut-in speed (m/s)	Cut-out speed (m/s)	Tower height (m)	Life span (yr)
1	10	13.8	3.4	0.05	4.06	7.0	138.5	615.4	2.5	45	18	25
2	20	13.8	6.3	0.09	7.60	10.0	178.5	1256.0	3.5	45	20	25
3	30	13.8	15.8	0.24	19.12	12.0	213.0	1808.6	3.5	45	22	25
4	50	13.8	33.5	0.50	40.45	16.4	311.1	3378.1	2.5	45	24	25
5	100	13.8	60.3	0.90	72.78	21.7	469.6	5914.4	3	20	30	25
6	200	13.8	113.7	1.71	137.23	25.0	590.6	7850.0	3.5	60	30	25
7	300	13.8	273.0	4.10	329.51	28.0	715.4	9847.0	3	60	32	25
8	500	13.8	450.0	6.75	543.14	31.2	864.2	12, 226.4	3	60	36	25
9	800	13.8	576.0	8.64	695.22	38.0	1233.5	18, 136.6	3	60	48	25
10	1500	13.8	750.0	11.25	905.23	70.0	3946.5	61, 544.0	3	60	60	25

a) IC: Initial investment.
b) O&M: Operation and maintenance cost.
c) CC: Comprehensive cost.

Table 2.3 PV panels parameters.

			Roof PV				Flexible PV				
Type	Rated power (kW)	NPV factor	IC[a] (CNY)	O&M[b] (CNY)	CC[c] (CNY)	Floor area (m^2)	IC[a] (CNY)	O&M[b] (CNY)	CC[c] (CNY)	Floor area (m^2)	Life span
1	0.25	13.80	1500	22.5	1810.47	2.00	2353.61	35.30	2840.76	2.50	25
2	0.40	13.80	2500	37.5	3017.45	3.50	3771.81	56.58	4552.50	4.00	25

a) IC: Initial investment.
b) O&M: Operation and maintenance cost.
c) CC: Comprehensive cost.

Table 2.4 Battery parameters.

Type	Rated power (kW)	NPV factor	Initial investment (CNY)	Operation and maintenance cost (CNY)	Comprehensive cost (CNY)	Life span (yr)	25 years' cost (CNY)
1	0.50	7.11	600.00	18.00	727.94	10	2183.82

As stated before, fuzzy numbers were applied to determine the uncertain parameters. For the electricity requirement for each unit of wastewater, it should be determined according to the treatment scale as there may be different likely and advanced values (Panepinto et al. 2016). Following the stipulation of energy consumption of municipal wastewater treatment of China (Beijing Municipal

Table 2.5 The electricity consumption parameters.

Category	Daily Treatment Scale (1×10^4 m^3/day)	Electricity data $\tilde{P}_k = (p_{k1}, p_{k2}, p_{k3}, p_{k4})$ (kWh/m^3)
I	[0.1,0.5)	(0.529, 0.570, 0.779, 0.795)
II	[0.5,2)	(0.529, 0.554, 0.779, 0.795)
III	[2,5)	(0.529, 0.540, 0.779, 0.795)
IV	[5,10)	(0.514, 0.524, 0.748, 0.763)
V	[10,20)	(0.497, 0.507, 0.722, 0.737)
VI	[20,50)	(0.475, 0.484, 0.690, 0.704)
VII	Above 50	(0.405, 0.463, 0.504, 0.514)

Table 2.6 Area coefficient parameters.

Fuzzy parameters	Treatment scale, area proportion ($1 \times 10^4 m^3$/day, –)	Estimation equation
Available idle area	(0.1, 10%), (100, 15%)	$\tilde{\lambda}_k^w = 0.0005 \cdot \mathrm{TS}_k + 0.1$
Roof area	(0.1, 30%), (100, 5%)	$\tilde{\lambda}_k^{pv1} = -0.0025 \cdot \mathrm{TS}_k + 0.300$
Wastewater treatment area	(0.1, 25%), (100, 50%)	$\tilde{\lambda}_k^{pv2} = 0.0025 \cdot \mathrm{TS}_k + 0.2497$

Bureau of Market Supervision 2021), the original electricity consumption values of WWTPs of different scales are summarized in Table 2.5.

The expected value of energy consumption can be estimated using Eq. (2.10). For the area coefficient parameters, fuzzy numbers and the interpolation method were applied to estimate Eq. (2.14). The results are shown in Table 2.6.

The category, location, treatment scale, maximum area for roof PV panels, maximum area for flexible PV panels, maximum area for wind turbines, annual full load hours of wind turbines, annual full load hours of PV panels, electricity consumption per m^3 of wastewater, and the annual total electricity consumption of each of the 24 WWTPs among the 1662 WWTPs are provided in Table 2.7.

The carbon emission coefficients per kWh of wind, PV, and thermal power generation were 0.034, 0.0499, and 0.96 kg/kWh, respectively (Nugent and Sovacool 2014). The industrial electricity price was set at the average level of 0.7 CNY/kWh.

2.4.3 Calculation Results and Analysis

This section comprehensively analyzes the optimal numbers of wind turbines and PV panels for WWTPs of a typical scale. Regional resultsinclude installed capacity and decarbonization potential, economic performance, and SSR.

2.4.3.1 Optimal Configurations Results

Using the proposed general model, the optimal results of the 24 typical WWTPs listed in Table 2.7 are explored. The optimal decision variables and objective values are

Table 2.7 Parameters of typical WWTPs.

		Parameters						
Category and Location	**TS 1 × 10⁴ m³/day**	S_r (m²)	S_f (m²)	S_w (m²)	FLH^w (h)	FLH^s (h)	$\tilde{p}^d$ (kWh/m³)	E^d (MWh)
I-JSLYG	0.1	600	500	200	1973	1191	0.668	244
I-JSLYG	0.15	900	750	300	1973	1191	0.668	366
I-JSLYG	0.2	1199	1001	400	1973	1191	0.668	488
I-JSLYG	0.25	1498	1252	501	1973	1191	0.668	610
I-JSLYG	0.3	1797	1503	601	1973	1191	0.668	732
I-SH	0.4	2394	2006	802	2065	1179	0.668	976
II-SH	0.5	2990	2510	1003	2065	1179	0.664	1212
II-JSLYG	1	5459	4625	1843	1973	1191	0.664	2425
II-JSLYG	1.5	7412	6338	2519	1973	1191	0.664	3637
III-SH	2	8857	7643	3030	2065	1179	0.661	4823
III-JSLYG	3	12295	10805	4263	1973	1191	0.661	7235
III-JSLYG	4	15092	13508	5304	1973	1191	0.661	9647
IV-SH	5	17264	15736	6150	2065	1179	0.637	11630
IV-JSLYG	6	19168	17792	6922	1973	1191	0.637	13956
IV-SH	8	21522	20718	7987	2065	1179	0.637	18608
V-SH	10	22018	21982	8400	2065	1179	0.616	22475
V-SH	12	24644	25516	9667	2065	1179	0.616	26970
V-JSWX	15	27585	30165	11288	1973	1070	0.616	33712
VI-SH	20	30024	35976	13200	2065	1179	0.588	42942
VI-JSNJ	30	38280	55220	19550	1973	1070	0.588	64413
VI-SH	40	42699	74635	25600	2065	1179	0.588	85885
VII-ZJHZ	60	42028	111972	36400	2090	986	0.472	103259
VII-SH	110	16500	165000	49500	2065	1179	0.472	189307
VII-SH	280	42000	420000	126000	2065	1179	0.472	481873

reported in Table 2.8, which includes the optimal rated power of the wind turbines and PV panels for each WWTP. The processing is conducted using MATLAB software, and the attitude parameters of the decision maker are set as $\alpha=\beta=0.8$.

Table 2.8 indicates that the appropriate power for the wind turbine model of WWTPs with different treatment scales ranges from 20 to 800 kW. As shown in Table 2.2, the hub height is between 20 and 48 m, and the cut-in wind speed is 2.5–3.5 m/s. Based on data from the Global Wind Atlas, the mean wind speed in the YRD region is around 3–7 m/s at a height of 10–50 m (Global Wind Atlas 2022). This indicates that the selected wind turbines are within the ideal range for the YRD region. The rated power of 0.25 kW for the PV panels is the best for all

Table 2.8 Results of typical WWTPs.

	Decision variables								Objective values		
Category and Location	**TS 1×10^4 m^3/day**	**P_w^* (kW)**	**P_{pv}^* (kW)**	***NW***	***NPV^r***	***NPV^f***	***NB***	**G^{main} (MWh)**	**LCOE (yuan)**	**IC (kW)**	**DC (t)**
I-JSLYG	0.1	20	0.25	1	200	300	29	56	0.349	145	172
I-JSLYG	0.15	30	0.25	1	300	449	44	84	0.358	217	258
I-JSLYG	0.2	50	0.25	1	400	599	60	92	0.342	300	362
I-JSLYG	0.25	100	0.25	1	500	749	83	41	0.272	412	521
I-JSLYG	0.3	200	0.25	1	601	898	115	−109	0.151	575	772
I-SH	0.4	300	0.25	1	802	1196	160	−233	0.148	799	1110
II-SH	0.5	500	0.25	1	1004	1494	225	−557	0.041	1124	1626
II-JSLYG	1	500	0.25	2	1849	2729	429	−911	0.085	2144	3067
II-JSLYG	1.5	800	0.25	2	2535	3706	633	−1378	0.042	3160	4614
III-SH	2	800	0.25	2	3057	4428	695	−687	0.154	3471	5068
III-JSLYG	3	800	0.25	3	4321	6147	1004	−617	0.186	5017	7221
III-JSLYG	4	800	0.25	3	5403	7546	1128	1057	0.283	5637	7893
IV-SH	5	800	0.25	4	6294	8632	1387	621	0.248	6932	10,124
IV-JSLYG	6	800	0.25	4	7116	9583	1475	2671	0.321	7375	10,371
IV-SH	8	800	0.25	5	8287	10,760	1753	4732	0.345	8762	12,760
V-SH	10	800	0.25	5	8791	11,009	1790	8377	0.402	8950	12,962
V-SH	12	800	0.25	6	10,206	12,322	2087	10,416	0.408	10,432	15,223
V-JSWX	15	800	0.25	6	12,066	13,792	2253	17,326	0.478	11,265	15,063
VI-SH	20	800	0.25	7	14,390	15,012	2591	22,709	0.478	12,951	18,598
VI-JSNJ	30	800	0.25	10	22,088	19,139	3662	37,603	0.511	18,307	24,651
VI-SH	40	800	0.25	12	29,852	21,348	4480	50,965	0.510	22,400	32,096
VII-ZJHZ	60	800	0.25	16	44,786	21,014	5850	60,284	0.512	29,250	39,536
VII-SH	110	800	0.25	19	66,000	8250	6753	136,028	0.570	33,763	48,989
VII-SH	280	800	0.25	47	168,000	21,000	16,970	348,505	0.572	84,850	122,613

WWTPs; this type of PV panel has been proved to be effective worldwide (Albadi et al. 2014). G^{main} is positive for the scale of 0.1–0.25 and above 4. The reason for the suitable power for wind turbines and PV panels increasing with the treatment scale of WWTPs is that small-scale WWTPs have a higher unit energy consumption, even though the renewable facilities that can be deployed are smaller in size. As the treatment scale increases, the total power consumption remains high even if the unit energy consumption is reduced, which means that energy still needs to be supplemented by the main power grid. In the treatment scale range of 0.3 to 3, G^{main} is negative, indicating that there is excess renewable energy that can be exported to the main grid. This results in a lower levelized cost of electricity (LCOE).

2.4.3.2 Economic Performance and Self-Sufficiency Ratio

This section involves a comparison of the economic performance and SSR) for each of the seven categories across Shanghai, Jiangsu, Zhejiang, and Anhui. After obtaining the optimal equipment results for 24 typical WWTPs (Table 2.8), the known values of P^{w*} and P^{pv*} are used in the calculation, and comprehensive datasets for 1662 WWTPs are computed using MATLAB, with attitude parameters of α=β =0.8.

By transforming the regional average LCOE of the wastewater treatment industry into a single-objective model, this study calculates the final overall LCOE = 0.3917 CNY/kWh. Compared to the average price of 0.7 CNY/kWh for electricity from the main grid, adopting renewable energy sources would be a cost-effective and efficient approach that would reduce the energy burden of the industry by 44.04%. Figure 2.6 demonstrates that the LCOE (histogram) and SSR (line chart) for each province and scale of WWTP exhibit similar trends. Among the different scales of WWTP, Category III, ranging from [2,5), has the lowest LCOE, indicating that adopting DRE in this category would be the most cost-effective. Categories I and II exhibit higher LCOE than Category III, and as the scale of WWTP increases, LCOE also gradually increases. This is because the I and II WWTPs are too small to make the installation of wind turbines economically feasible. On the other hand, for most Category III WWTPs, the optimal solution involves the use of both wind and solar energy, resulting in a significant reduction in LCOE. Nevertheless, as the scale of WWTP increases, more electricity is needed, which cannot be fulfilled solely by renewable energy sources. As a result, the reliance on the main power grid increases, leading to a corresponding rise in LCOE. The right axis of the line chart in Figure 2.6 indicates the SSR in the different WWTP scales and provinces. The variation patterns of the Shanghai, Jiangsu, Zhejiang, and Anhui are remarkably similar. Regarding each category, Category III exhibits the highest SSR, which is almost 100% in all four provinces. This trend is consistent with the changes in LCOE, which show an inverted V shape. Overall, Shanghai shows the best performance. This finding implies that for WWTPs situated in regions with abundant natural resources, priority should be given to the adoption of renewable energy infrastructure.

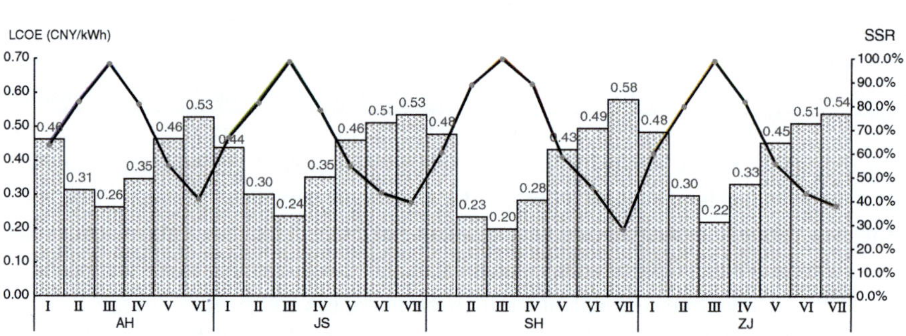

Figure 2.6 LCOE and SSR of each province and scale.

Table 2.9 Installed capacity and decarbonization potential.

Parameter	Unit	Category						
		I	II	III	IV	V	VI	VII
Installed capacity	GW	0.121	0.802	1.858	1.427	0.804	0.644	0.297
Decarbonization	Mt/year	0.127	1.056	2.572	1.967	1.1	0.87	0.418

2.4.3.3 Regional Decarbonization Potential

In this section, the installed capacity and decarbonization potential in the YRD region are analyzed. Taking LCOE as the objective, the installed capacity and decarbonization potential are transformed into constraints, and the calculation results are reported as follows.

Table 2.9 reports the calculation results of the installed capacity and annual carbon reduction of each of the seven categories. The installed capacity has a highly positive correlation with carbon reduction. The highest treatment scale is at Category III (31.21%) for the total installed capacity, followed by Categories IV (23.98%), V (13.50%), II (13.47%), VI (10.82%), VII (4.99%), and I (2.03%). To analyze the emission reduction effect of the renewable system, the difference between the carbon emissions of power generation using only coal and the carbon emissions of renewable energy is simulated. The total annual emission reduction amounts to 8.1 Mt, which is approximately equivalent to the total carbon dioxide absorbed by 8.1 million trees over 40 years (Tong et al. 2020). This highlights that the implementation of renewable energy in the wastewater treatment industry can significantly facilitate industrial decarbonization.

2.5 Comprehensive Discussions

This section presents a thorough discussion of the case study's calculation results, including scenario simulations and recommendations for strategy implementation.

2.5.1 Scenario Simulation

The base scenario assumes that the main grid is capable of accepting excess renewable energy. This is because in DRE systems, energy is primarily consumed locally, and the likelihood of injecting surplus energy into the main grid is relatively low. However, the simulation results of the YRD region's annual operation show that 34.9% of WWTPs generate excess renewable energy, which is exported to the main grid. However, in some instances, the national power grid may not permit the resale of renewable energy to ensure network stability (Numan et al. 2021). Therefore, it is necessary to take into account the issue of power grid congestion.

Here a scenario where export is prohibited is proposed for comparison. In the general model, G_k^{main} is unrestricted and can take any real number, with a negative

Table 2.10 Economy, technology, and environment results of two scenarios.

Category	LCOE (CNY/kWh)	Installed capacity (GW)	Decarbonization (Mt/year)
Export permitted (Base scenario)	0.3917	5.9525	8.11
Export not permitted	0.3926	5.6502	7.919

value indicating that the renewable energy is exported to the main power grid, and a positive value indicating that the renewable energy is insufficient, and the main power grid is used to replenish it. In the *Export not permitted* scenario, a restriction on export is added to the general model, ensuring that G_k^{main} is not less than 0. Table 2.10 reports the simulation results of the economy, technology, and environment of the studied region under the two scenarios. Table 2.10 indicates that with a limited export power, the regional installed capacity and annual carbon reduction decrease to 0.3023 GW and 0.191 Mt, respectively; and the average LCOE increases by 0.0009 CNY/kWh.

2.5.2 Management Recommendations

The optimal results from the multi-objective model show that the proposed grid-connected wind-solar storage system scheme for retrofitting existing WWTPs is economical for facility operators or local governments and useful in the facilitation of industrial decarbonization. These results provide quantitative evidence for the initiative of "encouraging wastewater treatment enterprises to build photovoltaic power generation projects", which was issued by the Chinese government in the "Guiding Opinions on Policies Related to Improving the Sewage Treatment Charging Mechanism in the Yangtze River Economic Zone (No. [2020] 561) (National Development and Reform Commission of China, Ministry of Finance China, Ministry of Housing and Urban-Rural Development China, Ministry of Ecology and Environment China, and Ministry of Water Resources China 2020)" in 2020. Further, the results also suggest that different models of wind turbines should be selected for WWTPs of different scales. In addition, WWTPs in Category III (daily treatment scale of 2×10^4 – 5×10^4 m^3/day) should prioritize the deployment of DRE systems as they have the lowest LCOE and highest SSR and can supply surplus renewable energy to the main grid for profit. Categories II and IV rank second. The varying performances with different WWTP scales can help decision makers formulate phased implementation strategies.

On the basis of the above propositions, a four-step renewable energy system implementation plan is suggested for the 1662 WWTPs in the study area: Step 1: Category III; Step 2: Categories II and IV; Step 3: Categories I and V; and Step 4: Categories VI and VII. The implementation of WWTP III promotes the regional installation of 2.572 GW of renewable energy infrastructure and reduces carbon emissions by 1.858 Mt per year. WWTP VI and VII are implemented last because of their large scale and significant impact on urban operations; therefore, implementation needs to take place when the technology and management are more mature.

References

Abbas, M.G., Wang, Z., Bashir, S. et al. (2021). Nexus between energy policy and environmental performance in China: the moderating role of green finance adopted firms. *Environmental Science and Pollution Research* 28: 63263–63277.

Albadi, M.H., Al Abri, R.S., Masoud, M.I. et al. (2014). Design of a 50 kW solar PV rooftop system. *International Journal of Smart Grid and Clean Energy* 3 (4): 402–409.

Alcaraz, J., Landete, M., Monge, J.F., and Sainz-Pardo, J.L. (2020). Multi-objective evolutionary algorithms for a reliability location problem. *European Journal of Operational Research* 283 (1): 83–93.

Ali, I., Shafiullah, G.M., and Urmee, T. (2018). A preliminary feasibility of roof-mounted solar PV systems in the Maldives. *Renewable and Sustainable Energy Reviews* 83: 18–32.

Babaelahi, M., Rafat, E., and Mofidipour, E. (2019). Emergy-based economic and environmental analysis and multi-objective optimization of a two-cascade solar gas turbine power plant. *Sustainable Production and Consumption* 20: 165–177.

Batista, A.C. and Batista, L.S. (2018). Demand side management using a multi-criteria ε-constraint based exact approach. *Expert Systems with Applications* 99: 180–192.

Beijing Municipal Bureau of Market Supervision (2021). Energy consumption quota for urban sewage treatment. http://scjgj.beijing.gov.cn/hdjl/myzj/bzzxdyjzj/202104/t20210425_2369595.html (accessed 10 October 2023).

Bey, M., Hamidat, A., and Nacer, T. (2021). Eco-energetic feasibility study of using grid-connected photovoltaic system in wastewater treatment plant. *Energy* 216: 119217.

Campana, P.E., Mainardis, M., Moretti, A., and Cottes, M. (2021). 100% renewable wastewater treatment plants: techno-economic assessment using a modelling and optimization approach. *Energy Conversion and Management* 239: 114214.

Campos, R.A., do Nascimento, L.R., and Rüther, R. (2020). The complementary nature between wind and photovoltaic generation in Brazil and the role of energy storage in utility-scale hybrid power plants. *Energy Conversion and Management* 221: 113160.

Cococcioni, M. and Fiaschi, L. (2021). The Big-M method with the numerical infinite *M*. *Optimization Letters* 15 (7): 2455–2468.

De Pee, A., Pinner, D., Roelofsen, O. et al. (2018). Decarbonization of industrial sectors: the next frontier. *McKinsey Global Institute* 66: 6–10.

Dewi, O.C., Ismoyo, A.D., Saskia, C.S. et al. (2019). Multifunction wastewater treatment plants as educational and recreational parks for the community. *IOP Conference Series: Earth and Environmental Science*, Volume 366, p. 012003. IOP Publishing.

Dong, S., Gong, Y., Wang, Z., and Incecik, A. (2019). Wind and wave energy resources assessment around the Yangtze River Delta. *Ocean Engineering* 182: 75–89.

Fernández-Marchante, C.M., Souza, F.L., Millán, M. et al. (2021). Improving sustainability of electrolytic wastewater treatment processes by green powering. *Science of the Total Environment* 754: 142230.

Global Wind Atlas (2022). https://globalwindatlas.info/ (accessed 10 October 2023).

Grant, S.B., Saphores, J.-D., Feldman, D.L. et al. (2012). Taking the "waste" out of "wastewater" for human water security and ecosystem sustainability. *Science* 337 (6095): 681–686.

Griggs, D., Stafford-Smith, M., Gaffney, O. et al. (2013). Sustainable development goals for people and planet. *Nature* 495 (7441): 305–307.

Helal, A., Ghoneim, W., and Halaby, A. (2013). Feasibility study for self-sustained wastewater treatment plants-using biogas CHP fuel cell, micro-turbine, PV and wind turbine systems. *Smart Grid and Renewable Energy* 4: 227–235.

Huang, R., Shen, Z., Wang, H. et al. (2021). Evaluating the energy efficiency of wastewater treatment plants in the Yangtze River Delta: perspectives on regional discrepancies. *Applied Energy* 297: 117087.

International Energy Agency (2020). World energy balances. https://www.iea.org/dataandstatistics/dataproduct/worldenergystatisticsandbalances (accessed 10 October 2023).

Jahangiri, M., Shamsabadi, A.A., Riahi, R. et al. (2020). Levelized cost of electricity for wind-solar power systems in Japan, a review. *Journal of Power Technologies* 100 (3): 188–210.

Jarvis, C.H. and Stuart, N. (2001). A comparison among strategies for interpolating maximum and minimum daily air temperatures. Part II: The interaction between number of guiding variables and the type of interpolation method. *Journal of Applied Meteorology* 40 (6): 1075–1084.

Kassem, Y., Gökçekuş, H., and Çamur, H. (2018). Economic assessment of renewable power generation based on wind speed and solar radiation in urban regions. *Global Journal of Environmental Science and Management* 4 (4): 465–482.

Kazemi, M. and Zareipour, H. (2017). Long-term scheduling of battery storage systems in energy and regulation markets considering battery's lifespan. *IEEE Transactions on Smart Grid* 9 (6): 6840–6849.

Mansouri Kouhestani, F., Byrne, J., Johnson, D. et al. (2020). Multi-criteria PSO-based optimal design of grid-connected hybrid renewable energy systems. *International Journal of Green Energy* 17 (11): 617–631.

Kumar, K.S. (2011). Computer aided design of waste water treatment plant with activated sludge process. *International Journal of Engineering Science and Technology* 3 (4): 3348–3356.

Lin, J., Fan, R., Tan, X., and Zhu, K. (2021). Dynamic decision and coordination in a low-carbon supply chain considering the retailer's social preference. *Socio-Economic Planning Sciences* 77: 101010.

Liu, J., Xilai, Z., Chao, G., and Lei, C. (2010). Study on land use, operation and construction cost of urban sewage treatment plant. *Journal of Environmental Engineering* 4: 2522–2526. in Chinese.

Lu, L., Guest, J.S., Peters, C.A. et al. (2018). Wastewater treatment for carbon capture and utilization. *Nature Sustainability* 1 (12): 750–758.

Mallapaty, S. (2020). How China could be carbon neutral by mid-century. *Nature* 586 (7830): 482–483.

McIlwaine, N., Foley, A.M., Morrow, D.J. et al. (2021). A state-of-the-art techno-economic review of distributed and embedded energy storage for energy systems. *Energy* 229: 120461.

McKuin, B., Zumkehr, A., Ta, J. et al. (2021). Energy and water co-benefits from covering canals with solar panels. *Nature Sustainability* 4: 609–617.

Ministry of Construction of the People's Republic of China (2001). Construction standard of urban sewage treatment project. http://mohurd.gov.cn/wjfb/200902/t20090211_185716.html (accessed 10 October 2023).

Ministry of Ecology and Environment of the People's Republic of China (2021). Government information disclosure. http://www.mee.gov.cn/xxgk2018/ (accessed 10 October 2023).

Ministry of Housing and Urban Rural Development of the People's Republic of China (2021). Design standard of wind power storage combined power station (GB/T51437-2021). http://www.mohurd.gov.cn/wjfb/202108/t20210818_251220.html (accessed 10 October 2023).

Nakkasunchi, S., Hewitt, N.J., Zoppi, C., and Brandoni, C. (2021). A review of energy optimization modelling tools for the decarbonisation of wastewater treatment plants. *Journal of Cleaner Production* 279: 123811.

National Bureau of Statistics of China (2020a). China Energy Statistics Yearbook 2020.

National Bureau of Statistics of China (2020b). China Environmental Statistics Yearbook 2020.

National Development and Reform Commission of China, Ministry of Finance China, Ministry of Housing and Urban-Rural Development China, Ministry of Ecology and Environment China, and Ministry of Water Resources China (2020). Guiding Opinions on Policies Related to Improving the Sewage Treatment Charging Mechanism in the Yangtze River Economic Zone. https://www.ndrc.gov.cn/xxgk/zcfb/tz/202004/t20200416_1225845.html (accessed 20 October 2023).

National Energy Adimistration (2020). Grid connected operation of wind power in 2019. http://www.nea.gov.cn/2020-02/28/c_138827910.htm (accessed 10 October 2023).

National Renewable Energy Laboratory of the US (2015). IEA Wind Task 26: Wind Technology, Cost, and Performance Trends in Denmark, Germany, Ireland, Norway, the European Union, and the United States: 2007–2012. https://www.nrel.gov/docs/fy15osti/64332.pdf (accessed 13 October 2023).

Nawarkar, C.J. and Salkar, V.D. (2019). Solar powered electrocoagulation system for municipal wastewater treatment. *Fuel* 237: 222–226.

Ng, C. and Ran, L. (2016). *Offshore Wind Farms: Technologies, Design and Operation*. Woodhead Publishing.

Nguyen, H.T., Safder, U., Nguyen, X.Q.N., and Yoo, C.K. (2020). Multi-objective decision-making and optimal sizing of a hybrid renewable energy system to meet the dynamic energy demands of a wastewater treatment plant. *Energy* 191: 116570.

Niu, K., Wu, J., Qi, L., and Niu, Q. (2019). Energy intensity of wastewater treatment plants and influencing factors in China. *Science of the Total Environment* 670: 961–970.

Nugent, D. and Sovacool, B.K. (2014). Assessing the lifecycle greenhouse gas emissions from solar PV and wind energy: a critical meta-survey. *Energy Policy* 65: 229–244.

Numan, M., Khan, A.Z., Asif, M. et al. (2021). Exploiting the inherent flexibility in transmission network for optimal scheduling, wind power utilization, and network congestion management. *IEEE Access* 9: 88746–88758.

O'Meara, S. (2020). China's plan to cut coal and boost green growth. *Nature* 584 (7822): S1–S1.

Our World in Data (2020). Emissions by sector. https://ourworldindata.org/emissions-by-sector (accessed 26 December 2022).

Panepinto, D., Fiore, S., Zappone, M. et al. (2016). Evaluation of the energy efficiency of a large wastewater treatment plant in Italy. *Applied Energy* 161: 404–411.

Patziger, M. (2021). Improving wastewater treatment plant performance by applying CFD models for design and operation: selected case studies. *Water Science and Technology* 84 (2): 323–332.

Ragheb, M. (2015). Economics of wind energy. *Wind Power Systems, Course NPRE*, 475.

Saleh, A. and Al-Nimr, M.'d.A. (2020). A novel vacuum wastewater treatment plant integrated with a solar absorption system. *International Journal of Energy Research* 44 (3): 1685–1697.

Schallenberg-Rodríguez, J. and Montesdeoca, N.G. (2018). Spatial planning to estimate the offshore wind energy potential in coastal regions and islands. Practical case: The Canary Islands. *Energy* 143: 91–103.

Seker, S. and Kahraman, C. (2021). Socio-economic evaluation model for sustainable solar PV panels using a novel integrated MCDM methodology: a case in Turkey. *Socio-Economic Planning Sciences* 77: 100998.

Shawon, M.J., El Chaar, L., and Lamont, L.A. (2013). Overview of wind energy and its cost in the middle east. *Sustainable Energy Technologies and Assessments* 2: 1–11.

Shojaei, A.H., Ghadimi, A.A., Miveh, M.R. et al. (2020). Multi-objective optimal reactive power planning under load demand and wind power generation uncertainties using ε-constraint method. *Applied Sciences* 10 (8): 2859.

Skouteris, G., Rodriguez-Garcia, G., Reinecke, S.F., and Hampel, U. (2020). The use of pure oxygen for aeration in aerobic wastewater treatment: a review of its potential and limitations. *Bioresource Technology* 312: 123595.

Spellman, F.R. (2013). *Water & Wastewater Infrastructure: Energy Efficiency and Sustainability*. CRC Press.

Spillias, S., Kareiva, P., Ruckelshaus, M., and McDonald-Madden, E. (2020). Renewable energy targets may undermine their sustainability. *Nature Climate Change* 10 (11): 974–976.

Strazzabosco, A., Kenway, S.J., and Lant, P.A. (2019). Solar PV adoption in wastewater treatment plants: a review of practice in California. *Journal of Environmental Management* 248: 109337.

Sun, B., Yu, Y., and Qin, C. (2017). Should China focus on the distributed development of wind and solar photovoltaic power generation? A comparative study. *Applied Energy* 185: 421–439.

Tong, X., Brandt, M., Yue, Y. et al. (2020). Forest management in southern China generates short term extensive carbon sequestration. *Nature Communications* 11 (1): 1–10.

Van Loosdrecht, M.C.M. and Brdjanovic, D. (2014). Anticipating the next century of wastewater treatment. *Science* 344 (6191): 1452–1453.

Wang, F., Xie, Y., and Xu, J. (2019). Reliable-economical equilibrium based short-term scheduling towards hybrid hydro-photovoltaic generation systems: case study from china. *Applied Energy* 253: 113559.

Wang, F., Xu, J., Liu, L. et al. (2021). Optimal design and operation of hybrid renewable energy system for drinking water treatment. *Energy* 219: 119673.

Wei, X., Duan, Y., Liu, Y. et al. (2019). Onshore-offshore wind energy resource evaluation based on synergetic use of multiple satellite data and meteorological stations in Jiangsu Province, China. *Frontiers of Earth Science* 13 (1): 132–150.

World Resources Institute (2020). Potential and vision of distributed renewable energy in Yangtze River Delta region. https://www.wri.org.cn/en/publication/potential-and-vision-distributed-renewable-energy-yangtze-river-delta-region (accessed 17 October 2023).

Wu, H., Locment, F., and Sechilariu, M. (2019). Experimental implementation of a flexible PV power control mechanism in a DC microgrid. *Energies* 12 (7): 1233.

Wu, W., Zhang, T., Xie, X., and Huang, Z. (2021). Regional low carbon development pathways for the Yangtze River Delta region in China. *Energy Policy* 151: 112172.

Xu, J. and Zhou, X. (2011). *Fuzzy-Like Multiple Objective Decision Making*, vol. 263. Springer.

Xu, J., Wang, F., Lv, C. et al. (2018a). Economic-environmental equilibrium based optimal scheduling strategy towards wind-solar-thermal power generation system under limited resources. *Applied Energy* 231: 355–371.

Xu, J., Wang, F., Lv, C., and Xie, H. (2018b). Carbon emission reduction and reliable power supply equilibrium based daily scheduling towards hydro-thermal-wind generation system: a perspective from China. *Energy Conversion and Management* 164: 1–14.

Zhang, Q., Wang, L., Li, G., and Liu, Y. (2020). A real-time energy management control strategy for battery and supercapacitor hybrid energy storage systems of pure electric vehicles. *Journal of Energy Storage* 31: 101721.

3

Sustainable Operation-Oriented Deployment of Hybrid Wind–Solar-Storage System

At present, a new round of global energy revolution and scientific and technological revolution has evolved in depth and is still in full swing, and the vigorous development of renewable energy has become a major strategic direction and a unanimous ambitious action in the global energy transition and the response to climate change.

Renewable Energy Development Plan for the 14th Five-Year Plan
National Development and Reform Commission of China, etc., Jun. 2022

With the increase in population and the acceleration of urbanization, urban drinking water supply has increased significantly. However, the industrial process of drinking water treatment requires intensive energy, leading to higher enterprise costs and increased carbon emissions carbon emission. Therefore, to ensure sustainable low polluting power supplies for drinking water treatment plants, this chapter proposes an optimal application of sustainability-oriented grid-connected hybrid wind–PV–battery system. A multi-objective nonlinear dynamic model with integer variables that considered both system planning and operations and included four sustainability dimensions: economic, technical, environmental, and social is constructed that allows decision makers to determine whether to deploy the proposed system. The ε-constraint method and system planner attitude parameters are introduced into the proposed model to transform it into its equivalent single objective form, which is then solved using LINGO software. A case study in China is given to assess the viability of the hybrid system components, and various electricity pricing strategies and natural resource changes are analyzed and compared. It is found that the grid-connected hybrid wind–solar-storage system was able to reduce the long-term electricity utilization costs at the drinking water treatment plant, with the feasible self-sufficiency ratio reaching 95%. Management recommendations for drinking water treatment power plants, the water treatment industry, and associated administration departments are also provided to explore the application of the proposed system.

Sustainable Hybrid Energy Systems: Carbon Neutral Approaches, Modeling, and Case Studies, First Edition. Jiuping Xu and Fengjuan Wang.

3.1 Background Review

Over the past few decades, there has been a significant increase in the global urban population, which has grown from an estimated 0.8 billion in 1950 (30%) to an estimated 4.2 billion in 2018 (55%) (Department of Economic and Social Affairs - the United Nations 2018). This population growth has led to an increase in urban drinking water supply due to the acceleration of urbanization index-urbanization (De Clercq et al. 2018). However, the industrial process of treating drinking water requires intensive energy, leading to higher enterprise costs and increased carbon emissions(García-Cáceres et al. 2019). For instance, in China, providing 83.81 billion cubic meters of drinking water in 2017 required 6.76 billion kilowatt-hours of electricity, resulting in 37.38 million tonnes of carbon dioxide emissions (National Bureau of Statistics of China 2019; International Energy Agency 2019b). With rising fuel prices and increasingly stringent water quality requirements, the future electricity use and carbon emissions associated with urban drinking water supply are likely to continue to increase (Gao et al. 2020). Traditionally, electricity is generated at large-scale power plants and distributed to water treatment plants through high- and low-voltage transmission and distribution networks (Luthander et al. 2015). However, since 70.41% of China's electricity is still generated from fossil fuels (China Electricity Council 2020) and long-distance transmissions cause unavoidable energy losses, there is a critical need for the deployment of decentralized renewable energy systems (Soshinskaya et al. 2014).

Out of all the currently developed distributed renewable sources, wind and solar have been found to be the most competitive due to their easy access, relative abundance, and extensive application prospects (Weschenfelder et al. 2020). According to Key World Energy Statistics 2019, the world's net installed capacity of wind and solar power was 514.9 and 397.7 GW separately, which ranked first and second in all non-hydro renewable sources (International Energy Agency 2019a). However, due to their randomness and intermittency, the utilization of a single source often leads to an excessive system size, increase in capital cost, and decrease in electricity reliability (Prebeg et al. 2016). Consequently, there has been significant useful research on the modeling, simulation, and optimization of integrated wind and solar resource utilization (Deng and Lv 2020). For example, Sanajaoba and Fernandez (2016) applied a Cuckoo Search algorithm to an off-grid hybrid energy system design problem and proved that the hybrid integration of photovoltaic, wind, and battery storage was the most reliable and cost-effective scheme for remote areas. Ahadi et al. (2016) proposed a wind, PV, and energy storage system optimization approach in diesel-free isolated communities and found that the wind turbine was an essential power component for renewable energy real installations. Ghorbani et al. (2018) conducted a sizing cost-reliability optimization for an off-grid house with photovoltaic panels, wind turbines, and batteries and found that the hybrid unit levelized cost was 0.502. However, compared with off-grid hybrid power systems that are widely used for remote and isolated areas, more and more urban utility customers are shifting toward a grid-connected hybrid energy system.

This system is not only environmentally friendly, but it also holds higher flexibility and reliability (Caballero et al. 2013).

As a result, a lot of research has shifted toward designing and operating grid-connected hybrid renewable energy systems, with a focus on analyzing their economic, technical, environmental, and social feasibility. Various studies have been conducted in this regard, such as a techno-economic analysis of PV-hydropower systems for a large office building in southeast France by Islam (2018), an analysis of optimal economic–environmental factors for a hybrid photovoltaic-wind-biomass power system in a sample township in central Catalonia by Gonzalez et al. (2018), and an analysis of the techno-economic-societal benefits for nine different integrated renewable energy systems for four remote and isolated communities in India by Chauhan and Saini (2016). Additionally, Baneshi and Hadianfard (2016) evaluated the techno-economic and environmental aspects of hybrid diesel-PV-wind-battery electricity generation systems for large non-residential electricity consumers based on the climate conditions in southern Iran. While these studies have made significant contributions toward investigating the feasibility of applying hybrid renewable energy systems to various objects, there has been little attention paid to the industrial process of urban drinking water treatment.

Previous studies on hybrid renewable energy systems for drinking water treatment plants (DWTPs) have primarily focused on assessing their technical and economic benefits. For instance, Soshinskaya et al. (2014) examined the potential for a renewable energy-based microgrid at a DWTP in the Netherlands and identified a grid-connected system with wind, PV, and demand response as the most cost-effective option, achieving 70% to 96% self-sufficiency. Similarly, García-Vaquero et al. (2014) evaluated the performance of a pilot wind and solar-powered nanofiltration plant compared to conventional DWTPs and concluded that the renewable energy system was suitable for sustainable and efficient drinking water treatment. Bukhary et al. (2017) investigated the feasibility of using photovoltaics to generate energy for an existing DWTP and found that solar PV could significantly reduce the dependence on the traditional electric grid. Although these studies have effectively explored the technical and economic benefits of hybrid renewable generation systems in DWTPs, a comprehensive analysis that also considers the environmental and social benefits has not been conducted yet.

Thus, this chapter aims to fill the research gap by proposing a sustainability-oriented hybrid renewable energy system for DWTPs. A grid-connected system integrating wind, PV, and storage, along with a rule-based operation strategy, will be developed for DWTPs. A multi-objective decision-making model will be created, taking into account four sustainability dimensions: economic sustainability, technical sustainability, environmental sustainability, and social sustainability. The decision objectives will be represented by levelized electricity cost, self-sufficiency ratio, carbon emissions, and job creation. A model transformation method based on the ε-constraint method and the decision-makers' atitude parameters will be introduced. To demonstrate the proposed optimization method's practicality and efficiency, a case study will be conducted in China. The results obtained from different hybrid system components, electricity price strategies, and changing natural resources will be compared, and management recommendations will be provided.

3.2 Main Issue Description

This section outlines the system components and operational strategy of the suggested hybrid renewable energy system for DWTPs, which consists of wind turbines, photovoltaic panels, and battery storage.

3.2.1 System Schematic

The proposed hybrid renewable energy system for DWTPs consists of wind turbines, solar PV arrays, and a battery storage system. The PV arrays and battery storage system are connected to a DC bus, while the wind turbines and main grid are connected to an AC bus. A power converter is used to connect the AC and DC buses, and the load at the drinking water treatment plant is supplied by the AC bus, as illustrated in Figure 3.1. The distribution of electricity in the proposed hybrid system for DWTPs is as follows: (i) Solar power, after being converted to AC, indirectly supplies the demand, with any surplus power sent to battery storage. (ii) Wind power directly meets demand, with any excess power converted to DC and sent to the battery. (iii) If the battery is full, any surplus electricity is exported to the grid. (iv) In case of insufficient electricity generated by the hybrid system, the shortage is first provided by the battery and then by the grid.

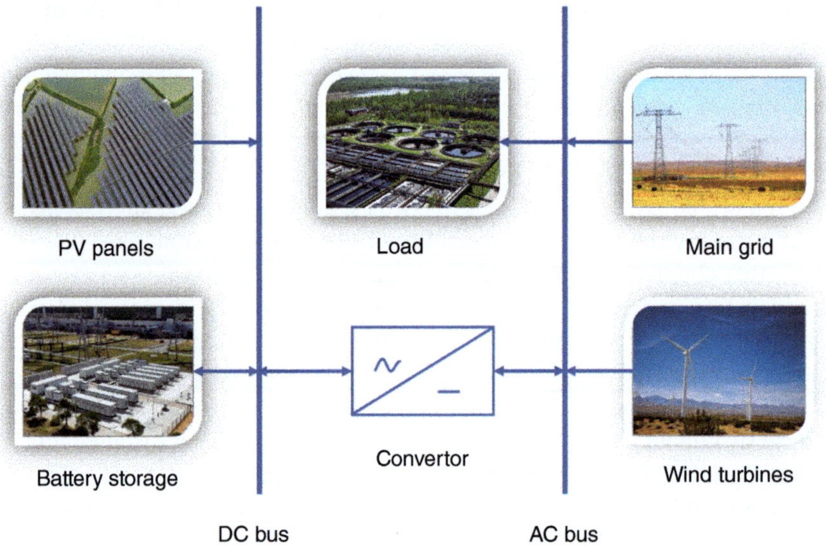

Figure 3.1 Component and structure of the proposed grid-connected hybrid wind–PV–battery system for drinking water treatment plant. Source: Kekyalyaynen/Adobe Stock (Load).

3.2.2 Operation Strategy

The operational procedure for the hybrid system is illustrated in Figure 3.2. At each time step t, the wind turbine and solar PV array power outputs are denoted as $P_{w,t}$ and $P_{pv,t}$, respectively. The power demand at time t is represented by D_t. The state-of-charge (SOC) of the battery at time step t, S_t, is constrained by its minimum and maximum values, S_{min} and S_{max}, respectively. Moreover, p_w and p_{pv} represent the wind and solar power prices, respectively. The operating strategy is as follows.

If the sum of the wind and solar power outputs at time t, $P_{w,t} + P_{pv,t}$, is greater than the power demand, D_t, then the excess power is used to meet the demand. If the battery state-of-charge, S_t, is equal to the upper bound S_{max}, the surplus electricity is sold to the main grid. If S_t is less than S_{max}, the excess power is used to charge the battery, giving priority to the power source with the lower on-grid price. However, if the battery is fully charged ($S_t = S_{max}$), all remaining power is sold to the main grid. If the battery is not fully charged, the source with the higher on-grid price is used to charge the system.

If the sum of the wind and solar power outputs, $P_{w,t} + P_{pv,t}$, is less than the power demand, D_t, the wind and solar power alone are unable to meet the demand. In this case, if the battery state-of-charge, S_t, is not at its lower bound, S_{min}, the required

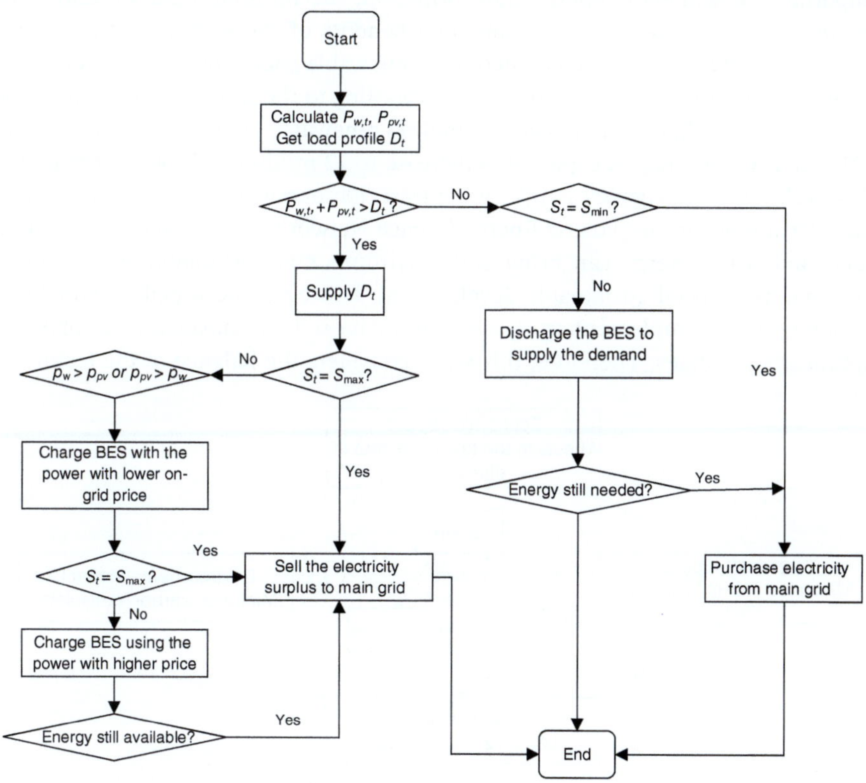

Figure 3.2 Operating strategy flowchart.

energy is supplied by the battery system. However, if $S_t = S_{min}$, the required energy is purchased from the main grid. Moreover, if the energy stored in the battery is insufficient and more energy is required to meet the demand, the shortfall is purchased from the main grid.

3.2.3 Optimization Scheme

The configuration of wind–solar-storage complementary systems needs to balance multiple objectives. The configuration of wind-solar-storage complementary systems needs to balance multiple objectives, as shown in Figure 3.3. From a technical and economic perspective, different levels of self-sufficiency will result in different configuration scales and therefore different electricity costs. Therefore, self-sufficiency rates need to be considered in the complementary system capacity decision-making process. From an environmental perspective, relying on distributed renewable energy systems in drinking water treatment parks can provide nearly 100% green electricity. However, electricity supplied by the main grid contains a certain proportion of thermal power, as China and other developing countries still rely on coal-fired power plants for over 60% of their electricity. Therefore, different amounts of electricity from the main grid will result in different carbon emissions. Considering the cleanliness of solar and wind resources is important for achieving energy transformation, the minimum carbon emissions for users can measure the environmental benefits of the system. From a social perspective, the large-scale construction of renewable energy projects also creates a significant number of job opportunities. According to the International Renewable Energy Agency, there were already 11 million renewable energy jobs worldwide in 2018, and this number is expected to increase to 22 million by 2050 (International Renewable Energy Agency 2018). In contrast, job opportunities in the fossil fuel industry are declining (Polaris Energy Storage Network 2021). This indicates that solar and wind energy can bring both environmental sustainability advantages and maintain social sustainable development. However, the social sustainability advantages of complementary systems have been less considered in previous optimization studies. Therefore, this study considers the balance of the technical,

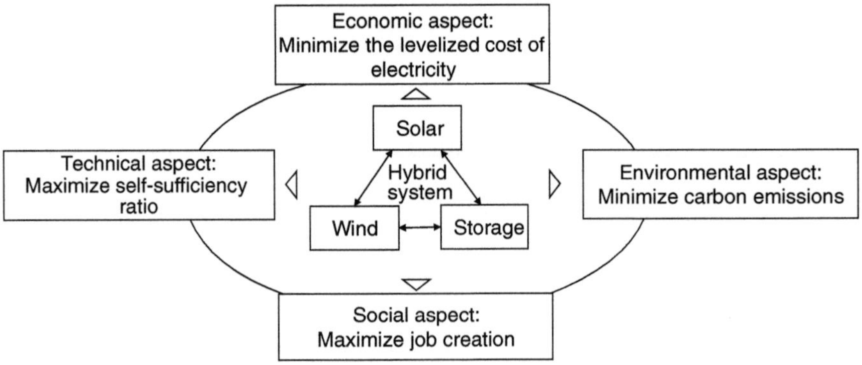

Figure 3.3 Optimization scheme of sustainable hybrid wind–solar-storage system.

economic, environmental, and social dimensions of the hybrid wind–solar-storage system (Figure 3.3).

3.3 Mathematical Modeling

3.3.1 Notations

Indices

i	Index of season, $i = 1 \ldots 4$.
t	Index of time period, $t = 1 \ldots 24$.
n	Index of year. $n = 1 \ldots 20$.

Decision variables

X_w	Number of wind turbines.
X_{pv}	Number of solar arrays.
X_b	Number of battery units.
Y_{it}^{ex}	Electricity exported to the main grid at time step t in season i (kWh).
Y_{it}^{in}	Electricity imported from the main grid at time step t in season i (kWh).
Z_{it}^{cha}	Battery charged power in season i time period t (kW).
Z_{it}^{dis}	Battery discharged power in season i time period t (kW).

Parameters

IC_w	Initial investment cost of a unit of wind turbine (CNY/kW).
IC_{pv}	Initial investment cost of a unit of photovoltaic array (CNY/kW).
IC_b	Initial investment cost of a unit of battery storage (CNY/kW · h).
Ba_{it}	Power of battery systems at time step t in season i (kW).
UC_w	Unit cost of wind turbine.
UC_{pv}	Unit cost of photovoltaic array.
UC_b	Unit cost of battery storage.
D_{it}	Electricity demand of drinking water treatment plant at time step t in season i (kW · h).
H	Height of hub (m).
H_0	Height of anemometer (m).
K_i	Number of days of season i.
MC_n^w	Operation and management cost of wind turbine in year n (CNY/kW).
MC_n^{pv}	Operation and management cost of photovoltaic array in year n (CNY/kW).
MC_n^b	Unit operation and management cost of battery storage in year n (CNY/kW · h).
A_b	Rated capacity of a unit of battery.
AP_{it}^{in}	Electricity price of the main grid at time step t in season i (CNY/kW · h).
AP_{it}^{pv}	Actual solar power output in season i at time period t (kW).
AP_{it}^{w}	Actual wind power output in season i at time period t (kW).

A_{pv}	Rated solar power output of a unit of photovoltaic array under standard test condition.
A_w	Rated power output of wind turbine.
r	Discount rate.
UR_n^w	Unit replacement cost of wind turbine in year n (CNY/kW).
UR_n^{pv}	Unit replacement cost of photovoltaic array in year n (CNY/kW).
UR_n^b	Unit replacement cost of battery storage in year n (CNY/kW · h).
s_0	Initial state of charge.
G_{it}	Actual solar radiation in season i at time period t (Wh/m^2).
G_s	Solar radiation under standard test condition (Wh/m^2).
$B_{\min}$	Minimum state-of-charge of the battery system.
$B_{\max}$	Maximum state-of-charge of the battery system.
T_{it}	Actual ambient temperature in season i at time period t (°C).
T_{sta}	Ambient temperature under standard test condition °C.
v_{it}	Actual wind speed at the height of hub in season i at time period t.
v_{it}^0	Wind speed at the height of anemometer in season i at time period t (m/s).
V_{in}	Cut-in wind speed of the turbine.
V_r	Rated wind speed of the turbine.
V_{out}	Cut-off wind speed of wind turbines.
α	Decision-maker's attitude toward carbon emission.
β	Decision-maker's attitude toward job creation.
ε	Decision-maker's attitude toward self-sufficiency ratio.
γ	Coefficient of hub and anemometer, often set as 0.143.
η	Charging and discharging efficiency of battery system.
μ	The ratio of thermal generation in the energy mix.
ϖ	Temperature coefficient, often set as −0.35%/°C.
σ	Self-discharging factor of battery system.
ζ	Carbon emission parameter of the coal burning generation process in the main grid (g/kWh).

3.3.2 Sustainable Deployment

The drinking water treatment plant power supply's sustainability is evaluated using four dimensions: economic, technical, environmental, and social sustainability. Indicators associated with each dimension are described below.

3.3.2.1 Economic Sustainability: Minimize the Levelized Cost of Electricity

The primary goal of decision makers responsible for hybrid generation system planning, such as drinking water treatment plant planners or third-party investors, is to establish a system that can meet the load demand at an acceptable cost. To determine the economic feasibility of the system, important indicators such as investment cost, operation and management cost, and equipment replacement cost must be evaluated (Lian et al. 2019). The most commonly used indicators for evaluating the economic feasibility of hybrid generation systems in literature include total annual cost (Yan et al. 2019), annualized system costs (Bilal et al.

2010), life cycle costs (Olcan 2015), levelized cost of energy (Barutha et al. 2019), and net present value (Zhang et al. 2017). This chapter utilizes the levelized cost of energy metric, as it is the most commonly used measure for comparing different electricity generation unit costs. To determine whether to implement a renewable system, decision makers compare the levelized costs of hybrid generation systems to the costs of purchasing electricity from the main grid, as shown mathematically in Eq. (3.1). The equation includes the levelized cost of electricity, f_1, and the costs of the wind turbine, solar array, and battery storage unit, denoted as UC_w, UC_{pv}, and UC_b, respectively. The number of wind turbines, PV arrays, and battery storage units to be installed, denoted as X_w, X_{pv}, and X_b, are decision variables that must also be determined during the system planning period.

$$\min f_1 = \frac{UC_w X_w + UC_{pv} X_{pv} + UC_b X_b + \sum_n \frac{\sum_i K_i (\sum_t Y_{it}^{in} AP_{it}^{in})}{(1+r)^n}}{\sum_n \frac{\sum_i K_i (\sum_t D_{it})}{(1+r)^n}} \tag{3.1}$$

To estimate the annual electricity demand, the daily demand is categorized into four groups to represent typical spring, summer, autumn, and winter loads, denoted as $i = 1, \dots, 4$. Each typical day is divided into 24 time steps of one hour, represented as $t = 1, \dots, 24$. The amount of electricity imported from the main grid at time step t during season i is denoted as Y_{it}^{in}, while p_{it}^{in} represents the price of purchasing electricity from the main grid at time step t during season i. The total cost of importing electricity during a typical day in season i is calculated as $\sum_t Y_{it}^{in} \cdot p_{it}^{in}$. In addition, K_i denotes the number of days in season i. The total annual cost of purchasing electricity is given by $\sum_i K_i \cdot \left(\sum_t Y_{it}^{in} \cdot p_{it}^{in}\right)$. Similarly, the annual electricity demand of the drinking water treatment plant can be calculated as $\sum_i K_i \cdot \left(\sum_t D_{it}\right)$. Since the levelized cost of electricity is a life cycle estimation, this chapter introduces the discount rate r and the lifetime of the proposed hybrid generation system n ($n = 1, \dots, 20$, without considering depreciation and residual value). The total cost of a unit of wind turbine, solar array, and battery storage includes the initial investment cost, operation and management cost, and equipment replacement cost, as shown in Eqs. (3.2)–(3.4).

$$UC_w = IC_w + \sum_n \frac{\left(MC_n^w + UR_n^w\right)}{(1+r)^n} \tag{3.2}$$

$$UC_{pv} = IC_{pv} + \sum_n \frac{\left(MC_n^{pv} + UR_n^{pv}\right)}{(1+r)^n} \tag{3.3}$$

$$UC_b = IC_b + \sum_n \frac{\left(MC_n^b + UR_n^b\right)}{(1+r)^n} \tag{3.4}$$

In Eqs. (3.2) to (3.4), IC_w, IC_{pv}, and IC_b represent the initial investment cost of a unit of wind turbine, photovoltaic array, and battery storage, respectively. Meanwhile, MC_n^w, MC_n^{pv}, and MC_n^b represent the unit operation and management cost of wind turbine, photovoltaic array, and battery storage in year n, while UR_n^w, UR_n^{pv}, and UR_n^b represent the unit replacement cost. In addition, r denotes the discount rate and n is the index of the year.

3.3.2.2 Technical Sustainability: Maximize Self-Sufficiency Ratio

The increasing energy costs and growing environmental concerns related to the use of fossil fuels have prompted the water treatment industry to reduce its power consumption and increase its on-site power generation for improved energy self-sufficiency (Macintosh et al. 2019). Combining energy from the grid and on-site sources offers new opportunities to enhance the flexibility (Schulze et al. 2019), security (Ozcan 2018), and sustainability (Sun et al. 2018) of energy-intensive processes. To determine reasonable self-sufficiency ratios (SSRs) for buildings (Chegari et al. 2020), industrial processes (Carvalho et al. 2020), and even entire cities (Jurasz et al. 2020), several studies have been conducted, and this chapter applies this metric as well. The SSR concept is used to describe the capability of covering the energy needs using local energy sources, with the SSR showing the percentage of the load that can be met by the locally generated energy, as shown in Eq. (3.5), where f_2 is the SSR, Y_{it}^{in} is the electricity imported from the main grid in season i time t, D_{it} is the power demand in season i time period t, and K_i is the number of days in season i.

$$\max f_2 = \frac{\sum_i K_i \left(\sum_t D_{it}\right) - \sum_i K_i \left(\sum_t Y_{it}^{in}\right)}{\sum_i K_i \left(\sum_t D_{it}\right)} \times 100\% \tag{3.5}$$

3.3.2.3 Environmental Sustainability: Minimize Carbon Emissions

Renewable energy sources offer a crucial advantage in that they do not release carbon emissions during generation, making them an attractive option for reducing air pollution and promoting environmental sustainability. However, despite this benefit, the latest Key World Energy Statistics reveal that 38.5% of the world's electricity is still generated through coal combustion (International Energy Agency 2019a), with some developing countries relying even more heavily on this fossil fuel, such as China, India, and Iran. In the case of hybrid systems that require electricity imports from the main grid, some carbon emissions are inevitable due to this energy mix. To ensure environmental sustainability, it is therefore imperative to minimize the carbon emissions associated with imported electricity from the main grid. Here, μ is the ratio of coal combustion generated electricity from the main grid, and ζ (g/kWh) is the average carbon emissions parameter for the coal burning generation process from the main grid; therefore, this objective is as shown in Eq. (3.6).

$$\min f_3 = \sum_i K_i \left(\sum_t Y_{it}^{in} \right) \mu\zeta \tag{3.6}$$

3.3.2.4 Social Sustainability: Maximize Job Creation

The social dimension is a key pillar for the sustainability measurement framework (Santoyo-Castelazo and Azapagic 2014). However, the social dimension has received less attention compared to the other dimensions (Cuesta et al. 2020). Past studies have mainly assessed the social performance of renewable energy systems at a national or city level from a macro perspective, using indicators such as the human development index, portfolio risk, and social acceptance (Lian et al. 2019). The contribution of a fixed hybrid generation system to social sustainability has been less explored. In this chapter, we use the job creation measuring method for hybrid generation systems proposed by Dufo-López et al. (2016) to evaluate the system's social sustainability. This method considers the job creation associated with component manufacturing, installation, operation, and maintenance simultaneously. Specifically, J_w and J_{pv} (job/kW) represent the number of jobs created per kWp of the wind turbine and PV generator, while J_b (job/kWh) represents the number of jobs created per kWh of the nominal battery storage bank. Then this objective can be written as Eq. (3.7), in which A_w, A_{pv}, and A_b are rated capacity of a unit of wind turbine, photovoltaic array, and battery storage.

$$\max f_4 = J_w X_w A_w + J_{pv} X_{pv} A_{pv} + J_b X_b A_b \tag{3.7}$$

3.3.2.5 Output of Solar Power

The output of photovoltaic power is affected by both solar radiation and ambient temperature, and can be accurately modeled using the equations developed by the National Renewable Energy Laboratory (Li and Qiu 2016). Specifically, the photovoltaic power output can be represented by Eq. (3.8).

$$AP_{it}^{pv} = X_{pv} \left\{ A_{pv} \frac{G_{it}}{G_s} \left[1 + \varpi \left(T_{it} - T_{sta}\right)\right] \right\} \tag{3.8}$$

AP_{it}^{pv} represents the actual solar power output at time t during season i, with X_{pv} being the number of photovoltaic arrays deployed. The solar radiation and ambient temperature at time t during season i are denoted by G_{it} and T_{it}, respectively, while A_{pv} is the rated solar power output of a photovoltaic unit under standard test conditions. The values of solar radiation and ambient temperature under standard test conditions are represented by G_s and T_{sta}, respectively, and ϖ represents the temperature coefficient which is usually set at −0.35.

3.3.2.6 Output of Wind Power

The power output of wind turbines can be described by a piecewise function that takes wind speed as input (Hemmati 2018). This function is represented by Eq. (3.9), where X_w denotes the number of installed wind turbines. The power output of a single turbine is given by $A_w \cdot \frac{v_{it}-V_{in}}{V_r-V_{in}}$ when the wind speed is within the cut-in (V_{in}) and rated (V_r) wind speed range of the turbine, and A_w is the rated power output of the wind turbine. The variables v_{it} and V_{out} correspond to the actual and cut-off wind speed of the wind turbines, respectively.

$$AP_{it}^{w} = \begin{cases} X_w \left(A_w \frac{v_{it}-V_{in}}{V_r-V_{in}}\right), V_{in} \leq v_{it} < V_r \\ X_w A_w, V_r \leq v_{it} < V_{out} \\ 0, v_{it} > V_{out} \text{ or } v_{it} < V_{in} \end{cases} \tag{3.9}$$

In addition, there have been recent suggestions to account for the impact of wind turbine installation height (Mostafaeipour 2010). This is due to the fact that wind speed is typically measured at the height of the anemometer, whereas the power output of wind turbines is calculated using the wind speed at the height of the hub. To address this discrepancy, wind speed is adjusted using Eq. (3.10), in which v_{it} and

v_{it}^{0} are the speed at the height of hub and anemometer, H and H_0 are the height of hub and anemometer, γ is a coefficient, often set as 0.143.

$$v_{it} = v_{it}^{0}\left(\frac{H}{H_0}\right)^{\gamma} \tag{3.10}$$

3.3.2.7 Balance of Battery Storage System

In order to mitigate the impact resulting from the unpredictability and intermittency of renewable energies, batteries are utilized as a backup energy source to store surplus energy generated by PV and wind systems. They are employed to provide power when the PV and wind power output is unable to meet the demand for electricity (Wang et al. 2020). The effect of battery bank can be modeled by Eqs. (3.11)–(3.12).

$$Ba_{it} = Ba_{i(t-1)}(1-\sigma) + Z_{it}^{cha}\eta - \frac{Z_{it}^{dis}}{\eta} \tag{3.11}$$

$$Ba_{i1} = s_0 X_b A_b \tag{3.12}$$

where Ba_{it} and $Ba_{i(t-1)}$ are power of battery systems at time step t and $t-1$ in season i, σ is the self-discharging factor, η is the charging and discharging efficiency, Z_{it}^{cha} and Z_{it}^{dis} are the battery charging and discharging power at season i time period t, s_0 is initial SOC. X_b is number of battery units, and A_b is the rated capacity of a unit of battery. Meanwhile, state-of-charge $S_t = \frac{Ba_t}{X_b \cdot A_b}$ needs to be restricted within $B_{\min}$ and $B_{\max}$ for safety reasons, as shown in Eq. (3.13).

$$B_{\min} \leq \frac{Ba_{it}}{X_b A_b} \leq B_{\max} \tag{3.13}$$

3.3.2.8 Balance of Demand and Supply

The energy balance of a hybrid system, which incorporates electricity generated from renewable sources, power imported from the grid, and discharged energy from batteries, is represented by Eq. (3.14) to ensure that the power supply meets the demand of the water treatment plant. Here, D_{it} denotes the power demand of the plant at time step t in season i, while Z_{it}^{cha} and Z_{it}^{dis}, respectively, represent the charged and discharged power from the battery system at time step t in season i, Y_{it}^{ex} and Y_{it}^{in} are power exported to and imported from the main grid, AP_{it}^{w} and AP_{it}^{pv} are the power output of wind turbines and solar arrays.

$$D_{it} + Z_{it}^{cha} + Y_{it}^{ex} = Y_{it}^{in} + Z_{it}^{dis} + X_w AP_{it}^{w} + X_{pv} AP_{it}^{pv} \tag{3.14}$$

3.3.2.9 Key Operation Constraints

The operational strategy dictates certain constraints to ensure that during a given time period t in season i, the batteries are either charging or discharging (Eq. (3.15)); the system is either importing or exporting electricity (Eq. (3.16)); and if the battery is discharging power, then no electricity is exported (Eq. (3.17)). These constraints are put in place to regulate the system.

$$Z_{it}^{cha} Z_{it}^{dis} = 0 \tag{3.15}$$

$$Y_{it}^{in} Y_{it}^{ex} = 0 \tag{3.16}$$

$$Z_{it}^{dis} Y_{it}^{ex} = 0 \tag{3.17}$$

3.3.3 Global Model

A multi-objective nonlinear dynamic model with integer variables is constructed for DWTPs to determine the feasibility of adopting a sustainable grid-connected hybrid electricity generation system based on the aforementioned steps. The economic viability objective is assessed using the levelized electricity cost, which reflects the total costs associated with installing and operating the power plant throughout its lifespan. This cost is calculated using a discount rate to account for the time value of money and is normalized over the total net electricity generated (Kabayo et al. 2019). The self-sufficiency ratio is used to evaluate the technical objective, which reflects the percentage of the load that is met by the hybrid system generated energy. A higher self-sufficiency ratio indicates a lower reliance on gray electricity, resulting in lower import costs, increased renewable energy contribution, and reduced environmental impacts. The system's environmental and social sustainability objectives aim to minimize carbon emissions from imported main-grid electricity and maximize job creation during system construction and operations. The decision variables in system planning period are numbers of wind turbines (X_w), PV arrays (X_{pv}), and battery storage units (X_b), which are all set as integer variables, while decision variables in operation period are electricity exported to/imported from the main grid (Y_{it}^{ex},Y_{it}^{in}) and power charged into/discharged from battery system (Z_{it}^{cha}, Z_{it}^{dis}). Additionally, since the amount of electricity stored in battery systems is time-dependent, it is challenging to obtain precise values for this quantity. Therefore, a dynamic equation is formulated to describe this relationship. The global model is shown in Eq. (3.18).

$$
\begin{aligned}
&\min\ f_1 = \frac{UC_wX_w+UC_{pv}X_{pv}+UC_bX_b+\sum_n \frac{\sum_i K_i\left(\sum_t Y_{it}^{in} p_{it}^{in}\right)}{(1+r)^n}}{\sum_n \frac{\sum_i K_i\left(\sum_t D_{it}\right)}{(1+r)^n}} \\
&\max\ f_2 = \frac{\sum_i K_i(\sum_t D_{it})-\sum_i K_i(\sum_t Y_{it}^{in})}{\sum_i K_i(\sum_t D_{it})} \times 100\% \\
&\min\ f_3 = \sum_i K_i \left(\sum_t Y_{it}^{in}\right) \mu\zeta \\
&\max\ f_4 = J_wX_wA_w + J_{pv}X_{pv}A_{pv} + J_bX_bA_b \\
&s.t. \begin{cases}
UC_w = IC_w + \sum_n \frac{(MC_n^w+UR_n^w)}{(1+r)^n} \\
UC_{pv} = IC_{pv} + \sum_n \frac{(MC_n^{pv}+UR_n^{pv})}{(1+r)^n} \\
UC_b = IC_b + \sum_n \frac{(MC_n^b+UR_n^b)}{(1+r)^n} \\
AP_{it}^{pv} = X_{pv}\left\{A_{pv}\frac{G_{it}}{G_s}\left[1+\varpi\left(T_{it}-T_{sta}\right)\right]\right\} \\
AP_{it}^{w} = \begin{cases} X_w\left(A_w\frac{v_{it}-V_{in}}{V_r-V_{in}}\right), V_{in} \le v_{it} < V_r \\ X_wA_w, V_r \le v_{it} < V_{out} \\ 0, v_{it} > V_{out} \text{ or } v_{it} < V_{in} \end{cases} \\
v_{it} = v_{it}^0\left(\frac{H}{H_0}\right)^{\gamma} \\
Ba_{it} = Ba_{i(t-1)}(1-\sigma) + Z_{it}^{cha}\eta - \frac{Z_{it}^{dis}}{\eta}, Ba_{i1} = s_0X_bA_b \\
SOC_{\min} \le \frac{Ba_{it}}{X_bA_b} \le SOC_{\max} \\
D_{it} + Z_{it}^{cha} + Y_{it}^{ex} = Y_{it}^{in} + Z_{it}^{dis} + X_wAP_{it}^{w} + X_{pv}AP_{it}^{pv} \\
Z_{it}^{cha}Z_{it}^{dis} = 0, Y_{it}^{in}Y_{it}^{ex} = 0, Z_{it}^{dis}Y_{it}^{ex} = 0
\end{cases}
\end{aligned}
\tag{3.18}
$$

The proposed model offers decision makers, such as DWTP planners or third-party investors, a comprehensive evaluation tool to assess the long-term technical, economic, environmental, and social sustainability of deploying on-site hybrid generation systems. To maximize economic advantages and integration reliability, the proposed system includes both wind and solar energy sources, and a battery storage system is incorporated as a backup resource to store excess energy and provide power during times of insufficient PV and wind power to meet load demand. The system is designed to be grid-connected, giving the operator the flexibility to operate on-grid or off-grid as needed, thereby increasing the system's flexibility and reliability.

3.3.4 Model Solving

When deciding whether to deploy the DWTP hybrid generation system, cost-effectiveness is typically the most important consideration. However, sustainable goals such as the self-sufficiency ratio, carbon emissions, and job creation cannot be ignored and must also be kept within acceptable ranges. To represent this situation, the ε-constraint method proposed by Haimes et al. (1971) and enhanced by Chankong and Haimes (1983) was applied to solve the proposed multi-objective model. Compared to alternative approaches (Mollaei et al. 2019)-(Jalil et al. 2019), this method prioritizes the maximization of one objective function, while the other objectives are treated as constraints that must remain within allowable levels (Guillén et al. 2005). The advantage of this approach is that it ensures an optimal value for one objective while taking into account the limitations imposed by the other objectives. To represent system decision-makers' attitude toward these objectives, three parameters are introduced (Wang et al. 2019): ε as the lower SSR bound, α as the attitude toward carbon emissions, and β as the lower job creation bound. Then the model is transformed as Eq. (3.19), where u_1 is DWTP's initial estimated carbon emissions from electricity utilization without the hybrid generation system.

$$\min f = \frac{UC_w X_w + UC_{pv} X_{pv} + UC_b X_b + \sum_n \frac{\sum_i K_i \left(\sum_t Y_{it}^{in} \cdot p_{it}^{in}\right)}{(1+r)^n}}{\sum_n \frac{\sum_i K_i \left(\sum_t D_{it}\right)}{(1+r)^n}}$$

$$s.t. \begin{cases} \frac{\sum_i K_i (\sum_t D_{it}) - \sum_i K_i (\sum_t Y_{it}^{in})}{\sum_i K_i (\sum_t D_{it})} \times 100\% \geq \varepsilon \\ \sum_i K_i \left(\sum_t Y_{it}^{in}\right) \mu\zeta \leq \alpha u_1 \\ J_w X_w A_w + J_{pv} X_{pv} A_{pv} + J_b X_b A_b \geq \beta \\ \text{Eqs. (3.8)–(3.17)} \end{cases} \tag{3.19}$$

After converting the proposed model into its equivalent single form with integer variables, dynamic equations, and nonlinear constraints, we utilized LINGO 13.0 to solve the model in Eq. (3.19). This software is widely used, easy to code, and has been found to be efficient in solving models with integer and nonlinear constraints. By adjusting the values of ε, α, and β according to the decision-maker's preferences, we obtained a set of optimized solutions for f. This approach allows for flexibility in considering the decision-maker's subjective attitudes toward each objective, resulting in a more personalized and practical solution.

3.4 Case Study

This section presents a case study of a DWTP located in Huili County in southwest China to illustrate the effectiveness and practicality of the proposed optimization method for hybrid renewable systems.

3.4.1 Case Description

Huili county, situated in southwest China (see Figure 3.4), has an urban population of 190,000. In 2018, the daily drinking water supply capacity of the county was 17,000 cubic meters, with an annual sales volume of 4922 thousand cubic meters, resulting in the consumption of 20.42 MWh of electricity (The Peoples' Government of Huili County 2020). However, the cost of raw materials such as disinfectants and coagulants increased, and there was a 10% rise in power costs (The Peoples' Government of Huili County 2020), resulting in the main drinking water supply net profit for 2018 being only 104 thousand CNY, which was 28.36% of the 2017 net profit. Consequently, due to the increasing water demand and stricter energy conservation policies, the managers tried several methods to reduce production costs, including equipment upgrades and peak load shifting. While these methods were somewhat successful, they had high equipment costs and a limited optimization scale.

Huili county, situated at an altitude of approximately 1800 meters above sea level, boasts abundant wind and solar resources. The average wind power density at a height of 70 meters is 865.82 w/m^2 (Yanchuan et al. 2017), while the yearly average global horizontal irradiation is 1645 kWh/m^2 (Solargis 2021). In line with Huili county's "13th Five Year Plan," which aims to increase the installed wind and solar capacity to 1000 MW by the end of 2020 (Website of Huili County Government 2016), the managers are exploring alternative methods to generate electricity for the urban DWTPs. To evaluate the cost-effectiveness of a hybrid generation system, the proposed nonlinear dynamic decision model was applied to the second DWTP, which has a daily treatment capacity of 5000 m^3/d and is one of the three urban DWTPs in Huili county.

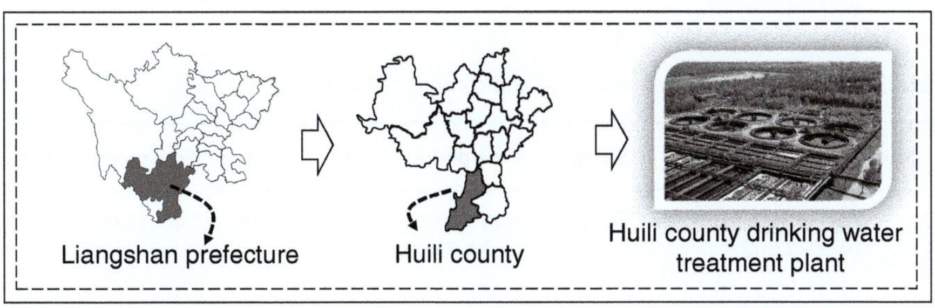

Figure 3.4 The location of the studied drinking water treatment plant. Source: Kekyalyaynen/Adobe Stock.

3.4.2 Data Collection

The wind turbine, PV array, and battery storage successfully application in Zhenkun et al. (2019) was chosen as the hybrid generation system, the cost parameters for which are shown in Table 3.1. The rated power of wind turbine is 1 kW, with the cut-in, rated, and cut-out speed as 3, 10, 30 m/s; the rated power of PV panels is also 1kW, the radiation and temperature of standard test condition are 100 W/m^2 and 25°C. The rated capacity of battery unit is kWh, the charging and discharging efficiency is 0.88, charge and discharge rate is 0.25, self-discharging rate is 0.1%, maximum and minimum SOC are 0.95 and 0.2 separately.

The industrial and commercial users' electricity prices in Sichuan Province, shown in Table 3.2, were collected from People's Government of Sichuan Province (2017) and Sichuan Provincial Development and Reform Commission (2019). Coal consumption for electricity generation in Huili county is 309 g/kWh (National Energy Administration 2018), and the associated carbon emissions of a gram of standard coal being 0.68 g. The job creation parameters for the wind and solar technologies were 2.7×10^{-3} and 1.1×10^{-3} job/kW (Wei et al. 2010), and a very conservative value of 10^{-5} was used for the battery bank (Dufo-López et al. 2016).

The typical hourly solar irradiation and air temperature data in Huili county in each season are shown in Figure 3.5a–b, which were collected from the website (Solargis 2021). The typical hourly wind speed data shown in Figure 3.5c were collected from the published paper (Yanchuan et al. 2017), and the demand data in Figure 3.5d were collected from Website of Huili County Government (2016).

Table 3.1 Cost parameters for each power supply component unit.

Parameters	Wind turbine	Solar array	Battery
Capacity	1 kW	1 kW	1 kW · h
Investment cost (CNY)	12,000	8000	800
Replacement cost (CNY)	10,200	6800	680
Annual O&M cost (CNY)	30	50	20
Utilization year	20 yr	20 yr	5 yr

Table 3.2 Electricity price of industrial and commercial users.

Season	Time period	Price	Season	Time period	Price	Season	Time period	Price
Dry	7:00–11:00, 19:00–23:00	0.9094	Normal	7:00–11:00, 19:00–23:00	0.8661	Flood	7:00–11:00, 19:00–23:00	0.8228
	23:00–7:00	0.3031		23:00–7:00	0.2887		23:00–7:00	0.2743
	11:00–19:00	0.6063		11:00–19:00	0.5774		11:00–19:00	0.5485

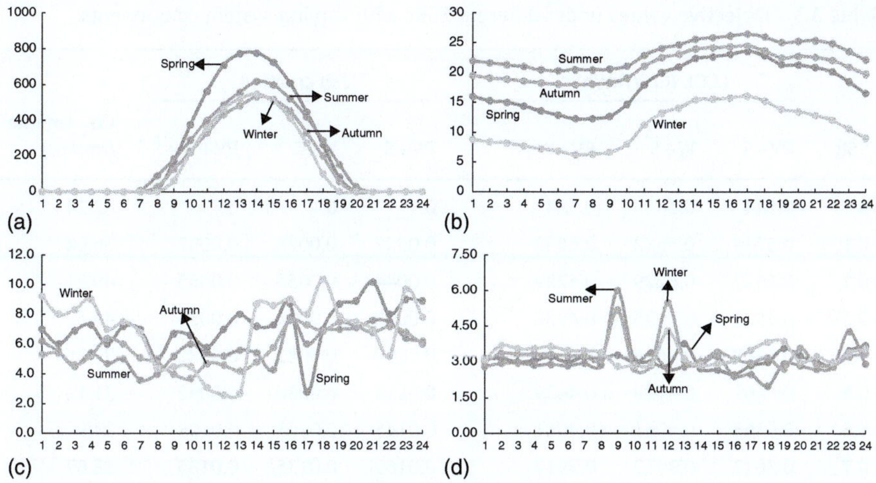

Figure 3.5 Model input data. (a) Global horizontal irradiation (Wh/m^2), (b) air temperature °C, (c) wind speed (m/s), and (d) electricity demand (kWh).

3.4.3 Calculation Results and Analysis

In this section, as the SSR step changes by 0.1, values of levelized cost of electricity (LCOE), created jobs and CO_2 emissions under the different hybrid system components (Section 3.4.3.1), the influence of different electricity price strategies (Section 3.4.4.1), and the changes in the natural resources on the LCOE (Section 3.4.4.2) are compared and presented.

3.4.3.1 Results Under Different Scenarios

Three hybrid system components: a wind+storage system, a PV+storage system, and a wind+PV+storage system, were compared for the DWTP. The calculation results for the other three objectives under different SSRs are shown in Table 3.3 and Figure 3.6. As can be seen in Table 3.3, as the SSR increases, the CO_2 emissions correspondingly decrease.

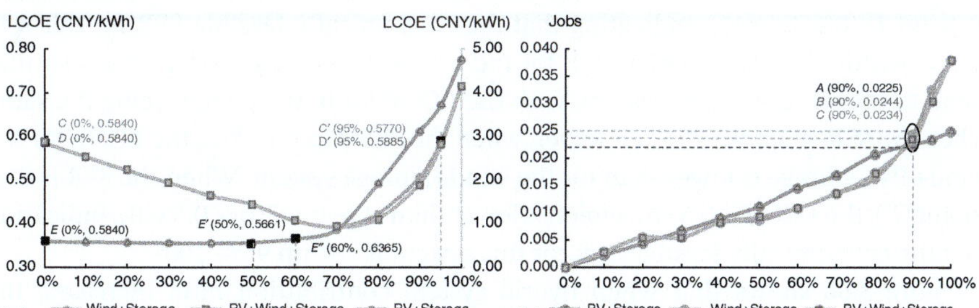

Figure 3.6 LCOE and created jobs under different SSRs with varying generation sources: wind+storage, PV+storage, and wind+PV+storage. (a) Levelized cost of electricity and (b) jobs created.

Table 3.3 Objective values under different SSRs with varying system components.

	LCOE (CNY/kWh)			Job creation			
SSR	PV+S	W+S	PV+W+S	PV+S	W+S	PV+W+S	CO_2 emission (tonnes)
0	0.584	0.584	0.584	0	0	0	62.24
0.1	0.5734	0.5532	0.5532	0.0022	0.0028	0.0027	56.04
0.2	0.5627	0.5229	0.5229	0.0044	0.0055	0.0055	49.81
0.3	0.5521	0.4935	0.4935	0.0066	0.0054	0.0054	43.57
0.4	0.5488	0.4634	0.4634	0.0089	0.0082	0.0082	37.35
0.5	0.5661	0.4429	0.4429	0.0111	0.0086	0.0092	31.12
0.6	0.6365	0.4061	0.4061	0.0145	0.0108	0.0108	24.9
0.7	0.8612	0.3917	0.3917	0.0169	0.0135	0.0135	18.67
0.8	1.9059	0.4218	0.4103	0.0207	0.0163	0.0173	12.45
0.9	3.1089	0.5078	0.492	0.0225	0.0244	0.0234	6.22
0.95	3.7143	0.5885	0.5799	0.0234	0.0325	0.0304	3.11
1	4.7486	0.7148	0.7148	0.0249	0.0379	0.0379	0

Figure 3.6a shows that the LCOE for PV+storage system first shows a very slight decrease when the SSR is within 40%, then gradually increases as the SSR changes from 40% to 70%; however, when the SSR exceeds 70%, it dramatically increases to 3.1089 (SSR=90%) and 4.7486(SSR=100%) correspondingly. The optimum SSR for the PV+storage system is 37.88%, which has an LCOE of 0.5403; as the LCOE at point E (0%, 0.5840) is within E' (50%, 0.5661) and E'' (60%, 0.6365), the economically feasible SSR for this system is within 50%–60%. For the wind+storage system, the LCOE drops gradually from 0.5840 to 0.3917 when the SSR is between 0% and 70%, then it rises quickly to 0.7418 when the SSR reaches 100%. The optimum SSR for the wind+storage system is 70.02%, which has a LCOE of 0.3930; however, when the SSR reaches 95% (point D'), the LCOE is 0.5885, which is a little higher than at point D (0%, 0.5840), indicating that the economically feasible SSR for this system is within 90%–95%. The LCOE for the wind+PV+storage system has a similar trend to the wind+storage system, with the LCOE for both systems being the same when the SSR is within 70%. However, when the SSR exceeds 70%, the LCOE for the wind+PV+storage is lower than for the wind+storage system. When the SSR is 95% (point C'), the LCOE is 0.5770, which is lower than at point C (0%, 0.5840), indicating that the economically feasible SSR for this system is within 95%–100%.

Figure 3.6b shows that when hybrid system contribution ratio increases, the related number of created jobs also increases. However, when the SSR is between 20% and 90%, the number of created jobs for the PV+storage system is higher than for the other two systems; however, when the SSR exceeds 90%, the result is exactly the opposite.

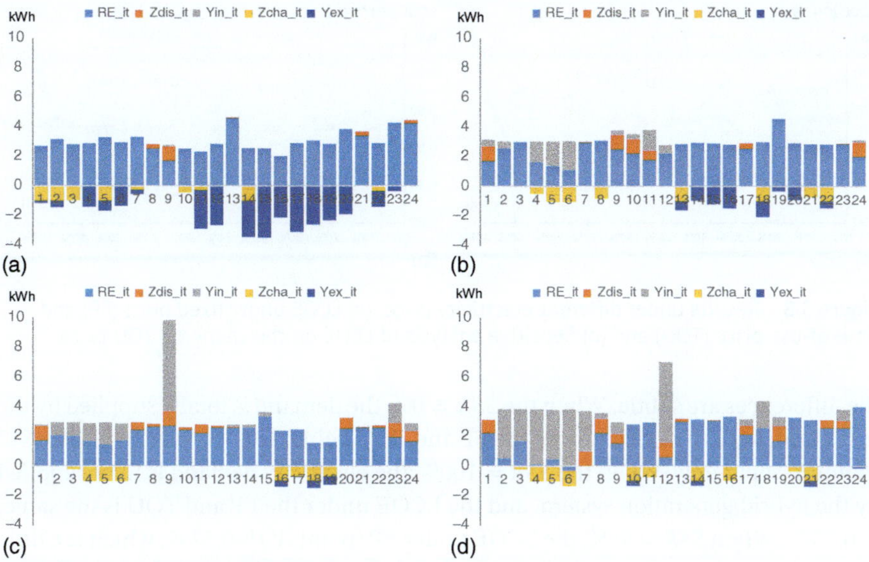

Figure 3.7 Typical daily energy balance results of the hybrid wind–solar-storage system. (a) Typical daily energy balance results – Spring, (b) typical daily energy balance results – Summer, (c) typical daily energy balance results – Autumn, and (d) typical daily energy balance results – Winter.

3.4.4 Results of Energy Balance

In this subsection, the grid connected wind–solar-storage system is presented as an example to take a deep look at this system's energy balance results, as shown in Figure 3.7. Re_{it} refers to wind and solar power output at season i time period t, $Zdis_{it}$ and $Zcha_{it}$ refer to battery discharged and charged power, Yin_{it} and Yex_{it} refer to electricity the hybrid wind–solar-storage system imported from and exported to the main grid. From the figure, it can be seen that the wind–solar-storage complementary system performs best in spring, covering not only the majority of electricity demand but also supplying a large amount of clean electricity to the main grid between 11:00 and 20:00. During the summer, the system requires support from the main grid from 1:00 to 12:00, but can export electricity to the main grid from 13:00 to 18:00. On a typical day in autumn, the system's reliance on the main grid increases, especially during periods of abnormal demand, such as at 9:00 in the figure. In winter, the system is most dependent on the main grid, particularly during the period from 2:00 to 7:00 when the photovoltaic system is unable to operate.

3.4.4.1 Influence of Electricity Price

In this subsection, the LCOE is compared under different SSR when the electricity price is fixed (FP) or based on time-of-use (TOU), and an LCOE sensitivity analysis conducted on the changing TOU price, as shown in Figure 3.8.

As can be seen in Figure 3.8a, the LCOE under both the FP and TOU first decreases then increases. When the SSR is within 60%, the LCOE under the FP is higher than for the TOU, but when the SSR exceeds 60%, the situation is the opposite, although

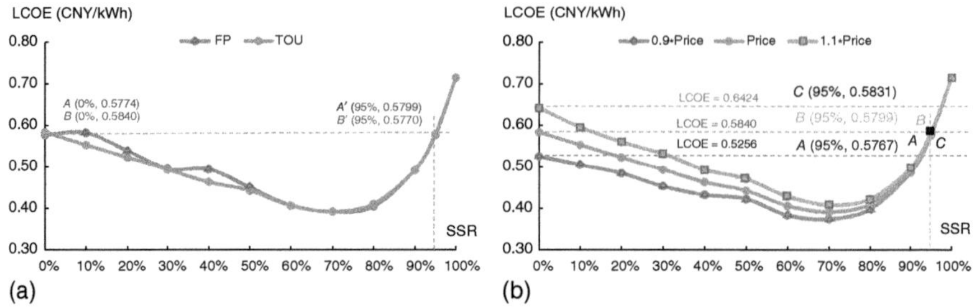

Figure 3.8 Results under different electricity price. (a) LCOE under fixed price (FP) and time-of-use price (TOU) and (b) Sensitive analysis of LCOE on the changing TOU price.

the differences are subtle. When the $SSR = 0\%$, the demand is totally supplied by the main-grid, with the LCOE under the FP and TOU being 0.5774 (point A) and 0.5840 (point B) separately. When the $SSR = 100\%$, the electricity demand is fully supplied by the hybrid generation system, and the LCOE under the FP and TOU is the same, at 0.7521. When $SSR = 95\%$, the LCOE under FP (point A') is 0.5799, which is a little higher than point A (0%, 0.5774), indicating that an SSR within 95% is cost-effective for the DWTP when the electricity price is fixed. Similarly, the LCOE under the TOU (point B') is 0.5770, which is less than point B (0%, 0.5840), indicating that the SSR can reach 95% if the electricity is based on TOU.

Figure 3.8b shows that while the LCOE has the same decreasing and increasing trends under different electricity price scenarios, a higher price corresponds to a higher economically feasible SSR. For example, when the SSR is 95%, the LCOE for the basic price scenario (the line with gray dots, point B) is 0.5799, which is lower than for the main grid supply (0%, 0.5840), meaning that the economically feasible SSR for this system is within 95% and 100%. However, when the electricity price is 1.1 times the basic case (the line with small squares), the LCOE is only 0.5831 (point C), far lower than for the main grid supply (0%, 0.6424)and when the electricity price is 0.9 times that of the basic case (the line with small triangles), the LCOE at point A is 0.5767, which is higher than when $SSR = 0$, that is 0.5256. Therefore, the higher the electricity importing price, the higher the economically feasible SSR in these three scenarios.

3.4.4.2 Influence of Natural Resources

In this subsection, a sensitivity analysis for the LCOE under changing wind speed and solar irradiation is conducted, with the results shown in Figure 3.9.

In Figure 3.9a, the LCOE has consistent trends under the different wind speed scenarios; however, better wind energy resources correspond to a lower LCOE and a higher SSR. For example, when the wind speed decreases to 0.9 times of the basic case (the blue line), the LCOE for the hybrid generation system when SSR is 90% (point A) is 0.5931, which is higher than 0.5840 ($SSR = 0\%$), indicating that the economically feasible SSR of this system is within 90%. When the wind speed is under the basic case (the orange line), the LCOE for the hybrid generation system when SSR is 95% (point B) is 0.5799, which is lower than 0.5840, indicating that

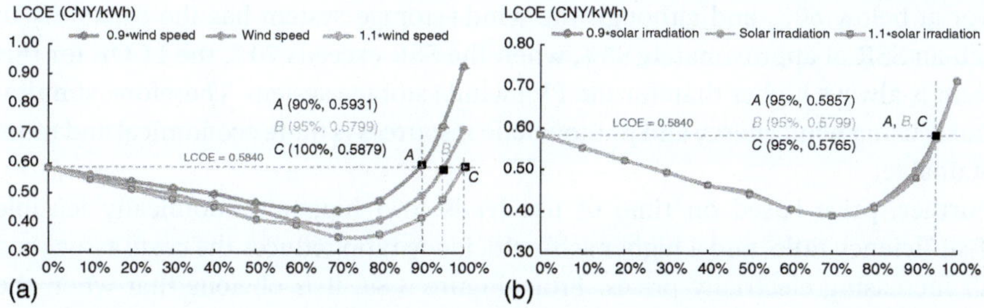

Figure 3.9 Results under changing natural resources. (a) LCOE under different wind speed scenarios and (b) LCOE under different solar irradiation scenarios.

the economically feasible SSR in this system can reach 95%. When the wind speed increases to 1.1 times of the basic case (the gray line), the LCOE for the hybrid generation system when the SSR is 100% (point *C*) is 0.5874, which is slightly higher than 0.5840, indicating that the economically feasible SSR for this system can reach approximately reaching 100%. Therefore, these three scenarios indicate that the better the wind resources, the higher the economically feasible SSR.

Figure 3.9b indicates that the solar irradiation changes do not have any obvious influences on the LCOE or the feasible SSR of the hybrid generation systems. Regardless of whether the solar irradiation is 0.9 or 1.1 times of basic scenario, the economically feasible SSR is approximately within 95% and the LCOE in these scenarios remains around the same.

3.5 Comprehensive Discussion

In this section, some related propositions are proposed and management recommendations are provided.

3.5.1 Related Propositions

Firstly, to reduce long-term electricity utilization cost at drinking water treatment plant, a grid-connected hybrid wind–solar-storage system is economically feasible. Figures 3.6–3.9 all indicate that when the SSR increases from 0% to 100%, the LCOE has a drop–rise trend; that is, when a hybrid generation system is employed, the optimized LCOE gradually decreases until the SSR is about 70% and then begins to increase, with the economically feasible SSR reaching 95%. However, to achieve a 100% renewable energy supply was shown to be a poor choice because the cost of the electricity utilization would be higher than buying all the electricity from the main grid.

Second, the complementary use of both wind and solar resources in a hybrid system reduces the levelized cost of electricity and increases the self-sufficiency ratio. Figure 3.6 shows that compared with the PV+wind+storage system, which can reach an SSR of 95%, the economically feasible SSR for the PV+storage system was far lower at below 60%, and although the wind+storage system has the capability to reach an SSR of approximately 95%, when the SSR exceeds 70%, the LCOE for this system is always higher than for the PV+wind+storage system. Therefore, simultaneously using wind energy and photovoltaic resources is more economical and more sustainable.

Further, price based on time of use leads to a higher economically feasible self-sufficiency ratio, and a higher self-sufficiency ratio reduces the costs associated with increasing electricity prices. From Figure 3.8a, it is obvious that when the electricity price is fixed, the economically feasible SSR is within 95%, whereas the TOU price strategy exceeds 95%; therefore, a TOU price strategy could assist in attaining a higher SSR for the DWTP. Figure 3.8b shows that as the SSR increases, the LCOE differences become smaller; for example, when the $SSR = 04$, the LCOEs for "0.9*price," "price," and "1.1*price" are 0.6214, 0.5824, and 0.5526, a difference about 0.3–0.4 CNY/kWh; however, when the SSR reaches 95%, the LCOEs for the "0.9*price," "price," and "1.1*price" are only 0.5831, 0.5799, and 0.5767, a difference of about 0.003 CNY/kWh.

Finally, better wind energy resources correspond to a lower levelized cost of electricity and a higher self-sufficiency ratio, while changes in the solar irradiation do not have any obvious influences on these values. Taking $SSR = 90\%$ as example (Figure 3.9a), when the wind speed changes from "0.9*wind speed" to "1.1*wind speed," the LCOE drops from 0.5931 to 0.4165, a difference of about 0.18 CNY/kWh; however, as the wind speed increases, the boundary for the economically feasible SSR becomes wider, from within 90% to 95% and finally reaches approximately 100%. However, changes in the solar irradiation (Figure 3.9b) have little influence on the LCOE and SSR boundary. The LCOE for the different solar irradiation condition is almost the same, and the economically feasible SSR is always 95% in three scenarios.

3.5.2 Management Recommendations

For the drinking water treatment power plant, the proposed hybrid wind-PV-storage generation system and related operation strategy should be considered as an alternative method to achieve sustainable power supply. The analyses demonstrated that the wind–solar-storage system was economically feasible, and that the SSR could achieve 95%. User-side electricity generation saves electricity transmission costs, and the storage and main-grid connections as back up sources ensure the power supply reliability. Further, when the renewable sources are abundant, the system can send the surplus power to the main grid, which generates extra revenue for the DWTP. Therefore, the deployment of the hybrid generation system not only contributes to electricity utilization sector carbon emissions reductions, but also provides job opportunities for the whole society.

For the water treatment industry, encouraging the establishment of on-site hybrid generation systems for power supply can competitively pursue sustainable development. This is because the deployment of a micro-grid optimally uses the available renewable power as much as possibleand ensures environmentally-friendly operations. Although it may not be economical to achieve a 100% renewable energy supply, an economically feasible SSR as high as 95% can both positively promote the industry's energy transformation and efficiently contribute to carbon emissions reduction. This finding provides an important reference for the Chinese government's initiative to encourage water supply companies to develop distributed photovoltaics in the "Notice on the Issuance of the Implementation Plan for the Conservation and Intensive Utilization of Water Resources in the Yellow River Basin (No. [2021]13) (National Development and Reform Commission of China 2021)".

For the administration department, electricity pricing policies are vital for motivating the uptake of renewable energy. In DWTPs that have renewable energy SSRs less than 70%, an electricity pricing policy based on time-of-use could act as an incentive because of the lower levelized electricity costs; however, for those DWTPs that already have renewable energy SSRs over 70%, a fixed price policy would be more effective. The administration can also adjust the industry's average renewable energy SSR by adjusting the electricity prices to a certain degree as the results indicated that higher prices can encourage industry parks to increase their renewable energy utilization.

References

Ahadi, A., Kang, S.-K., and Lee, J.-H. (2016). A novel approach for optimal combinations of wind, PV, and energy storage system in diesel-free isolated communities. *Applied Energy* 170: 101–115.

Baneshi, M. and Hadianfard, F. (2016). Techno-economic feasibility of hybrid diesel/PV/wind/battery electricity generation systems for non-residential large electricity consumers under southern Iran climate conditions. *Energy Conversion and Management* 127: 233–244.

Barutha, P., Nahvi, A., Cai, B. et al. (2019). Evaluating commercial feasibility of a new tall wind tower design concept using a stochastic levelized cost of energy model. *Journal of Cleaner Production* 240: 118001.

Bilal, B.O., Sambou, V., Ndiaye, P.A. et al. (2010). Optimal design of a hybrid solar-wind-battery system using the minimization of the annualized cost system and the minimization of the loss of power supply probability (LPSP). *Renewable Energy* 35 (10): 2388–2390.

Bukhary, S., Batista, J., and Ahmad, S. (2017). Evaluating the feasibility of photovoltaic-based plant for potable water treatment. *World Environmental and Water Resources Congress*, 256–262.

Caballero, F., Sauma, E., and Yanine, F. (2013). Business optimal design of a grid-connected hybrid PV-wind energy system without energy storage for an Easter Island's block. *Energy* 61: 248–261.

Carvalho, D.J., Moretti, R.R., Colodette, J.L., and Bizzo, W.A. (2020). Assessment of the self-sustained energy generation of an integrated first and second generation ethanol production from sugarcane through the characterization of the hydrolysis process residues. *Energy Conversion and Management* 203: 112267.

Chankong, V. and Haimes, Y.Y. (1983). *Multiobjective Decision Making: Theory and Methodology*. Elsevier.

Chauhan, A. and Saini, R.P. (2016). Techno-economic feasibility study on integrated renewable energy system for an isolated community of India. *Renewable and Sustainable Energy Reviews* 59: 388–405.

Chegari, B., Tabaa, M., Moutaouakkil, F. et al. (2020). Local energy self-sufficiency for passive buildings: case study of a typical Moroccan building. *Journal of Building Engineering* 29: 101164.

China Electricity Council (2020). List of basic data of electric power statistics in 2018. http://www.cec.org.cn/guihuayutongji/tongjxinxi/niandushuju/2020-01-21/197086.html (accessed 17 October 2023).

Cuesta, M.A., Castillo-Calzadilla, T., and Borges, C.E. (2020). A critical analysis on hybrid renewable energy modeling tools: an emerging opportunity to include social indicators to optimise systems in small communities. *Renewable and Sustainable Energy Reviews* 122: 109691.

De Clercq, D., Smith, K., Chou, B. et al. (2018). Identification of urban drinking water supply patterns across 627 cities in china based on supervised and unsupervised statistical learning. *Journal of Environmental Management* 223: 658–667.

Deng, X. and Lv, T. (2020). Power system planning with increasing variable renewable energy: a review of optimization models. *Journal of Cleaner Production* 246: 118962.

Department of Economic and Social Affairs - the United Nations (2018). World urbanization prospects: the 2018 revision. https://population.un.org/wup/Publications/Files/WUP2018-Report.pdf (accessed 14 December 2023).

Dufo-López, R., Cristóbal-Monreal, I.R., and Yusta, J.M. (2016). Optimisation of PV-wind-diesel-battery stand-alone systems to minimise cost and maximise human development index and job creation. *Renewable Energy* 94: 280–293.

Gao, J., Zhao, J., Zhang, J. et al. (2020). Preparation of a new low-cost substrate prepared from drinking water treatment sludge (DWTS)/bentonite/zeolite/fly ash for rapid phosphorus removal in constructed wetlands. *Journal of Cleaner Production* 261: 121110.

García-Cáceres, R.G., Galvis, M.T.C., and Suárez-Fajardo, J.F. (2019). Towards an efficient and sustainable planning of the drinking water supply chain. *Journal of Cleaner Production* 230: 394–401.

García-Vaquero, N., Lee, E., Castañeda, R.J. et al. (2014). Comparison of drinking water pollutant removal using a nanofiltration pilot plant powered by renewable energy and a conventional treatment facility. *Desalination* 347: 94–102.

Ghorbani, N., Kasaeian, A., Toopshekan, A. et al. (2018). Optimizing a hybrid wind-PV-battery system using GA-PSO and MOPSO for reducing cost and increasing reliabvility. *Energy* 154: 581–591.

Gonzalez, A., Riba, J.-R., Esteban, B., and Rius, A. (2018). Environmental and cost optimal design of a biomass-wind-PV electricity generation system. *Renewable Energy* 126: 420–430.

Guillén, G., Mele, F.D., Bagajewicz, M.J. et al. (2005). Multiobjective supply chain design under uncertainty. *Chemical Engineering Science* 60 (6): 1535–1553.

Haimes, Y.Y., Lasdon, L.S., and Wismer, D.A. (1971). On a bicriterion formulation of the problems of integrated system identification and system optimization. *IEEE Transactions on Systems, Man, and Cybernetics* SMC-1 (3): 296–297.

Hemmati, R. (2018). Optimal design and operation of energy storage systems and generators in the network installed with wind turbines considering practical characteristics of storage units as design variable. *Journal of Cleaner Production* 185: 680–693.

International Energy Agency (2019a). Key world energy statistics. https://www.oecd-ilibrary.org/energy/key-world-energystatistics-2019_71b3ce84-en (accessed 14 December 2023).

International Energy Agency (2019b). CO_2 emissions from fuel combustion 2019-highlight. https://webstore.iea.org/co2-emissions-from-fuel-combustion-2019.

International Renewable Energy Agency (2018). Global renewable energy jobs already reached 11 million million in 2018. https://www.irena.org/-/media/Files/IRENA/Agency/Press-Release/2019/Jun/PR_IRENA-JOBS_2019_ZH.pdf?la=en&hash=1D632DBB37B40206920445BBC78CE9F7774DD1A6 (accessed 17 October 2023).

Islam, M.S. (2018). A techno-economic feasibility analysis of hybrid renewable energy supply options for a grid-connected large office building in southeastern part of France. *Sustainable Cities and Society* 38: 492–508.

Jalil, S.A., Hashmi, N., Asim, Z., and Javaid, S. (2019). A de-centralized bi-level multi-objective model for integrated production and transportation problems in closed-loop supply chain networks. *International Journal of Management Science and Engineering Management* 14 (3): 206–217.

Jurasz, J.K., Dąbek, P.B., and Campana, P.E. (2020). Can a city reach energy self-sufficiency by means of rooftop photovoltaics? Case study from Poland. *Journal of Cleaner Production* 245: 118813.

Kabayo, J., Marques, P., Garcia, R., and Freire, F. (2019). Life-cycle sustainability assessment of key electricity generation systems in Portugal. *Energy* 176: 131–142.

Li, F. and Qiu, J. (2016). Multi-objective optimization for integrated hydro-photovoltaic power system. *Applied Energy* 167: 377–384.

Lian, J., Zhang, Y., Ma, C. et al. (2019). A review on recent sizing methodologies of hybrid renewable energy systems. *Energy Conversion and Management* 199: 112027.

Luthander, R., Widén, J., Nilsson, D., and Palm, J. (2015). Photovoltaic self-consumption in buildings: a review. *Applied Energy* 142: 80–94.

Macintosh, C., Astals, S., Sembera, C. et al. (2019). Successful strategies for increasing energy self-sufficiency at Grüneck wastewater treatment plant in Germany by food waste co-digestion and improved aeration. *Applied Energy* 242: 797–808.

Mollaei, A., Mohammadi, M., and Naderi, B. (2019). A bi-objective MILP model for blocking hybrid flexible flow shop scheduling problem: robust possibilistic programming approach. *International Journal of Management Science and Engineering Management* 14 (2): 137–146.

Mostafaeipour, A. (2010). Feasibility study of harnessing wind energy for turbine installation in province of Yazd in Iran. *Renewable and Sustainable Energy Reviews* 14 (1): 93–111.

National Bureau of Statistics of China (2019). Annual data of 2018. https://www.stats.gov.cn/sj/ndsj/2019/indexch.htm (accessed 14 December 2023).

National Development and Reform Commission of China (2021). Notice on the issuance of the implementation plan for the conservation and intensive utilization of water resources in the Yellow River Basin. https://www.gov.cn/zhengce/zhengceku/2021-12/17/content_5661519.htm (accessed 17 October 2023).

National Energy Administration (2018). National power reliability annual report, 2017. http://zfxxgk.nea.gov.cn/auto79/201806/t20180606_3191.htm (accessed 17 October 2023).

Olcan, C. (2015). Multi-objective analytical model for optimal sizing of stand-alone photovoltaic water pumping systems. *Energy Conversion and Management* 100: 358–369.

Ozcan, M. (2018). The role of renewables in increasing Turkey's self-sufficiency in electrical energy. *Renewable and Sustainable Energy Reviews* 82: 2629–2639.

People's Government of Sichuan Province (2017). Notice of the Sichuan Provincial Development and Reform Commission on relevant issues concerning the adjustment of the electricity price policy for the peak and valley of Sichuan power grid. http://www.sc.gov.cn/10462/11857/13305/13312/2017/12/1/10440883.shtml (accessed 17 October 2023).

Polaris Energy Storage Network (2021). Research report: renewable energy jobs will increase fivefold by 2050. https://chuneng.bjx.com.cn/news/20210804/1167631.shtml (in Chinese) (accessed 17 October 2023).

Prebeg, P., Gasparovic, G., Krajacic, G., and Duic, N. (2016). Long-term energy planning of croatian power system using multi-objective optimization with focus on renewable energy and integration of electric vehicles. *Applied Energy* 18: 1493–1507.

Sanajaoba, S. and Fernandez, E. (2016). Maiden application of Cuckoo Search algorithm for optimal sizing of a remote hybrid renewable energy system. *Renewable Energy* 96: 1–10.

Santoyo-Castelazo, E. and Azapagic, A. (2014). Sustainability assessment of energy systems: integrating environmental, economic and social aspects. *Journal of Cleaner Production* 80: 119–138.

Schulze, C., Blume, S., Siemon, L. et al. (2019). Towards energy flexible and energy self-sufficient manufacturing systems. *Procedia CIRP* 81: 683–688. 52nd CIRP Conference on Manufacturing Systems.

Sichuan Provincial Development and Reform Commission (2019). Notice on reducing the general industrial and commercial electricity price of the Sichuan power grid. http://fgw.sc.gov.cn//sfgw/jgtzgg/2019/5/31/20b5872a34914075a078d7ccce420419.shtml (accessed 17 October 2023).

Solargis (2021). https://apps.solargis.com/prospect/map?s=26.646987,102.256345&c=27.273246,102.23671,11 (accessed 17 October 2023).

Soshinskaya, M., Crijns-Graus, W.H.J., van der Meer, J., and Guerrero, J.M. (2014). Application of a microgrid with renewables for a water treatment plant. *Applied Energy* 134: 20–34.

Sun, S.I., Kiaee, M., Norman, S., and Wills, R.G.A. (2018). Self-sufficiency ratio: an insufficient metric for domestic PV-battery systems? *Energy Procedia* 151: 150–157.

The Peoples' Government of Huili County (2020). Publicity of the conclusion information of the supervision and examination of the price and cost of urban water supply in Huili County. http://www.huili.gov.cn/zfxxgk/zfxxgknr/tzgg/202001/t20200115_1450696.html (accessed 17 October 2023).

Wang, F., Xie, Y., and Xu, J. (2019). Reliable-economical equilibrium based short-term scheduling towards hybrid hydro-photovoltaic generation systems: case study from China. *Applied Energy* 253: 113559.

Wang, R., Xiong, J., He, M.-f. et al. (2020). Multi-objective optimal design of hybrid renewable energy system under multiple scenarios. *Renewable Energy* 151: 226–237.

Website of Huili County Government (2016). Outline of the 13th Five Year Plan for national economic and social development of Huili County. http://www.ahmhxc.com/shisanwu/12588.html (accessed 17 October 2023).

Wei, M., Patadia, S., and Kammen, D.M. (2010). Putting renewables and energy efficiency to work: how many jobs can the clean energy industry generate in the US? *Energy Policy* 38 (2): 919–931.

Weschenfelder, F., de Novaes Pires Leite, G., da Costa, A.C.A. et al. (2020). A review on the complementarity between grid-connected solar and wind power systems. *Journal of Cleaner Production* 257: 120617.

Yan, Y., Zhang, H., Long, Y. et al. (2019). Multi-objective design optimization of combined cooling, heating and power system for cruise ship application. *Journal of Cleaner Production* 233: 264–279.

Yanchuan, Z., Zhengfeng, M., Haiyan, G. et al. (2017). Analysis of wind energy resources in Huili. *Plateau and Mountain Meteorology Research* 37 (2): 72–77. (in Chinese).

Zhang, Y., Campana, P.E., Lundblad, A., and Yan, J. (2017). Comparative study of hydrogen storage and battery storage in grid connected photovoltaic system: storage sizing and rule-based operation. *Applied Energy* 201: 397–411.

Zhenkun, L., Kai, H., Qun, L., and Yang, F. (2019). Power configuration and optimal operation of grid-connected microgrid in electricity selling market. *Electric Power Automation Equipment* 39 (11): 41–49. https://doi.org/10.16081/j.epae.201909042.

4

Disaster Resilience-Oriented Deployment of Hybrid Wind–Solar-Storage-Gas System

Climate change and extreme weather events increasingly impact all components of the energy system. They affect the availability of primary energy sources, in particular renewable energy sources, the transformation, transmission, distribution and storage of energy, and energy demand. It is crucial that these impacts are considered in the clean energy transition.

Adaptation Challenges and Opportunities for the European Energy System
The European Environment Agency, Jan. 2019

Because extreme events and disasters can seriously threaten power system safety and resilience, there has been an increased research focus on power system resilience. This chapter proposes a bi-level multi-objective model for an industrial park deploying a hybrid wind–solar-storage-gas system to optimize configuration in dealing with extreme events, which considers the interactions between the authority and the industrial park in a leader–follower decision process, and seeks to trade-off between economic cost, environmental protection, and power system resilience. An interactive algorithm is designed, and the ε-constraint method is used to convert the model into a bi-level single objective model to solve the proposed complex model. A case example from an industrial park in Jiangsu Province, China, is given to demonstrate the practicality and efficiency of the proposed optimization method. The joint analysis of the power supply and financial losses found that a distributed energy system with an optimal resilience configuration could guarantee 95.85% of demand for three consecutive hours after an extreme event, maintain more than 50% of demand at any time, and reduce power outage financial losses at any time to a minimum of 0.232 million CNY. Based on the analyses and discussion, this model was proven to provide a reasonable and practical strategy for resilience–economy–environment trade-off in industrial park-distributed energy systems.

Sustainable Hybrid Energy Systems: Carbon Neutral Approaches, Modeling, and Case Studies,
First Edition. Jiuping Xu and Fengjuan Wang.

4.1 Background Review

Key infrastructure systems are crucial for maintaining the security, stability, socio-economic prosperity, and sustainability of modern society (Rinaldi et al. 2001). Among these systems, power networks are particularly important (Zou and Chen 2019), given the global trend toward electrified and digital economies. Today, many critical infrastructure sectors, including communications, water, food, defense, and healthcare, depend on a reliable power supply, whether directly or indirectly (Mishra et al. 2020). Climate change has resulted in an increase in the frequency and severity of extreme weather events, which often lead to power disruptions (Hussain et al. 2019). In the past 15 years, power outages in the US have amounted to approximately 161,462 hours, impacting about 157 million customers and causing a loss of 445 GW in demand (Masrur et al. 2021). Hurricane Michael, which occurred in 2018, caused 1.7 million power outages in the southeastern region of the US (Vedullapalli et al. 2020). In February 2021, the Texas freeze led to electricity disruptions lasting several days, affecting roughly 10 million people due to a combination of insufficient power generation and high electricity demand (Wu et al. 2021). Although current power systems are typically reliable during normal operating conditions, maintaining a continuous power supply during unexpected and high-impact events remains a significant challenge (Panteli and Mancarella 2015).

Resilience of power systems is essential for sustainable development goals and future energy systems (Wang et al. 2015). In the face of extreme weather events or other disruptions, a system's resilience refers to its ability to withstand and adapt to the consequences of such events and recover quickly to an acceptable level (Arghandeh et al. 2016). One way to improve power system resilience is to modify the physical infrastructure of the power supply using traditional methods (Ma et al. 2018). For example, strengthening components by adding guy wires or using steel or composite poles, reinforcing selected lines leading to a population center or other critical loads, installing additional lines, breakers, and transformers, and undergrounding overhead distribution systems (Yuan et al. 2016). However, users who rely on main grid power supplies need to increase their resilience using other means as well. The increased use of distributed energy sources such as photovoltaic, wind turbines, and gas turbines has led to the diversification of methods needed to improve power system resilience (Li et al. 2022).

Centralized power systems that rely on large networks of overhead power lines are vulnerable to extreme weather events, making them prone to single-point network faults. Decentralized grids, on the other hand, can reduce the impact of such faults by bringing the energy source closer to the demand. Microgrids (MG) are particularly effective at maintaining renewable energy penetration and can be useful for meeting critical load requirements, especially in isolated areas. Therefore, the use of distributed generators (DG) and microgrids (MG) can enhance power system resilience by providing a more distributed architecture (Bennett et al. 2021). Numerous studies have investigated distributed generators and microgrids for resilience, including approaches to optimize restoration times and allocate distributed energy grid resources (Poudel and Dubey 2018), and resilience-oriented optimization for hybrid energy storage system microgrids (Tobajas et al. 2022). However, while these studies have focused on day-ahead scheduling and performance optimization, they have not fully considered extreme events and policy impacts on the economic, environmental, and resilient aspects of distributed energy configuration optimization (Griffin 2020).

Taking into account the effects of extreme disasters and policies on the optimal configuration of distributed energy sources and microgrids, local emergency power supplies could be developed for important power users to ensure the supply of key loads. Distributed energy based on clean energy sources could also promote the use of cleaner energy and provide economic benefits to these power users (Schwaegerl and Tao 2013). In its 14th Five-Year Plan for Modern Energy Systems, China proposed to strengthen power security in key regions, promote the construction of local emergency power supplies, encourage important qualified users to develop distributed energy and microgrids, and improve emergency self-provided power supply configurations (National Development and Reform Commission of China 2022). China has more than 2500 national and provincial industrial parks, which contribute more than 50% of the country's industrial output (National Development and Reform Commission of China 2018). Since industrial energy consumption still accounts for more than 70% of the country's total energy consumption(Kuby et al. 2011), promoting the construction of green industrial parks and ensuring power security in key regions are essential for deploying distributed energy systems based on clean energy in industrial parks.

After considering the preceding discussion, this chapter proposes a bi-level multi-objective model to expedite the utilization of clean energy in industrial parks under government direction, improve the emergency self-provided power supply configurations and power system resilience, and create green and resilient industrial parks through distributed energy systems and microgrids. The proposed model involves the authority searching for an optimal clean energy proportion for the industrial park, while the park responds sensibly to optimize its goals based on the authority's decisions. Therefore, the allocation of clean energy proportion relies on the interdependent relationship between the authority and the industrial park. When the industrial park adopts a distributed energy system, it takes into account several objectives, including economic cost, carbon emissions, and power system resilience. Given that these objectives are unlikely to be achieved simultaneously, multi-objective optimization is utilized to balance the competing objectives, and fuzzy theory with uncertain fuzzy parameters is utilized to define the distributed energy power supply capacity under extreme events and estimate the ideal clean energy proportion.

4.2 Main Issue Description

Before establishing the mathematical model, the basic system descriptions are given.

4.2.1 System Schematic

Due to the increase in natural disasters in recent years, there is an urgent need to enhance the resilience of power systems. Although diesel generators have commonly been utilized as backup power sources during emergencies, they have certain limitations, including low reliability, high emissions, and dependence on fuel resupplies. A study by Anderson et al. (2022) revealed that diesel generators that are poorly maintained have only a 50% probability of operating for 48 hours or longer, which has led facilities managers to search for alternatives, such as hybrid microgrids. Microgrids are small-scale, self-contained power systems that consist of distributed power generators, loads, energy storage components, and control units (Barra et al. 2020). They can operate autonomously, providing supplementary power and/or maintaining standard services in a defined area in both grid-isolated or grid-connected modes (Hossain et al. 2019). Maryland's Energy Administration recommended enhancing the state's resilience by adopting microgrids in a report (Maryland Energy Administration 2014), and Martini suggested integrating distributed generators into California's distribution (De Martini 2014). This would help mitigate the impacts of future natural disasters.

Research on microgrids and their potential to improve power system resilience during extreme events has been extensive, with a majority of studies concentrating on microgrid control strategies and operational modes. For example, one study by Amirioun et al. (2018) explored the use of proactive operational scheduling in multicarrier microgrids to improve operational resilience against hurricanes. Another study by Liang et al. (2016) proposed a control strategy for electric springs to enhance the resilience of microgrids during operation. Furthermore, Dragičević et al. (2015) outlined three approaches to control microgrid operations, namely centralized or hierarchical control, decentralized control, and distributed control. However, there has been a limited amount of research focused on the initial planning of distributed generators, energy storage equipment, and other resources for microgrids.

This chapter focuses on optimizing the initial planning of distributed energy configurations in coastal areas prone to typhoon events, using the microgrid configuration illustrated in Figure 4.1. The microgrid consists of two distributed power generation sources: photovoltaic power generation and gas power generation. Photovoltaic power generation is selected as one of the power sources because of the black start capabilities of renewable energy (Schneider et al. 2016), which can ensure critical load operations during extreme situations. To maintain a stable energy system, gas turbines are also included as a distributed power source (He et al. 2020). Gas turbines have fast start-up, shut-down, and load regulation characteristics (Martinez-Mares and Fuerte-Esquivel 2012) and can use various fuels, such as natural gas, biogas, waste gas, flare gas, liquefied petroleum gas, and other fuels, to overcome fuel shortage problems during extreme weather or other events (Capstone Turbine Corporation 2020). Energy storage is added to the microgrid to alleviate the intermittency of photovoltaic power generation (Kök et al. 2018) and provide reliable power (Li and Wang 2012), while also mitigating the PV impact on grid system stability (Zhao et al. 2015).

4.2.2 Resilience Characterization

The decision-making environment in this context is uncertain and imprecise due to potential typhoons and their impacts on equipment performance. Various techniques are available to model uncertainty, such as possibilistic, probabilistic,

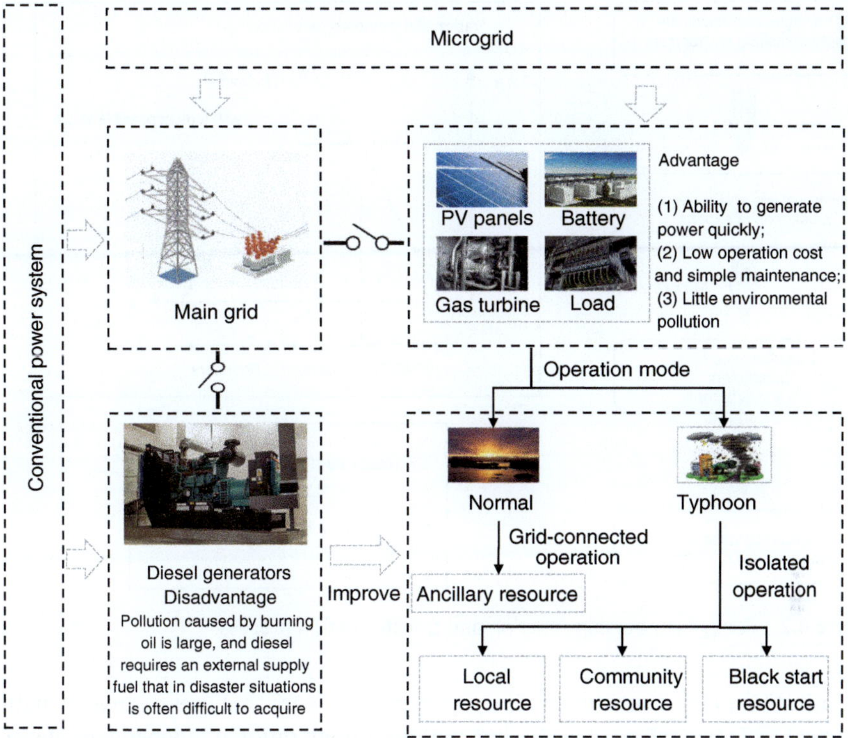

Figure 4.1 Structure, advantages, and operation modes of the proposed hybrid energy system. Source: Deyan Georgiev/Adobe Stock (PV); Val Thoermer/Adobe Stock (Battery); Num/Adobe Stock (Diesel generators Disadvantage); Petersjanchristoph/Pixabay (Normal).

and hybrid possibilistic–probabilistic approaches, robust optimization, information gap decision theory, and interval analyses (Khaloie et al. 2019). Among these, the possibilistic approach is commonly used to represent the uncertainty of variables in the absence of accurate information (Aien et al. 2016). Therefore, the possibilistic approach was chosen in this study to assess power system resilience under uncertainty. Possibilistic uncertainty modeling and fuzzy set theory, proposed by Zadeh, are used to define the boundaries of uncertain parameters and convert them into certain parameters (Zadeh 1978), as shown in Figure 4.2. Many studies have applied fuzzy set theory to solve problems under uncertainty, such as integrating multi-criteria decision-making with fuzzy set theory to address outranking problems (Çalı and Balaman 2019), and solving biomass-coal co-firing uncertainty problems using fuzzy set theory (Xu et al. 2018). Therefore, inspired by these studies, fuzzy set theory is used in this study to model uncertain resilience parameters.

4.2.3 Optimization Scheme

Because both the local authority and the industrial park are critical decision makers, game theory can be used to develop the installed capacity distributed hybrid energy

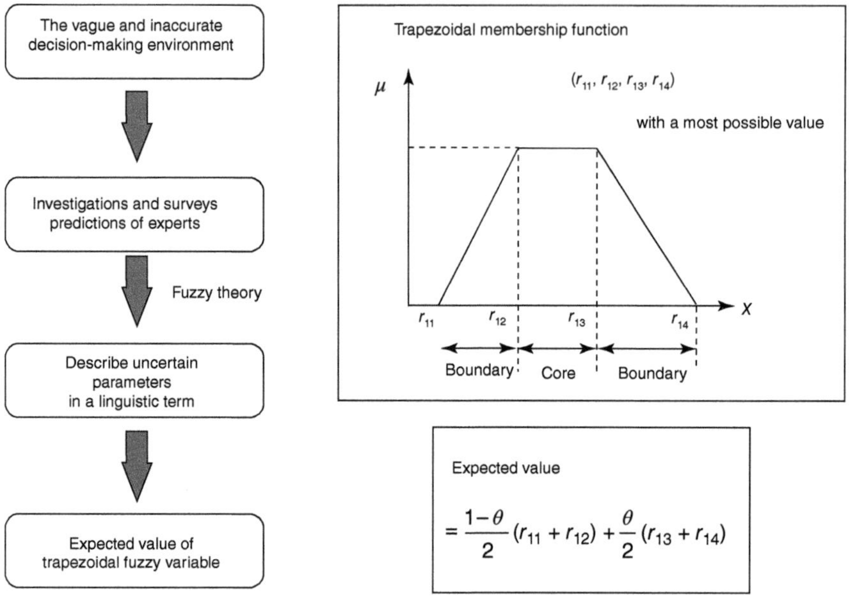

Figure 4.2 Fuzzy membership function and transformation process.

optimization plans. Game theory is a research framework that uses a range of mathematical tools to analyze complex interactions among rational independent players (Tang et al. 2018). It is widely used and can be applied to situations where multiple participants have different or conflicting goals (Su and Huang 2014). The interaction between the authority and the industrial parks can be described as a cooperative game with both collective and individual rationality elements. The main objective of this chapter is to highlight the collective rationality that cooperation between the two parties can bring about the transformation of the national energy system and enhance the resilience of the regional power system. At the same time, the individual rationality of the authority is to meet its environmental protection and clean energy objectives by collaborating with the industrial park, while the individual rationality of the industrial park is to satisfy its production and operational needs with effective and stable power supply through collaboration with the authority.

As the authority holds the power to set the minimum requirement for the use of clean energy, which directly impacts the industrial park's installed strategy, it has higher priority and makes the initial decision. However, the industrial park can adjust its distributed energy system strategy based on its power needs and cost limitations. Thus, the authority and industrial park seek to maximize their respective and collective interests through cooperation, as expressed in mathematical form in Figure 4.3. Bi-level programming, which has been extensively used in various fields, is effective in resolving the trade-off between decision makers on different levels. For instance, Lv et al. (2016) proposed a bi-level multi-objective equilibrium strategy-based coal blending method to optimize the relationship between the authority and coal-fired power plants, while Zhang and Vesselinov (2016) proposed an interactive fuzzy optimization methodology to balance the energy–water nexus

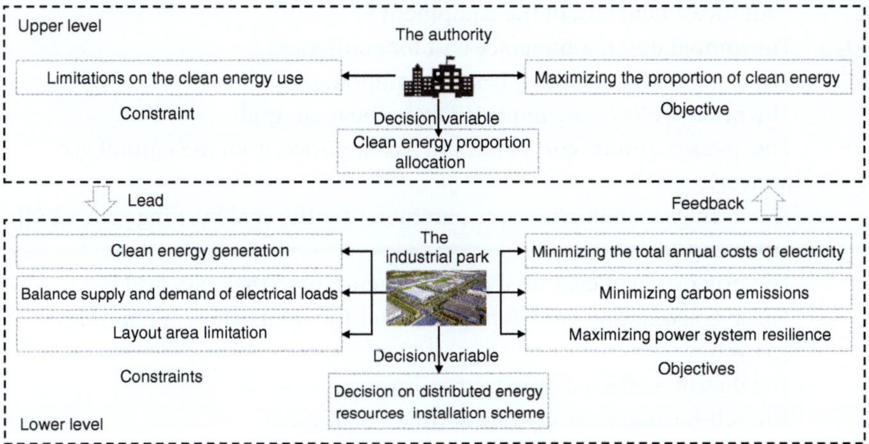

Figure 4.3 Flowchart of the bi-level structure. Source: peteri/Adobe Stock (The industrial park).

management decision-making levels, effectively quantifying the trade-off between decision makers. Therefore, motivated by these studies, this chapter applies a bi-level programming method to determine the optimal relationship between the local authority and the industrial park.

4.3 Mathematical Modeling

4.3.1 Notations

Indices

i	Index of day, $i = 1 \ldots 365$.
t	Index of time period, $t = 1 \ldots 24$.
k	Index of type of the device, $v = 1 \ldots 4$.

Decision variables

y	The proportion of clean energy use allocated to industrial park.
x_{pv}	Number of solar arrays.
x_b	Number of battery storage systems.
x_{gt}	Number of gas turbines.
E_{it}^{m}	The electricity purchased imported from the main grid of industrial park at time t in day i.
E_{it}^{g}	The amount of natural gas consumed by industrial park at time t in day i.

Parameters

$\underline{Y}$	The proportion of clean energy used in industrial park's existing production operations.
$\overline{Y}$	The maximum proportion of clean energy use when all of industrial park's electricity demand is generated by clean energy.
α_k	Annual equivalent investment conversion factor for the equipment k.

UC_k	Unit investment cost of the equipment k.
AM_k	The annual unit maintenance cost for equipment k.
AO_k	The annual unit operating cost for equipment k.
C^e_{it}	The price of electricity imported from the main grid.
AC^g	The average annual cost of natural gas imported from the natural gas network.
δ_1	The average carbon emissions parameter for the coal-burning generation process (g/kWh).
η_1	The proportion of coal-fired power generation in the main grid.
δ_2	The average carbon emissions parameter for combustion of natural gas (g/kWh).
h	The duration affected by extreme disasters.
R_{it}	The self-healing recovery rate at time t in day i.
D_{it}	Electricity demand at time t in day i.
AP^{pv}_{it}	Power generation of unit photovoltaic panel at time t in day i.
AP^g_{it}	Power generation per unit gas turbine at time t in day i.
S^0	The maximum area in an industrial park that can be used for distributed energy sources installation.
A^{pv}	The rated capacity of PV array.
D_{pv}	The PV derating factor (percent), often set as 0.88.
G_{it}	The solar irradiation observed on the PV array at time t in day i (kW/m^2).
G_s	The solar irradiation incident at standard condition (1 kW/m^2).
α_{pv}	The temperature coefficient of power (%/°C).
T_{it}	The PV cell temperature at time t in day i (°C).
T_{sta}	The PV cell temperature under standard test conditions (STCs) (°C).
E^{ba}_{it}	The power of battery systems at time t in day i.
Z^{cha}_{it}	The battery charging power at time t in day i.
Z^{dis}_{it}	The battery discharging power at time t in day i.
$P^{dis}_{\max}$	Maximum battery discharge value.
$P^{cha}_{\max}$	Maximum battery charge value.
χ	The charging and discharging efficiency.
λ	The self-discharging factor.
ε_{gt}	The conversion efficiency of gas turbine.
H^g	The calorific value of burning natural gas.
RU^b	The upward climbing rate of energy storage battery.
RD^b	The downward climbing rate of energy storage battery.
RU^g	The upward climbing rate of gas turbine.
GT_{rd}	The downward climbing rate of gas turbine.

Uncertain parameters

$\widetilde{E_{ba}}$	The storage power of energy storage battery under extreme disaster.
$\widetilde{E^g_t}$	The power generated by gas turbine under extreme disaster.
$\widetilde{S_k}$	The required installation area per unit of energy equipment k.

4.3.2 Resilient Deployment

The objectives and constraints of the disaster resilience-oriented hybrid system deployment problem will be presented in this section.

4.3.2.1 The Upper-Level Decision Maker: To Maximize the Use of Clean Energy

The authority aims to achieve carbon neutrality and carbon peak targets as soon as possible, which requires the transformation of energy systems and increased use of clean energy by companies in the industrial park. To ensure this, a proportion y of clean energy use must be allocated to the industrial park, where the clean energy generation proportion of the annual electricity consumption of the industrial park should not be less than y. The clean energy proportion used by the industrial park reflects the degree to which the regional energy transition has been achieved. Given the authority's promotion of clean energy use, this objective can be expressed mathematically as shown in Eq. (4.1).

$$\max f_1 = y \tag{4.1}$$

where the clean energy installations should be no less than the lower bound $\underline{Y}$ and no more than the upper bound $\overline{Y}$. This limitation can be described as Eq. (4.2).

$$\underline{Y} \leq y \leq \overline{Y} \tag{4.2}$$

For the lower-level decision maker, there are two objectives considered, as described in the following contents.

4.3.2.2 To Minimize the Total Annual Power Costs

Annual total costs, annual system costs, life cycle costs, energy equilibrium costs, and net present value are widely used economic indicators in studies. In this research, the economic objectives are expressed as the annual total costs, which consist of investment, maintenance, operation, and electricity and gas purchase costs. The mathematical expression of the economic objective is presented in Eq. (4.3).

$$\min f_2 = \sum_k x_k \left(\alpha_k UC_k + AM_k + AO_k\right) + C^e_{it} \sum_i \sum_t E^m_{it} + AC^g \sum_i \sum_t E^g_{it} \tag{4.3}$$

αk represents the annual equivalent investment conversion factor for equipment k, where n denotes the equipment life and r denotes the annual interest rate. Specifically, $\alpha_k = \frac{r(1+r)^n}{(1+r)^n-1}$. The unit investment costs of the equipment k are represented by UC_k, while x_{jk} denotes the capacity of installed equipment k. The annual unit maintenance costs for equipment k are represented by M_k, and AO_k represents the annual unit operating costs for equipment k. C^e_{it} denotes the price of electricity imported from the main grid, and C^{gas}_{in} represents the average annual costs for natural gas imported from the natural gas network. E^{ge}_{it} represents the purchase costs of the electricity imported from the main industrial park grid at time t on day i, while F^{gas}_{it} is the natural gas consumed at the industrial park at time t on day i.

4.3.2.3 To Minimize Carbon Emissions

In order to improve environmental quality and decrease carbon emissions, the development goals of the industrial park established by the authority necessitate the adoption of clean energy. However, using natural gas as a fuel source results in carbon emissions, and the electricity sourced from the main grid is often generated through thermal power. Thus, the objective of minimizing carbon emissions can be expressed as presented in Eq. (4.4).

$$\min f_3 = \sum_i \sum_t E_{it}^m \eta_1 \delta_1 + \sum_i \sum_t E_{it}^g \delta_2 \tag{4.4}$$

δ_1 denotes the average carbon emissions parameter for the coal-burning generation process (g/kWh), while η_1 is the proportion of coal-fired power generation in the main grid. On the other hand, δ_2 is the average carbon emissions parameter for natural gas combustion (g/kWh). E_{it}^m denotes the power imported from the main grid at time t on day i, and E_{it}^g is the natural gas consumed at the same time.

4.3.2.4 To Maximize Power System Resilience

In order to ensure the continuity of key load operations and maintain normal production and operation activities for the users of the industrial park, the power system resilience needs to be improved due to the occurrence of frequent extreme events. The ability of the distributed energy system to meet power recovery requirements during such events can be represented by the variable Q_{it}. The objective of this improvement can be mathematically expressed as shown in Eq. (4.5).

$$\max f_4 = \sum_i \sum_t Q_{it} \tag{4.5}$$

Q_{it} can be further expressed by Eq. (4.6), R_{it} is the self-recovery rate of the hybrid system at time t in day i.

$$Q_{it} = \begin{cases} R_{it}, & R_{it} < 1 \\ 1, & R_{it} \geq 1 \end{cases} \tag{4.6}$$

Self-recovery rate index (R_{it}) can be measured using Eq. (4.7). A higher self-recovery rate indicates greater resilience of the hybrid energy system. In this equation, $\widetilde{E_{ba}}$ represents the energy storage battery storage capacity during extreme disasters, $\widetilde{E_t^g}$ represents the power generated by the gas turbine during extreme events, h represents the duration of the extreme event, and $\sum_t^{t+h} D_{it}$ represents the total power demand of the industrial park from time i to time $i + h$.

$$R_{it} = \frac{E\left[\widetilde{E_{ba}}\right] x_{ba} + E\left[\widetilde{E_t^g}\right] x_{gt} h}{\sum_t^{t+h} D_{it}} \tag{4.7}$$

$\widetilde{E_{ba}}$ and $\widetilde{E_t^g}$ (written as $\widetilde{E_k}$ here after) are the electricity that can be provided per unit of energy equipment k. Influenced by some objective factors, the distributed energy operation in the event of a major event and its ability to withstand such events cannot be exactly valued. However, through field research, it was determined that this value was within a certain range, with the most likely value being in the relatively smaller range. Therefore, parameter $\widetilde{e_k}$ is a typical trapezoidal fuzzy number, which can be written as $\widetilde{e_k} = (e_{k1}, e_{k2}, e_{k3}, e_{k4})$, where $e_{k1} \leq e_{k2} \leq e_{k3} \leq e_{k4}$. This indicates that the maximum value for $\widetilde{E_k}$ is e_{k4}, the minimum value is e_{k1}, and the most likely value is between e_{k2} and e_{k3}. Because this parameter cannot be directly calculated, the expected value operator method for trapezoidal fuzzy numbers proposed by Xu and Zhou (2011) was used. Xu's method determines the trapezoidal fuzzy number's corresponding expected value using Eq. (4.8).

$$\widetilde{E_k} \rightarrow E\left[\widetilde{E_k}\right] = \frac{1-\varsigma}{2}\left(e_{k1} + e_{k2}\right) + \frac{\varsigma}{2}\left(e_{k3} + e_{k4}\right) \tag{4.8}$$

4.3.2.5 Clean Energy Use Restrictions

When planning for the installation of clean energy, the industrial park must consider the decisions of the authority, as it is in a subordinate position in the decision-making process. The proportion of electricity generated from the clean energy installed by the industrial park must not be lower than the clean energy proportion allocated by the authority annually, or else the industrial park may be penalized and lose its right to supply electricity. This constraint is expressed in Eq. (4.9).

$$\frac{\sum_i \sum_t \left(AP_{it}^{pv} x_{pv} + AP_{it}^{g} x_{gt}\right)}{\sum_i \sum_t D_{it}} \geq y \tag{4.9}$$

4.3.2.6 Installation Area Restriction

Limited land availability in the industrial park restricts the installation space for photovoltaic arrays, energy storage batteries, and gas turbines, as illustrated in Eq. (4.10).

$$\sum_k E\left[\widetilde{S_k}\right] x_k \leq S^0 \tag{4.10}$$

where $E[\widetilde{S_k}]$ is the expected value of $\widetilde{S_k}$, which is the required installation area per unit of energy equipment k, and S^0 is the maximum area in the industrial park that can be used for the distributed energy source installation.

4.3.2.7 PV Panel Operation

PV generation is a physical process in which solar energy is transformed directly into electrical energy (Al Garni et al. 2018). The PV panel power output was obtained using Eq. (4.11).

$$AP_{it}^{pv} = A^{pv} D_{pv} \frac{G_{it}}{G_s}\left[1 + \alpha_{pv}\left(T_{it} - T_{sta}\right)\right] \tag{4.11}$$

where A^{pv} is the rated PV array capacity, D_{pv} is the PV derating factor (percent), which is often set at 0.88, G_{it} is the solar irradiation observed on the PV array at time t (kW/m^2), G_s is the solar irradiation standard condition incident (1 kW/m^2), α_{pv} is the power temperature coefficient (%/°C), which is often set as 0.35%/°C. T_{it} is PV cell temperature at time t and T_{sta} is the PV cell temperature under STCs.

4.3.2.8 Energy Storage System Operation

To mitigate the unpredictability and intermittent nature of renewable energy, enhance the efficiency of gas turbines, stabilize the load, and maintain the photovoltaic power generation within a specific range, the installation of energy storage batteries is necessary (Pingkuo and Zhongfu 2016). This would also decrease the frequency and intensity of gas turbine peak shaving (International Renewable Energy Agency 2019). The impact of the battery bank can be expressed using the equation given in Eq. (4.12).

$$E^{ba}_{i,t+1} = E^{ba}_{it}(1-\lambda) + Z^{cha}_{it}\chi - \frac{Z^{dis}_{it}}{\chi} \tag{4.12}$$

where $E^{ba}_{i,t+1}$ and E^{ba}_{it} are the battery system power at time step $t+1$ and t in day i, λ is the self-discharging factor, Z^{cha}_{it} and Z^{dis}_{it} are the charging and discharging battery power at time t on day i, χ is the charging and discharging efficiency.

4.3.2.9 Battery State Restrictions

For safety reasons, the battery state needs to meet the limitation shown in Eqs. (4.13)–(4.16), in which $P^{cha}_{\max}$ and $P^{dis}_{\max}$ are the maximum charging and discharging energy storage battery power, RU^b and RD^b are the upward and downward energy storage battery climbing rates.

$$0 \le Z^{cha}_{it} \le P^{cha}_{\max} \tag{4.13}$$

$$0 \le Z^{dis}_{it} \le P^{dis}_{\max} \tag{4.14}$$

$$Z^{cha}_{it} Z^{dis}_{it} = 0 \tag{4.15}$$

$$-RD^b \le E^{ba}_{i+1,t} - E^{ba}_{it} \le RU^b \tag{4.16}$$

4.3.2.10 Gas Turbine Operation

The gas turbine is essential in ensuring the stability of the distributed energy system's power supply. The conversion efficiency of the gas turbine, denoted as ε_{gt}, is expressed in Eq. (4.17), where E^{g}_{it} represents the natural gas consumed by the gas turbine at time t on day i, and H^g is the calorific value obtained from burning natural gas. The gas turbine output is subject to constraints, as shown in Eqs. (4.18) and (4.19). Here, RU^g and RD^g are the upward and downward climbing rates of the gas turbine, while $P^{gt}_{\max}$ represents the maximum gas turbine power output. Here, RU^g and RD^g denote the gas turbine's upward and downward climbing rates, respectively, and $P^{gt}_{\max}$ represents the maximum gas turbine power output.

$$AP^{g}_{it} = \varepsilon_{gt} F^{gas}_{it} H^g \tag{4.17}$$

$$0 \le AP^{g}_{it} \le P^{gt}_{\max} \tag{4.18}$$

$$-RD^g \le P^{gt}_{i+1,t} - AP^{g}_{it} \le RU^g \tag{4.19}$$

4.3.2.11 Power Supply and Demand Balance

The provision of sufficient electricity is crucial for the production and operation activities of the industrial park users, as depicted in Eq. (4.20).

$$AP_{it}^{pv}x_{pv} + AP_{it}^{g}x_{gt} + Z_{it}^{dis} = D_{it} + Z_{it}^{cha} \tag{4.20}$$

4.3.3 Global Model

In order to achieve sustainable socioeconomic development and environmental protection, China has set the goal of carbon peaking and carbon neutralization, which requires optimization of energy and industrial structures. Additionally, due to the frequency of extreme events, it is essential for industrial parks to strengthen their local emergency power supply. To address these challenges, a bi-level model for a distributed energy source installation scheme under uncertainty has been developed. The objective of the authority is to achieve carbon neutrality and carbon peaking by maximizing the proportion of clean energy in the industrial park, as indicated by the maximum and minimum proportions in Eq. (4.2). After the required clean energy proportion is determined by the authority, the industrial park develops a suitable distributed energy source installation scheme (Eq. (4.3)), considering the area and demand balance constraints (Eqs. (4.9)–(4.20)). The industrial park presents its proposed scheme to the authority, which then adjusts its initial decisions to align better with the industrial park's proposal. Subsequently, the authority sends an improved clean energy allocation scheme to the industrial park, following which the industrial park modifies its previous distributed energy source installation scheme and submits it to the authority. This iterative process continues until all stakeholders reach a consensus on an equilibrium solution. The global bi-level model can be expressed as Eq. (4.21).

$$\begin{aligned}
&\max f_1 = y\\
&\left\{\begin{aligned}
&\underline{Y} \le y \le \overline{Y}\\
&\min f_2 = \textstyle\sum_k x_{jk}\left(\alpha_k UC_k + AM_k + AO_k\right) + C_{it}^{in,e}\sum_i\sum_t E_{it}^{m} + AC^{g}\sum_i\sum_t E_{it}^{g}\\
&\min f_3 = \textstyle\sum_i\sum_t E_{it}^{m}\eta_1\delta_1 + \sum_i\sum_t E_{it}^{g}\delta_2\\
&\max f_4 = \textstyle\sum_i\sum_t Q_{it}\\
&\left\{\begin{aligned}
&\frac{\sum_i\sum_t(AP_{it}^{pv}x_{pv} + AP_{it}^{g}x_{gt})}{\sum_i\sum_t D_{it}} \ge y\\
&\textstyle\sum_{k=1}^{K} E\left[\widetilde{S_k}\right] x_k \le S_{\max}\\
&AP_{it}^{pv} = A^{pv}D_{pv}\frac{G_{it}}{G_s}\left[1 + \alpha_{pv}\left(T_{it} - T_{sta}\right)\right]\\
&E_{i+1,t}^{ba} = E_{it}^{ba}(1-\lambda) + Z_{it}^{cha}\chi - \frac{Z_{it}^{dis}}{\chi}\\
&0 \le Z_{it}^{cha} \le P_{\max}^{cha}\\
&0 \le Z_{it}^{dis} \le P_{\max}^{dis}\\
&Z_{it}^{cha}Z_{it}^{dis} = 0\\
&-RD^{b} \le E_{i+1,t}^{ba} - E_{it}^{ba} \le RU^{b}AP_{it}^{g} = \varepsilon_{gt}F_{itk}^{in}H^{g}\\
&P_{\min}^{gt} \le AP_{it}^{g} \le P_{\max}^{gt}\\
&-P_{rd} \le P_{i+1,t}^{gt} - AP_{it}^{g} \le P_{ru}AP_{it}^{pv}x_{pv} + AP_{it}^{g}x_{gt} + Z_{it}^{dis} = D_{it} + Z_{it}^{cha}
\end{aligned}\right.
\end{aligned}\right.
\end{aligned} \tag{4.21}$$

A bi-level model is created to identify the optimal clean energy utilization plan for the industrial park, aligned with the national energy system transformation objectives and enhancing the regional power system's resilience. Unlike earlier schemes, this model incorporates multiple decision-makers, making it more practical for industrial parks and the policy-making process of the authority. With flexibility in adjusting policy control parameters, both the authority and the industrial park can reach a mutually acceptable decision based on the situation.

4.3.4 Model Solving

To deal with the trade-off between the resilience, economy, environment factors and to determine the optimal relationship between the local authority and the industrial park, a bi-level multi-objective model is established. The ε-constraint method, which guarantees that the most important objective reaches its ideal value while also considering the other objectives, has proven to be a very valuable method for solving multi-objective optimization (Haimes 1971; Chankong and Haimes 2008). Since the economic objective is the most important objective for the industrial park, it is given priority. However, the environmental and resilience objectives are also essential, and they must be met within predetermined acceptable ranges. To represent the industrial park's attitude toward these objectives, two parameters are introduced: a, representing the attitude toward carbon emissions, and b, representing the attitude toward resilience. The proposed model is then transformed into an equivalent single objective form, as presented in Eq. (4.22).

$$
\begin{aligned}
&\max f_1 = y \\
&\begin{cases}
\overline{Y} \leq y \leq \underline{Y}, \\
\min f_2 = \sum_k x_{jk} \left(\alpha_k UC_k + AM_k + AO_k \right) + C_{it}^{in,e} \sum_i \sum_t E_{it}^m + AC^g \sum_i \sum_t E_{it}^g \\
\sum_i \sum_t E_{it}^m \eta_1 \delta_1 + \sum_i \sum_t E_{it}^g \delta_2 \leq aL \\
\sum_i \sum_t Q_{it} \geq bU \\
\text{Eqs. (4.6)–(4.20)}
\end{cases}
\end{aligned}
\tag{4.22}
$$

Studies have demonstrated that even the simplest form of bi-level models is an NP-hard problem due to the interactions between the objectives and constraints of the upper and lower levels (Colson et al. 2007). Interactive algorithms, which evolve toward an optimal solution through multiple iterations between the two levels, offer a simple and fast approach for solving bi-level optimization problems (Emam 2013). Consequently, an interactive algorithm was devised to solve the bi-level model. The solution processes of the model are depicted in Figure 4.4.

The clean energy utilization proportion before and after termination differs by less than 1%. Specifically, if the difference between each proportion in two adjacent cycles is less than 1%, the interactive algorithm is considered complete, and the result from the final interaction is the final solution. While this may not be the optimal

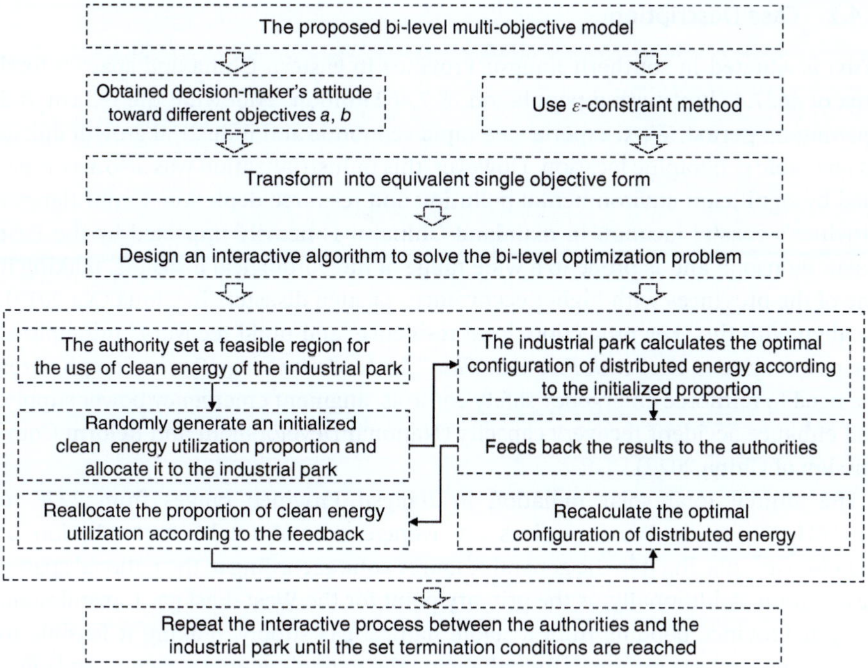

Figure 4.4 Model solving and associated results analysis strategy.

solution, this algorithm is practical as it saves time and yields results very close to the optimal solution, given the allowable error set at 1%.

4.4 Case Study

In this section, a practical example from an industrial park in Wuxi, Jiangsu Province, China (Figure 4.5), is given to demonstrate the practicality and efficiency of the proposed optimization model.

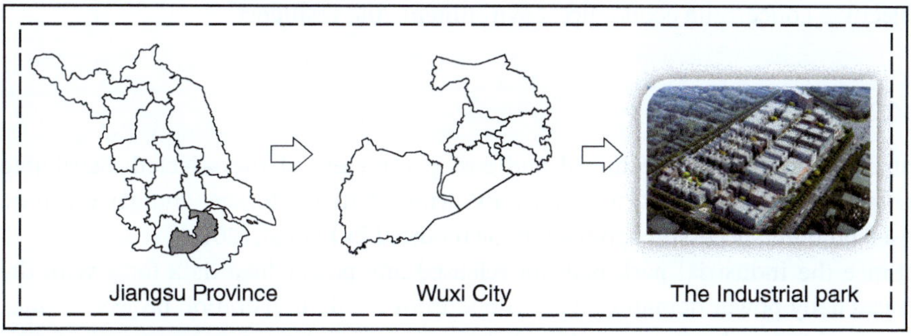

Figure 4.5 Industrial park location.

4.4.1 Case Description

Wuxi is situated in southern Jiangsu Province in eastern China and spans a total area of 4627.47 km^2 with a population of 7.462 million. Following the reform and opening-up period, Wuxi experienced rapid economic and industrial growth due to its favorable geographic location. However, this industrialization was also accompanied by significant environmental pollution and resource depletion. Given Jiangsu Province's coastal location in mainland China, it is heavily impacted by the East Asian monsoon and is prone to a wide range of meteorological disasters, making it one of the provinces with higher occurrences of such disasters in China (Xu 2019). To align with the "key protection, local resilience, and rapid recovery" principle of the power system, the *"14th Five-Year Plan" for Modern Energy Systems* emphasizes the need to reinforce emergency safety controls, augment emergency power supply, and enhance accident recovery capacity (National Development and Reform Commission of China 2022).

The annual total solar radiation in Jiangsu Province ranges from 2978 to 5657 MJ/m^3, while Wuxi receives an average annual total solar radiation of 4514 MJ/m^3, placing it in the general resource zone according to the national spatial distribution. Additionally, as the primary point for the West–East gas transmission, Jiangsu Province benefits from a stable natural gas supply, making it feasible to leverage natural gas resources for enhancing the region's energy structure and environmental sustainability (Central People's Government of the People's Republic of China 2006). As part of the province's "14th Five-Year Plan" (Jiangsu Development & Reform Commission 2021), the installation of photovoltaic power generation capacity is set to reach 26 million kilowatt, with natural gas utilization targeted to exceed 32 billion cubic meter by 2025. Accelerating the transformation of green industrial structure is imperative to establish a green, low-carbon industrial system and foster the development of green factories, parks, and supply chains in the manufacturing industries and industrial parks throughout the province (Department of Industry and Information Technology of Jiangsu Province 2021). Consequently, the authorities are investigating alternative installations of distributed energy sources in energy-intensive industrial parks to achieve these objectives. To this end, the proposed bi-level multi-objective model was employed in an industrial park located in Wuxi to determine if the designed system could effectively fulfill multiple objectives concerning economics, environment, and resilience.

4.4.2 Data Collection

Table 4.1 displays the technical and cost parameters of the selected distributed energy source candidates, which are PV arrays, battery storage, and gas turbines, based on their successful applications as reported in Li et al. (2019).

Since the industrial park had not released any power load data for a year, the typical daily power demand was obtained from the demonstration project records. Previous studies have demonstrated the validity of using data from typical days when actual data is unavailable or hard to calculate (Zhou et al. 2020). For instance,

Table 4.1 Technical and cost parameters of the hybrid system components.

	Technical parameters			Cost parameters		
Components	**Parameters**	**Value**	**Unit**	**Parameters**	**Value**	**Unit**
PV	STC radiation	100	W/m^2	Initial investment cost	12,000	CNY/kW
	STC temperature	25		Operation and maintenance cost	30	CNY/kW
	Apv	−0.0035%	—	Expected lifetime	20	Year
Battery	Charge/discharge efficiency	0.88	—	Initial investment cost	7786	CNY/kW
	Charge/discharge rate	0.25	—	Operation and maintenance cost	615.094	CNY/kW
	Self-discharging rate	0.001	—	Expected lifetime	15	Year
	Max, min, and initial SOC	0.95, 0.2, 0.25	—			
Gas turbine	Power generation efficiency	0.84	—	Initial investment cost	800	CNY/kWh
	Natural gas consumption	2.86	m^2/kW	Operation and maintenance cost	20	CNY/kWh
				Expected lifetime	5	Year

SOC, state of charge; STC, standard test condition.

Shen et al. (2020) simulated interprovincial hydropower system operations using three typical inflow scenarios representing dry, normal, and wet years, and Wang et al. (2018) conducted an optimal scheduling problem case study using integrated energy system operations data from a typical summer day. Therefore, following existing research and an analysis of the industrial park's power demand across typical days in spring, summer, autumn, and winter in one year, the typical daily power demand, solar irradiation, and ambient temperature data were established and are presented in Table 4.2.

The industrial electricity prices and the photovoltaic and gas-fired power generation feed-in tariffs were collected from the Jiangsu Provincial Price Bureau. The price of natural gas 2.86 CNY/m^3, buying electricity from the main grid follows Time-of-Use price, which is 0.9647 CNY/kWh during peak hours from 8:00 to 12:00 and 17:00 to 12:00; 0.5568 CNY/kWh from 21:00 to 24:00, and 0.2289 CNY/kWh during off-peak hours from 0:00 to 8:00. Price of selling PV and gas turbine generated power to the main grid is 0.98 CNY/kWh and 0.449 CNY/kWh separately.

4.4.3 Calculation Results and Analysis

In this section, maximum resilience emission results, comparison of different scenarios, operation under normal modes, and operation under extreme disasters are displayed.

Table 4.2 Typical daily power demand (MW), solar irradiation (kW/m^2), and ambient temperature (°C) data of the studied industrial park located in Wuxi city.

Typical day	Parameter	Time series																							
		1 h	2 h	3 h	4 h	5 h	6 h	7 h	8 h	9 h	10 h	11 h	12 h	13 h	14 h	15 h	16 h	17 h	18 h	19 h	20 h	21 h	22 h	23 h	24 h
Typical spring day	Demand (MW)	22.45	21.33	20.20	20.77	21.89	24.13	26.38	29.18	35.92	56.12	52.75	44.90	42.65	44.90	40.97	31.99	29.18	26.94	29.18	28.06	26.94	25.82	23.57	22.45
	Solar radiation (kW/m^2)	0	0	0	0	0	23	125	245	376	497	584	625	609	562	466	344	209	95	10	0	0	0	0	0
	Ambient temperature (°C)	19.8	19.4	19.3	19.1	18.8	19.1	19.9	20.2	20.6	21.5	22.9	23.6	24.1	24.7	24.8	24.8	24.5	23.8	22.6	22.1	21.7	21.2	20.4	20.1
Typical summer day	Demand (MW)	40.18	38.17	36.16	37.16	39.17	43.19	47.21	52.23	64.29	100.45	94.42	80.36	76.34	80.36	73.33	57.25	52.23	48.21	52.23	50.22	48.21	46.21	42.19	40.18
	Solar radiation (kW/m^2)	0	0	0	0	0	8	104	230	359	480	558	589	570	531	446	338	211	94	8	0	0	0	0	0
	Ambient temperature (°C)	27	26.8	26.7	26.5	26.3	26.6	27.1	27.4	27.9	28.6	29.8	30.3	30.6	31	31.1	31	30.6	29.9	28.9	28.6	28.3	28	27.5	27.3
Typical autumn day	Demand (MW)	19.52	18.54	17.56	18.05	19.03	20.98	22.93	25.37	31.22	48.79	45.86	39.03	37.08	39.03	35.62	27.81	25.37	23.42	25.37	24.39	23.42	22.44	20.49	19.52
	Solar radiation (kW/m^2)	0	0	0	0	0	0	7	93	208	309	379	406	391	343	256	146	36	0	0	0	0	0	0	0
	Ambient temperature (°C)	11.3	11	11	10.8	10.5	10.5	10.7	11	11.7	12.7	14.4	15	15.4	15.9	16	15.7	15	14.1	13.2	13	12.7	12.4	11.9	11.6
Typical winter day	Demand (MW)	12.76	12.12	11.48	11.80	12.44	13.71	14.99	16.58	20.41	31.89	29.97	25.51	24.23	25.51	23.28	18.18	16.58	15.31	16.58	15.94	15.31	14.67	13.39	12.76
	Solar radiation (kW/m^2)	0	0	0	0	0	0	0	57	166	272	351	400	413	378	311	214	101	10	0	0	0	0	0	0
	Ambient temperature (°C)	5.1	4.9	4.8	4.6	4.3	4.2	4.1	4.3	4.9	5.9	7.5	8.2	8.7	9.6	9.4	9.2	8.5	7.7	6.9	6.7	6.5	6.2	5.7	5.4

4.4.3.1 Maximum Resilience Emission Results

In this section, the two groups of results under different decision-maker attitudes are discussed.

Group1: a = 1, b Is Changing In this group, the industrial park's attitude toward carbon emissions reductions was fixed at 1 ($a = 1$), and the attitude parameter toward resilience (b) is changing, $a = 1$ means there is no restriction on the amount of emission reduction. As shown in Table 4.2, the sensitivity analysis step length was set at 0.1. When $b = 0.1$, the cost is 1.7169×10^8 CNY and resilience index is 24.37%; the highest resilience index (b) is 0.955, which means the distributed energy resilience in the industrial park reaches as high as 95.85%, and the annual power cost is 2.1909×10^8 CNY.

Group2: b = 0, a Is Changing In this group, the industrial park's resilience attitude (b) is fixed at 0, and the attitude toward carbon emissions (a) is changing, $b = 0$ means the industrial park is not keen on establishing a resilient power system. As shown in

Table 4.3 Sensitivity analysis on the industrial park's attitude toward resilience targets.

Parameter		Decision variables			Objective values		
		N_{pv} (MW)	N_{gt} (MW)	N_{ba} (MW)	Cost (10^8 CNY)	Emission (10^5 tonnes)	Resilience index (%)
$a = 1$[a]	$b = 0.1$	70	0	20	1.7169	1.3387	24.37
	$b = 0.2$	70	4	20	1.7277	1.2907	38.99
	$b = 0.3$	70	5	20	1.7284	1.3008	42.64
	$b = 0.4$	70	8	18	1.7375	1.2774	51.16
	$b = 0.5$	70	11	20	1.7464	1.2287	62.96
	$b = 0.6$	70	18	13	1.7811	1.1634	74.24
	$b = 0.7$	70	25	12	1.8512	1.0992	84.16
	$b = 0.8$	70	26	12	1.9485	1.0387	85.21
	$b = 0.9$	70	38	12	2.0797	0.9762	94.32
	$b = 0.955$	70	40	18	2.1909	0.9620	95.85
$b = 0$[b]	$a = 1$	70	0	20	1.7169	1.3387	24.37
	$a = 0.9$	70	4	20	1.7277	1.2907	38.99
	$a = 0.8$	70	11	20	1.7464	1.2287	62.96
	$a = 0.7$	70	18	13	1.7811	1.1634	74.24
	$a = 0.6$	70	26	11	1.9485	1.0387	85.21
	$a = 0.5$	70	37	12	2.0750	0.9767	93.84
	$a = 0.493$	70	40	18	2.1909	0.9620	95.85

a) $a = 1$: The industrial park's attitude toward emissions reductions was set at 1, there is no emission reduction pressure;

b) $b = 0$: The industrial park's resilience attitude is set at 0, there is no resilience improvement pressure.

Table 4.3, the sensitivity analysis step length was set at 0.1. When $a = 1$, the annual power cost is 1.7169×10^8 CNY and the carbon emissions are 1.3387×10^5 tonnes. The lowest carbon emissions ratio (a) can reach 0.493, indicating that when $b = 0$, the total carbon emissions reduce by 50.7% compared with the total power supplies from the main grid, the annual power cost is 2.1909×10^8 CNY, and the resilience index is 95.85%.

4.4.3.2 Comparison of Different Scenarios

In this section, results under four different scenarios are obtained and compared, as shown in Table 4.4,

Scenario 1 To rely only on the main grid. In this scenario, the industrial park relies on main grid for power supply, as shown in Table 4.4, the value of resilience index is 0, which means that when extreme events occur, a power system that only relies on the main grid lacks resilience.

Scenario 2 To minimize annual power cost. In this scenario, a distributed energy system is established to improve power supply resilience, and given the optimal economic cost, the system resilience could be improved by installing a distributed 70 MW photovoltaic and a 20 MW energy storage system. Because of the severe weather conditions that often accompany extreme events, photovoltaic panels are often unable to generate electricity. Therefore, only the discharge provided by the energy storage can supply the power demand, at which time the resilience index is 24.36%.

Scenario 3 To obtain the economically feasible resilience index. In this scenario, the annual power cost of deploying distributed energy is equal to scenario 1, so that the economically feasible resilience index can be obtained. As shown in Table 4.4, a 66 MW photovoltaic, a 19 MW energy storage facility, and a 29 MW gas turbine could be installed, which would provide 89.34% of the resilience index.

Scenario 4 To maximize power supply resilience. In this scenario, the maximum resilience index of deploying distributed energy sources is obtained. Under this situation, a 40 MW gas turbine is added to a 70 MW photovoltaic and an 18 MW energy storage, which together raise the resilience index to 95.85%.

4.4.3.3 Operation Under Normal Modes

Although the aim is to improve the resilience of the power system through the installation of distributed energy, the distributed energy system mainly operates under daily scenarios because of the low frequency of extreme disasters. In this section, the normal operation of the distributed energy system under the maximum resilience installation plan is discussed.

Figure 4.6a illustrates the power demand for typical spring days, which is primarily fulfilled by gas power generation. From 1:00 to 2:00, the main grid is required to provide some power as the gas turbine's start-up ramp rate is limited. The excess

Table 4.4 Installation plans under different scenarios.

		Decision variables			Objective values		
Scenario	**Description**	N_{pv} (MW)	N_{gt} (MW)	N_{ba} (MW)	Cost (10^8 CNY)	Carbon emission (10^5 tonnes)	Resilience lindex (%)
Scenario 1	Relying only on the main grid	0	0	0	1.8456	1.9533	0
Scenario 2	Minimum annual power cost for ldeploying distributed energy sources	70	0	20	1.7169	1.3386	24.36
Scenario 3	The annual power cost of deploying ldistributed energy is equal to scenario 1	66	29	19	1.8456	1.2188	89.34
Scenario 4	Deploying distributed energy sourcesl to generate maximum resilience	70	40	18	2.1947	0.9532	95.85

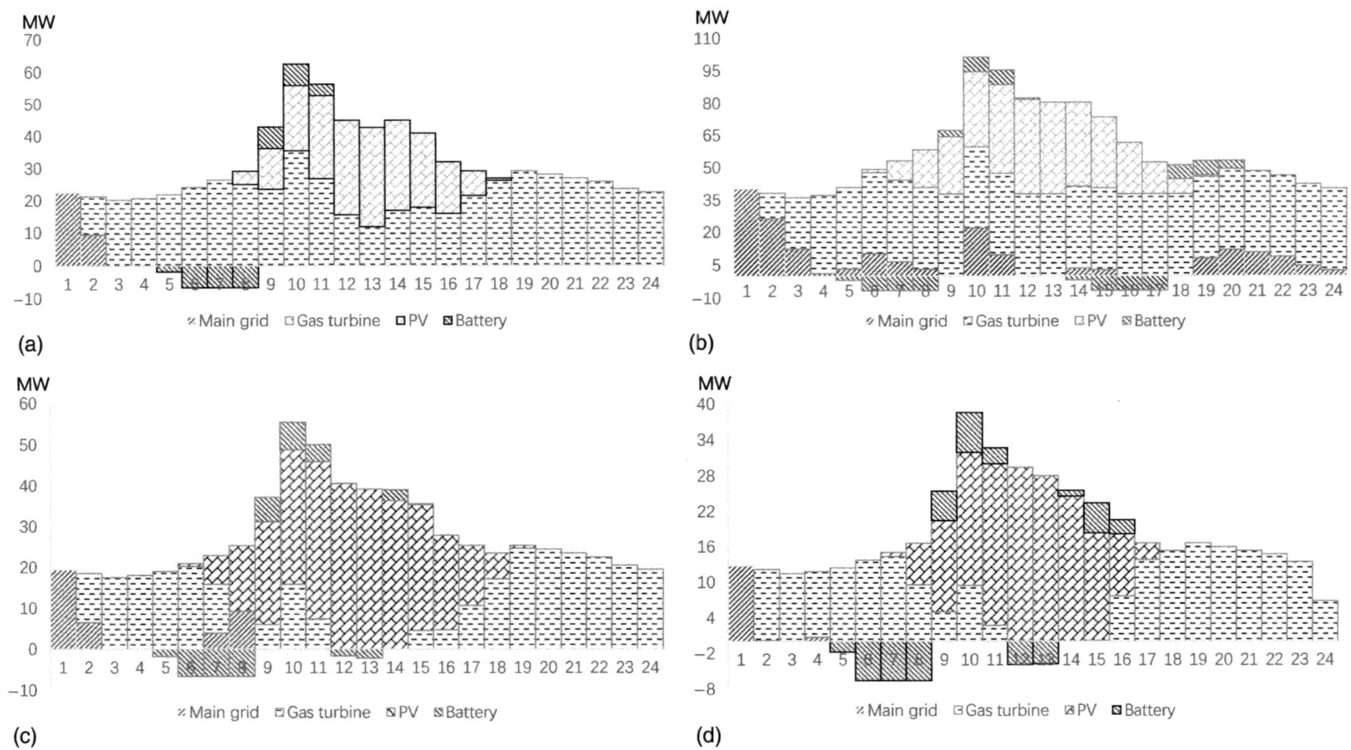

Figure 4.6 The typical power supply in spring, summer, autumn, and winter. (a) Power supply on typical days in spring. (b) Power supply on typical days in summer. (c) Power supply on typical days in autumn. (d) Power supply on typical days in winter.

power generated by the gas turbine is stored in the energy storage battery from 5:00 to 8:00. The battery discharges From 9:00 to 11:00 to meet the power supply requirements. Between 8:00 and 17:00, the photovoltaic panels begin to generate electricity as the solar radiation meets the photovoltaic power generation requirements. As a result, photovoltaic power generation fulfills around 50% of the electricity demand between 11:00 and 16:00. This optimal power supply and demand schedule ensures the efficient use of all the distributed energy sources and minimizes the reliance on the main grid.

Figure 4.6b shows that the power demand on typical summer days is still mainly provided by gas power generation. Even from 10:00 to 16:00, when the photovoltaic power generation is the highest, it is not able to meet the summer power demand, which means the gas turbine is still needed to supplement the power supply. From 10:00 to 11:00, the main grid provides part of the power supply, as there is no photovoltaic power generation from 6:00 to 7:00 and from 19:00 to 24:00, only the gas turbine supplies the main power, and the energy storage battery and main grid supply the auxiliary power.

Figure 4.6c,d shows that the power supply on typical autumn and winter days is similar. Because of the small electricity demand and the good solar radiation conditions, the photovoltaic power generation can replace the gas power generation as the main power supply source from 9:00 to 17:00. However, in all other periods, the gas turbine is still the main power supply, and the main grid and the energy storage battery are the auxiliary power supplies.

4.4.3.4 Operation Under Extreme Disasters

Assuming that extreme events can happen at any time, the distributed energy resilience in the three hours after an extreme event was analyzed and shown in Figure 4.7, and economic loss of industrial parks facing extreme disasters under different scenarios is shown in Figure 4.8, for which the power outage costs of the industrial users (34,900 CNY/MW) in Bao et al. (2019) were selected.

Power Supply Recovery Rate Under Different Scenarios Figure 4.7a–d shows that the installed distributed energy is unable to fully meet the power demand for three consecutive hours following extreme events. Among the four typical days analyzed, winter provides the best supply conditions, with up to 47% of off-peak hour power demand being met. In contrast, summer poses the greatest challenge, as only 19% of the power demand during peak hours can be met by the installed distributed energy.

Figure 4.7e–h shows that in Scenario 3 the installed distributed energy can provide more than 50% of the power supply at most times, except for the period between 8:00 and 13:00 in summer. In winter, the installed distributed energy is capable of fully meeting the power demand for three consecutive hours at any time. In autumn, a 100% power supply cannot be guaranteed for the peak power consumption period between 9:00 and 13:00. In summer, approximately 40% of the power supply can be guaranteed between 8:00 and 14:00, but at other times, more than 60% of the power

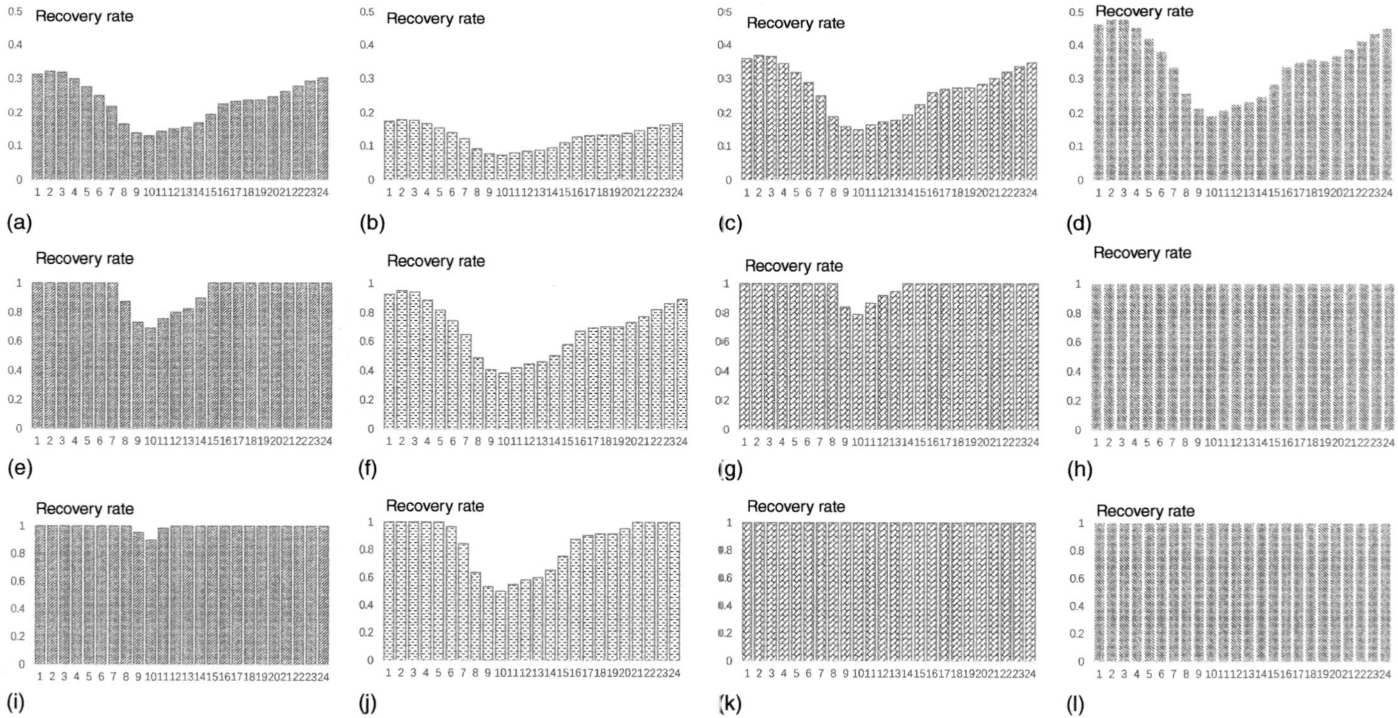

Figure 4.7 The typical power supply in spring, summer, autumn, winter under different scenarios. (a) Scenario 2: Recovery rate on typical days in spring. (b) Scenario 2: Recovery rate on typical days in summer. (c) Scenario 2: Recovery rate on typical days in autumn. (d) Scenario 2: Recovery rate on typical days in winter. (e) Scenario 3: Recovery rate on typical days in spring. (f) Scenario 3: Recovery rate on typical days in summer. (g) Scenario 3: Recovery rate on typical days in autumn. (h) Scenario 3: Recovery rate on typical days in winter. (i) Scenario 4: Recovery rate on typical days in spring. (j) Scenario 4: Recovery rate on typical days in summer. (k) Scenario 4: Recovery rate on typical days in autumn. (l) Scenario 4: Recovery rate on typical days in winter.

supply is guaranteed, with the maximum power supply reaching 90%. In spring, only 70–85% of the power can be supplied between 8:00 and 14:00.

Figure 4.7i–l shows that in Scenario 4 the installed distributed energy is capable of fully supplying the power demand at any time for three consecutive hours after an extreme event in autumn and winter. In spring, except for the period between 9:00 and 11:00, the distributed energy can fully meet the power demand within three hours after an extreme event. When extreme events occur at 9:00, the distributed energy can meet 95.3% of the power demand, while for events at 10:00 and 11:00, the distributed energy can meet 89.7% and 98.35% of the power demand, respectively. However, in summer, when extreme events occur between 8:00 and 15:00, the distributed energy can only meet around 50% of the power demand. On the other hand, when extreme events occur between 6:00 and 7:00 or from 16:00 to 20:00, the distributed energy can meet more than 85% of the power demand, and it can fully meet the power demand at all other times.

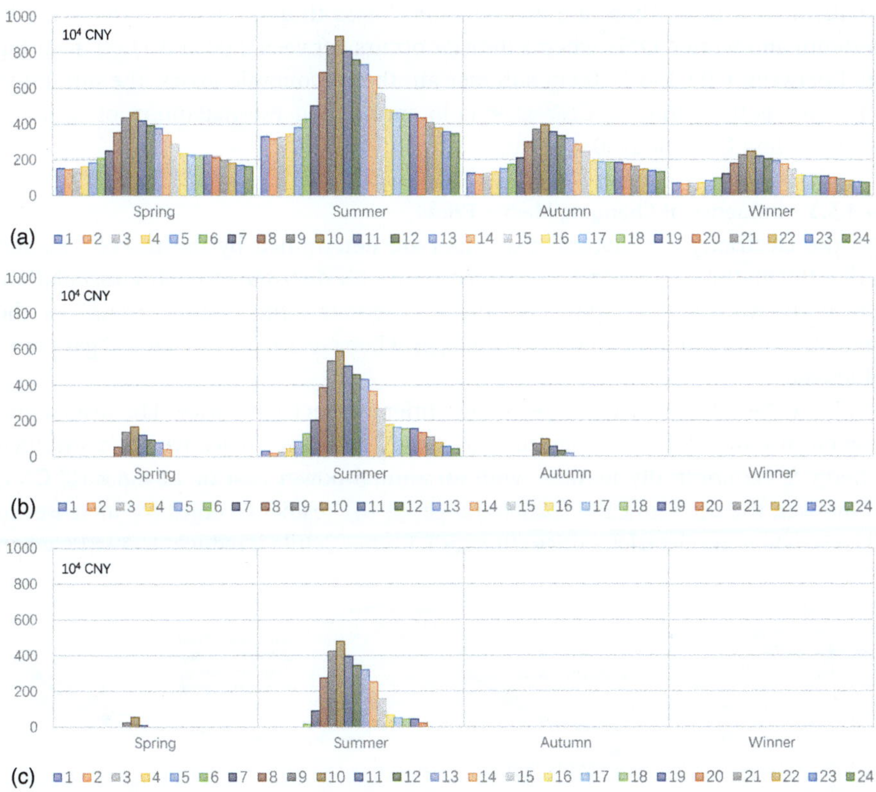

Figure 4.8 Financial losses of industrial parks facing extreme disasters under different scenarios (10^4 CNY). (a) Financial losses of industrial parks facing extreme disasters under scenario 2. (b) Financial losses of industrial parks facing extreme disasters under scenario 3. (c) Financial losses of industrial parks facing extreme disasters under scenario 4.

Financial Losses Under Different Scenarios Figure 4.8a shows the economic losses from extreme events in the industrial park in Scenario 2. To show the results more clearly, a yellow-red gradient was used to indicate the economic losses. Extreme events at any time on the four typical days result in economic losses, the largest of which are from 8:00 to 14:00; the maximum economic losses will occur at 10:00 in summer, with a loss value of 8.907 million CNY, and the minimum economic losses occur at 2:00 in winter, with a loss value of 0.644 million CNY.

Figure 4.8b shows the economic losses from extreme events in the industrial park in Scenario 3. To present the economic loss results more clearly, a white-blue gradient was used. It can be seen that extreme disasters do not cause economic losses in winter, but cause economic losses at any time of the day in summer. However, in spring and autumn, only when extreme events occur at the 8:00–14:00 power consumption peak are there economic losses. The maximum economic loss caused by extreme events at 10:00 in summer is 5.906 million CNY, and the minimum is 0.192 million CNY at 2:00 in summer.

Figure 4.8c shows the economic losses from extreme events in the industrial park in Scenario 4. To present the economic loss results more clearly, a white-green gradient was used. There are no economic losses three hours after the event in autumn and winter. Only when a disaster occurs between 9:00 and 11:00 in spring and between 6:00 and 10:00 in summer are there economic losses, the minimum of which is 0.237 million CNY at 9:00 in spring, and the maximum of which is 4.789 million CNY at 10:00 in summer.

4.4.3.5 Influence of Changing Market Prices

As the electricity and natural gas prices are determined by the market, market price fluctuations can affect the industrial park's distributed energy installation, power production, and purchase plans. Therefore, this section examines the industrial park's annual power cost changes when electricity and natural gas prices fluctuate.

Figure 4.9a shows cost changes under different electricity price. The results presented in point *B* demonstrate that an 89.34% resilience objective for distributed energy is economically feasible, with an annual power cost of 1.8456×10^8 CNY. Moreover, the economically feasible range for the resilience objective increases to 95.85% when the electricity price changes 1.1 times, while it reduces to 59.81% when

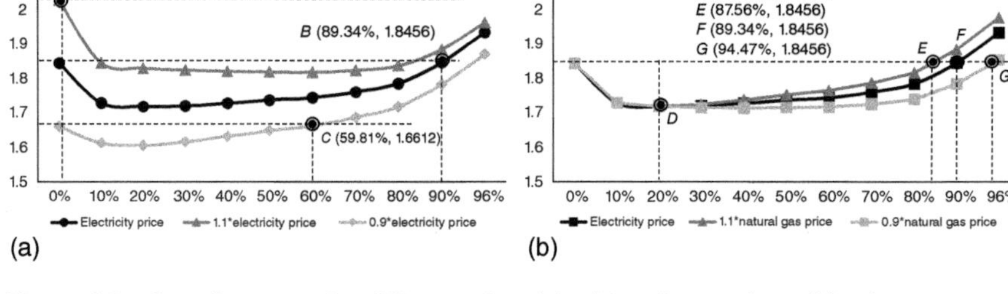

Figure 4.9 Cost change under different electricity (a) and natural gas (b) prices.

the electricity price changes 0.9 times. It is worth noting that, under different electricity prices, the annual power costs initially decrease and then increase as the system's resilience objective increases. This trend is due to the selection of lower-cost photovoltaic and energy storage systems to achieve the resilience enhancement goal. As the costs decrease, the objective can be achieved. However, when photovoltaic and energy storage systems reach their upper limits, gas turbines must be installed to achieve the objective. Since the gas turbine power generation costs are high, increasing the resilience objective leads to more gas turbine installations and, consequently, higher annual power costs.

Figure 4.9b shows cost changes under different natural gas price. As shown in point *D*, the cost curves overlap when the resilience is lower than 20.68%, indicating that natural gas price changes have no impact on the cost. This is because only photovoltaic energy and energy storage are needed to achieve the resilience objective, and gas turbines are not required. Point *E* demonstrates that an 87.65% resilience objective can be achieved economically using distributed energy systems when the natural gas price increases by 1.1 times. Similarly, point *G* indicates that a distributed energy resilience objective of 94.47% is economically feasible when the natural gas price decreases to 0.9 times the original value.

4.5 Comprehensive Discussion

In this section, some propositions obtained from the results analysis are presented, and management recommendations are provided.

4.5.1 Related Propositions

Firstly, deploying hybrid wind–solar-storage-gas system makes a positive contribution to carbon emissions reduction and is economically feasible. Table 4.3 indicates that an increase in parameter *b* leads to an enhancement of power system resilience and a reduction in generated power consumption carbon emissions. Specifically, the deployment of a hybrid wind–solar-storage-gas system resulted in a reduction of carbon emissions by 0.991×10^5 tonnes compared to reliance solely on the main grid power supply. Additionally, Table 4.4 demonstrates that the installation of a hybrid wind–solar-storage-gas system has a positive effect on power system resilience and economic feasibility. The annual power costs are reduced from 1.8456×10^8 CNY to 1.7169×10^8 CNY, corresponding to a decrease of 6.97%, when a hybrid system is deployed and the power system resilience index is 24.36%. In summary, the results of this study suggest that deploying a hybrid wind–solar-storage-gas system is both economically feasible and beneficial in terms of enhancing power system resilience and reducing carbon emissions in industrial parks.

Secondly, deploying hybrid wind–solar-storage-gas system improves the ability of industrial parks to withstand extreme events and improves power system resilience. The findings in Figure 4.7i–l indicate that the distributed energy can fully guarantee power supply for three consecutive hours after an extreme event occurs on typical days in autumn and winter. In spring, when disasters occur from 8:00 to 10:00, the distributed energy can supply over 90% of the power demand. Even during the summer season when electricity demand is at its peak, the distributed energy system can ensure more than 50% of the power demand. Furthermore, the financial analysis presented in Figure 4.8 demonstrates that deploying a hybrid energy system with optimal distributed energy resilience configuration can minimize the economic losses from power failure to 0.232 million CNY. As a result, the deployment of hybrid energy systems can significantly improve the ability of industrial parks to withstand extreme events and enhance power system resilience.

Finally, deploying hybrid wind–solar-storage-gas system can help industrial parks cope with the risk of future changes in electricity and natural gas market prices. As shown in Figure 4.9a, the hybrid energy system remains a more cost-effective option than relying solely on the main grid, even when electricity prices increase by 1.1 times the current level. The system can achieve a maximum resilience index of 95.85%. Similarly, Figure 4.9b shows that the annual power costs of the hybrid energy system remain lower than relying solely on the main grid, even if natural gas prices increase by 1.1 times the current level. The hybrid system can achieve a resilience index of 87.65%. Therefore, the deployment of a hybrid wind–solar-storage-gas system can significantly enhance the ability of industrial park users to manage changes in electricity and natural gas prices. Moreover, it is important to note that as long as appropriate control parameters are maintained to improve the power system resilience, it is always more cost-effective to install a distributed energy system than relying solely on the main grid.

4.5.2 Management Recommendations

Based on the above discussion and analysis, this section gives some management recommendations as the foundation for distributed energy system optimization.

The optimization model and analysis method presented in this chapter can serve as a valuable analytical tool for authorities in guiding the installation and optimization of hybrid energy systems in industrial parks and other significant power consumers. Hence, the implementation of clean energy requirements by authorities can encourage industrial parks and other important power users to undertake energy transformations to achieve clean production and establish local emergency power supply systems. This finding provides an important reference for the action plan of "increasing the proportion of green power consumption and optimizing the allocation of power resources" proposed in the Peak Carbon Implementation Plan for China's Industrial Sector (Ministry of Industry and Information Technology of China, China Development and Reform Commission, Ministry of Ecology and Environment of China 2022).

In the face of the potential occurrence of extreme events, it is advisable for significant power users to establish hybrid energy systems that possess complementary capabilities in providing local emergency power supplies. In this study, the selected configuration consisting of photovoltaics, gas turbines, and energy storage batteries exhibit excellent complementary abilities and can provide consistent power supplies. Therefore, power users should develop customized plans based on their respective geographical locations, natural resources, and other relevant factors, and subsequently design hybrid energy systems that possess complementary capabilities to ensure the stability of power supply.

Industrial parks can flexibly adjust their distributed energy installation plans under the guidance of the government. Table 4.3 shows that appropriate distributed energy installations can assist industrial parks to improve their energy structures and power system resilience. Therefore, industrial parks need to adjust their distributed energy installation and operational plans based on the current economic situation and their external conditions. When extreme events are frequent, industrial parks need to increase their distributed energy installations and utilization to strengthen their power security capabilities. If the extreme event frequencies decrease, and the economic conditions are poor, the industrial parks could choose the most economically distributed energy utilization mode.

References

Aien, M., Hajebrahimi, A., and Fotuhi-Firuzabad, M. (2016). A comprehensive review on uncertainty modeling techniques in power system studies. *Renewable and Sustainable Energy Reviews* 57: 1077–1089.

Al Garni, H.Z., Awasthi, A., and Ramli, M.A.M. (2018). Optimal design and analysis of grid-connected photovoltaic under different tracking systems using HOMER. *Energy Conversion and Management* 155: 42–57.

Amirioun, M.H., Aminifar, F., and Shahidehpour, M. (2018). Resilience-promoting proactive scheduling against hurricanes in multiple energy carrier microgrids. *IEEE Transactions on Power Systems* 34(3): 2160–2168.

Anderson, K., Farthing, A., Elgqvist, E., and Warren, A. (2022). Looking beyond bill savings to equity in renewable energy microgrid deployment. *Renewable Energy Focus* 41: 15–32.

Arghandeh, R., Von Meier, A., Mehrmanesh, L., and Mili, L. (2016). On the definition of cyber-physical resilience in power systems. *Renewable and Sustainable Energy Reviews* 58: 1060–1069.

Bao, M., Ding, Y., Singh, C., and Shao, C. (2019). A multi-state model for reliability assessment of integrated gas and power systems utilizing universal generating function techniques. *IEEE Transactions on Smart Grid* 10(6): 6271–6283.

Barra, P.H.A., Coury, D.V., and Fernandes, R.A.S. (2020). A survey on adaptive protection of microgrids and distribution systems with distributed generators. *Renewable and Sustainable Energy Reviews* 118: 109524.

Bennett, J.A., Trevisan, C.N., DeCarolis, J.F. et al. (2021). Extending energy system modelling to include extreme weather risks and application to hurricane events in Puerto Rico. *Nature Energy* 6(3): 240–249.

Çalı, S. and Balaman, Ş.Y. (2019). A novel outranking based multi criteria group decision making methodology integrating ELECTRE and VIKOR under intuitionistic fuzzy environment. *Expert Systems with Applications* 119: 36–50.

Capstone Turbine Corporation (2020). The energy landscape has changed. https://www.capstoneturbine.com/ (accessed 18 October 2023).

Central People's Government of the People's Republic of China (2006). The great significance of the West-East Gas Transmission project. https://www.gov.cn/ztzl/2006-01/02/content_145310.htm (accessed 14 December 2023).

Chankong, V. and Haimes, Y.Y. (2008). *Multiobjective Decision Making: Theory and Methodology*. Courier Dover Publications.

Colson, B., Marcotte, P., and Savard, G. (2007). An overview of bilevel optimization. *Annals of Operations Research* 153(1): 235–256.

De Martini, P. (2014). More than smart: a framework to make the distribution grid more open, efficient and resilient. https://gridworks.org/wp-content/uploads/2015/06/More-Than-Smart-Report-by-GTLG-and-Caltech-08.11.14.pdf (accessed 18 October 2023).

Department of Industry and Information Technology of Jiangsu Province (2021). Notice on the issuance of Jiangsu Province's "14th Five-Year Plan" for industrial green development. https://gxt.jiangsu.gov.cn/art/2021/8/25/art_83673_10000511.html (accessed 18 October 2023).

Dragičević, T., Lu, X., Vasquez, J.C., and Guerrero, J.M. (2015). DC microgrids–Part I: A review of control strategies and stabilization techniques. *IEEE Transactions on Power Electronics* 31(7): 4876–4891.

Emam, O.E. (2013). Interactive approach to bi-level integer multi-objective fractional programming problem. *Applied Mathematics and Computation* 223: 17–24.

Griffin, P.A. (2020). Energy finance must account for extreme weather risk. *Nature Energy* 5(2): 98–100.

Haimes, Y. (1971). On a bicriterion formulation of the problems of integrated system identification and system optimization. *IEEE Transactions on Systems, Man, and Cybernetics* 1(3): 296–297.

He, G.X., Yan, H., Chen, L., and Tao, W.Q. (2020). Economic dispatch analysis of regional electricity–gas system integrated with distributed gas injection. *Energy* 201: 117512.

Hossain, M.A., Pota, H.R., Hossain, M.J., and Blaabjerg, F. (2019). Evolution of microgrids with converter-interfaced generations: challenges and opportunities. *International Journal of Electrical Power & Energy Systems* 109: 160–186.

Hussain, A., Bui, V.H., and Kim, H.M. (2019). Microgrids as a resilience resource and strategies used by microgrids for enhancing resilience. *Applied Energy* 240: 56–72.

International Renewable Energy Agency (2019). NDCs in 2020: advancing renewables in the power sector and beyond.

Jiangsu Development & Reform Commission (2021). Circular of Jiangsu Province on the Fourteenth Five Year Plan for National Economic and Social Development and the Outline of the Long term Goals for the Year 2035. http://www.jiangsu.gov.cn/art/2021/3/2/art_46143_9684719.html (accessed 18 October 2023).

Khaloie, H., Abdollahi, A., Rashidinejad, M., and Siano, P. (2019). Risk-based probabilistic-possibilistic self-scheduling considering high-impact low-probability events uncertainty. *International Journal of Electrical Power & Energy Systems* 110: 598–612.

Kök, A.G., Shang, K., and Yücel, Ş. (2018). Impact of electricity pricing policies on renewable energy investments and carbon emissions. *Management Science* 64(1): 131–148.

Kuby, M., He, C., Trapido-Lurie, B., and Moore, N. (2011). The changing structure of energy supply, demand, and CO_2 emissions in China. *Annals of the Association of American Geographers* 101(4): 795–805.

Li, C. and Wang, R.Z. (2012). Building integrated energy storage opportunities in China. *Renewable and Sustainable Energy Reviews* 16(8): 6191–6211.

Li, Z., He, K., Lu, Q., and Fu, Y. (2019). Power configuration and optimal operation of grid-connected microgrid in electricity selling market. *Electric Power Automation Equipment* 39(11): 41–49.

Li, X., Du, X., Jiang, T. et al. (2022). Coordinating multi-energy to improve urban integrated energy system resilience against extreme weather events. *Applied Energy* 309: 118455.

Liang, L., Hou, Y., Hill, D.J., and Hui, S.Y.R. (2016). Enhancing resilience of microgrids with electric springs. *IEEE Transactions on Smart Grid* 9(3): 2235–2247.

Lv, C., Xu, J., Xie, H. et al. (2016). Equilibrium strategy based coal blending method for combined carbon and PM10 emissions reductions. *Applied Energy* 183: 1035–1052.

Ma, S., Su, L., Wang, Z. et al. (2018). Resilience enhancement of distribution grids against extreme weather events. *IEEE Transactions on Power Systems* 33(5): 4842–4853.

Martinez-Mares, A. and Fuerte-Esquivel, C.R. (2012). A unified gas and power flow analysis in natural gas and electricity coupled networks. *IEEE Transactions on Power Systems* 27(4): 2156–2166.

Maryland Energy Administration (2014). Resiliency through microgrids task force established. https://energy.maryland.gov/Documents/MarylandResiliencyThroughMicrogridsTaskForceReport_000.pdf (accessed 18 October 2023).

Masrur, H., Sharifi, A., Islam, M.R. et al. (2021). Optimal and economic operation of microgrids to leverage resilience benefits during grid outages. *International Journal of Electrical Power & Energy Systems* 132: 107137.

Ministry of Industry and Information Technology of China, China Development and Reform Commission, Ministry of Ecology and Environment of China (2022). Peak carbon implementation plan for China's industrial sector. https://www.gov.cn/gongbao/content/2022/content_5717004.htm (accessed 20 October 2023).

Mishra, S., Anderson, K., Miller, B. et al. (2020). Microgrid resilience: a holistic approach for assessing threats, identifying vulnerabilities, and designing corresponding mitigation strategies. *Applied Energy* 264: 114726.

National Development and Reform Commission of China (2018). Catalog of Chinese industrial parks. https://www.ndrc.gov.cn/fggz/lywzjw/zcfg/201803/t20180302_1047056.html (accessed 14 December 2023).

National Development and Reform Commission of China (2022). The 14th Five Year Plan for Modern Energy System. https://www.ndrc.gov.cn/xwdt/tzgg/202203/t20220322_1320017.html (accessed 18 October 2023).

Panteli, M. and Mancarella, P. (2015). Modeling and evaluating the resilience of critical electrical power infrastructure to extreme weather events. *IEEE Systems Journal* 11(3): 1733–1742.

Pingkuo, L. and Zhongfu, T. (2016). How to develop distributed generation in China: in the context of the reformation of electric power system. *Renewable and Sustainable Energy Reviews* 66: 10–26.

Poudel, S. and Dubey, A. (2018). Critical load restoration using distributed energy resources for resilient power distribution system. *IEEE Transactions on Power Systems* 34(1): 52–63.

Rinaldi, S.M., Peerenboom, J.P., and Kelly, T.K. (2001). Identifying, understanding, and analyzing critical infrastructure interdependencies. *IEEE Control Systems Magazine* 21(6): 11–25.

Schneider, K.P., Tuffner, F.K., Elizondo, M.A. et al. (2016). Evaluating the feasibility to use microgrids as a resiliency resource. *IEEE Transactions on Smart Grid* 8(2): 687–696.

Schwaegerl, C. and Tao, L. (2013). Quantification of technical, economic, environmental and social benefits of microgrid operation. *Microgrids: Architectures and Control* 275–313. https://onlinelibrary.wiley.com/doi/epdf/10.1002/9781118720677.ch07.

Shen, J., Cheng, C., Wang, S. et al. (2020). Multiobjective optimal operations for an interprovincial hydropower system considering peak-shaving demands. *Renewable and Sustainable Energy Reviews* 120: 109617.

Su, W. and Huang, A.Q. (2014). A game theoretic framework for a next-generation retail electricity market with high penetration of distributed residential electricity suppliers. *Applied Energy* 119: 341–350.

Tang, Y., Zhang, Q., Mclellan, B., and Li, H. (2018). Study on the impacts of sharing business models on economic performance of distributed PV-battery systems. *Energy* 161: 544–558.

Tobajas, J., Garcia-Torres, F., Roncero-Sánchez, P. et al. (2022). Resilience-oriented schedule of microgrids with hybrid energy storage system using model predictive control. *Applied Energy* 306: 118092.

Vedullapalli, D.T., Hadidi, R., and Bozeman, L.S. (2020). Distribution system restoration using graph theory after multiple faults. *IET Energy Systems Integration* 2(3): 235–242.

Wang, Y., Chen, C., Wang, J., and Baldick, R. (2015). Research on resilience of power systems under natural disasters: a review. *IEEE Transactions on Power Systems* 31(2): 1604–1613.

Wang, C., Lv, C., Li, P. et al. (2018). Modeling and optimal operation of community integrated energy systems: a case study from China. *Applied Energy* 230: 1242–1254.

Wu, D., Zheng, X., Xu, Y. et al. (2021). An open-source extendable model and corrective measure assessment of the 2021 texas power outage. *Advances in Applied Energy* 4: 100056.

Xu, L.A.Z. (2019). Temporal and spatial distribution and evolution characteristics of major meteorological disasters in Jiangsu Province. *Chinese Agricultural Science Bulletin* 35(19): 112.

Xu, J. and Zhou, X. (2011). *Fuzzy-Like Multiple Objective Decision Making*, vol. 263. Springer.

Xu, J., Huang, Q., Lv, C. et al. (2018). Carbon emissions reductions oriented dynamic equilibrium strategy using biomass-coal co-firing. *Energy Policy* 123: 184–197.

Yuan, W., Wang, J., Qiu, F. et al. (2016). Robust optimization-based resilient distribution network planning against natural disasters. *IEEE Transactions on Smart Grid* 7(6): 2817–2826.

Zadeh, L.A. (1978). Fuzzy sets as a basis for a theory of possibility. *Fuzzy Sets and Systems* 1(1): 3–28.

Zhang, X. and Vesselinov, V.V. (2016). Energy-water nexus: balancing the tradeoffs between two-level decision makers. *Applied Energy* 183: 77–87.

Zhao, X., Zeng, Y., and Zhao, D. (2015). Distributed solar photovoltaics in China: policies and economic performance. *Energy* 88: 572–583.

Zhou, J., Wu, Y., Dong, H. et al. (2020). Proposal and comprehensive analysis of gas-wind-photovoltaic-hydrogen integrated energy system considering multi-participant interest preference. *Journal of Cleaner Production* 265: 121679.

Zou, Q. and Chen, S. (2019). Enhancing resilience of interdependent traffic-electric power system. *Reliability Engineering & System Safety* 191: 106557.

5

Bi-level Emission Quota Allocation Toward Coal and Biomass Co-combustion

Blending various grades of coal with biomass presents a credible approach for reducing carbon dioxide emissions and producing alternative jet fuel.

Coal, Biomass Mix May Be in Military Jet Fuel Future
The U.S. Defense Logistics Agency etc., 2015

Coal–biomass co-combustion has been demonstrated to be an acceptable option for energy production and has been found to significantly reduce the environmental impact of coal-fired power plants (CPPs) by reducing not only net greenhouse gas emissions but also SO_2 and NO_x emissions due to biomass' lower carbon, sulfur, and fuel-bound nitrogen contents. This chapter suggests a dynamic equilibrium strategy for the hybrid coal–biomass co-combustion system to reduce carbon emissions, where a bi-level programming approach is used to create a cooperative relationship between the authority and CPPs, a dynamic programming approach is used to handle conflicts in biomass availability times, and a multi-objective programming approach is used to seek the trade-off between economic development and environmental protection. The introduction of uncertainty theory also addresses vague uncertain parameters. The suggested model is then used to analyze a case study in Jiangsu, China, to show how effective and useful it is. As part of the solution, an interactive algorithm is created to reflect the goals and constraints of various stakeholders. Based on analyses and discussions under various scenarios, the suggested approach can accomplish sustainable growth and establish economic–environmental coordination. Additionally, a framework for a competitive allocation of carbon emissions is advised. With only minor tweaks to specific factors, this methodology could potentially be applied to different nations and business sectors.

5.1 Background Review

Reliance on coal has resulted in excessive greenhouse gas emissions (Li et al. 2009). According to BP Energy Outlook, global carbon emissions have been growing

Sustainable Hybrid Energy Systems: Carbon Neutral Approaches, Modeling, and Case Studies, First Edition. Jiuping Xu and Fengjuan Wang.

relentlessly for nearly 50 years, although the growth has gradually slowed in recent years (BP Global 2018). Two-thirds of the world's electricity is generated by fossil fuel power plants, of which CPP is one of largest sources of carbon emissions (BP Global 2015; Service 2017). So carbon emissions from CPPs need to be urgently reduced. Studies on reducing carbon emissions from CPPs have focused on either hard-technology or soft-technology (Lv et al. 2016). Hard-technology concentrates on technological innovation or investment to reduce carbon emissions, For example, coal washing, retrofitting the stream turbine flow passage, and carbon capture and storage/sequestration (Mao et al. 2014). But it is expensive and beyond the ability of most developing countries (Xu et al. 2017). So soft-technology (co-firing, carbon tax, and Cap-and-Trade (C&T)) (Dai et al. 2014; Yearwood-Lee 2015; Van Den Broek et al. 1996) has been the more favored option (Goto et al. 2013; Xu et al. 2015; Li 2012). Lin and Li (2011) claimed that although carbon tax was an immediate carbon price signal, it also had some disadvantages as the mitigation impacts were uncertain and rational carbon tax rates were hard to determine. Keohane (2009) and Grubb (2012) believed that C&T had some advantages over carbon tax, but its procedures are complex and cost is relatively high. So coal–biomass co-combustion method provides the simplest, most practical, and most cost-effective method for carbon emissions reductions (Li et al. 2012; Liu et al. 2016; Tilman et al. 2009).

Co-firing biomass and coal has been thoroughly studied. Because of biomass' lower levels of carbon, sulfur, and fuel-bound nitrogen, this method has been shown to be a reasonable option for power generation and has been found to significantly reduce the environmental footprint of CPPs (Agbor et al. 2014; Sami et al. 2001; Weldu 2017; Narayanan and Natarajan 2007; Beagle and Belmont 2016). From the viewpoint of CPPs, Eksioglu and Karimi (2014) created a non-linear optimization model that took into account the extra expenses, potential savings, and loss of process efficiency. Additionally, Yilmaz and Selim (2015) created a fuzzy multi-objective strategy-based decision-making model to create the biomass supply chain with the highest profit margin. A number of co-firing technologies have matured; for instance, Al-Mansour and Zuwala (2010) outlined three mature technological approaches: "direct co-firing," in which biomass and coal were burned in the same or separate mills and burners; "indirect co-firing," in which biomass was converted into the coal furnace using a biomass gasifier; and "parallel co-firing," in which separate biomass boiler and coal boiler were used. Basu et al. (2011) examined the capital and operating expenses of co-firing technologies in five CPPs and discovered that the direct co-firing strategy at lower ratios was the simplest and least expensive option. The major issues with direct co-firing of biomass and coal, however, were the co-firing ratio and biomass storage (Agbor et al. 2014; Rentizelas et al. 2009; Allen et al. 1998). According to Sahu et al. (2014) and Moon et al. (2013), the maximum allowable mixing ratio for direct co-firing without any unresolvable difficulties was a 10% biomass co-firing ratio. In order to get the most affordable storage, Rentizelas et al. (2009) suggested merging various biomass supply chains to deal with the seasonal availability of biomass storage. Although these studies have helped reduce carbon emissions in some ways, they frequently concentrate only on one decision maker, one aspect of the economy, or one aspect of biomass storage from a short-term viewpoint. The reality is more intricate, and additional advancements are still required.

In this chapter, a bi-level, multi-objective, multistage coal–biomass co-combustion approach is suggested as a means of lowering carbon emissions. The authorities and CPP are two of the decision-makers actually involved in reducing carbon emissions. The CPPs then make decisions in order to satisfy the authority's decisions and its goal of maximizing profits after the authority first makes an initial decision based on optimizing itself and anticipating CPPs' actions. The authority is informed of the CPPs' decisions because they could have an impact on their goals. The authority then modifies its first decision, and the CPPs then decide once more and offer input. Until a final plan that is acceptable to the authorities and all CPPs is decided, this procedure is repeated (Lv et al. 2016; Zhang et al. 2015). In the process, as the authority is the leader and CPPs are the followers, the relationship is similar to Stackelberg game. Therefore, bi-level programming method is adopted to seek a Stackelberg–Nash equilibrium (Colson et al. 2007; Sinha et al. 2014). The authority has a broad goal of sustainable development and places a high value on both environmental preservation and economic growth. Therefore, the trade-off between the economic and the environment is sought using the multi-objective programming method. As a result of their significant contribution to global climate change, carbon emissions play the function of environmental protection in this chapter, while revenue maximization plays the part of economic development because it can ensure the stability and growth of the economy and society. Storage of biomass that is secure and long-term is essential for CPPs. In order to ensure that CPPs can optimize profits over the long run, a dynamic programming method is used to identify the best choices at each stage as seasonal access to biomass storage occurs. In order to determine the accurate value of unknown parameters like emissions factors, which are unpredictable due to the unstable combustion process, uncertainty theory is also used (Xu and Zhou 2011; Liu et al. 2015).

Based on the foregoing discussion, this chapter suggests a thorough methodology that combines a bi-level programming method, multi-objective programming method, and dynamic programming method to reduce carbon emissions. This methodology aims to achieve a cooperative relationship between the authority and the CPPs, balance the trade-off between environmental protection and economic development, and resolve conflicts over biomass storage time in an uncertain environment. The following part uses a mathematical modeling approach to explain this bi-level, multi-objective, multistage coal–biomass co-combustion system that is based on a dynamic equilibrium strategy.

5.2 Main Issues' Description

In this section, three main issues associated with the coal and biomass co-combustion hybrid energy system optimization process will be discussed.

5.2.1 System Schematic

The process of directly burning biomass in a coal furnace is known as direct co-firing. This can be accomplished in one of two ways: either by milling the raw solid biomass and then immediately injecting it into the pulverized coal fire system, or by pre-mixing it with the coal in the coal-handling system, as shown in Figure 5.1. Raw solid biomass is typically in granular, pelletized, or dust form. Direct co-firing is considered to be the simplest, cheapest, and most widespread method of co-firing biomass with coal in a boiler. By establishing coal–biomass co-combusted hybrid energy system, the coal used in power plants can be partly replaced by biomass, and the associated greenhouse gas emissions can be reduced; At the same time, it can help diversify energy sources and reduce dependence on fossil fuels.

5.2.2 Uncertain Decision-Making Environment

The decision of the hybrid system generation in a uncertain environment is also a problem for system operators. Influenced by some objective factors, some parameters, for example, carbon emissions caused by burning a unit of fuel, and conversion parameters from a unit of fuel to power, cannot be valued exactly. However, through field research, it can be found that this value can be estimated to be within a certain range, with the most likely value being in a relatively smaller range. For instance, $\widetilde{P}_{ij}$ is an uncertain parameter, the certain range of which is from the minimum value r_{11} to the maximum value r_{14}, and the most likely value of which is between r_{12} and r_{13}, so it can be written as a trapezoidal fuzzy number $\widetilde{P}_{ij} = (r_{11}, r_{12}, r_{13}, r_{14})$, where $r_{11} \leq r_{12} \leq r_{13} \leq r_{14}$. However, numbers in this form cannot be calculated directly, so the expected value operator method of trapezoidal fuzzy numbers proposed by Xu and Zhou (2011) is adopted so that the uncertain decision, making environment can be handled. The flowchart of this process is shown in Figure 5.2.

5.2.3 Bi-level Decision-Making Structure

The government decides on the carbon emissions quota to be distributed to each CPP after taking into account its revenue goals and carbon emissions reduction

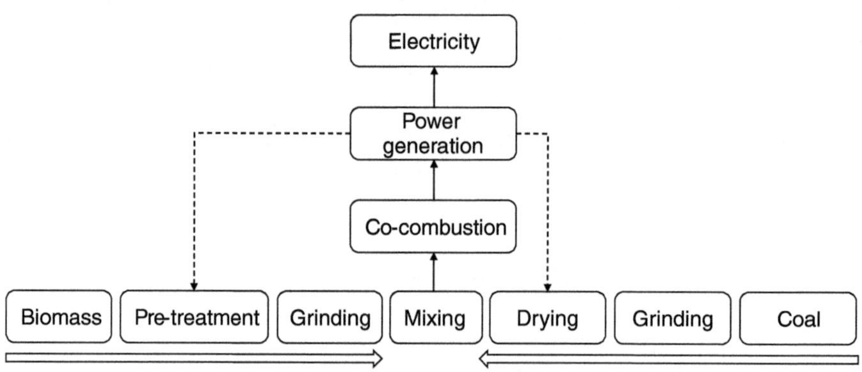

Figure 5.1 Schematic of coal and biomass co-combusted hybrid energy system.

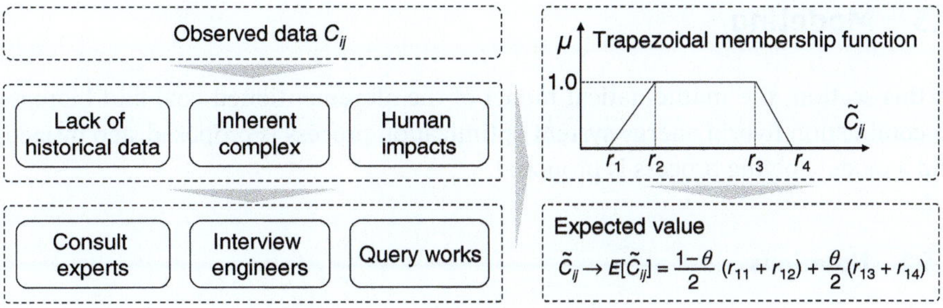

Figure 5.2 Process of handling uncertain decision-making environment.

target. The CPPs then decide on their own fuel-using and fuel-purchasing strategies based on outside environmental pressures and internal corporate innovations. The process for adjusting emission quotas serves as a link between environmental preservation and economic growth. A larger quota, in the eyes of the CPPs, enables greater generation and greater potential profits. The CPPs will attempt to modify their coal–biomass co-combustion power generation plans in an effort to improve their performances and strive for the maximum amount of allowance, though, as the government places a strong emphasis on emissions control, CPPs with better low-carbon performances will obtain more allowances. Due to the fact that the CPPs' decisions have an impact on the government's objectives through the natural feedback loop, the government must modify its allocation choices, which in turn requires the CPPs to modify their choices once more. Until a satisfactory solution is found, this leader–follower decision-making interaction is repeated. The bi-level decision-making structure is shown in Figure 5.3.

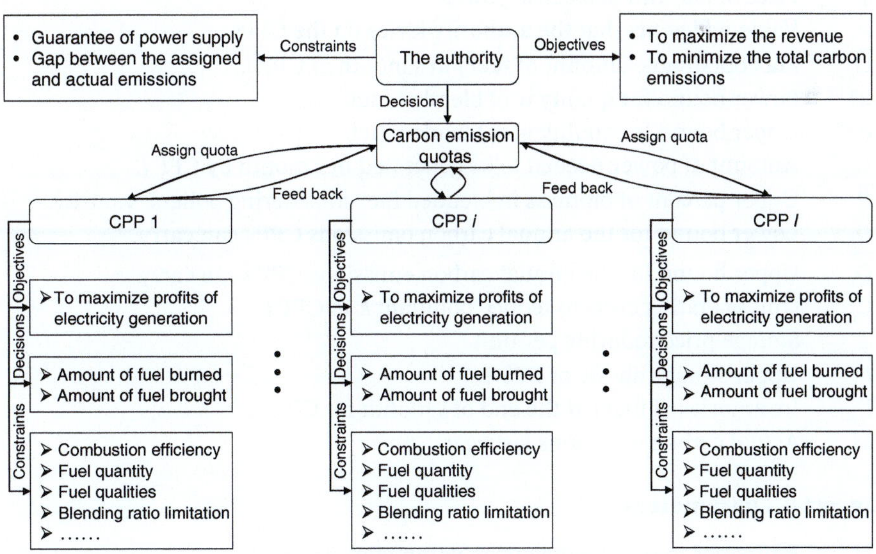

Figure 5.3 Bi-level decision-making structure of coal and biomass co-combusted hybrid energy system.

5.3 Modeling

In this section, the mathematical model of the aforementioned coal and biomass co-combustion hybrid energy system optimization process is proposed step by step, and a model-solving process is proposed.

5.3.1 Notations

Indices

i	Index of coal-fired power plants, $i \in \vartheta = 1, 2, \ldots, I$.
j	Index of fuels, $j \in \gamma = 1, 2, \ldots, j_b, \ldots, J$, where 1–j_b represents biomass fuel, j_b–n represents coal fuel.
k	Pollutant index, $k \in \psi = 1, 2, \ldots, K$.
w	Fuel quality index, $w \in \varsigma = 1, 2, \ldots, W$.
s	Month index, $s \in \sigma = 1, 2, \ldots, S$.

Decision variables

x_{ijs}	Amount of fuel j used in s month for CPP i.
z_{ijs}	Amount of fuel j bought in s month for CPP i.
y_i	Annual carbon emissions allocation quota for CPP i allocated by the authority.

Crisp parameter

u	Price of per unit of carbon emissions quota.
η	Price of per unit of electric power.
π	Value-added tax that the authority levies on the CPPs.
$\overline{O}_{ijs}$	Max obtainable amount of fuel j in s month at CPP i.
$u_w^{\min}$	Lower bound for quality w of blended fuel.
$u_w^{\max}$	Upper bound for quality w of blended fuel.
D_{is}	Amount of power needed to be generated in s month by CPP i.
$\overline{B}$	Upper percent of biomass in blended fuel considering boilers' security.
$\underline{Q_i}$	Lower bound for the annual carbon emissions CPP i can carry.
$\overline{\overline{Q_i}}$	Upper bound for the annual carbon emissions CPP i can carry.
C_{ik}^T	Unit operating costs to reduce pollutant k in CPP i.
Z	Storage price monthly per unit.
$\overline{I_i}$	Upper storage ability of CPP i.
I_{ijs}	Inventories of fuel j at the end of s month for CPP i.
M_c	Actual carbon emissions last year.

Uncertain parameters

$\widetilde{H}_i$	Emissions to power conversion parameters in CPP i.
$\widetilde{P}_{ijs}$	Unit purchasing price of fuel j in s month for CPP i.
$\widetilde{P}_{ij}$	Conversion parameters from a unit of fuel j to power in CPP i.

$\widetilde{S}_{jk}$	Amount of pollutant emissions k caused by burning a unit of fuel j.
$\widetilde{W}_{ij}$	Carbon emissions caused by burning a unit of fuel j in CPP i.
$\widetilde{B}_{jw}$	Quality w of fuel j.
$\widetilde{\eta}_j$	Burn-out fraction of fuel j char.

Policy control parameters

α	Attitude of the authority toward carbon emissions reduction.
λ	Attitude of the authority toward the gap between the assigned quota and actual value.
β	Combustion efficiency.

5.3.2 Perspective from the Local Authority

5.3.2.1 To Maximize the Revenue

The authority imposes value-added tax and fees on the carbon emission quotas in order to ensure the stability and growth of the economy and society. Let π be value-added tax rate, $\widetilde{P}_{ij}$ be the conversion parameters from a unit of fuel j to power in CPP i, then the annual profits of CPP i can be represented as $\sum_j \sum_s \eta E[\widetilde{P}_{ij}]x_{ijs}$; therefore, the total value-added tax received from the CPPs is $\sum_i \sum_j \sum_s \pi\eta E[\widetilde{P}_{ij}]x_{ijs}$. Let u be the price of per unit carbon emissions allocation quota and $\sum_i y_i$ be the total amount of carbon emissions allowances allocated to all the CPPs by the authority, then fees on the carbon emissions quotas are represented by $u\sum_i y_i$. Therefore, the total revenue is formulated as Eq. (5.1).

$$\max F_1 = \sum_i \sum_j \sum_s \pi\eta E[\widetilde{P}_{ij}]x_{ijs} + u\sum_i y_i \tag{5.1}$$

5.3.2.2 To Minimize the Total Carbon Emissions

The authority is responsible to the public for environmental protection planning. Let y_i be carbon emissions quota allocated by the authority to CPP i. Then the total carbon emissions of the CPPs generation industry is $\sum_i y_i$, which should be minimized, as shown in Eq. (5.2).

$$\min F_2 = \sum_i y_i \tag{5.2}$$

5.3.2.3 Limitations on Each CPP's Carbon Emissions

Each CPP is required to pay taxes; thus, the authority must guarantee its fundamental rights, such as sustaining basic operations. Let $\underline{Q_i}$ represent the lower bound of carbon emissions quantity produced by CPP i when expenses balance receipts. y_i is carbon emissions allocation quota for CPP i, then y_i should be no less than $\underline{Q_i}$, namely, $\underline{Q_i} \leq y_i$. The allocation of carbon emissions quotas that each CPP can carry is required as the authority assesses taxes on carbon emissions quotas. Let $\overline{Q_i}$ represent the upper bound of carbon emissions in CPP i which is obtained under full-load production, then the assigned quota should not exceed the bound, namely, $y_i \leq \overline{Q_i}$. These two constraints can be written as Eq. (5.3).

$$\underline{Q_i} \leq y_i \leq \overline{Q_i} \tag{5.3}$$

5.3.2.4 Guarantee of Power Supply

The authority has to guarantee the region's power requirements have been satisfied. Let $\widetilde{H}_i$ be the conversion parameter for a unit of carbon emissions to power in CPP i, and $E[\widetilde{H}_i]y_i$ denote electric power generated by CPP i. The sum of all the CPPs-generated power can be represented as $\sum_i E[\widetilde{H}_i]y_i$, the total electricity generated by all CPPs should not exceed the amount of power needed in the region $\left(\sum_i \sum_s D_{is}\right)$, as shown in Eq. (5.4).

$$\sum_i E[\widetilde{H}_i]y_i \geq \sum_i \sum_s D_{is} \tag{5.4}$$

5.3.2.5 Gap Between the Assigned Emission Quota and the Actual Emissions

The authority must consider each CPP's requirements in order to allocate carbon emissions quotas in a rational manner as the carbon emissions quota may, on occasion, be far from reality. Let $\widetilde{W}_{ij}$ represent the carbon emissions caused by burning a unit of fuel j in CPP i, and x_{ijs} is the amount of fuel j used in s month for CPP i. Then the actual carbon emissions can be represented as $\sum_j \sum_s E[\widetilde{W}_{ij}]x_{ijs}$, and this constraint can be written as Eq. (5.5).

$$1 - \frac{\sum_j \sum_s E[\widetilde{W}_{ij}]x_{ijs}}{y_i} \leq \lambda \tag{5.5}$$

5.3.3 Perspective from the CPPs

5.3.3.1 To Maximize Profits of Electricity Generation

The regional authority allots a specific carbon emissions quota to the CPPs. With respect to this capped carbon emission quota, the CPPs create monthly production schedules in order to maximize earnings. In this chapter, for CPP i, the income is from power sales and the cost can be divided into fuel purchase costs, fuel storage costs, pollutants treatment costs, taxes and fees on carbon emissions allowances; the objective of maximizing profits can be represented as Eq. (5.6)

$$\max f = f^1 - f^2 - f^3 - f^4 - f^5 \tag{5.6}$$

where f^1 is the annual income of selling electricity and can be further explained as Eq. (5.7), in which η is the unit price of electric power and $\sum_j \sum_s E[\widetilde{P}_{ij}]x_{ijs}$ denotes the electric power generated by CPP i

$$f^1 = \eta \sum_j \sum_s E[\widetilde{P}_{ij}]x_{ijs} \tag{5.7}$$

f^2 is fuel purchase costs and can be further explained as Eq. (5.8). As biomass prices change from month to month, CPP i decides on the quantity z_{ijs} to buy depending on the month, and $\widetilde{P}_{ijs}$ denotes the unit purchase price for fuel j in s month.

$$f^2 = \sum_j \sum_s E[\widetilde{P}_{ijs}] z_{ijs} \tag{5.8}$$

f^3 is fuel storage costs and can be further explained as Eq. (5.9), where Z is the per unit monthly storage price, and I_{ijs} is the closing stock of the previous month plus the present month's purchase.

$$f^3 = Z \sum_j \sum_s I_{ijs} \tag{5.9}$$

f^4 is pollutants treatment costs and can be further explained as Eq. (5.10), in which $\widetilde{S}_{jk}$ is the pollutant emissions k resulting from burning a unit of fuel j, C^T_{ik} is the unit operating costs at CPP i for treating pollutant k.

$$f^4 = \sum_j \sum_s \sum_k E[\widetilde{S}_{jk}] C^T_{ik} x_{ijs} \tag{5.10}$$

f^5 is taxes and fees on carbon emissions allowances and can be further explained as Eq. (5.11), where π is value-added tax, and tax cost is $\pi\eta \sum_s \sum_j E[\widetilde{P}_{ij}] x_{ijs}$. The cost of purchasing the carbon emissions quota is uy_i.

$$f^5 = \pi\eta \sum_s \sum_j E[\widetilde{P}_{ij}] x_{ijs} + uy_i \tag{5.11}$$

5.3.3.2 Combustion Efficiency

Because fuel blends with biomass and coal typically have low calorific values and high moisture content, it's possible that this lowers overall efficiency (Dumortier 2013). Let $\widetilde{B}_{jw}(w = 1)$ be the volatile matter content and $\widetilde{\eta}_j$ be the burn-out fraction of fuel j; therefore, $\sum_j E[\widetilde{B}_{jw}] x_{ijs} + \sum_j E[\widetilde{\eta}_j](1 - E[\widetilde{B}_{jw}]) x_{ijs}$ is what can be burnt in the fuel blends (Sami et al. 2001). The co-combustion efficiency of the fuel blends should be higher than the threshold values β, as shown in Eq. (5.12).

$$\frac{\sum_j E[\widetilde{B}_{jw}] x_{ijs} + \sum_j E[\widetilde{\eta}_j](1 - E[\widetilde{B}_{jw}]) x_{ijs}}{\sum_j x_{ijs}} \geq \beta \tag{5.12}$$

5.3.3.3 Fuel Quantity Requirements

There is a maximum supply quantity $\overline{O}_{ijs}$ that can be purchased each month, as shown in Eq. (5.13).

$$0 \leq z_{ijs} \leq \overline{O}_{ijs} \tag{5.13}$$

5.3.3.4 Fuel Qualities' Requirements

The characteristics of biomass feedstock can be very diverse, and biomass fuels differ greatly from coal in terms of their composition, physical and chemical characteristics, and energy content. The majority of biomass fuels have larger and a wider variety of particle sizes, more volatile compounds, more ash, less sulfur and nitrogen, less heating value, and less heating value than coal. The co-firing process may suffer if the fuel blends being burned do not adhere to the boiler's criteria, which could lead to more serious issues. Therefore, it is essential to concentrate on the characteristics of the fuel blend in order to ensure that the biomass and coal blends could be co-fired in the current boilers. Let tw be the fuel quality, $u_w^{\min}$ and $u_w^{\max}$ are the lower and upper bounds for quality w of blended fuel, then the fuel qualities requirements can be found in Eq. (5.14).

$$E\left[\tilde{u}_w^{\min}\right] \le \frac{\sum_j E[\widetilde{B}_{jw}]x_{ijs}}{\sum_j x_{ijs}} \le E\left[\tilde{u}_w^{\max}\right] \tag{5.14}$$

where $w = 1$ represents the volatile matter content, $w = 2$ represents the heat rate, $w = 3$ represents the ash content, $w = 4$ represents the moisture content, and $w = 5$ represents the sulfur content (Lv et al. 2016).

5.3.3.5 Blending Ratio Limitation of Biomass

In comparison to coal, biomass fuels have far less carbon and a higher proportion of hydrogen and oxygen, which results in a significantly lower energy density. Typical biomass has only approximately one-tenth the overall fuel density of coal. The increased ignition characteristics and the devolatilization in the volatile reaction area from the synergetic effects in coal–biomass co-combustion have been reported to improve the combustion efficiency of coal (Agbor et al. 2014). Let 1–j_b be the biomass fuel and j_b–J be the coal fuel with $\overline{B}$ being upper percent of biomass. Then this constraint can be written as Eq. (5.15).

$$\frac{\sum_{i=1}^{j_b} x_{ijs}}{\sum_j x_{ijs}} \le \overline{B} \tag{5.15}$$

5.3.3.6 Responsibility to Ensure Power Supply

Profits are important, but CPPs should also place a high value on social responsibility. In today's world, electricity is a requirement for living. Each CPP benefits from social resources as a member of society and is required to produce a specific amount of electricity due to its capacities. Let D_{is} be the power demands in s month by CPP i, then this constraint can be written as Eq. (5.1).

$$\sum_j E[\widetilde{P}_{ij}]x_{ijs} \ge D_{is} \tag{5.16}$$

5.3.3.7 Emissions Quota Constraints

Since the authority and CPPs have a leader–follower relationship, the authority creates an adequate framework for carbon emission quotas. The actual carbon emissions of each CPP cannot go beyond the quotas because they are located on the lowest level. Set $\widetilde{W}_{ij}$ as the carbon emissions caused by burning a unit of fuel j. Therefore the mathematical form is described as Eq. (5.17).

$$\sum_j \sum_s E[\widetilde{W}_{ij}]x_{ijs} \le y_i \tag{5.17}$$

5.3.3.8 Dynamic Fuel Storage

Biomass has been considered a desirable renewable fuel to replace coal since it is a renewable carbon dioxide–neutral energy source that can lower pollutants. However, time has a significant impact on the availability of feedstock for power generation. Therefore, a dynamic multi-stage model is needed to describe fuel resources storage, as shown in Eq. (5.18).

$$I_{ijs} = I_{ji(s-1)} + z_{ijs} - x_{ijs} \tag{5.18}$$

5.3.3.9 Logistic Constraint on Fuel Storage

The CPPs must check their monthly fuel purchase strategies and cumulative fuel resource storage quantities; the terminal's storage quantity cannot be negative, as shown in Eq. (5.19).

$$0 \leq x_{ijs} \leq z_{ijs} + I_{ji(s-1)} \tag{5.19}$$

5.3.3.10 Limitation of Warehousing Ability

The warehouse capacity also has a maximum quantity limit, set as $\overline{I_i}$. Beyond these limitations, CPPs are unable to save or purchase; hence, the amount stored at the terminal in the previous month and the amount purchased in the current month are limited as Eq. (5.20).

$$\sum_j I_{ji(s-1)} + \sum_j z_{ijs} \leq \overline{I_i} \tag{5.20}$$

5.3.4 Global Model

In order to handle the relationship between the authorities and CPPs, the trade-off between economic development and environmental preservation, and biomass availability timing conflicts, a dynamic equilibrium strategy-based bi-level multi-objective multistage decision-making model is created. To maximize revenue and reduce carbon emissions, the authority at the upper level of the decision-making system first decides how to allocate carbon emissions quotas, With full knowledge of the authority's decision, CPPs work to maximize their profits while developing coal–biomass co-combustion ratio schemes that take into account technical constraints, carbon emissions allowances, social responsibility requirements, and fuel resource quantity and quality restrictions. After the authority updates the carbon emission allowances, the CPPs provide feedback on their decisions, reevaluate the strategies for the coal–biomass co-combustion ratio, and provide feedback once more to the authority. From this interactive process, the global satisfactory solution is obtained. Therefore, from an integration of the constraints and objective functions as Eq. (5.1)–(5.20), the global multi-objective bi-level dynamic model Eq. (5.21) is formulated as follows.

$$\max F_1 = \sum_i \sum_j \sum_s \pi \eta E[\widetilde{P}_{ij}] x_{ijs} + u \sum_i y_i$$

$$\min F_2 = \sum_i y_i$$

$$
\text{s.t.}\begin{cases}
\underline{Q_i} \le y_i \le \overline{Q_i} \\
\sum_i E[\widetilde{H}_i] y_i \ge \sum_i \sum_s D_{is} \\
1 - \frac{\sum_j \sum_s E[\widetilde{W}_{ij}] x_{ijs}}{y_i} \le \lambda, \forall i \in \vartheta \\
\max f = f^1 - f^2 - f^3 - f^4 - f^5 \\
\text{s.t.}\begin{cases}
f^1 = \eta \sum_j \sum_s E[\widetilde{P}_{ij}] x_{ijs} \\
f^2 = \sum_j \sum_s E[\widetilde{P}_{ijs}] z_{ijs} \\
f^3 = Z \sum_j \sum_s I_{ijs} \\
f^4 = \sum_j \sum_s \sum_k E[\widetilde{S}_{jk}] C^T_{ik} x_{ijs} \\
f^5 = \pi \eta \sum_s \sum_j E[\widetilde{P}_{ij}] x_{ijs} + u y_i \\
\frac{\sum_j E[\widetilde{B}_{jw}] x_{ijs} + \sum_j E[\tilde{\eta}_j](1 - E[\widetilde{B}_{jw}]) x_{ijs}}{\sum_j x_{ijs}} \ge \beta, \quad \forall s \in \sigma \\
0 \le z_{ijs} \le \overline{O}_{ijs}, \quad \forall j \in \gamma, \quad s \in \sigma \\
E\left[\tilde{u}_w^{\min}\right] \le \frac{\sum_j E[\widetilde{B}_{jw}] x_{ijs}}{\sum_j x_{ijs}} \le E\left[\tilde{u}_w^{\max}\right], \quad \forall s \in \sigma, \quad w \in \varsigma \\
\frac{\sum_{i=1}^{j_b} x_{ijs}}{\sum_j x_{ijs}} \le \overline{B}, \quad \forall s \in \sigma \\
\sum_j E[\widetilde{P}_{ij}] x_{ijs} \ge D_{is}, \quad \forall s \in \sigma \\
\sum_j \sum_s E[\widetilde{W}_{ij}] x_{ijs} \le y_i \\
I_{ijs} = I_{ji(s-1)} + z_{ijs} - x_{ijs}, \quad \forall j \in \gamma \\
0 \le x_{ijs} \le z_{ijs} + I_{ji(s-1)} \\
\sum_j I_{ji(s-1)} + \sum_j z_{ijs} \le \overline{I}_i, \quad \forall s \in \sigma
\end{cases}
\end{cases}
\tag{5.21}
$$

The aforementioned bi-level dynamic multi-objective biomass and coal co-firing model was created to make sure that CPPs' carbon emissions are reduced. This model is better suited for actual production than earlier coal–biomass co-combustion models because it takes into account various decision-makers and multiple conflicts. As a result, it can help CPPs create reasonable production plans that adhere to the authority's policy. The authority has the capacity to modify the policy control parameters in order to get the best outcomes in various situations. Additionally, this strategy is workable when the authority has complete control over the CPPs. At that time, the authority's relationship with CPPs is comparable to that of a general corporation and its subsidiaries. The suggested approach can be utilized for the distribution of carbon emissions allowances by the general corporation to its subsidiaries under the constrained carbon emissions resources (Lu et al. 2006).

5.3.5 Model Solving

Generally, authority needs to guarantee further economic development, and gives top priority to revenue when making decisions. But it also needs to consider environmental protection to ensure that the economy develops within its environmental carrying capacity. Therefore, in this chapter, a parameter (β) representing the attitude of the authority toward carbon emissions reduction is introduced, and the carbon emissions minimization objective is transformed into a constraint controlled within a range of acceptance (Lv et al. 2016). Therefore, Eq. (5.21) can be transformed as Eq. (5.22).

$$\max F = \sum_i \sum_j \sum_s \pi\eta E[\widetilde{P}_{ij}]x_{ijs} + u\sum_i y_i$$

$$\text{s.t.}\begin{cases} \sum_{j=1}^{m} y_i \leq \alpha M_c \\ \text{Eqs. (5.2)–(5.5)} \\ \max f = f^1 - f^2 - f^3 - f^4 - f^5 \\ \text{s.t. \{Eqs. (5.7)–(5.20)} \end{cases} \tag{5.22}$$

The aforementioned method transforms the proposed multi-objective bi-level global model into its equivalent single-objective bi-level model, which can be solved in Matlab software using the following algorithm. The authority (the leader) and the CPPs (the followers) in this hierarchical optimization system are required to maximize their own goals while taking into account the goals of others. The following approach is used to find the optimal solution to this linear bi-level programming model. By achieving the balance between economic growth and environmental conservation, this algorithm hopes to end conflicts between the government and CPPs. First, the feasible range of the upper level model is found and randomly choose a y_i^1 that maximizes the revenue of the upper-level authorities. Then, this initial upper-level optimal solution is put into the lower-level model and an optimal solution of the lower level is calculated. Next, the results of the lower-level model are fed back into the upper-level model to determine the subsequent optimal solution. This process is repeated until an optimal solution to the bi-level programming is found (Lv et al. 2016). The procedure for this solution approach is shown in Figure 5.4.

5.4 Case Study

In this section, an actual case in Jiangsu Province, China, is chosen to explore the practical and effectiveness of the proposed optimization model.

5.4.1 Case Description

China is entering a post-coal growth period where it has the largest coal market in the world and coal consumption is expected to peak, accounting for nearly 50% of

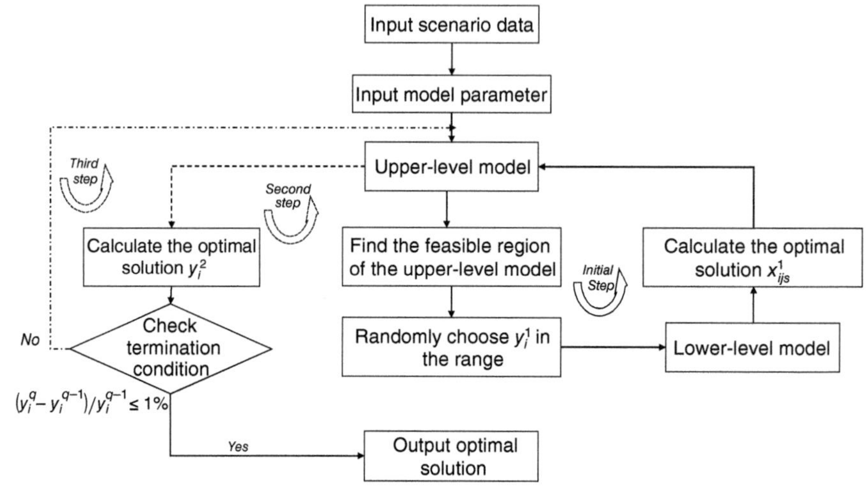

Figure 5.4 Algorithmic flowchart.

global coal consumption in 2035, along with many carbon emissions (Qi et al. 2016). Jiangsu province in southeastern China is not only a booming economy and the second largest of all Chinese provinces, it is also a large population province with an estimated population of more than 78.66 million in 2016. Due to its large population and economic advantages, Jiangsu Province is a major emitter of carbon in China, emitting 20.3 million tonnes per year. CPP is the most important end-user of coal and has become a significant source of regional carbon emissions (Zhao et al. 2008). Therefore, Jiangsu authorities are urgently seeking to reduce carbon emissions while maintaining economic growth.

Under economic and environmental pressures, there is a push for sustainable development and many policies are being implemented with the aim of developing cleaner and lower carbon fuel sources. The Chinese government's strategic action plan for energy development aims to reduce the share of coal in the energy mix to less than 62% by 2020 (BP Global 2017). Fortunately, Jiangsu Province is rich in agricultural bioenergy. As of 2009, the province was producing around 40 million tonnes of bioenergy annually, and it had previously thought about using coal–biomass co-combustion to address the anticipated increase in energy demand and lower carbon emissions (Zhang et al. 2013). Since 2003, a number of coal–biomass co-combustion CPPs have been constructed in Jiangsu Province's north and center due to the abundance of biomass resources there.

To reduce the computation burden, only three main coal–biomass co-combustion CPPs are examined in this case: Fengxian CPP, Ganyu CPP, and Baoying CPP (Figure 5.5), all of which use straw (S), wood waste (W), and coal as fuel. As coal from different places has different properties, Yaoqiao coal (Y) and Zhangshuanghe coal (Z) are used in these CPPs.

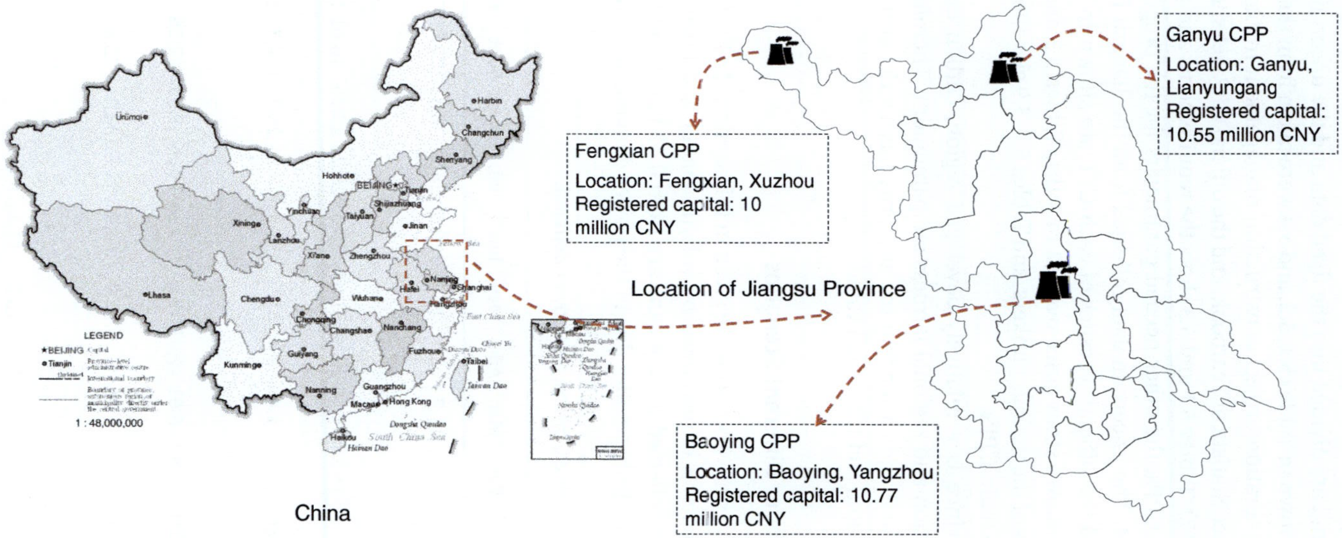

Figure 5.5 Case region and CPP locations.

5.4.2 Data Collection

The basic necessary data for Fengxian CPP, Ganyu CPP, and Baoying CPP are shown in Tables 5.1–5.7 and are divided into the fixed data and the uncertain data.

The fixed data shown in Tables 5.1, 5.3, and 5.4 were taken from the White chapter on straw power generation project construction management in Jiangsu Province, the Jiangsu Province Statistical Yearbook, and the official CPP websites (Jiangsu Statistical Bureau 2017). Tables 5.1 and 5.3 show the requirements and responsibilities at the three main CPPs, the other needed parameters for the proposed model are shown in Table 5.4. The uncertain data collected from the Jiangsu Province Statistical Yearbook and the official websites of each CPP are shown in Tables 5.2 and 5.5–5.9. Table 5.2 shows straw price over 12 months, Tables 5.5 and 5.6 show the fuel characteristics of the various bunkers, and Tables 5.7–5.9 give the boiler requirements at the three main CPPs.

As Jiangsu Province is a relatively developed region, carbon emissions quota allocation satisfaction and regional fairness are paramount. Therefore, the control parameter λ is set at 0, indicating that the authority has a strict attitude toward the gap between the quota and the actual value.

5.4.3 Results Under Different Scenarios

The results under various scenarios are given in this section. As Eq. (5.12) indicates, combustion efficiency β is restrained by volatile matter and burn-out fraction of char. Based on the collected data, β can vary from 0.42 to 0.49; however, the variation range for β is a little smaller due to the power demand and carbon emissions

Table 5.1 Straw resource quantities, prices, and power demand.

CPPs	Parameter	Jan.	Feb.	Mar.	Apr.	May.	Jun.	Jul.	Aug.	Sept.	Oct.	Nov.	Dec.
Fengxian CPP	Quantity ($\overline{O}_{ijs}$) 10^4 tonnes	4.2	3.7	11.7	18.5	29.6	116.5	38.2	36.8	177.5	77.2	22.1	10.3
	Power (D_{is})10^7 kWh	22.4	15.24	21.56	20.26	21.97	22.3	25.23	23.26	16.61	17.34	21.5	22.57
Ganyu CPP	Quantity ($\overline{O}_{ijs}$) 10^4 tonnes	0.6	1.1	1.4	2	81.9	91.3	19.9	28.7	142.3	58.2	9.5	1.3
	Power (D_{is})10^7 kWh	10.07	5.8	8.61	7.97	9.07	9.38	11.06	9.93	7.36	7.98	10.82	11.95
Baoying CPP	Quantity ($\overline{O}_{ijs}$) 10^4 tonnes	0.6	0.9	1.2	2	92.9	95.9	65.7	34.3	132.6	60.2	11.3	1.7
	Power (D_{is})10^7 kWh	14.47	8.46	13.1	12.52	13.3	13.44	16.78	15.18	9.67	11.28	13.64	14.78

Table 5.2 Fuzzy straw prices.

CPPs	Monthly straw prices			
Fengxian	Jan.	Feb.	Mar.	Apr.
CPP $\widetilde{P}_{ijs}$	(715,725,735,740)	(710,720,730,740)	(705,715,725,735)	(690,695,700,705)
(CNY/tonne)	May.	Jun.	Jul.	Aug.
	(663,670,676,685)	(630,635,640,650)	(640,645,656,660)	(630,635,645,650)
	Sept.	Oct.	Nov.	Dec.
	(665,675,685,790)	(615,620,630,635)	(645,655,665,670)	(705,710,720,730)
Ganyu CPP	Jan.	Feb.	Mar.	Apr.
$\widetilde{P}_{ijs}$ (CNY/	(725,730,740,750)	(690,695,705,710)	(670,675,680,685)	(670,680,690,705)
tonne)	May.	Jun.	Jul.	Aug.
	(625,630,640,645)	(590,598,606,615)	(633,638,645,650)	(620,626,632,638)
	Sept.	Oct.	Nov.	Dec.
	(577,585,594,600)	(615,620,625,631)	(652,657,665,670)	(690,695,705,710)
Baoying CPP	Jan.	Feb.	Mar.	Apr.
$\widetilde{P}_{ijs}$ (CNY/	(727,732,748,755)	(720,725,735,740)	(710,715,720,725)	(685,695,705,710)
tonne)	May.	Jun.	Jul.	Aug.
	(565,570,575,580)	(528,535,543,650)	(630,635,640,645)	(655,660,665,670)
	Sept.	Oct.	Nov.	Dec.
	(580,590,600,610)	(620,635,645,660)	(675,685,695,700)	(710,715,720,725)

Table 5.3 Crisp parameters of CPPs.

CPPs	Least carbon emissions $\underline{Q}_i$ (tonnes)	Utmost carbon emissions $\overline{Q}_i$ (tonnes)	Max storage ability $\overline{I}_i$ (tonnes)	Procurement price V_k (CNY/kg)	
				For SO_2	For NO_x
Fengxian CPP	5.76×10^6	1.05×10^7	7.85×10^5	1.8	14.4
Ganyu CPP	4.67×10^6	1.034×10^7	5.91×10^5	2	14.7
Baoying CPP	6.47×10^6	1.13×10^7	6.15×10^5	2.2	14.5

quota constraints. To examine and verify the variation range for β, α was set at 1 and then Matlab was used to calculate the results, from which it was found that when β changes from 0.43 to 0.44; the results of the proposed model were acceptable.

Therefore, three scenarios are presented to demonstrate the effectiveness of the model. Each scenario changes the value of α, which stands for the attitude toward reducing carbon emissions, from 1 to 0.75; the lower the value, the stricter the attitude. For comparison, the model is also computed without the carbon emission

Table 5.4 Other crisp parameters.

Parameters	Value	Parameters	Value
Carbon emissions quota price u (CNY/tonne)	30	Value-added tax S (%)	1
Amount of power needed in a region U (kWh)	5.1686×10^9	Price of unit electric η (CNY/kWh)	0.45
Storage price monthly per unit K(CNY/10^4 tonnes)	0.125	Carbon emissions last year M_c (tonnes)	2.37×10^7
Upper percentage of biomass in fuel blend $\overline{B}$	0.1		

Table 5.5 Uncertain fuel properties in fuzzy form ($\widetilde{B}_{jw}$).

Parameters	Units	Straw	Wood waste	Yaoqiao coal	Zhangshuanghe coal
Volatile matter	(wt%)	(64.2,64.7, 65.4,65.9)	(44.7,45.0, 45.3,45.6)	(25.5,25.8, 26.2,26.5)	(31.6,31.9, 32.2,32.5)
Heat rate	(Gj/tonne)	(19.3,19.6, 20.2,20.9)	(18.3,18.7, 19.1,19.81)	(27.8,28.2, 28.5,28.8)	(21.8,22.2, 22.75,23.1)
Ash content	(wt%)	(27.3,28.1, 28.4,28.9)	(19.8,20, 20.4,20.56)	(8.86,9.1, 9.5,9.8)	(19,19.3, 19.7,20)
Moisture content	(wt%)	(12.5,12.8, 13.2,13.5)	(8.6,8.93, 9.2,9.55)	(6.75,6.9, 7.2,7.4)	(4.6,4.8, 5.22,5.45)
Sulfur content	(w%)	(0.17,0.18, 0.21,0.24)	(0.09,0.1, 0.12,0.14)	(0.33,0.4, 0.47,0.52)	(0.59,0.62, 0.66,0.69)

Table 5.6 Pollutant emission factors and char burn-out fraction.

Parameters		Straw	Wood waste	Yaoqiao coal	Zhangshuanghe coal
Emission factor $\widetilde{E}_{ik}$ (kg/tonne)	For SO_2	(1.5,1.8, 2.13,2.4)	(3.95,4.1, 4.4,4.72)	(2.7,3,3.35, 3.7)	(8.25,8.6, 8.9,9.2)
	For NO_x	(4.6,4.8, 5.1,5.3)	(0.9,1.1, 1.3,1.45)	(9.3,9.6, 9.7,10.2)	(9.9,10.2, 10.5,10.7)
Char burn-out fraction: $\widetilde{\eta}_j$ (%)		(44,48, 52,56)	(36,38, 40,43)	(17,19, 21,23)	(17,20, 22,24)

Table 5.7 Fuel to carbon emissions parameters in fuzzy form $\widetilde{W}_{ij}$ (kg/tonne).

CPPs	Straw	Wood waste	Yaoqiao coal	Zhangshuanghe coal
Fengxian CPP	(1375,1500, 1600,1750)	(1700,1850, 2070,2290)	(2325,2450, 2500,2600)	(1970,2030, 2070,2100)
Ganyu CPP	(1200,1350, 1500,1650)	(1750,1900, 2000,2150)	(2125,2350, 2500,2630)	(2150,2200, 2375,2450)
Baoying CPP	(1350,1400, 1500,1600)	(1350,1500, 1700,1850)	(2250,2300, 2450,2600)	(2150,2250, 2300,2400)

Table 5.8 Fuel to power parameters in fuzzy form $\widetilde{P}_{ij}$ (kWh/tonne).

CPPs	Straw	Wood waste	Yaoqiao coal	Zhangshuanghe coal
Fengxian CPP	(1250,1400, 1550,1700)	(1550,1700, 1900,2050)	(2050,2200, 2350,2500)	(1950,2100, 2250,2350)
Ganyu CPP	(1500,1650, 1850,1950)	(1450,1650, 1870,2090)	(2100,2200, 2350,3450)	(1900,2050, 2130,2250)
Baoying CPP	(1300,1500, 1700,1900)	(1200,1400, 1550,1750)	(2200,2300, 2450,2550)	(2010,2150, 2300,2450)

restriction. Table 5.8 illustrates Scenario 1, in which β is set at 0.44 and α fluctuates between 1 and 0.75. Table 5.9 contains Scenario 2 with β fixed at 0.435 and α changing from 1 to 0.75. In Scenario 3, β is set at 0.43 and α shifts from 1 to 0.75 in Table 5.10.

Based on these results, the following conclusions are drawn. First, revenue, overall carbon emissions, and CPP profits all decline when α decreases. As a result of the authority's tighter stance toward carbon emissions, the total amount of emissions decreases when α declines. CPPs produce less power and make less money when using resources with looser limits on carbon emissions. As tax and carbon emission allowance fees from CPPs decrease, revenue likewise decreases. Second, without environmental protection, CPPs would be able to generate as much power as possible, emit CO_2 at full production capacity, and make greater profits. And the authorities have greater revenues and higher total carbon emissions. Finally, even though CPP's profits change as combustion efficiency changes, carbon emission allowances do not change. This is because combustion efficiency changes the co-firing ratio of biomass and coal, which in turn affects profits. However, because combustion efficiency does not change the total maximum acceptable carbon emissions, emission allowances remain the same.

Table 5.9 Fuel quality requirements and carbon to power parameter in fuzzy form.

Parameters		Volatile matter (wt%)	Heat rate (Gj/tonne)	Ash content (wt%)
Fengxian CPP	Lower bound	(5.5,5.9, 6.2,6.6)	(20.9,21.3, 21.6,21.9)	—
	Upper bound	(35.9,36.2, 36.5,36.9)	—	(25.6,26, 26.3,26.5)
Ganyu CPP	Lower bound	(7.1,7.3, 7.6,7.8)	(21.6,21.9, 22.1,22.4)	—
	Upper bound	(37.6,37.9, 38.1,38.4)	—	(27.7,27.9, 28.2,28.4)
Baoying CPP	Lower bound	(8.7,8.9, 9.2,9.4)	(21.4,21.7, 21.9,22.1)	—
	Upper bound	(38.6,38.8, 39,39.3)		(28.6,29, 29.2,29.5)
Parameters		**Moisture content (wt%)**	**Sulfur content (wt%)**	**Carbon to power $\widetilde{H}_i$ (kWh/tonne)**
Fengxian CPP	Lower bound	—	—	(935,945, 955,965)
	Upper bound	(8.6,8.9, 9.1,9.4)	(0.91,0.96, 0.99, 1.15)	
Ganyu CPP	Lower bound	—	—	(1003,1007, 1013,1016)
	Upper bound	(9.6,9.9, 10.1,10.4)	(1.1,1.3, 1.8,2.2)	
Baoying CPP	Lower bound	—	—	(960,966,978,993)
	Upper bound	(12.6,12.9, 13.2,13.7)	(1.7,1.9, 2.2,2.5)	

5.5 Discussion

In this section, results with changing α and β under different scenarios are compared, and policy implementations are proposed.

5.5.1 Propositions and Analyses

First, carbon emissions can be reduced by using a coal–biomass co-combustion approach based on a dynamic equilibrium strategy. This chapter proposes a dynamic equilibrium strategy-based coal–biomass co-combustion method. It is clear from Figure 5.6 that the coal–biomass co-combustion strategy significantly lowers carbon emissions. The lowest possible amount of carbon emissions, α, which symbolizes the government's attitude toward reducing carbon emissions, can be modified to

Table 5.10 Results when $\beta = 0.440$ and α is changing.

Decision maker	Calculated Results	$\alpha = 1.00$	$\alpha = 0.95$	$\alpha = 0.90$	$\alpha = 0.85$	$\alpha = 0.80$	$\alpha = 0.75$	No regulation
Authority	Revenue (10^7CNY)	82.41	78.33	74.25	70.18	66.06	61.95	83.73
Fengxian CPP	Profits (10^7CNY)	445.18	430.65	414.81	393.92	391.42	390.68	460.31
	Assigned quota (10^5 tonnes)	75.61	72.61	69.35	65.10	64.59	64.44	78.75
Ganyu CPP	Profits (10^7CNY)	416.82	372.68	342.11	333.25	255.43	255.27	413.36
	Assigned quota (10^5 tonnes)	77.47	69.00	63.19	61.50	46.74	46.71	77.55
Baoying CPP	Profits (10^7CNY)	443.63	441.98	429.87	399.34	417.02	356.63	447.18
	Assigned quota (10^5 tonnes)	83.92	83.54	80.76	74.85	78.27	66.59	84.75

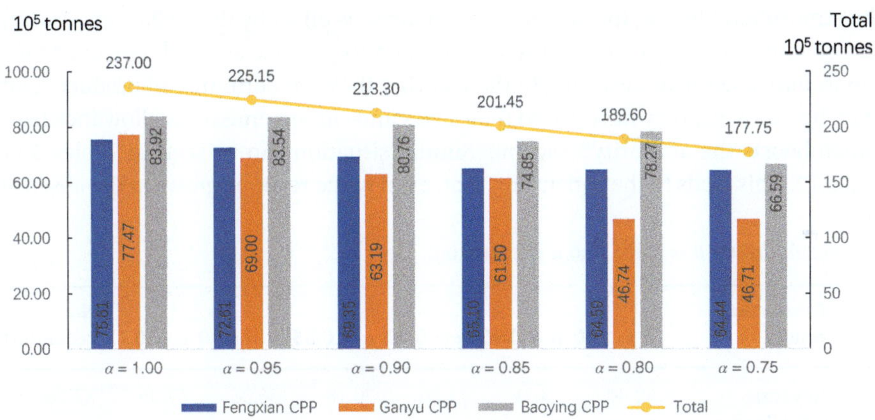

Figure 5.6 Carbon emissions under different α values when β is 0.440.

0.75 based on the findings in Tables 5.10–5.12. This clearly demonstrates that the coal–biomass co-combustion method, which is based on a dynamic balancing strategy, has the potential to reduce carbon emissions by around 25% compared to the previous production period. In the proposed model, a coal–biomass co-combustion approach based on a dynamic equilibrium strategy is used, and the authorities allocate carbon emission allowances to multiple subordinate CPPs, giving higher allocations to CPPs with better carbon reduction performance. When CPPs are assigned limited carbon emission quota, they need to adjust the coal–biomass co-combustion ratio and develop fuel purchase plans to improve carbon reduction performance in order to obtain more carbon emission allocations. Therefore, a

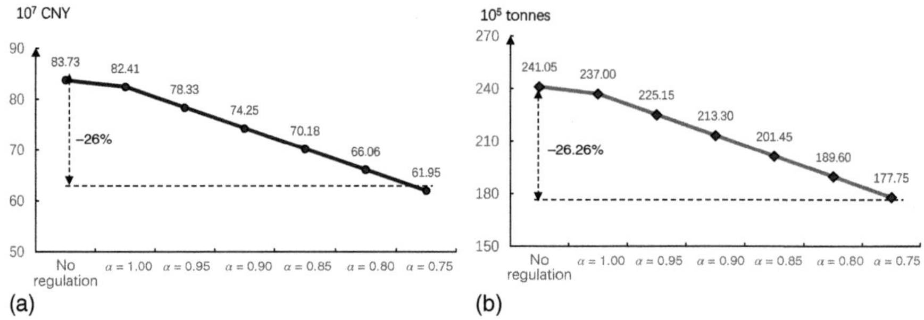

Figure 5.7 Revenue and carbon emissions when $\beta = 0.44$. (a) Authority revenue and (b) total emission.

coal–biomass co-combustion approach based on a dynamic equilibrium strategy can reduce carbon emissions.

Second, the sustainability of development would be harmed by a lack of environmental protection restrictions. If there are no restrictions on environmental protection, as shown in Figure 5.7, the carbon emissions are higher. For instance, if there is no environmental protection when $\beta = 0.44$, the authority receives 26% higher money; nevertheless, an additional 26.26% carbon dioxide is released into the atmosphere; hence, the 26% revenue increase is offset by the 26.26% increase in carbon emissions. All CPPs in this scenario aim to produce as much power as they can to maximize revenues. Due to the fact that they are permitted to produce a lot of carbon dioxide, they have raised their taxes and carbon emissions allowance fees, which boosts the authority's income. Similar situation can be seen in Tables 5.11 and 5.12. This leads to the conclusion that, even while revenue grows in the absence

Table 5.11 Results when $\beta = 0.435$ and α is changing.

Decision maker	Calculated results	$\alpha = 1.00$	$\alpha = 0.95$	$\alpha = 0.90$	$\alpha = 0.85$	$\alpha = 0.80$	$\alpha = 0.75$	No regulation
Authority	Revenue (10^7CNY)	82.40	78.31	74.26	70.18	66.06	61.95	83.76
Fengxian CPP	Profits (10^7CNY)	445.44	430.99	415.22	394.48	391.99	391.25	460.50
	Assigned quota (10^5 tonnes)	75.61	72.61	69.35	65.10	64.59	64.44	78.75
Ganyu CPP	Profits (10^7CNY)	418.40	373.96	343.28	334.39	256.28	256.13	416.25
	Assigned quota (10^5 tonnes)	77.47	69.00	63.19	61.50	46.74	46.71	77.55
Baoying CPP	Profits (10^7CNY)	445.26	443.61	431.68	401.72	419.49	358.76	448.81
	Assigned quota (10^5 tonnes)	83.92	83.54	80.76	74.85	78.27	66.59	84.75

Table 5.12 Results when $\beta = 0.43$ and α is changing.

Decision maker	Calculated results	$\alpha = 1.00$	$\alpha = 0.95$	$\alpha = 0.90$	$\alpha = 0.85$	$\alpha = 0.80$	$\alpha = 0.75$	No regulation
Authority	Revenue (10^7CNY)	82.38	78.30	74.24	70.19	66.07	61.96	83.76
Fengxian CPP	Profits (10^7CNY)	445.44	430.99	415.22	394.48	391.99	391.25	460.50
	Assigned quota (10^5 tonnes)	75.61	72.61	69.35	65.10	64.59	64.44	78.75
Ganyu CPP	Profits (10^7CNY)	419.63	375.19	344.42	335.49	257.10	255.73	418.17
	Assigned quota (10^5 tonnes)	77.47	69.00	63.19	61.50	46.74	46.71	77.55
Baoying CPP	Profits (10^7CNY)	446.66	445.01	433.07	404.04	421.92	360.83	450.22
	Assigned quota (10^5 tonnes)	83.92	83.54	80.76	74.85	78.27	66.59	84.75

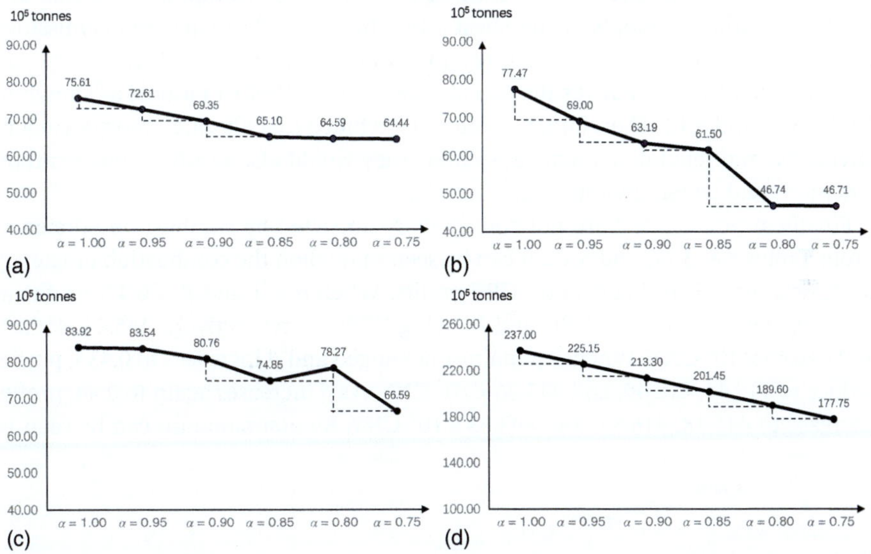

Figure 5.8 Revenue and carbon emissions when $\beta = 0.44$. (a) Fengxian CPP, (b) Ganyu CPP, (c) Baoying CPP, and (d) total.

of environmental protection constraints, the higher carbon dioxide emissions that result are detrimental to sustainable development.

Further, different CPPs respond differently to various environmental protection limitations. It is clear from Figure 5.8 that the CPPs respond differently to shifting carbon emission limitations. For instance, the carbon emissions allowances decrease for all three CPPs when α changes from 1 to 0.95; Fengxian CPP by 3×10^5 tonnes, Ganyu CPP by 8.47×10^5 tonnes, and Baoying CPP by 0.38×10^5 tonnes. The reduction ratios are typically different; thus, the carbon emission permits are decreased by 4%, 10.9%, and 0.5%, respectively. Therefore, while vying for carbon emissions allowances, Ganyu CPP has a benefit over the other two CPPs. When α changes from 0.95 to 0.9 and from 0.85 to 0.8, similar events occur. However, the carbon emission allowances the three CPPs obtain are lowered by 6.1%, 2.7%, and 7.3%, respectively, when α varies from 0.9 to 0.85, indicating that Baoying CPP has an advantage over the other two CPPs when bidding for carbon emission allowances. Similar results are also obtained when α ranges between 0.8 and 0.75. We can figure out from Eq. (5.4) that the CPP with higher $\tilde{E}_j$, exactly as Ganyu CPP and Baoying CPP, has high power generation and low carbon emissions and is preferred to be allocated higher carbon emissions allowance by the authority.

What's more, when the carbon emission constraint is tighter, the marginal rate of decrease in emissions is a little higher than revenue. As illustrated in Figure 5.9 tougher carbon emission regulations result in a loss in revenue as well as overall carbon emissions. However, the marginal rate of decrease in carbon emissions is slightly higher than the revenue rate. The overall carbon emissions allowance is 189.60×10^5 tonnes when $\alpha = 0.8$, and the revenue is 66.06×10^7 CNY when the authority has an attitude toward reducing carbon emissions that is somewhat lax. When the authority adopts the harshest stance (α=0.75), the total carbon emissions drop to 177.75×10^5 tonnes and the revenue drops to 61.95×10^7 CNY, or 6.7% and 6.6%, respectively. Similar results can also be seen in other situations, as shown in Tables 5.11 and 5.12. Therefore, it could be concluded that although more stringent environmental regulations reduce revenue, they would also result in lower carbon emissions and better environmental benefits.

Finally, profits at CPP are not considerably impacted by combustion efficiency. From Tables 5.8, 5.11, and 5.12, it can be seen that when the combustion efficiency increases, there is little effect on CPP profits. When $\alpha = 1$ and $\beta = 0.43$, profits at the Fengxian CPP, Ganyu CPP, and Baoying CPP are, respectively, 445.44, 419.63, and 446.66×10^7 CNY. When α remains unchanged and β increases to 0.435, profits reduce to 445.44, 418.40, and 445.26×10^7 CNY. As β increases again to 0.44, profits decrease to 445.18, 416.82, and 443.63×10^7 CNY. Similar situation can be seen in

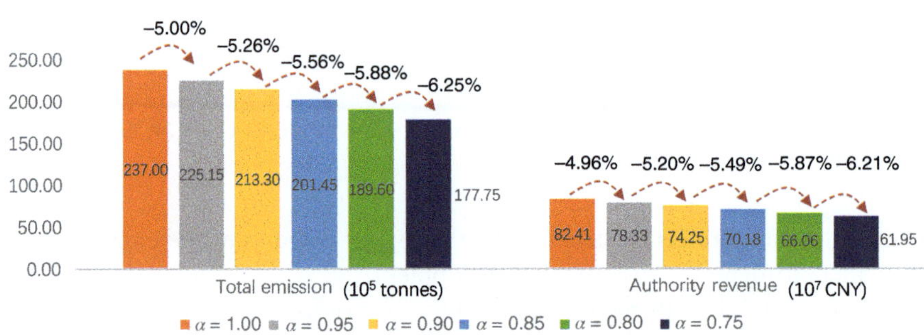

Figure 5.9 Revenue and carbon emissions when $\beta = 0.44$ and α is changing.

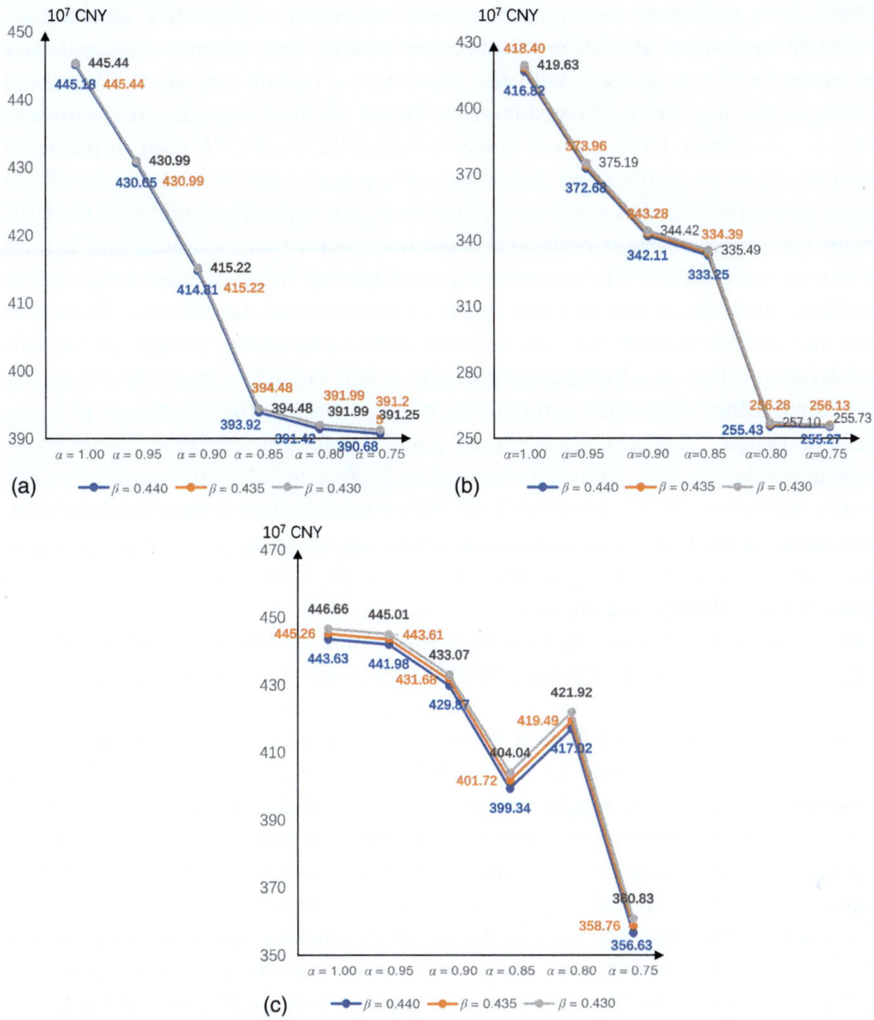

Figure 5.10 Carbon emissions under different α values and β are changing. (a) Fengxian CPP, (b) Ganyu CPP, and (c) Baoying CPP.

other situation. The reason for this is that combustion efficiency does not change carbon allowances, which are a key factor in profitability. It only affects the percentage of biomass and coal co-firing, and slightly affects profits. Therefore, CPPs are willing to sacrifice a small drop in profits to improve combustion efficiency and reduce emissions (Figure 5.10).

5.5.2 Policy Implications

Based on the above discussions and analyses, in this section, some policy implications are given.

First, it is necessary to set up a carbon emissions competition mechanism. A carbon emissions allocation competition system that involves the authority and several CPPs is advised, with the allocation of carbon emissions allowances serving as the key strategic consideration. Based on their response, the authority assigns subordinate CPPs carbon emissions permits. Each CPP then implements a coal–biomass co-combustion ratio strategy to ensure environmentally beneficial power generation. As was previously mentioned, the authority would give the CPP, which has greater carbon emissions reduction performances, additional carbon emission permits. The CPPs compete with one another to obtain additional carbon emissions allowances due to these strict carbon emission restrictions. Therefore, with mechanism, it is easy to achieve lower carbon emissions. As opposed to C&T, which has cumbersome transaction procedures and expensive transaction fees, the carbon emissions allocation competition mechanism has the advantage of being easily accessible. This strategy, however, has the potential to reduce carbon emissions more significantly than a carbon price. More crucially, this mechanism has the power to spur all CPPs to switch to cleaner manufacturing through competition. As a result, a method for carbon emissions allocation competition is proposed in order to lower the overall carbon emissions from the CPPs and foster cooperation between the authority and the CPPs.

Second, it is important to approach environmental conservation with flexibility. It is advised that the authority use a flexible approach to reducing carbon emissions depending on the level of economic development. The authority should pick the best environmental protection stance while taking into account the current circumstances. As discussed in Section 5.5.1, if there were no limits on carbon emissions, the authority would earn more money, which would be financed by higher carbon emissions, which would cause more environmental harm. Therefore, to promote better economic development when the economy is in a relatively poor state, the government could adopt a more permissive attitude toward environmental preservation. The authority may, however, adopt a harder stance as the economy matures over time; in other words, in succeeding periods, they could gradually tighten the policy control parameter until the harshest conditions were reached. Sustainable growth is ensured and the harmony between the economy, society, and ecology system is realized in this way.

Finally, the combustion efficiency should gradually increase. For energy conservation and emissions reduction, combustion efficiency is crucial. Volatile matter and the degree of char combustion in the coal–biomass co-combustion method in this chapter are the main factors limiting combustion efficiency. Greater amounts of biomass are needed to attain high combustion efficiency, which is very advantageous for the environment. According to Section 5.5.1, high combustion efficiency only results in minor losses in CPP revenues, which may be accepted by CPPs and do not jeopardize the authority and CPPs' cooperative relationship. As a result, increasing combustion efficiency gradually is advised in order to better safeguard the environment. This finding provides important support for China's National Development and Reform Commission to include "coal-coupled biomass power generation" as an encouraged industry in the "Guidance Catalog for Industrial Structure Adjustment 2019" (National Development and Reform Commission of China 2019).

References

Agbor, E., Zhang, X., and Kumar, A. (2014). A review of biomass co-firing in North America. *Renewable and Sustainable Energy Reviews* 40: 930–943.

Allen, J., Browne, M., Hunter, A. et al. (1998). Logistics management and costs of biomass fuel supply. *International Journal of Physical Distribution & Logistics* 28 (6): 463–477.

Al-Mansour, F. and Zuwala, J. (2010). An evaluation of biomass co-firing in Europe. *Biomass and Bioenergy* 34 (5): 620–629.

Basu, P., Butler, J., and Leon, M.A. (2011). Biomass co-firing options on the emission reduction and electricity generation costs in coal-fired power plants. *Renewable Energy* 36 (1): 282–288.

Beagle, E. and Belmont, E. (2016). Technoeconomic assessment of beetle kill biomass co-firing in existing coal fired power plants in the Western United States. *Energy Policy* 97: 429–438.

BP Global (2015). BP Energy Outlook.

BP Global (2017). BP Energy Outlook.

BP Global (2018). BP Energy Outlook.

Colson, B., Marcotte, P., and Savard, G. (2007). An overview of bilevel optimization. *Annals of Operations Research* 153 (1): 235–256.

Dai, C., Cai, X.H., Cai, Y.P., and Huang, G.H. (2014). A simulation-based fuzzy possibilistic programming model for coal blending management with consideration of human health risk under uncertainty. *Applied Energy* 133 (10): 1–13.

Dumortier, J. (2013). Co-firing in coal power plants and its impact on biomass feedstock availability. *Energy Policy* 60 (5): 396–405.

Eksioglu, S.D. and Karimi, H. (2014). *An Optimization Model in Support of Biomass Co-firing Decisions in Coal Fired Power Plants*. Springer International Publishing.

Goto, K., Yogo, K., and Higashii, T. (2013). A review of efficiency penalty in a coal-fired power plant with post-combustion CO2 capture. *Applied Energy* 111 (11): 710–720.

Grubb, M. (2012). Cap and trade finds new energy. *Nature* 491 (7426): 666–667.

Jiangsu Statistical Bureau (2017). https://tj.jiangsu.gov.cn/col/col86297/index.html.

Keohane, N.O. (2009). Cap and trade, rehabilitated: using tradable permits to control U.S. greenhouse gases. *Review of Environmental Economics and Policy* 3 (1): 42–62.

Li, Y. (2012). Dynamics of clean coal-fired power generation development in China. *Energy Policy* 51 (6): 138–142.

Li, S., Xu, T., Hui, S., and Wei, X. (2009). NOx emission and thermal efficiency of a 300 MWe utility boiler retrofitted by air staging. *Applied Energy* 86 (9): 1797–1803.

Li, J., Brzdekiewicz, A., Yang, W., and Blasiak, W. (2012). Co-firing based on biomass torrefaction in a pulverized coal boiler with aim of 100% fuel switching. *Applied Energy* 99 (6): 344–354.

Lin, B. and Li, X. (2011). The effect of carbon tax on per capita co emissions. *Energy Policy* 39 (9): 5137–5146.

Liu, Z., Guan, D., Wei, W. et al. (2015). Reduced carbon emission estimates from fossil fuel combustion and cement production in China. *Nature* 524 (7565): 335.

Liu, X., Chen, M., and Wei, Y. (2016). Assessment on oxygen enriched air co-combustion performance of biomass/bituminous coal. *Renewable Energy* 92: 428–436.

Lu, J., Shi, C., and Zhang, G. (2006). *On Bilevel Multi-Follower Decision Making: General Framework and Solutions*. Elsevier.

Lv, C., Xu, J., Xie, H. et al. (2016). Equilibrium strategy based coal blending method for combined carbon and PM_{10} emissions reductions. *Applied Energy* 183: 1035–1052.

Mao, X.Q., Zeng, A., Hu, T. et al. (2014). Co-control of local air pollutants and CO_2 from the Chinese coal-fired power industry. *Journal of Cleaner Production* 67 (67): 220–227.

Moon, C., Sung, Y., Ahn, S. et al. (2013). Effect of blending ratio on combustion performance in blends of biomass and coals of different ranks. *Experimental Thermal and Fluid Science* 47 (5): 232–240.

Narayanan, K.V. and Natarajan, E. (2007). Experimental studies on cofiring of coal and biomass blends in India. *Renewable Energy* 32 (15): 2548–2558.

National Development and Reform Commission of China (2019). Guidance Catalog for Industrial Structure Adjustment 2019. https://www.gov.cn/zhengce/2021-11/30/content_5713262.htm (accessed 20 October 2023).

Qi, Y., Stern, N., Wu, T. et al. (2016). China's post-coal growth. *Nature Geoscience* 9: 564–566.

Rentizelas, A.A., Tolis, A.J., and Tatsiopoulos, I.P. (2009). Logistics issues of biomass: the storage problem and the multi-biomass supply chain. *Renewable and Sustainable Energy Reviews* 13 (4): 887–894.

Sahu, S.G., Chakraborty, N., and Sarkar, P. (2014). Coal-biomass co-combustion: an overview. *Renewable and Sustainable Energy Reviews* 39 (6): 575–586.

Sami, M., Annamalai, K., and Wooldridge, M. (2001). Co-firing of coal and biomass fuel blends. *Progress in Energy and Combustion Science* 27 (2): 171–214.

Service, R.F. (2017). Cleaning up coal—cost-effectively. *Science* 356: 798.

Sinha, A., Malo, P., Frantsev, A., and Deb, K. (2014). Finding optimal strategies in a multi-period multi-leader-follower Stackelberg game using an evolutionary algorithm. *Computers & Operations Research* 41 (1): 374–385.

Tilman, D., Socolow, R., Foley, J.A. et al. (2009). Energy. Beneficial biofuels–the food, energy, and environment trilemma. *Science* 325 (5938): 270.

Van Den Broek, R., Faaij, A., and Van Wijk, A. (1996). Biomass combustion for power generation. *Biomass and Bioenergy* 11 (4): 271–281.

Weldu, Y.W. (2017). Life cycle human health and ecosystem quality implication of biomass-based strategies to climate change mitigation. *Renewable Energy* 108: 11–18.

Xu, J. and Zhou, X. (2011). *Fuzzy-Like Multiple Objective Decision Making*. Berlin Heidelberg: Springer-Verlag.

Xu, J., Song, X., Wu, Y., and Zeng, Z. (2015). GIS-modelling based coal-fired power plant site identification and selection. *Applied Energy* 159: 520–539.

Xu, J., Ma, N., and Xie, H. (2017). Ecological coal mining based dynamic equilibrium strategy to reduce pollution emissions and energy consumption. *Journal of Cleaner Production* 167: 514–529.

Prepared By Yearwood-Lee, E. (2015). Carbon emission trading.

Yilmaz, B. and Selim, H. (2015). A decision model for cost effective design of biomass based green energy supply chains. *Bioresource Technology* 191: 97–109.

Zhang, Q., Zhou, D., Zhou, P., and Ding, H. (2013). Cost analysis of straw-based power generation in Jiangsu Province, China. *Applied Energy* 102 (2): 785–793.

Zhang, G., Lu, J., and Gao, Y. (2015). *Multi-Level Decision Making*. Berlin Heidelberg: Springer-Verlag.

Zhao, Y., Wang, S., Duan, L. et al. (2008). Primary air pollutant emissions of coal-fired power plants in China: current status and future prediction. *Atmospheric Environment* 42 (36): 8442–8452.

6

Bi-Level Emission Quota Allocation Toward Coal and Municipal Solid Waste Co-combustion

Reducing global carbon dioxide emissions to net zero by 2050 is consistent with efforts to limit the long-term increase in average global temperatures to 1.5°C. This calls for nothing less than a complete transformation of how we produce, transport and consume energy.

Net Zero by 2050: A Roadmap for the Global Energy Sector
International Energy Agency, Oct. 2021

Co-combustion of coal and municipal solid waste (MSW) has been considered an important method for establishing a co-combustion hybrid energy system, which not only reduces greenhouse gas emissions and coal costs compared to burning coal alone but also reduces the amount of waste entering landfills and extends the life of existing landfills. This chapter presents an optimized leader–follower approach for achieving economic and environmental equilibrium in coal and MSW co-combustion power plants. The approach incorporates a carbon emission quota allocation scheme that operates within combustion and pollutant limitations. In this leader–follower game, the upper-level regional authority determines carbon allocation quotas and sets environmental targets, while the lower-level power plants devise strategies to maximize revenue within the constraints imposed by the upper level. By incorporating uncertain parameters that account for fluctuations in carbon and power conversion, this approach more accurately reflects the characteristics of coal–waste incineration in the context of balancing carbon emissions and economic considerations. The robustness and practicality of the proposed methodology were validated through a case study. Scenario analysis under varying political parameters demonstrates that the methodology can assist authorities in achieving a carbon-economy trade-off. In situations where strict carbon controls are in place, the methodology can encourage power plants to reduce their blended coal ratios and invest in low-carbon incineration technology. The chapter also provides managerial insights for the authority and relevant practitioners regarding further industrial developments.

Sustainable Hybrid Energy Systems: Carbon Neutral Approaches, Modeling, and Case Studies,
First Edition. Jiuping Xu and Fengjuan Wang.

6.1 Background Review

Co-combustion of coal and MSW has provided another significant alternative for establishing a co-combustion hybrid energy system, which not only reduces greenhouse gas emissions and coal costs compared to burning coal alone but also reduces the amount of waste entering landfills and extends the life of existing landfills. MSW refers to the refuse from residential, commercial, institutional, and demolition activities. It includes food waste, paper, plastics, glass, textiles, metals, and wood (Sha'Ato et al. 2007). The most commonly used disposal method of MSW is sanitary landfill (more than 68%). While this method has low operational difficulty and allows for high waste processing, it causes serious soil and groundwater pollution (Liu et al. 2019). Compared to landfill, waste-to-energy technology, especially MSW incineration power generation, can achieve more than a 90% waste volume reduction with no further decomposition needed (Li et al. 2015). These technologies are widely developed. By the end of 2019, as many as 418 waste incineration power plants (IPPs) burning blended coal and solid waste for power generation had been built in China (He 2012; Wang et al. 2018).

However, the incineration of 1 Mg of municipal waste in incinerators is also associated with the production/release of about 0.7–1.2 Mg of CO_2 (Lee et al. 2019). As a result, a significant amount of work has been devoted to addressing the carbon emissions issue in coal–waste co-combustion (Jouhara et al. 2017; Durán et al. 2017). For example, McPhail et al. (2014) developed a model to assess profits and emission costs under varying co-combustion ratios. Zhao et al. (2012) used a life cycle assessment model to evaluate the environmental impacts of MSW and coal incineration. Lu et al. (2016) used computational modeling to simulate co-combusting coal with MSW, in which the feasibility of co-combustion with different MSW blending ratios was assessed, and results indicated that up to 30% MSW co-combusted with coal was viable without significant adjustments to coal-fired reactors. While these studies have all made significant contributions to the co-combustion of MSW and coal, they only focused on single decision-makers from power plants, and did not consider the essential role of local authorities.

As coal–waste co-combustion projects are public projects in many countries, their authority-driven and social-influenced nature requires special policy attention (Parker and Karlsson 2018). Political regulations can effectively reduce emissions at a low cost to the authority. However, in developing countries, MSW IPPs may cause concern for climate change. It is crucial for regional authorities and IPP owners to collaborate on developing interactive carbon mitigation strategies for coal–waste co-combustion. Carbon allowance allocation schemes (CAAS) have been identified as one of the most practical and efficient instruments for mitigating carbon emissions (Zhang and Duan 2020). And there have been extensive studies on emission allocation schemes and associated provisions for material usage and pollutant emissions standards (Xu et al. 2017b; Zhou et al. 2010; Huang et al. 2013; Han et al. 2017).

However, most of them tend to take carbon reduction from the macroeconomic perspective based on single-optimization approaches (Cho and Kim 2019; Ilic and

Ödlund 2018). Nonetheless, they fail to explore the interactive leader–follower relationship between regional authorities and IPPs. In reality, when implementing a CAAS to reduce carbon emissions, there are multiple stakeholders with conflicting objectives. The regional authority (upper-level decision-makers) prioritizes economic and environmental sustainability, while waste IPPs (lower-level decision-makers) pursue the highest possible profit. Therefore, the key to achieving a successful trade-off between IPP profits and emission reductions lies in finding an appropriate balance through the allocation of emission allowances (Choi 2005). To reflect the interactive relationship in the IPP-CAAS and optimize carbon allocations from a game-theoretic perspective, the leader–follower (bi-level) approach has been widely accepted as a robust modeling technique for reconciling hierarchical conflicts, and has been successfully applied to water resource optimization (Xu et al. 2017a), policy suggestion (Lee 2019), and product sales strategy (Wei et al. 2020).

This chapter addresses the current IPP-CAAS knowledge gap by adopting a leader–follower optimized approach. The approach models the carbon-economy equilibrium of multiple stakeholders. The upper and lower levels are represented by the regional authority (leader) and IPPs (followers), respectively. The goals of this research are to investigate the CAAS decision-making relationship in the MSW incineration industry. This is achieved by integrating horizontal coordination and vertical interactions. A global framework is established under an uncertain decision-making environment. The equilibrium between carbon emission reductions and economic sustainability is determined by policy-makers. The specific implications between MSW transportation, combustion, and pollutant limitations in the IPP-CAAS problem are holistically assessed. The results are analyzed to deepen understanding of CAAS environmental performances and provide suggestions for the MSW incineration industry.

6.2 Main Issue Description

Before establishing the mathematical model, specific details about basic background and descriptions are as follows.

6.2.1 System Schematic

Due to its heterogeneous nature, high moisture content, and the presence of incombustible materials, MSW is not suitable for combustion in large-scale coal power plants without prior transformation into a more homogeneous and calorific feedstock. Various processes have been developed to convert MSW into combustible fuels. These processes typically involve sorting and mechanical separation of waste materials, size reduction through shredding, chipping, and milling, separation and screening, and blending, as well as drying and densification (Vekemans and Chaouki 2016). In a coal–MSW co-combustion power plant, coal is pulverized into a fine powder and put into a feeder. Conbustable MSW is put into another feeder, as shown in

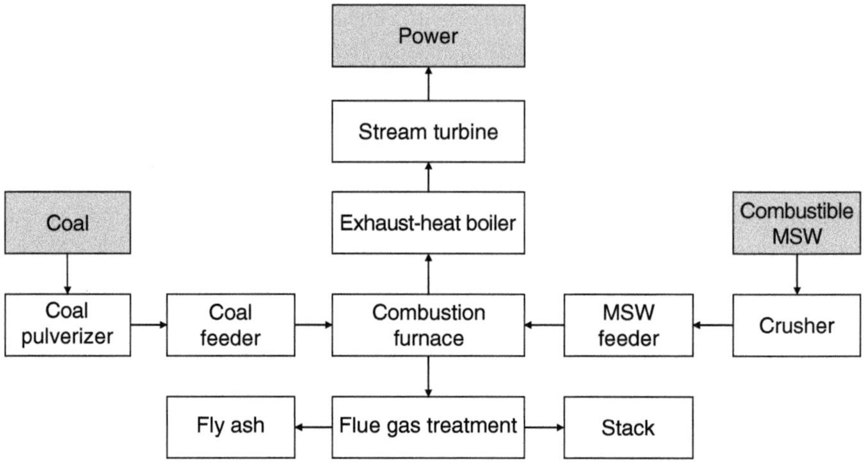

Figure 6.1 Schematic of coal and municipal solid waste co-combustion hybrid energy system.

Figure 6.1. The coal and MSW are then transported to the furnace where they are burned together. The heat generated from the combustion process is used to heat water in an exhaust-heat boiler to produce steam. The steam will drive a steam turbine so that electricity can be generated. Establishing a co-combustion hybrid energy system through co-firing coal and MSW holds several benefits; it allows for the use of MSW as a valuable energy source while reducing the volume of waste disposal in landfills. It also provides power plant operators with flexibility in their fuel supply and can help reduce emissions of pollutants such as NO_x, SO_2, and CO_2.

6.2.2 Uncertain Decision-Making

The carbon allowances help establish an interactive collaborative bi-level decision IPP-CAAS. In real decision-making environments, some factors are uncertain (Mian et al. 2017), such as the carbon emissions coefficients and the electricity transformation during incineration; therefore, in this chapter, they are dealt with using a trapezoidal fuzzy function method (Van Broekhoven and De Baets 2006). Take the waste-to-power coefficient as an example. Because of differences in equipment and waste conditions, the waste-to-power coefficient cannot be directly calculated; however, it can be estimated to be within a certain range, with the most likely value being in a relatively small range. Therefore, this coefficient is a typical trapezoidal fuzzy number and can be written as $\widetilde{M}_{ij} = r_{1j}; r_{2j}; r_{3j}; r_{4j}$, where $r_{1j} \leq r_{2j} \leq r_{3j} \leq r_{4j}$, and the most likely value being between r_{2j} and r_{3j}. Then, an uncertain conversion factor can be calculated using the expected value operator method for trapezoidal fuzzy numbers proposed by Xu and Zhou (2011) and changed to a direct value. The process of observed data going through fuzzification inference and defuzzification process is shown in Figure 6.2.

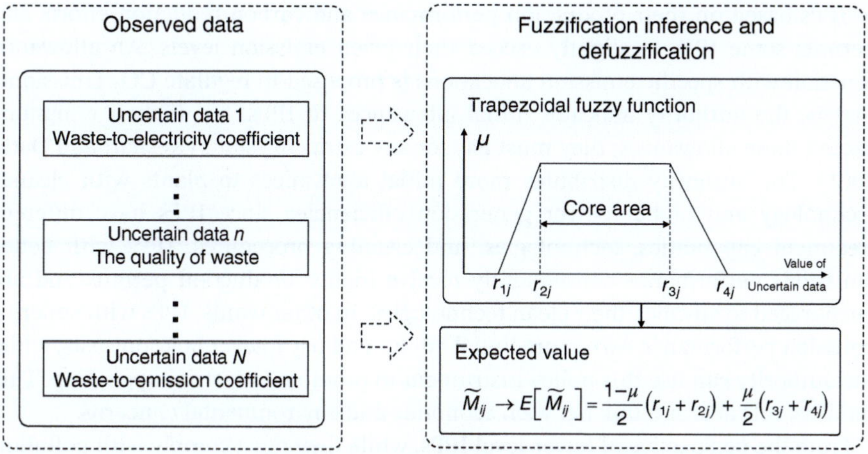

Figure 6.2 Process to handle uncertain parameters with fuzzy theory.

6.2.3 Bi-Level Relationship

In this decision problem, the regional authority at the upper level, acting on behalf of the public, takes into consideration factors such as public interest, social welfare, regional economic development, and environmental protection, as shown in (Figure 6.3). Given that air pollution from MSW IPPs (e.g. CO_2 and other polluting gases) affects society as a whole (He et al. 2004), the regional authority must supervise and control air pollution emission ceilings at incineration plants to ensure ecological balance and community health in relevant areas. To encourage production, the authority provides subsidies and allowances to each IPP (He and Lin 2019). However, the emission allowances allocated to each IPP may not accurately reflect their actual production, potentially resulting in imbalanced resource usage and energy output. To address this issue, the authority initially allocates allowances

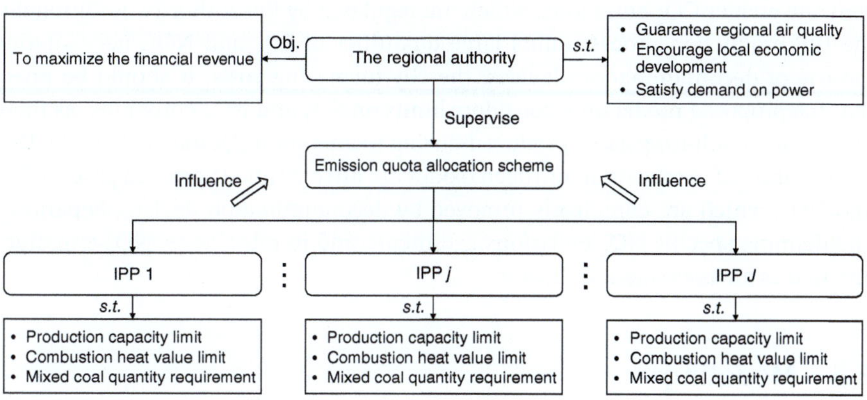

Figure 6.3 Hierarchical structure of each participant.

to IPPs based on their production performance and carbon reduction efforts and permits some IPPs to slightly exceed their given emission levels. An allowance function with specific emission allocations is proposed to regulate CO_2 emissions; that is, the authority allocates initial allowances to IPPs, but if their emissions exceed these allowances, they must pay for any extra emissions (Keshari and Datta 2001). The authority distributes more initial allowances to plants with cleaner technology and higher power generation efficiencies since IPPs have different treatment capabilities, technologies, and cleaning procedures. IPPs with better emissions performance consequently receive higher production permits and are encouraged to advance their clean technologies. In other words, IPPs with superior emission performance earn more initial quotas and pay lower emission costs, while the authority can use this policy instrument to promote industrial upgrading. This punitive role acts as a link between economic and environmental concerns.

From the perspective of lower-level IPPs, while they must comply with pollution emission regulations, they retain the ability to make decisions aimed at maximizing their revenue. As subordinates in the hierarchy, each IPP occupies the same decision-making position and must compete for allocated allowances. However, they can also influence the authority's initial allocations by developing their own incineration plans. In China, MSW classification treatment is not well developed, with MSW to be incinerated divided roughly into high-heat MSW (e.g. plastic and paper waste) and low-heat MSW (e.g. kitchen and other mixed waste), resulting in significant variation in combustion emissions. For instance, high-heat MSW has a higher heat value and produces fewer polluting gases but generates more CO_2 during incineration (Tolis et al. 2010), while low-heat MSW has a lower energy value and produces less power when combusted but generates more polluting gases and fewer CO_2 emissions (Kuo et al. 2011). In an effort to increase their revenue, IPPs aim to utilize their allocated carbon allowances to produce more power, leading them to prefer higher-heat-value MSW (e.g. waste paper and plastic waste). However, the pre-incineration collection and recycling costs for high-heat-value waste are higher than for low-heat value waste, and high-heat-value waste tends to generate greater CO_2 emissions, which are regulated by the authority. Additionally, the IPP production process emits large quantities of SO_2 and NO_x, necessitating the use of decontamination devices, thereby increasing costs. It should be noted that the proposed model only considers limits on SO_2 and NO_x emissions, as these are the main polluting gases produced during incineration (Quina et al. 2011). Due to the nature of combustion, nitrogen oxides are often emitted as mixed gases of NO and NO_2, which are completely removed by decontamination devices. Separating and defining specific NO_x emissions is difficult, and in most cases, NO_x emissions are used as an assessment indicator.

6.3 Modeling

In this section, the bi-level programming for the above-described IPP-CAAS is constructed. The mathematical description is given in the following.

6.3.1 Notations

To facilitate the problem description, the basic parameter notations and definitions are introduced.

Indices

i	Index of IPPs, $i = 1, 2, \dots, I$.
j	Index of fuels, $j = 1, 2, \dots, \theta, \dots J$, where θ-J represent municipal solid waste and others represent coal.
k	Index of pollutant emissions, $k = 1, 2, \dots, K$.
w	Index of waste quality, $w = 1, 2, \dots, W$.

Decision variables

y_i	Allocated threshold value for emission k at IPP i.
Q_{ij}	The amount of fuel j used at IPP i, where $1 - \theta$ represent coal and $\theta - J$ represent municipal solid waste.

Certain parameters

π	Tax that IPP i pays for each unit of sold power.
η	Unit price of electric power at IPP i.
θ	Unit subsidy from the regional authority for the treated waste.
M_c	The actual carbon emissions from the last production period.
D	Electric power demand from the waste incineration industry in the local district.
P	Penalty coefficient for a carbon emission unit when exceeding the threshold.
C_{ij}^P	Procurement costs for a unit of combustible fuel j at IPP i.
C_i^O	Operating costs for burning a combustible fuel unit at IPP i.
C_{ik}^T	Treatment costs for pollutant emissions k at IPP i.
$\overline{CA_i}$	Maximum electricity production capacity at IPP i.
$\underline{CA_i}$	Minimal electricity production capacity at IPP i.
MP	Limits of coal ratio in blended fuels.

Uncertain parameters

$\widetilde{P}_{ij}$	Power generated by per unit of fuel j at power plant i.
$\widetilde{W}_{ij}$	Carbon emission coefficient for treated combustible material i at IPP i.
$\widetilde{q}_{jw}$	The wth quality of fuel j.
$\widetilde{u}_{iw}^{\min}$	Lower waste quality bound w to meet operations at IPP i.
$\widetilde{u}_{iw}^{\max}$	Upper waste quality bound w to meet operations in IPP i.
$\widetilde{S}_{ik}$	Pollutant emission coefficient for treating a unit of mixed waste

Policy control parameters

α	The parameter of total carbon emission reduction level.
β	The parameter of carbon emission intensity reduction target.
γ	Attitude of the authority toward the power supply demand risk.

6.3.2 Modeling Description for Regional Authority

The objective of maximizing revenues and related constraints for the regional authority in the emission quota allocation problem are presented here in detail.

6.3.2.1 To Maximize Revenue

Since the authority acts on behalf of the public, it has priority in decision-making. In order to ensure a combination of carbon emission control and stable regional economic development, the authorities need to ensure an increase in revenues and regulation of pollution emissions. The revenue from IPPs can be divided into two components: taxes and additional revenue from penalties for CO_2 emissions; the authorities incur expenses because of the need to subsidize the IPPs' waste disposal practices. The electricity power generated by the IPPs is $\sum_{i=1}^{I}\sum_{j=1}^{J} Q_{ij}E[\widetilde{P}_{ij}]$, and let π be the tax the authority levies on the IPPs. The authority's income from tax can be written as $\pi\sum_{i=1}^{I}\sum_{j=1}^{J} Q_{ij}\widetilde{P}_{ij}$. Each IPP i has to pay for more allowances if discharging more CO_2 than the given carbon allocation quota y_i, so the total authority income of the carbon emission penalty in the region is $\sum_{i=1}^{I} g(Q_{ij}, y_i)$. Let parameter θ be the unit subsidy price, then the subsidies spending on waste treatment for the authority is $\theta\sum_{i=1}^{I}\sum_{j=1}^{J} Q_{ij}$. Combining the two parts (revenue and expenditure), the authority's objective can be finally written as Eq. (6.1).

$$\max\ G = \pi\sum_{i=1}^{I}\sum_{j=1}^{J} Q_{ij}E[\widetilde{P}_{ij}] + \sum_{i=1}^{I} g(Q_{ij}, y_i) - \theta\sum_{i=1}^{I}\sum_{j=1}^{J} Q_{ij} \tag{6.1}$$

Carbon emission penalty $g(Q_{ij}, y_i)$ is formulated as Eq. (6.1), where $\widetilde{W}_{ij}$ represents the carbon emission range for a combustible unit. Considering the inherent carbon emission uncertainty associated with temperature, technology, and incineration furnace conditions in the waste incineration process, the carbon coefficient $\widetilde{W}_{ij}$ is represented as fuzzy number and evaluated using an expected value operator method.

$$g(Q_{ij}, y_i) = \begin{cases} P(E[\widetilde{W}_{ij}]Q_{ij} - y_i); & \text{when } E[\widetilde{W}_{ij}]Q_{ij} \geq y_i, \\ 0; & \textit{others} \end{cases} \tag{6.2}$$

6.3.2.2 Emission Quota Limitation

Each IPP is granted a starting allotment of carbon emissions, or y_i, by the authority. Some IPPs may exceed their allotted amount and be required to pay for additional emissions. But the total emissions $\sum_{i=1}^{I} E[\widetilde{H}_{ij}]Q_{ij}$ should be no more than the total carbon allocation quotas $(\sum_{i=1}^{I} y_i)$ in the IPP industry, as shown in Eq. (6.3).

$$\sum_{j=1}^{J} E[\widetilde{H}_{ij}]Q_{ij} \leq \sum_{i=1}^{I} y_i \tag{6.3}$$

6.3.2.3 Total Emissions Limitation

Additionally, the regional authority must ensure that the total emission limits are below a specified percentage of the highest emissions from the previous year in order to achieve carbon reduction, as shown in Eq. (6.4), where α is used as an adjustable parameter indicating the authority's attitude.

$$\sum_{i=1}^{I} y_i \leq \alpha M_c \tag{6.4}$$

6.3.2.4 Power Supply and Demand Risk

The authority is responsible for ensuring a sufficient and consistent supply of electricity because it is a requirement for continuing economic and social development. However, because of the inherent uncertainties and randomness in the incineration power generation process, there are always risks (Chen and Christensen 2010). To resolve these risks, the fuzzy chance constraint method is used in this chapter to describe the power supply demand risk control, as shown in Eq. (6.5).

$$Pos\left\{\sum_{i=1}^{I}\sum_{j=1}^{J} Q_{ij}\widetilde{P}_{ij} \geq D\right\} \geq \gamma \tag{6.5}$$

where *Pos* is the possibility measure introduced in Dubois and Prade (1987), and D is the demand for the incineration power in the region. This fuzzy constraint can be transformed into its crisp form using the method in Xu and Zhou (2011), as shown in Eq. (6.6), in which r_{1j} and r_{2j} are the random parameters in the member function for the trapezoidal fuzzy number $\widetilde{P}_{ij}$.

$$(1-\gamma)\sum_{i=1}^{I}\sum_{j=1}^{J} r_{1j}Q_{ij} + \gamma\sum_{i=1}^{I}\sum_{j=1}^{J} r_{2j}Q_{ij} - D \geq 0 \tag{6.6}$$

6.3.3 Modeling Description for Each IPP

The profit maximization objective and associated limitations for each IPP in the emission quota allocation problem are described in detail here.

6.3.3.1 To Maximize Profits

Although they are in a lower-level position, each IPP is an individual decision-maker and makes its own production plans based on the initial decision values from the upper authority, the objectives of which are to ensure the highest profits by disposing of as much waste as possible. However, the IPPs can only function within the decision range of the regional authority. Each IPP's profit objective consists of internal revenue, internal costs, external revenue, and external costs, as shown in Eq. (6.7).

The external parts of the model depend on the interaction with the authority, which are subsidies and extra allowance fares.

$$\max\ F_i = F_i^1 - F_i^2 - F_i^3 + F_i^4 \tag{6.7}$$

The internal revenue is the income received from selling electric power without any governmental subsidies ($\eta \sum_{j=1}^{J} Q_{ij}E[\widetilde{P}_{ij}]$). Let η and π be the electricity power price and power tax, respectively. The revenue from selling electric power is calculated as the power income minus the corresponding taxation, as shown in Eq. (6.8).

$$F_i^1 = (\eta - \pi) \sum_{j=1}^{J} Q_{ij}E[\widetilde{P}_{ij}] \tag{6.8}$$

The internal costs consist of the basic operational cost ($Q_{ij}C_i^O$) and pollutant treatment cost ($Q_{ij}S_{ik}C_{ik}^T$) that only incurred during the production cycle, as shown in Eq. (6.9), where C_i^O is the unit cost for burning waste in plant i, S_{ik} is the pollutant emissions k from burning a unit of mixed waste, and C_{ik}^T is the unit treatment cost in IPP i for disposing of pollutant k.

$$F_i^2 = Q_{ij}C_i^O + Q_{ij}S_{ik}C_{ik}^T \tag{6.9}$$

External costs are associated with the carbon emission quota. When the IPP discharges its given carbon emission quota, it must pay for any extra emissions; therefore, the extra emission costs for IPP i are shown in Eq. (6.10).

$$F_i^3 = g(Q_{ij}, y_i) \tag{6.10}$$

External revenue is associated with the subsidy. Under the regional authority's subsidy policy, IPPs receive some funds to encourage waste incineration; therefore, let parameter θ be the unit subsidy that each IPP receives for disposing of a unit of waste, then the external revenue is shown in Eq. (6.11).

$$F_i^4 = \theta \sum_{i=1}^{I} Q_{ij} \tag{6.11}$$

6.3.3.2 Available Capacity Limitations of Power Plants

The amount of waste that each IPP can handle should be smaller than the designed maximum treatment standard *overlineCA*$_i$, and in order to retain its fundamental temperature, the incineration apparatus must also ensure the lowest furnace working quantity ($\underline{CA_i}$); therefore, this constraint is written as Eq. (6.12).

$$\underline{CA_i} \leq \sum_{j=1}^{J} Q_{ij} \leq \overline{CA_i} \tag{6.12}$$

6.3.3.3 Quality Requirements of Fuels

The incinerator for each IPP is predetermined and cannot be changed in the short term. The incinerator has several requirements for the water content and calorific value of MSW, and any fuel mix that does not meet these requirements can impair the efficiency of the incinerator in terms of power generation or even have serious consequences. Therefore, to improve the effectiveness of waste incineration, the IPP needs to ensure the quality of the fuel. In this context, three MSW characteristics are considered in the model to describe the quality of the blend: w_1 for water content, w_2 for heat value, and w_3 for biomass content. The comprehensive condition for the blended fuel is therefore expressed as Eq. (6.13).

$$Q' = \frac{\sum_{j=1}^{J} E[\widetilde{q}_{jw}]Q_{ij}}{\sum_{j=1}^{J} Q_{ij}} \tag{6.13}$$

where $E[\widetilde{q}_{jw}]$ is the expected value and $\widetilde{q}_{jw}$ is the quality characteristic w for fuel j. The upper and lower expected values for the wth characteristic in IPP i are $\widetilde{u}_{iw}^{\min}$ and $\tilde{u}_{iw}^{\min}$, respectively. The comprehensive conditions for the waste must be in the permitted range without any deviation, as shown in the following constraint:

$$E\left[\tilde{u}_{iw}^{\min}\right] \leq Q' \leq E\left[\tilde{u}_{iw}^{\min}\right] \tag{6.14}$$

6.3.3.4 Combustion Technical Requirements

Although co-firing a portion of MSW with coal can increase combustion efficiency (Agbor et al. 2014), the amount of coal in the waste mixture must be controlled to prevent the IPP from acting as a mini coal power plant and to guarantee the MSW disposal quantities (Srivastava et al. 2020), as shown in constraint (6.15), where Q_{i1} is the quantity of coal that is being burned with the other MSW.

$$\frac{Q_{i1}}{\sum_{j=1}^{J} Q_{ij}} \leq MP \tag{6.15}$$

6.3.4 Global Model

The global optimization model shown in Eq. (6.16) can be used to summarize decisions made by the upper and lower levels. The regional authority first determines the initial carbon allocation quotas for the IPPs in accordance with their historical emission performance, technological developments, and regional and personal emission laws. The IPPs then create their own combustion plans based on these allocations, taking into account the total costs of their operating capabilities, waste quality, and government regulations. The IPPs' plans are then submitted to the upper-level authority, which reconsiders its position and resubmits its plans to the IPPs. Due to the conflicts between economic targets and carbon reduction, each IPP will reconsider their own combustion plans based on the new emission regulations and adjusted allocation schemes and resubmit these plans to the authority.

$$
\max\ G = \pi \sum_{i=1}^{I}\sum_{j=1}^{J} E[\widetilde{P}_{ij}]Q_{ij} + \sum_{i=1}^{I} g(Q_{ij}, y_i) - \sum_{i=1}^{I}\sum_{j=1}^{J} Q_{ij}\theta
$$

$$
s.t.\begin{cases}
g(Q_{ij}, y_i) = \begin{cases} P(E[\widetilde{W}_{ij}]Q_{ij} - y_i)\,; & when\, E[\widetilde{W}_{ij}]Q_{ij} \geq y_i \\ 0\,; & others \end{cases} \\
\sum_{i=1}^{I}\sum_{i=1}^{I} E[\widetilde{W}_{ij}]Q_{ij} \leq \sum_{j=1}^{J} y_i \leq \alpha M_c \\
\sum_{j=1}^{J} E[\widetilde{H}_{ij}]Q_{ij} \leq \sum_{i=1}^{I} y_i \\
\sum_{i=1}^{I} y_i \leq \alpha M_c \\
(1-\gamma)\sum_{i=1}^{I}\sum_{j=1}^{J} r_j^1 Q_{ij} + \gamma \sum_{i=1}^{I}\sum_{j=1}^{J} r_j^2 Q_{ij} - D \geq 0 \\
\max\ F_i = F_i^1 - F_i^2 - F_i^3 + F_i^4 \\
s.t.\begin{cases}
F_i^1 = (\eta - \pi)\sum_{j=1}^{J} Q_{ij}E[\widetilde{P}_{ij}] \\
F_i^2 = Q_{ij}C_i^O + Q_{ij}S_{ik}C_{ik}^T \\
F_i^3 = g(Q_{ij}, y_i) \\
F_i^4 = \theta \sum_{i=1}^{I} Q_{ij} \\
\underline{CA_i} \leq \sum_{j=1}^{J} Q_{ij} \leq \overline{CA_i} \\
E\left[u_{iw}^{\min}\right] \leq \frac{\sum_{j=1}^{J} E[\widetilde{q}_{jw}]Q_{ij}}{\sum_{j=1}^{J} Q_{ij}} \leq E\left[\widetilde{u}_{iw}^{\min}\right] \\
\frac{Q_{i1}}{\sum_{j=1}^{J} Q_{ij}} \leq MP
\end{cases}
\end{cases}
\tag{6.16}
$$

If an IPP is prepared to pay for the additional emissions, it is possible for them to emit more than their initial quotas. IPPs are rivals in the non-cooperative game and are not aware of each other's decisions during the lower-level decision-making process. Without any interactions, each IPP solely decides how to best achieve its own economic objectives. The authority, as opposed to lower-level decision-makers, has access to all IPP production and emissions plans and must coordinate these conflicts in order to achieve its own socio-economic objectives. The interactions happen between each IPP and the upper-level authority. Requests for additional emissions could be approved if the total additional emissions do not exceed the established maximum carbon allowance. If exceeded, the regional authority will need to analyze the carbon efficiency of each IPP's electricity production and the potential impact on the authority's goals if the total additional emissions exceed the maximum carbon allowance. This interactive process will continue until a multi-consensus is reached. By integrating Eqs. (6.1)–(6.15), the environmental IPP-CAAS problem process is formulated as the global bi-level model shown in Eq. (6.16).

6.3.5 Solution Approach

A bi-level mathematical model of the IPP-CAAS problem optimization was developed (6.16). Bi-level programming problems are NP-hard problems (Isaloo and Paydar 2020), so it is actually challenging to find solutions. The hierarchical problem would still be a non-convex programming problem even if all decision-makers' objective functions and the common constraint functions were linear (Boyd and

Vandenberghe 2004). This greatly increases the complexity and length of time required to find an ideal solution (Ganesan et al. 2014). The Karush–Kuhn–Tucker (KKT) conditions are one of the many meta-heuristics that have been presented in previous research studies to address these kinds of difficult bi-level programming challenges. The fundamental idea behind the KKT approach is that it appends the leader-level problem after replacing the lower-level problem with suitable KKT conditions (Hanson 1981). During the transformation of the proposed model, a Lagrange multiplier $v_r(r = 1, 2 \ldots 6)$ is imported, with the Lagrange function for each IPP objective being $L(Q_{ij}, y_i)$. Based on the KKT condition theory, $L(Q_{ij}, y_i)$ is composed as Eq. (6.17).

$$L(Q_{ij}, y_i) = F(Q_{ij}, y_i) + \sum_{r=1}^{6} v_r g_r(Q_{ij}, y_i), \tag{6.17}$$

where function $g_r(Q_{ij}, y_i)$ $(r = 1, 2 \ldots, 6)$ are the non-negative forms for the inequalities in the lower-level constraints, the mathematical expressions for $g_r(Q_{ij}, y_i)$ $(r = 1, 2 \ldots, 6)$ are shown in Eqs. (6.18)–(6.23):

$$g_1(Q_{ij}, y_i) = \sum_{j=1}^{J} Q_{ij} - \underline{CA}_i \geq 0 \tag{6.18}$$

$$g_2(Q_{ij}, y_i) = \overline{CA}_i - \sum_{j=1}^{J} Q_{ij} \geq 0 \tag{6.19}$$

$$g_3(Q_{ij}, y_i) = y_i(\lambda - 1) - \sum_{j=1}^{J} E[\widetilde{W}_{ij}]Q_{ij} \geq 0 \tag{6.20}$$

$$g_4(Q_{ij}, y_i) = E[\widetilde{q}_{iw}]Q_{ij} - E\left[\tilde{u}_{iw}^{\min}\right] \sum_{j=1}^{I} Q_{ij} \geq 0 \tag{6.21}$$

$$g_5(Q_{ij}, y_i) = E[\tilde{u}_{iw}^{\max}] \sum_{j=1}^{J} Q_{ij} - E[\widetilde{q}_{iw}]Q_{ij} \geq 0 \tag{6.22}$$

$$g_6(Q_{ij}, y_i) = MP \sum_{j=1}^{J} Q_{ij} - Q_{ij} \geq 0 \tag{6.23}$$

What is more, based on the discussion presented by Zhang et al. (2008), the KKT solution can only be established if the additional conditions shown in Eqs. (6.24)–(6.27) are satisfied.

$$\nabla_{Q_{ij}} L(Q_{ij}, y_i) = 0 \tag{6.24}$$

$$\nabla_{y_i} L(Q_{ij}, y_i) = 0 \tag{6.25}$$

$$\sum_{r=1}^{6} g_r v_r = 0 \tag{6.26}$$

$$v_r(r = 1, 2, \ldots 6) \geq 0 \tag{6.27}$$

The computation flowchart is as follows, as shown in Figure 6.4. The upper-level decision-maker first determines the initial feasible carbon quota solution y_i, and

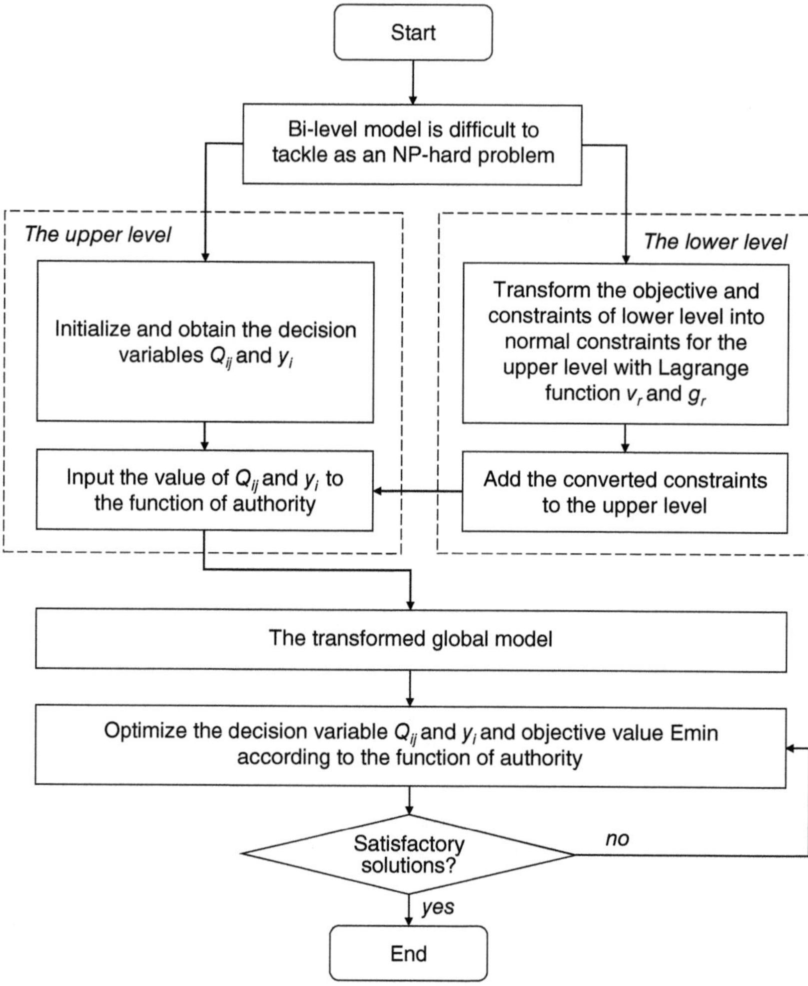

Figure 6.4 The flowchart of the computation process.

then lower-level decision-makers determine their own acceptable decision-making variables Q_{ij} that satisfy the KKT transformation requirements (Sakawa et al. 2015). These lower-level determinations are then returned to the upper-level decision-maker, who derives temporary values for y_i and max G. After the information is fed back, the original decision is adjusted under the original lower-level constraint limitations, which are regarded as normal constraints by the upper-level decision-maker. As the lower-level decisions may not be entirely suitable, the local authority checks whether these new values are optimal and decides on their desired output variables, after which the optimal solutions to the global model are determined. Based on the above processing, the original conflicts and influences between the leader (authority) and the followers (IPPs) can be coordinated and resolved when the KKT conditions are satisfied. In this way, as long as the authority's objective is optimized, a global optimization solution can be found. To find the solution to the transformed model, the fmincon function is introduced and calculated using Matlab.

6.4 Case Study

To demonstrate the efficiency of the proposed leader–follower IPP-CAAS model, a real-world case study is given in this section.

6.4.1 Case Presentation

Sichuan Province is located in central southwest China. It is one of the most populous provinces in China and has the sixth largest provincial economy, with more than US$ 6000 per capita GDP in 2019. However, the growth of the provincial cities has brought forth significant MSW issues. The tremendous economic and population growth in Chengdu, the capital of Sichuan, has resulted in a dramatic increase in trash output during the past few decades. Therefore, it has been a top priority for local governments to address the urgent need for waste treatment while working with limited capital and infrastructural assistance. In 2018, there were roughly 16,500 tonnes of waste produced per day on average, or 1.04 kg per person (NBS Sichuan 2018).

Chinese authorities have called for the construction of IPPs and the use of advanced incineration technologies to treat waste. As one of the 33 pilot cities in China for the recycling and environmentally sound treatment of kitchen waste (Liqiang 2019), Chengdu has planned and built several municipal waste incineration facilities in recent years(Wang et al. 2016). However, the large amount of CO_2 and polluting gases produced by these plants can have a negative impact on the local environment. Since Sichuan Province has a lower air quality than the national average and a total carbon emission of over 40 million tonnes, ranking ninth in China (Liu 2016), Chengdu is representative of MSW treatment and environmental management, so it was selected as a case study area to investigate the practicality and efficiency of the proposed bi-level optimization approach.

Since waste classification in China is not precise, to reduce the difficulty of calculation, we divided MSW into four classifications, accounting for about 80% of all waste, and considered only four major incineration plants: the Luodai IPP, the Jiujiang IPP, the Xiangfu IPP, and the Wanxing IPP. The locations of the chosen IPPs are shown in Figure 6.5.

6.4.2 Data Collection

The parameters for the fuel properties, such as heat value, water content, and ash content, as well as some IPP operational data, such as furnace refuse requirements, carbon emissions, and electricity transmission coefficients, were determined using a combination of expert interviews and historical data. These parameters are presented in Tables 6.6–6.4. The steps are as follows: first, surveys were carried

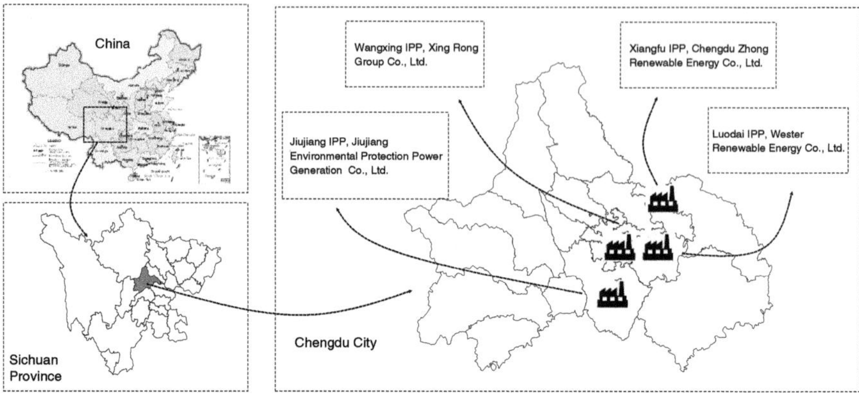

Figure 6.5 The geographical position of the case study region.

Table 6.1 Parameters of fuels in fuzzy form.

Parameters	Coal	Plastic waste	Kitchen waste	Mixed waste
Properties of MSW: $\tilde{q}_{jw}$				
Heat value (Gj/tonne)	(21.5, 21.8, 22.2, 22.9)	(5.8, 6.3, 6.7, 7.12)	(3.35, 3.65, 4.2, 4.42)	(4.48, 4.79, 5.09, 5.41)
water content (%)	—	(11.4, 14, 18.2, 20.3)	(47.4, 51.8, 56.2, 62.3)	(33.5, 36.8, 38.2, 44.3)
Ash content (%)	(20.8, 21.3, 21.7, 22.1)	(9.82, 10.17, 11.9, 16.7)	(25.27, 33.35, 38.4, 48.56)	(24.43, 29.1, 37.8, 39.2)
Pollutants emission coefficients: $\tilde{S}_{ik}$				
SO_2 (kg/tonne)	8.7	13.6	7.6	9.36
NO_x (kg/tonne)	6.3	8.2	7.25	7.13

out at waste transfer stations and at each IPP in the case region, and experts were asked to provide ranges for the necessary data based on historical data and their own judgments; second, the initial numbers were then analyzed and the extremes were deleted that were outside the range of the average values; third, the minimum value was established from the lower bound (i.e. r_{ij}^1) data and the maximum value set as the upper bound (i.e. r_{ij}^4) for the fuzzy numbers; Then, it was considered that the range of values from r_{ij}^2 to r_{ij}^3, which had the highest frequencies, was the most likely range for the uncertain parameters.

Operational management parameters such as the production capacity limits, the unit operational costs, and the pollution treatment technology and relevant costs were collected by the research team, and are also shown in Table 6.5.

Table 6.2 Fuel properties (u_{iw}^{min}, $\tilde{u}_{iw}^{min}$).

Fuel properties		Luodai IPP	Jiujiang IPP	Xiangfu IPP	Wanxing IPP
Heat value (Gj/tonne)	$\tilde{u}_{i1}^{min}$	(5, 5.4, 5.6, 5.7)	(4.8, 5.2, 5.3, 5.5)	(5.1, 5.2, 5.45, 5.6)	(4.1, 4.5, 4.7, 5.0)
	$\tilde{u}_{i1}^{max}$	(5.6, 6, 6.5, 6.8)	(5.5, 5.8, 6.0, 6.2)	(5.9, 6.15, 6.4, 6.5)	(6.0, 6.6, 6.9, 7.2)
Water content (%)	$\tilde{u}_{i2}^{min}$	—	—	—	—
	$\tilde{u}_{i2}^{max}$	(38, 40, 42, 48)	(35, 39, 44, 46)	(38, 44, 46, 50)	(40, 45, 49, 53)
Ash content (%)	$\tilde{u}_{i3}^{min}$	—	—	—	—
	$\tilde{u}_{i3}^{max}$	(21.8, 22.4, 22.6, 23.5)	(18.3, 20.48, 21, 21.5)	(21.1, 24.32, 28.45, 30.72)	(22.3, 24.87, 26.4, 27.9)

Table 6.3 Carbon emission coefficient $\widetilde{W}_{ij}$ ($kgCO_2$/tonne).

Fuels	Luodai IPP	Jiujiang IPP	Xiangfu IPP	Wanxing IPP
Coal	(1952, 1965, 1978, 1986)	(1965, 1976, 1985, 1998)	(1940, 1955, 1967, 1980)	(1965, 1980, 1993, 2005)
Plastic waste	(330, 365, 380, 385)	(386, 402, 415, 420)	(378, 394, 410, 429)	(385, 400, 434, 442)
Kitchen waste	(240, 265, 290, 300)	(245, 260, 278, 294)	(257, 280, 297, 300)	(250, 284, 300, 315)
Mixed waste	(292, 360, 373, 382)	(315, 370, 385, 394)	(355, 378, 391, 414)	(400, 428, 435, 437)

Table 6.4 Electric transmission parameter $\widetilde{T}_{ij}$ (kWh/tonne).

Fuels	Luodai IPP	Jiujiang IPP	Xiangfu IPP	Wanxing IPP
Coal	(2205, 2210, 2230, 2250)	(2200, 2220, 2250, 2280)	(2190, 2210, 2260, 2290)	(2095, 2130, 2160, 2200)
Plastic waste	(1087, 1198, 1290, 1584)	(993, 1032, 1105, 1174)	(1164, 1235, 1327, 1380)	(1337, 1450, 1471, 1534)
Kitchen waste	(568, 610, 623, 635)	(595, 624, 636, 648)	(653, 662, 672, 688)	(653, 670, 680, 686)
Mixed waste	(794, 827, 830, 846)	(828, 890, 904, 910)	(908, 947, 956, 960)	(901, 932, 950, 995)

Table 6.5 Operational management parameters.

Operational management parameters		**Luodai IPP**	**Jiujiang IPP**	**Xiangfu IPP**	**Wanxing IPP**
Emission reduction measure	SO_2	SDA	SDA	SDA	SDA
	NO_x	SNCR	SNCR	SCR	SCR
Emission reduction efficiency	SO_2(%)	95.5	95.7	95.6	97.0
	NO_x(%)	84.5	85.1	84.5	85.5
Pollution treatment cost	SO_2(CNY/kg)	2	1.8	2.3	2.2
	NO_x(CNY/kg)	14.5	14.7	15	14.8
Operation cost	(CNY/tonne)	102	100	105	110
Lower production capacity	(tonne/d)	420	600	840	640
Upper production capacity	(tonne/d)	1200	1800	2200	1800

SDA, spray drying absorption; SNCR, selective non-catalytic reduction; SCR, selective catalytic reduction.

Table 6.6 Macro data used in the proposed model.

Parameter	**Value**	**Parameter**	**Value**
Price of unit electric (CNY/kWh)	0.65	Waste treatment subsidy coefficient (CNY/tonne)	70
Added-value tax: π (CNY/kWh)	0.001	Penalty coefficient of carbon emission (CNY/tonne)	30
Local carbon emission amount in the last production cycle (tonne)			3.2×10^7

Macro data used in the proposed model are shown in Table 6.6. The actual carbon emissions cap in the previous production period, the tax, the waste treatment subsidy, and the penalty coefficient were taken from the Statistical Yearbook of Sichuan and the China and Chinese Southwest Electric Power Design Institute. As Sichuan Province is currently undergoing rapid development and has a crucial electricity power supply, the policy control parameter *gamma* in this chapter, which represents the risk control attitude of the authority toward ensuring the electric power supply, was set at 1, indicating that the local authority had a resolute attitude toward the guarantee of power supply demand.

6.4.3 Calculation Results

In this section, results under six changing α are displayed, and comparisons among these scenarios are conducted.

6.4.4 Results of Different Scenarios

As this was a static problem for a single period, the IPP industry production cycle was set at one quarter, with the quantities for coal, the residues, and the corresponding CO_2 emissions being calculated for three months. The carbon emissions parameter α was designed to examine how the results changed under various carbon and economic preference conditions. The step length for the sensitivity analysis was set at 0.05, and the lowest α was 0.75. Results under the six scenarios are presented in Tables 6.7–6.12.

6.4.4.1 S0: Baseline Scenario, α=1

The initial calculation outcomes for this model are represented in this scenario. The local authority's revenue is $-43{,}151 \times 10^7$ CNY, which is negative because, in this scenario, the regional authority is encouraging the development of the incineration power industry and, as a result, allocating a higher government subsidy to each IPP.

Table 6.7 Results under different authority attitudes (S0: $\alpha = 1$).

Power plants	Allocated emission quota y_i	Coal burned Q_{i1} (10^3tonnes)	Plastic waste burned Q_{i2} (10^3tonnes)	Kitchen waste burned Q_{i3} (10^3tonnes)	Mixed waste burned Q_{i4} (10^3tonnes)	Total emission (10^3tonnes)	Benefits F_j (10^7CNY)
Luodai IPP	73	18.45	27.90	41.22	36.45	72.90	6.8569
Jiujiang IPP	97	25.20	36.99	49.50	59.04	99.626	9.2856
Xiangfu IPP	140	35.10	61.65	72.72	81.00	136.95	15.2433
Wanxing IPP	115	27.00	42.30	53.37	65.07	114.56	11.1757

Table 6.8 Results under different authority attitudes (S1: $\alpha = 0.95$).

Power plants	Allocated emission quota y_i	Coal burned Q_{i1} (10^3tonnes)	Plastic waste burned Q_{i2} (10^3tonnes)	Kitchen waste burned Q_{i3} (10^3tonnes)	Mixed waste burned Q_{i4} (10^3tonnes)	Total emission (10^3tonnes)	Benefits F_j (10^7CNY)
Luodai IPP	73	17.37	27.00	37.80	39.42	70.51	6.8319
Jiujiang IPP	93	21.96	40.23	44.73	56.70	92.25	9.000
Xiangfu IPP	132	32.22	55.8	70.20	82.62	129.21	14.6653
Wanxing IPP	105	25.20	45.45	48.42	64.80	110.72	11.2739

Table 6.9 Results under different authority attitudes (S2: $\alpha = 0.9$).

Power plants	Allocated emission quota y_i	Coal burned Q_{i1} (10^3tonnes)	Plastic waste burned Q_{i2} (10^3tonnes)	Kitchen waste burned Q_{i3} (10^3tonnes)	Mixed waste burned Q_{i4} (10^3tonnes)	Total emission (10^3tonnes)	Benefits F_j (10^7CNY)
Luodai IPP	71	16.20	27.90	34.20	41.85	68.43	6.7467
Jiujiang IPP	190	20.88	40.50	42.30	43.02	84.31	8.1793
Xiangfu IPP	122	29.88	61.47	62.82	77.94	123.47	14.2676
Wanxing IPP	98	22.95	47.16	45.90	60.48	104.39	10.8531

Table 6.10 Results under different authority attitudes (S3: $\alpha = 0.85$).

Power plants	Allocated emission quota y_i	Coal burned Q_{i1} (10^3tonnes)	Plastic waste burned Q_{i2} (10^3tonnes)	Kitchen waste burned Q_{i3} (10^3tonnes)	Mixed waste burned Q_{i4} (10^3tonnes)	Total emission (10^3tonnes)	Benefits F_j (10^7CNY)
Luodai IPP	68	15.30	24.21	37.80	38.43	64.92	6.2108
Jiujiang IPP	84	21.60	41.04	46.35	57.24	92.49	8.9110
Xiangfu IPP	115	25.38	39.33	55.53	74.07	103.18	12.5421
Wanxing IPP	94	23.40	37.80	50.40	54.90	100.36	9.8159

Table 6.11 Results under different authority attitudes (S4: $\alpha = 0.8$).

Power plants	Allocated emission quota y_i	Coal burned Q_{i1} (10^3tonnes)	Plastic waste burned Q_{i2} (10^3tonnes)	Kitchen waste burned Q_{i3} (10^3tonnes)	Mixed waste burned Q_{i4} (10^3tonnes)	Total emission (10^3tonnes)	Benefits F_j (10^7CNY)
Luodai IPP	64	13.50	28.26	33.30	41.13	62.71	6.3758
Jiujiang IPP	82	18.00	37.80	49.23	50.49	82.38	8.2060
Xiangfu IPP	102	25.20	56.16	60.75	74.07	100.79	12.0017
Wanxing IPP	90	16.20	40.14	44.37	47.07	81.95	8.8049

The total carbon allocation standard is the same as that in the previous production cycle (Song et al. 2015). The Luodai IPP, Jiujiang IPP, Xiangfu IPP, and Wanxing IPP each have a carbon allocation of 73, 97, 140, and 115×10^3 tonnes, with Xiangfu IPP being in charge of disposing of the majority of the coal and residue, as shown in Table 6.7. The benefits of the power plants ranged from 15.2433×10^7 to 6.8569×10^7 CNY, whereas the carbon emissions from each IPP ranged from 136.95×10^3 tonnes to 72.9×10^3 tonnes.

Table 6.12 Results under different authority attitudes (S5: $\alpha = 0.75$).

Power plants	Allocated emission quota y_i	Coal burned Q_{i1} (10^3tonnes)	Plastic waste burned Q_{i2} (10^3tonnes)	Kitchen waste burned Q_{i3} (10^3tonnes)	Mixed waste burned Q_{i4} (10^3tonnes)	Total emission (10^3tonnes)	Benefits F_j (10^7CNY)
Luodai IPP	60	11.34	21.24	30.42	43.29	55.56	5.6179
Jiujiang IPP	74	13.41	36.99	47.61	45.99	70.87	7.3090
Xiangfu IPP	98	20.25	47.88	60.84	72.00	97.66	11.6752
Wanxing IPP	80	15.12	44.19	47.79	53.82	85.38	9.4599

6.4.4.2 S1: Initial Curb Scenario, α=0.9

In this scenario, the regional authority gives economic concerns a priority and has a conciliatory approach toward reducing carbon emissions, with the carbon emission allowance parameter weighting being 95%. In this scenario, the authority's overall benefits significantly outweigh S0 at -3.4898×10^7 CNY. As seen in Table 6.8, the carbon allocations decrease to 73, 93, 132, and 105 × 103 tonnes. Actual emissions total 70.51×10^3 tonnes, 92×10^3 tonnes, 129×10^3 tonnes, and 110×10^3 tonnes, respectively. The Wanxing IPP is responsible for paying for the 5.72×10^3 excess tonnes of CO_2 that it generates beyond its allotted amount. The benefits at four IPPs (Luodai IPP, Jiujiang IPP, Xiangfu IPP, and Wanxing IPP) gradually decline with the Xiangfu having the largest declining trend.

6.4.4.3 S2: Moderate Curb Scenario, α=0.9

The regional authority's attitude toward excess emission level is raised a further 5%, a group of results is determined. The authority's fiscal benefit is -3.3157×10^7 CNY. As can be seen in Table 6.9, the actual emissions and enterprise benefits for each IPP are 68.43×10^3 tonnes and 6.7467×10^7 CNY (Luodai), 84.31×10^3 tonnes and 8.1793×10^7 CNY (Jiujiang), 123.47×10^3 tonnes and 14.2676×10^7 CNY (Xiangfu), 104.3×10^3 tonnes and 10.8531×10^7 CNY (Wanxing), the total carbon allocations for the incineration industry are 481×10^3 tonnes (71, 190, 122, 98×10^3 tonnes).

6.4.4.4 S3: Serious Curb Scenario, α=0.85

The regional authority places more emphasis on reducing carbon emissions for sustainability, leading to in a 15% reduction in CO_2 emissions. As a result, the carbon allocations are correspondingly 68×10^3, 84×10^3, 115×10^3, and 94×10^3 tonnes, with the difference in the overall allocations between S2 and S3 being bigger than in the earlier scenarios at 120×10^3 tonnes of CO_2. The IPPs earn 6.2108, 8.911, 12.5421, and 9.8159×10^7 CNY individually, and their respective real carbon emissions are 64.92, 92.49, 103.18, and 100.36×10^3 tonnes. The authority's revenue is -3.2804×10^7 CNY in this scenario.

6.4.4.5 S4: Vigorous Curb Scenario, α=0.8

Under this scenario, the regional authority has limited its attitude toward emissions to 80% of the most recent production cycle. The benefits to the authority is -3.221×10^{-7} CNY. As indicated in Table 6.11, the carbon allocations for the four IPPs are 64×10^3 tonnes (Luodai IPP), 82×10^3 tonnes (Jiujiang IPP), 102×10^3 tonnes (Xiangfu IPP), and 90×10^3 tonnes (Wanxing IPP). Correspondingly, the actual carbon emissions decrease considerably: the Xiangfu IPP emits the most CO_2 (110.79×10^3 tonnes), Jiujiang and Wanxing IPPs emit at 82.38 and 81.95×10^3 tonnes, and Luodai IPP only produces 62.71×10^3 tonnes. The respective enterprise benefits are 6.3758, 8.206, 12.0017, and 8.8049×10^7 CNY.

6.4.4.6 S5: Maximal Limitation Scenario, α=0.75

The CO_2 emissions parameter α is vigorously curbed at 0.75 compared to the baseline. With the total CO_2 emissions reduced by 25%, the regional authority's revenue is -3.2152×10^7 CNY, which increases slightly by about 0.0158×10^7 CNY compared with the base scenario. But the IPPs' benefits are the lowest among all the six scenarios, which are 5.6179, 7.309, 11.6753 and 9.4599×10^7 CNY, respectively. Furthermore, the associated actual carbon emissions also fall to 55.56, 70.87, 97.66, and 85.38×10^3 tonnes, respectively.

In general, the amount of coal decreased from S0 to S5 by an average of 10,000 tonnes when α decreased. The IPPs burned as much coal as was permitted under the authority's MSW incineration coal mixing ratio in the beginning (S0). The proportion of coal decreases across the scenarios from roughly one-sixth to less than one-ninth since coal can create significantly more energy than waste, resulting in higher carbon emissions. Particularly, the amount of coal produced by Jiujiang IPP drops by more than 5000 tonnes from S4 to S5. The majority of municipal waste gradually decreases, with mixed waste accounting for about 40% of all waste, followed by roughly 34% of kitchen waste and 26% of plastic waste. Due to its greater operational capacity, Xiangfu IPP outperformed the other three IPPs in terms of waste disposal. When dealing with kitchen waste, extra coal is mixed in to ensure that the incineration criteria are met because kitchen waste has a high water and ash content. As a result, less kitchen trash is processed when the amount of coal is decreased. Plastic trash burns naturally in the incinerator because of its high heat value and low water content; thus, any reduction in coal has no impact on it. As a result, the amount of plastic waste at Wanxing IPP increases from S3 to S5.

6.4.5 Scenario Results Comparison

In this section, the carbon emission results, revenue, costs, and profits of each power plant will be presented can compared, and the influence of subsidy variation on profits trend will be discussed in detail.

6.4.5.1 Comparison of Total Carbon Emissions at Each Power Plant

As can be seen from Figure 6.6, the mixed carbon allocation method has a significant effect on reducing the carbon emissions. Changes in the regional authority's

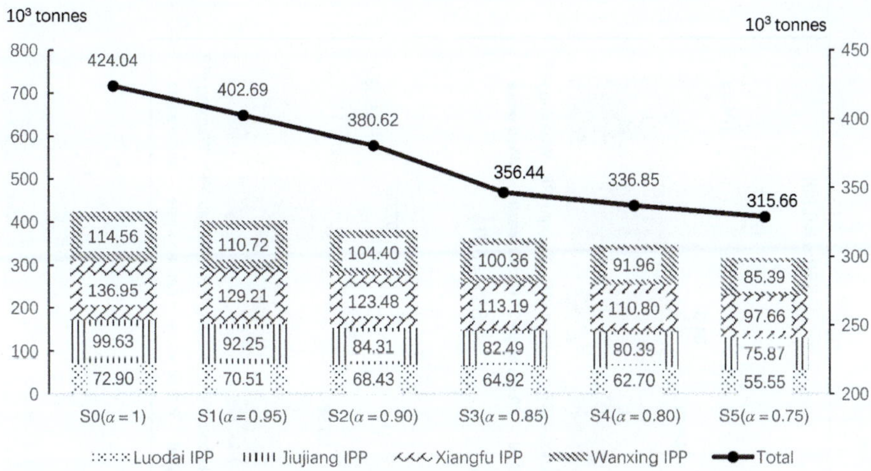

Figure 6.6 The quantity of carbon emissions in IPPs (unit: $\times 10^3$ tonnes).

attitude cause a reduction in carbon emissions of 25.55%, from 424.04×10^3 tonnes (S0) to 315.68×10^3 tonnes (S5). The regional authority should control the carbon emissions at about 85% to better foster carbon reduction measures in the IPP industry and achieve its carbon-efficiency goal, as can be seen from the line graph, which shows a slight inflection at S3 that indicates the trend of carbon emission reduction varies in S3. Figure 6.6 also shows that IPPs have diverse sensitivities to the changing carbon emission constraints. For instance, when α varies from 1 to 0.95, the carbon emissions decrease at Luodai IPP by 2.397×10^3 tonnes, Jiujiang IPP by 7.376×10^3 tonnes, Xiangfu IPP by 7.74×10^3 tonnes, and Wanxing IPP by 3.84×10^3 tonnes, with corresponding reduction ratios of 2.08%, 7.94%, 5.73%, and 6.32%. Therefore, in this situation, Xiangfu IPP has an advantage over the other three IPPs when competing for the carbon emission allowances. However, there is a different trend from S3 (α=0.85) to S4 (α=0.8), with the carbon emissions at Wanxing IPP decreasing by 8.4% compared to S1 and S2, but those at Xiangfu IPP dropping by only 2.39%. Xiangfu IPP has greater carbon efficiency than Jiujiang IPP when the government sets stronger limits on carbon emissions, but finds it more challenging to reduce its initial emissions.

6.4.5.2 Carbon Emissions from Different Fuels at Each Power Plant

As shown in Figure 6.7, although coal is obviously burned in smaller amounts than other fuels, its carbon emissions nonetheless account for a sizeable share of all IPPs' total emissions, at around 40% on average. As can be seen in S0 in Table 6.7, quantity of coal is limited in the model since it emits five times as much CO_2 as plastic or kitchen trash due to it's high carbon content. IPPs prefer to burn as much coal as they can at first since coal has a significantly higher energy output. In S0, total emissions from Xiangfu IPP are 136.95×10^3 tonnes, of which 65.28×10^3 tonnes come from coal, 22.67×10^3 tonnes from mixed waste, 19.27×10^3 tonnes from kitchen waste, and 24.04×10^3 tonnes from plastic waste. Wanxing IPP is the

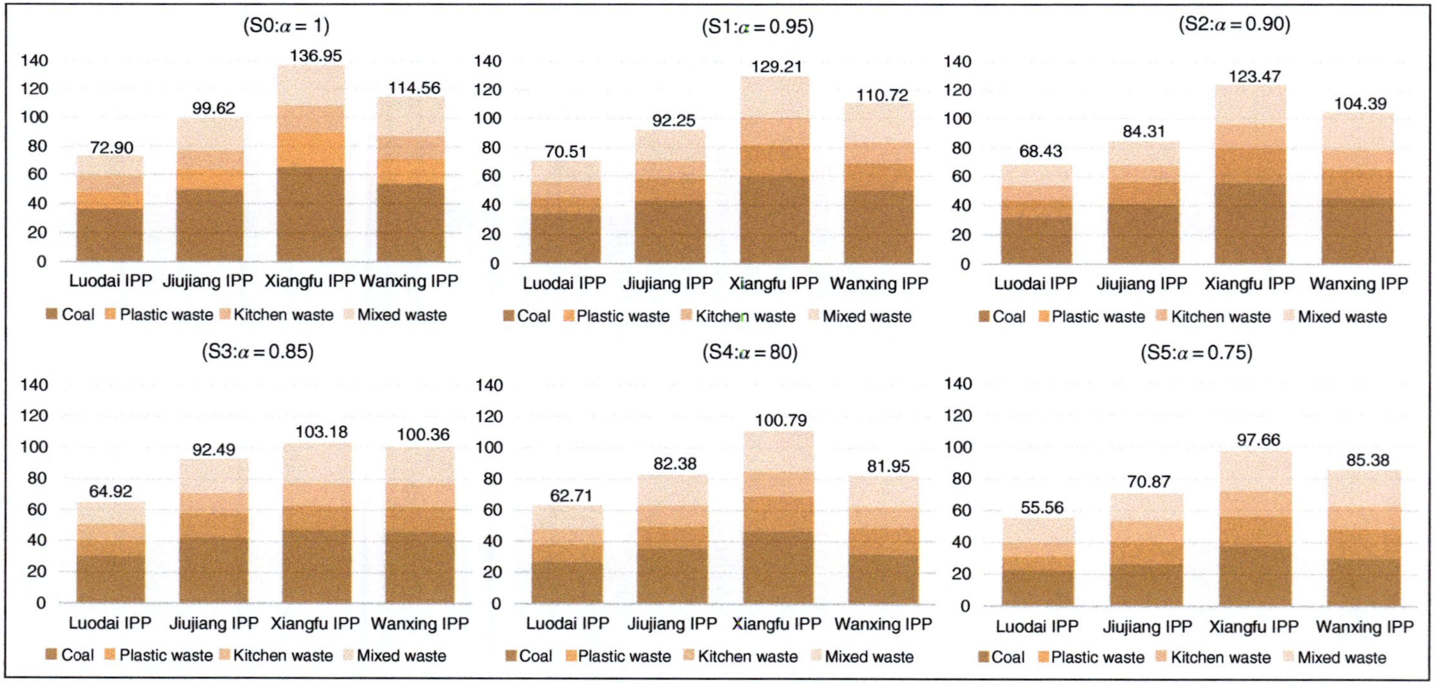

Figure 6.7 The carbon emissions of different wastes in IPPs (unit: $\times 10^3$ tonne).

source of 114.56 × 10^3 tonnes of emissions overall, followed by Jiujiang IPP (99.626 ×10^3 tonnes) and Luodai IPP (72.9 ×10^3 tonnes). From S0 to S5, as α decreases, coal emissions also decrease, but distributions of other waste essentially remain steady. To maintain a stable incinerator and satisfy their social obligations, IPPs must dispose of the necessary city MSW, so they must manage the coal volumes to meet carbon reduction targets. From S1 to S5, the emissions of plastic waste decrease by 2.33 × 10^3 tonnes (Luodai IPP), 1.20 × 10^3 tonnes (Jiujiang IPP), 3.08 × 10^3 tonnes (Xiangfu IPP), and 0.51 × 10^3 tonnes (Wanxing IPP), the emissions of kitchen waste decrease by 2.08 × 10^3 tonnes, −0.78 × 10^3 tonnes, 2.47 × 10^3 tonnes, and 0.18 × 10^3 tonnes, and the emissions of mixed waste decrease by −1.41 × 10^3 tonnes, 4.11 × 10^3 tonnes, 3.71 × 10^3 tonnes and 4.66 × 10^3 tonnes. The fluctuating amounts of MSW collected from surrounding residents are the cause of the negative figures for the kitchen waste emissions at Jiujiang IPP (−0.78) and the mixed waste emissions at Luodai IPP (−1.41). Because greater amounts are collected, mixed waste tends to generate the highest CO_2 of the three MSW types.

6.4.5.3 Comparison of Revenue, Costs, and Profits at Each Power Plant

Figure 6.8 shows the four IPPs' performances in terms of revenue, costs, and profit when the regional authority gradually introduces a stricter carbon-curbing policy. From S0 to S5, the four IPPs all see a decline in revenue and profits; however, because the Luodai IPP has the smallest operational size, it continues to generate income under 10×10 CNY, with the maximum value coming in at 9.50×10 CNY. However, as it spends less on environmental preservation and carbon allocations than the other IPPs, its costs are also lower. As shown in Figure 6.8c, the performance of the Xiangfu IPP is significantly better than that of the other IPPs: 20.14 × 10^7 CNY revenue and 15.24 × 10^7 CNY profits compared to Jiujiang IPP's 12.87 × 10^7 CNY revenue and 9.29 × 10^7 CNY profits, and Wanxing IPP's 15.47 × 10^7 CNY revenue and 11.18 × 10^7 CNY profits. However, Xiangfu IPP's declining tendency is equally noteworthy, as the revenue falls by 4.87 × 10^7 CNY compared with a decrease of only 1.85 × 10^7 CNY at Luodai IPP. The internal and external costs also affect the net profits. The IPP costs gradually decrease from S0 to S2, suggesting that the inner unit cost of operation and treatment costs for SO_2, NO_x are lower at the start of the carbon-curb policy. However, from S2 to S3, the costs rise moderately because of the extra external carbon emissions costs, and from S3 to S4, the costs at Jiujiang IPP and Xingfu IPP rise to 3.87 × 10^7 and 2.93 × 10^7 CNY, respectively. The IPPs' overall prices rise as a result of the requirement to purchase more emission allocations due to the regional authority's tougher enforcement of carbon emission objectives.

6.4.5.4 Influence of Subsidy Variation on Profits Trend

The authority's entire revenue is impacted by its subsidy strategy to promote the waste combustion development. The results show that the power plants in deficit due to the substantial subsidies given for the treated waste. While this situation is common in the early days of new technologies such as incineration power (Song et al. 2015), the authority will need to adjust the unit subsidies in the future. The current government subsidy is 70 CNY per unit of treated waste. The revenue/profit patterns

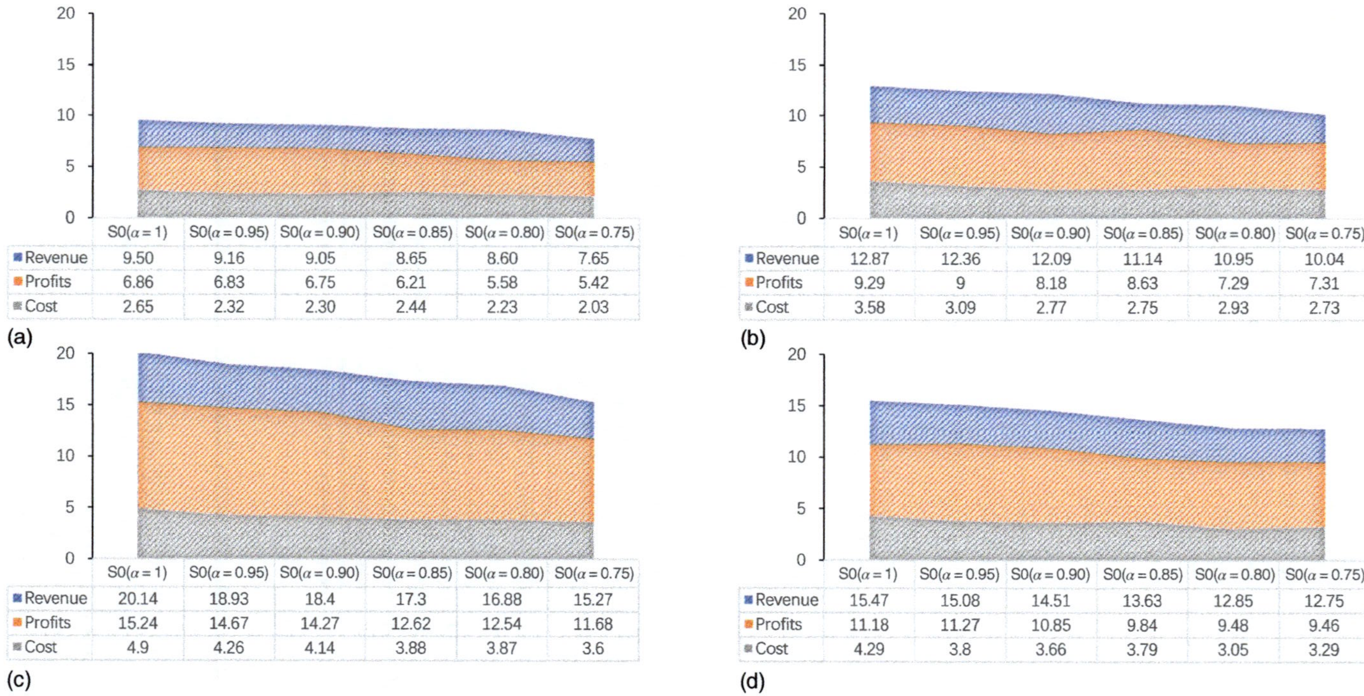

Figure 6.8 IPPs' economic performances (unit: $\times10^7$ CNY). (a) Luodai IPP, (b) Jiujiang IPP, (c) Xiangfu IPP, and (d) Wanxing IPP.

for the authorities and the IPPs when the subsidy decreases by steps of 5 CNY under the various scenarios are shown in Figure 6.9. In the first line graph (S0, $\alpha = 1$), the authority's initial revenue is -4.3151×10^7 CNY due to the preferential tax. As the subsidy declines, the revenue increases substantially and peaks at 0.8129×10^7 CNY, with a subsidy of 15 CNY being the point at which the revenue is positive (0.22×10^7 CNY). IPP earnings decline as the subsidy decreases, but the authority's revenue rises. The authority has more options to balance the subsidies and its own revenue when the carbon allocation allowances policy is tougher. However, the profitability of IPPs alters dramatically since production capacity is constrained by a strict carbon reduction requirement, making the IPP with lower unit carbon emissions more competitive because it can process more trash and receive more subsidies.

6.5 Comprehensive Discussion

Based on the previous discussion, policy implications and industrial management recommendations are provided for better exploration of coal–MSW co-combustion.

6.5.1 Policy Implications

A carbon emission allocation mechanism should be established. Under the basic scenario (S0), the CO_2 emissions reduced from 424.04×10^3 tonnes to 315.68×10^3 tonnes (S5), suggesting that the CAAS is a practical strategy for lowering IPP carbon emissions. Without such a plan, IPPs would be less inclined to continuously lower their emissions and would instead be more inclined to pursue the largest profits and the lowest emissions performance.

To meet the carbon reduction goals, the authority should adopt a proactive attitude to guarantee that the carbon emission allowances are kept within a reasonable range. As the carbon reduction targets tighten, actual carbon emissions decrease, allowing the authority to thoroughly assess their current condition and select their preferred carbon reduction strategy. It is advised that authorities should gradually tighten their stance on environmental preservation in less developed regions while maintaining the harshest stance on carbon reduction in more developed regions.

The authority needs to support research and development for cleaner, more effective coal–waste co-combustion. As energy costs rise and renewable energy sources take up a greater market share, advanced incineration systems should become more competitive. The taxes and subsidy policies encourage the power plants to dispose of more waste, which motivates them to modernize their furnaces and production technology, even though the authority initially has negative revenue. The authority can alter the preferential tax and lower the subsidy as the market and advances in technology strive to strike a balance between economic growth and environmental responsibility. This finding provides an important reference for the Chinese government to include "coal-biomass co-combustion power generation" technology into the "14th Five-Year Plan" for Scientific and Technological Innovation in the Field of Energy (Chinese Government 2022).

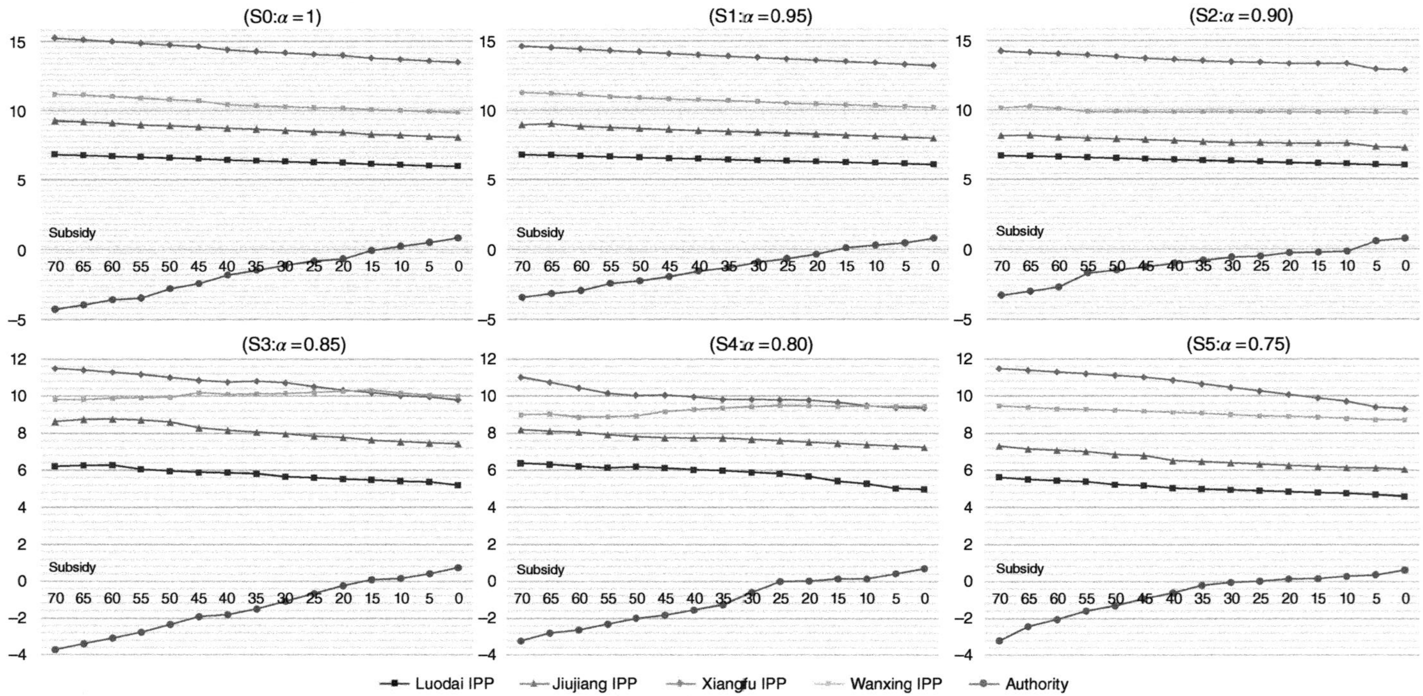

Figure 6.9 The profit trend of subsidy variations (unit: $\times 10^7$ CNY).

6.5.2 Industrial Management Recommendations

The validity of CAAS and initial carbon quota needs to be ensured. Low carbon production in the MSW sector is a gradual development process that requires continuous optimization and improvement. During this process, some IPPs still have to maintain high carbon emission levels to meet the growing regional waste treatment requirements, which means that low-carbon, efficient IPPs should also be given appropriate base waste treatment quotas at the beginning of the production cycle and encouraged to improve their production technologies. If they are required to significantly reduce their production, then regional waste treatment capacity and industrial profits will be affected. Therefore, authorities need to consider both industrial emission reduction targets and market benefits.

In this study, IPPs are found to have different sensitivities to changes in environmental constraints, and IPPs' competitive strength shifts from high production priority to low carbon priority (starting from S4 and S5); that is, the original high production strategy (in S0 and S1) becomes less feasible when low carbon emission quotas are available, which means that IPPs need to reform their carbon-intensive production activities and proactively update their production technologies to reduce unit carbon emission factor, which will strengthen their competitive advantage when low carbon emission quota policies are available.

The proportion of coal blended with MSW should be reasonably controlled. When the authorities change their attitude toward carbon emission limits under certain circumstances, the proportion of blended coal in each IPP will fluctuate, and even if the differences are small, these fluctuations will have a significant impact on total carbon emissions. IPP owners should try to control the amount of blended coal, especially coal with a high carbon factor, in order to reduce environmental pollution.

References

Agbor, E., Zhang, X., and Kumar, A. (2014). A review of biomass co-firing in North America. *Renewable & Sustainable Energy Reviews* 40: 930–943.

Boyd, S. and Vandenberghe, L. (2004). *Convex Optimization*, 244. Cambridge: Cambridge University Press.

Chen, D. and Christensen, T.H. (2010). Life-cycle assessment (EASEWASTE) of two municipal solid waste incineration technologies in China. *Waste Management & Research* 28 (6): 508–519.

Chinese Government (2022). "14th Five-Year Plan" for Scientific and Technological Innovation in the Field of Energy. https://www.gov.cn/zhengce/zhengceku/2022-04/03/5683361/files/489a4522c1da4a7d88c4194c6b4a0933.pdf (accessed 20 October 2023).

Cho, S. and Kim, J. (2019). Multi-site and multi-period optimization model for strategic planning of a renewable hydrogen energy network from biomass waste and energy crops. *Energy* 185: 527–540.

Choi, I. (2005). Global climate change and the use of economic approaches: the ideal design features of domestic greenhouse gas emissions trading with an analysis of the European Union's CO_2 emissions trading directive and the climate stewardship act. *Natural Resources Journal* 45: 865.

Dubois, D. and Prade, H. (1987). Properties of measures of information in evidence and possibility theories. *Fuzzy Sets and Systems* 24 (2): 161–182.

Durán, I., Rubiera, F., and Pevida, C. (2017). Separation of CO_2 in a solid waste management incineration facility using activated carbon derived from pine sawdust. *Energies* 10 (6): 827.

Ganesan, T., Elamvazuthi, I., Shaari, K.Z.K., and Vasant, P. (2014). Hopfield differential evolution for multi-objective optimization of a cement-bonded sand mould system. *International Journal of Management Science and Engineering Management* 9 (1): 40–47.

Han, R., Yu, B.Y., Tang, B.J. et al. (2017). Carbon emissions quotas in the Chinese road transport sector: a carbon trading perspective. *Energy Policy* 106: 298–309.

Hanson, M.A. (1981). On sufficiency of the Kuhn-Tucker conditions. *Journal of Mathematical Analysis and Applications* 80 (2): 545–550.

He, P.J. (2012). Municipal solid waste in rural areas of developing country: do we need special treatment mode? *Waste Management* 32 (7): 1289–1290.

He, J. and Lin, B. (2019). Assessment of waste incineration power with considerations of subsidies and emissions in China. *Energy Policy* 126: 190–199.

He, P.J., Zhang, H., Zhang, C.G., and Lee, D.J. (2004). Characteristics of air pollution control residues of MSW incineration plant in Shanghai. *Journal of Hazardous Materials* 116 (3): 229–237.

Huang, J., Xue, F., and Song, X. (2013). Simulation analysis on policy interaction effects between emission trading and renewable energy subsidy. *Journal of Modern Power Systems and Clean Energy* 1 (2): 195–201.

Ilic, D.D. and Ödlund, L. (2018). Method for allocation of carbon dioxide emissions from waste incineration which includes energy recovery. *Energy Procedia* 149: 400–409.

Isaloo, F. and Paydar, M.M. (2020). Optimizing a robust bi-objective supply chain network considering environmental aspects: a case study in plastic injection industry. *International Journal of Management Science and Engineering Management* 15 (1): 26–38.

Jouhara, H., Nannou, T.K., Anguilano, L. et al. (2017). Heat pipe based municipal waste treatment unit for home energy recovery. *Energy* 139: 1210–1230.

Keshari, A.K. and Datta, B. (2001). A combined use of direct search algorithms and exterior penalty function method for groundwater pollution management. *Journal of Porous Media* 4 (3): 259–270.

Kuo, J.H., Lin, C.L., Chen, J.C. et al. (2011). Emission of carbon dioxide in municipal solid waste incineration in Taiwan: a comparison with thermal power plants. *International Journal of Greenhouse Gas Control* 5 (4): 889–898.

Lee, J. (2019). Operational decision model with carbon cap allocation and carbon trading price. *Journal of Open Innovation: Technology, Market, and Complexity* 5 (1): 11.

Lee, E., Irving, B., Martinsen, T., and Mareckova, K. (2019). *Emissions from Waste Incineration*. Cambridge University Press.

Li, Y., Zhao, X., Li, Y., and Li, X. (2015). Waste incineration industry and development policies in China. *Waste Management* 46: 234–241.

Liqiang, H. (2019). Chinadaily. http://www.chinadaily.com.cn/a/201905/14/WS5cd9fe7ca3104842260bb625.html (accessed 19 October 2023).

Liu, Z. (2016). China's carbon emissions report 2016. Working Paper. Belfer Center for Science and International Affairs, Harvard Kennedy School.

Liu, X., Liu, L., Leng, P., and Hu, Z. (2019). Feasible and effective reuse of municipal sludge for vegetation restoration: physiochemical characteristics and microbial diversity. *Scientific Reports* 9 (1): 879.

Lu, L., Ismail, T.M., Jin, Y. et al. (2016). Numerical and experimental investigation on co-combustion characteristics of hydrothermally treated municipal solid waste with coal in a fluidized bed. *Fuel Processing Technology* 154: 52–65.

McPhail, A., Griffin, R., El-Halwagi, M. et al. (2014). Environmental, economic, and energy assessment of the ultimate analysis and moisture content of municipal solid waste in a parallel co-combustion process. *Energy & Fuels* 28 (2): 1453–1462.

Mian, M.M., Zeng, X., Nasry, A.A.N.B., and Al-Hamadani, S.M.Z.F. (2017). Municipal solid waste management in China: a comparative analysis. *Journal of Material Cycles and Waste Management* 19 (3): 1127–1135.

NBS Sichuan (2018). National Buereau of Statistics of Sichuan Province. http://tjj.sc.gov.cn// (accessed 19 October 2023).

Parker, C.F. and Karlsson, C. (2018). The UN climate change negotiations and the role of the United States: assessing American leadership from Copenhagen to Paris. *Environmental Politics* 27 (3): 519–540.

Quina, M.J., Bordado, J.C.M., and Quinta-Ferreira, R.M. (2011). Air pollution control in municipal solid waste incinerators. In: *The Impact of Air Pollution on Health, Economy, Environment and Agricultural Sources*, 331–358.

Sakawa, M., Nishizaki, I., Uemura, Y., and Kubota, K. (2015). Interactive fuzzy programming for multilevel linear programming problems with fuzzy parameters. *Journal of Japan Society for Fuzzy Theory & Systems* 83 (6): 1–9.

Sha'Ato, R., Aboho, S.Y., Oketunde, F.O. et al. (2007). Survey of solid waste generation and composition in a rapidly growing urban area in Central Nigeria. *Waste Management* 27 (3): 352–358.

Song, J., Song, D., and Zhang, D. (2015). Modeling the concession period and subsidy for bot waste-to-energy incineration projects. *Journal of Construction Engineering and Management* 141 (10): 04015033.

Srivastava, V., Goel, G., Thakur, V.K. et al. (2020). Analysis and advanced characterization of municipal solid waste vermicompost maturity for a green environment. *Journal of Environmental Management* 255: 109914.

Tolis, A., Rentizelas, A., Aravossis, K., and Tatsiopoulos, I. (2010). Electricity and combined heat and power from municipal solid waste; theoretically optimal investment decision time and emissions trading implications. *Waste Management & Research the Journal of the International Solid Wastes & Public Cleansing Association Iswa* 28 (11): 985.

Van Broekhoven, E. and De Baets, B. (2006). Fast and accurate center of gravity defuzzification of fuzzy system outputs defined on trapezoidal fuzzy partitions. *Fuzzy Sets and Systems* 157 (7): 904–918.

Vekemans, O. and Chaouki, J. (2016). Municipal solid waste cofiring in coal power plants: combustion performance. In: *Developments in Combustion Technology* (ed. K.G. Kyprianidis and J. Skvaril). InTech. ISBN: 978-953-51-2668-3 978-953-51-2669-0.

Wang, H., He, J., Kim, Y., and Kamata, T. (2016). Municipal solid waste management in small towns: an economic analysis conducted in Yunnan, China. *Social Science Electronic Publishing* 31: 151–165.

Wang, G., Deng, J., Ma, Z. et al. (2018). Characteristics of filterable and condensable particulate matter emitted from two waste incineration power plants in China. *Science of the Total Environment* 639: 695–704.

Wei, J., Lu, J., and Zhao, J. (2020). Interactions of competing manufacturers' leader-follower relationship and sales format on online platforms. *European Journal of Operational Research* 280 (2): 508–522.

Xu, J. and Zhou, X. (2011). *Fuzzy-Like Multiple Objective Decision Making*. Berlin Heidelberg: Springer-Verlag.

Xu, J., Hou, S., Yao, L., and Li, C. (2017a). Integrated waste load allocation for river water pollution control under uncertainty: a case study of Tuojiang River, China. *Environmental Science and Pollution Research* 24 (21): 17741–17759.

Xu, X., Zhang, W., He, P., and Xu, X. (2017b). Production and pricing problems in make-to-order supply chain with cap-and-trade regulation. *Omega* 66: 248–257.

Zhang, H. and Duan, M. (2020). China's pilot emissions trading schemes and competitiveness: an empirical analysis of the provincial industrial sub-sectors. *Journal of Environmental Management* 258: 109997.

Zhang, G., Lu, J., and Dillon, T. (2008). Solution concepts and an approximation Kuhn–Tucker approach for fuzzy multiobjective linear bilevel programming. In: *Pareto Optimality, Game Theory and Equilibria*, Springer Optimization and its Applications, vol. 17 (ed. A. Chinchuluun, P.M. Pardalos, A. Migdalas, and L. Pitsoulis), 457–480. New York, NY: Springer.

Zhao, Y., Xing, W., Lu, W. et al. (2012). Environmental impact assessment of the incineration of municipal solid waste with auxiliary coal in China. *Waste Management* 32 (10): 1989–1998.

Zhou, X., James, G., Liebman, A. et al. (2010). Partial carbon permits allocation of potential emission trading scheme in Australian electricity market. *IEEE Transactions on Power Systems* 25 (1): 543–553.

7

Bi-level Multi-objective Emission Quota Allocation Toward Coal and Sewage Co-combustion

The power sector lies at the heart of this transition, which requires increased electrification of end uses and the adoption of variable renewable energy such as wind and solar PV as the main sources of electricity. In this context, it is essential to establish robust structures to guide the procurement of electricity and ensure the flexibility required for a just and sustainable renewable era.

RE-organising Power Systems for the Transition
International Renewable Energy Agency, Jun. 2022

Co-combustion of coal and sewage sludge not only addresses the environmental damage of sewage sludge, but also reduces carbon emissions and coal consumption. This chapter presents a bi-level multi-objective model for the allocation of carbon emission allowances for the co-combustion of coal and sewage sludge in an uncertain environment. It considers the interaction of the decision-making process with the authorities as leaders and coal-fired power plants as followers, and seeks trade-offs between economic development, environmental protection, and renewable energy use. In addition, fuzzy theory, an uncertainty modeling technique, is used to describe vague and inaccurate decision-making environments. In order to solve the proposed complex model, this chapter converts the developed bi-level multi-objective model into a single-level single-objective model using the weighted sum method and the Karush–Kuhn–Tucker (KKT) optimization condition. The practicality and effectiveness of the optimization model is demonstrated through a real-life case study from Zhejiang Province, China. Based on the analysis and discussion, the model provides authorities with a reasonable and practical strategy for allocating carbon emission allowances to coal–biomass co-combustion.

Sustainable Hybrid Energy Systems: Carbon Neutral Approaches, Modeling, and Case Studies, First Edition. Jiuping Xu and Fengjuan Wang.

7.1 Background Review

Co-combustion of coal and sewage sludge has provided an applicable way to establish a co-combustion hybrid energy system and reduce carbon emissions from coal-fired power plants. Sewage sludge is the by-product of wastewater treatment processes. As it contains a great quantity of heavy metals, polycyclic aromatic hydrocarbons, and different kinds of viruses (Lei et al. 2020; Akdağ et al. 2018), it is widely accepted as a hazardous waste. It is predicted by Shanmugam et al. (2022) that the global sewage sludge production is highly likely to reach 127.5 million tonnes by 2030. Sewage sludge has traditionally been disposed of through landfills, incineration, composting, and agricultural use (Tyagi and Lo 2013). Sewage sludge incineration is considered to be more sustainable compared to other disposal methods because it both reduces sludge volume and recovers the contained calorific value (Folgueras et al. 2005). Combustion with coal is even more advantageous. First, sewage sludge can replace part of the coal used for power generation, which is important to counteract the risk of fossil fuel depletion (Hong et al. 2013). Furthermore, the co-firing of sewage sludge with coal does not require much modifications to existing coal-fired power plants and can be implemented quickly. Therefore, the mixed combustion of sewage sludge and coal-fired power generation has been widely used.

The co-combustion of sewage sludge and coal has been studied extensively. For example, Wang et al. (2018) conducted experiments to analyze the effect of the mass ratio of sewage sludge to pulverized coal on co-combustion characteristics, and found that the ignition performance improved significantly as the proportion of sewage sludge increased. Yu et al. (2018) developed a dynamic full-field synergy model and applied it to a 220 tonnes per hour co-combustion coal and sludge boiler to control pollutant emissions. Holik et al. (2019) studied the coal–sludge co-combustion characteristics in a 100 MW coal-fired utility boiler, and found that the annual profit could reach 4.2–5.5 million CNY for the power plant. These laboratory-scale studies and practical applications have provided abundant theoretical and practical basis for co-combustion of coal and sewage sludge. However, few of them considered applying the policy instruments of carbon emission quota allocation to better explore the performance of co-combustion of coal and sewage sludge.

Research on carbon emission quota allocation has attracted significant attention, and various allocation methods have been proposed. Zhou and Wang (2016) classified the allocation methods into indicator, optimization, game theoretic, and hybrid approaches. Zhang et al. (2014) employed the Shapley value method to allocate the quotas among regions under the consideration of collaboration. The previous studies focused on carbon emission quota allocation at different levels, for example, burden sharing among countries and regional allocation of carbon emission reduction targets, but paid little attention to allocation at an operational level in a specific industry such as co-combustion of coal and sewage sludge. Therefore, a comprehensive study that integrates carbon emission quota allocation into the combustion of coal and sewage sludge is urgently needed.

This chapter presents a bi-level multi-objective model for the allocation of carbon emission allowances for coal and sewage sludge co-combustion in an uncertain environment. In the proposed model, the authority tries to find an optimal strategy based on its objectives, while the coal-fired power plant reacts in a rational manner and seeks to optimize its objectives in response to the authority's decision. Thus, the allocation involves the authorities and the coal-fired power plants and depends on the interaction. In this chapter, the allocation of carbon emission allowances for sewage sludge and coal co-firing is considered as a bi-level decision problem, with the authorities as leaders and coal-fired power plants as followers. Furthermore, under the guidance of the UN development goal of "affordable and clean energy" (The United Nations 2019), multiple goals of economic development, environmental protection, and renewable energy use are considered. Since these objectives cannot be achieved simultaneously, a multi-objective optimization approach was used to seek trade-offs between conflicting objectives. In addition, fuzzy theory is employed to describe uncertain, imprecise, and fuzzy parameters and to estimate exact values. For this bi-level multi-objective model, the weighted sum method and KKT optimization conditions are used to calculate the results. Then, the practicality of the model is demonstrated for Zhejiang Province, and the results show that the model provides a reasonable and practical carbon emission allowance allocation strategy for the government.

7.2 Main Issue Description

Before establishing the mathematical model, specific details about basic background and descriptions are as follows.

7.2.1 System Schematic

A coal and sludge co-combusted hybrid energy system involves mixing and burning sludge in coal-fired power plant boilers to achieve a large-scale synergistic treatment of sludge while using the heat generated by sludge combustion for power generation. In essence, sludge with a moisture content of 60–85% from a sewage treatment plant is transported to the power plant in an enclosed vehicle, dried to a moisture content of 30% or less using power plant steam to meet the fueling requirements, and then blended with coal in a certain proportion and sent to the power plant boiler, where it is burned at a high temperature of over 1200 °C; see scheme shown in Figure 7.1. Not only does co-combustion significantly reduce the cost of sludge disposal and eliminate environmental pollution from sludge stockpiles, but it is also an effective way to transform and upgrade thermal power plants to achieve circular growth and business extension.

7.2.2 Uncertain Decision Environment

Decision-makers are faced with ambiguous and imprecise decision-making environments because of the uncertainty inherent in various system components. Uncertainty modeling techniques usually include probabilistic approach, possibilistic approach, and fuzzy theory, etc. (Khaloie et al. 2019). In the absence of sufficient historical data (Aien et al. 2016), the possibilistic approach has been widely used to characterize uncertainty in variables. Zadeh first introduced the concept of

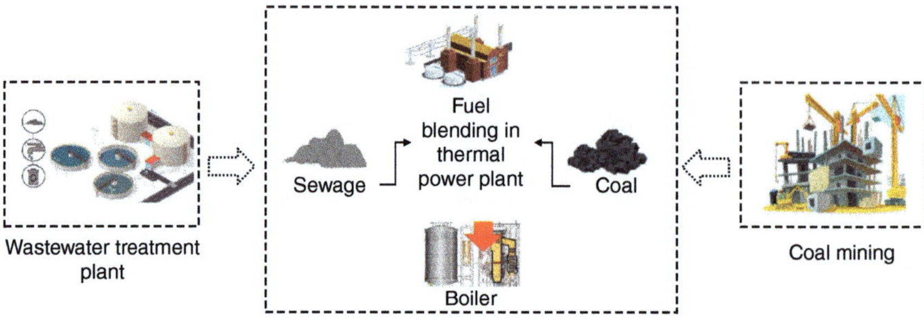

Figure 7.1 Schematic of coal and sludge co-combusted hybrid energy system.

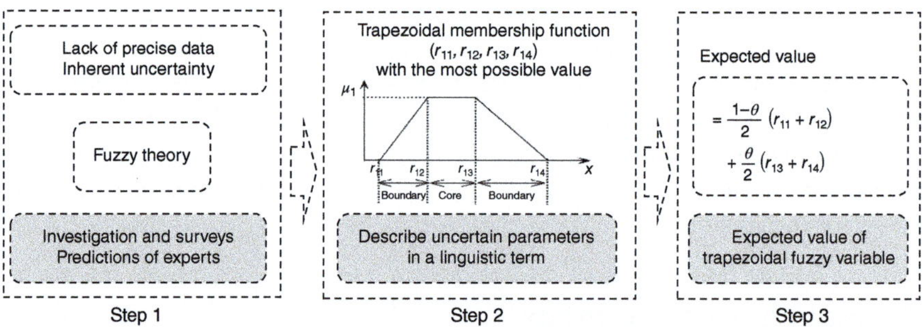

Figure 7.2 Fuzzy membership function and transformation process.

probabilistic uncertainty modeling and proposed fuzzy theory, which has explicit bounds on uncertain parameters and is capable of converting uncertain parameters into definite ones (Zadeh 1978). Fuzzy theory has been widely used in many studies to deal with uncertain factors. For example, Xu et al. (2018) used fuzzy theory to describe the physical and chemical properties of biomass, and Çalı and Balaman (2019) combined multi-criteria decision-making and fuzzy theory to effectively solve the ranking problem under uncertainty. Inspired by these applications, this chapter uses fuzzy theory to characterize uncertain parameters and obtain their exact values. Figure 7.2 shows the flowchart for the construction of the fuzzy membership functions.

7.2.3 Optimization Scheme

Two decision-makers are involved in the allocation of carbon emission quotas for the co-combustion of coal and sewage sludge: the authorities and the coal-fired power plants. The authority with the higher decision priority makes the first decisions and develops the carbon emission quota allocation. Each coal-fired power plant has its own response to the decision of the authority. With the allocated carbon emission quotas, the coal-fired power plant develops a coal and sewage sludge blending strategy to achieve the target. If there are insufficient carbon emission quotas, coal-fired power plants need to buy the required quotas from the carbon trading market; conversely, they can sell additional quotas. Since the authority's decision is also affected by the response of the coal-fired power plant, the relationship between

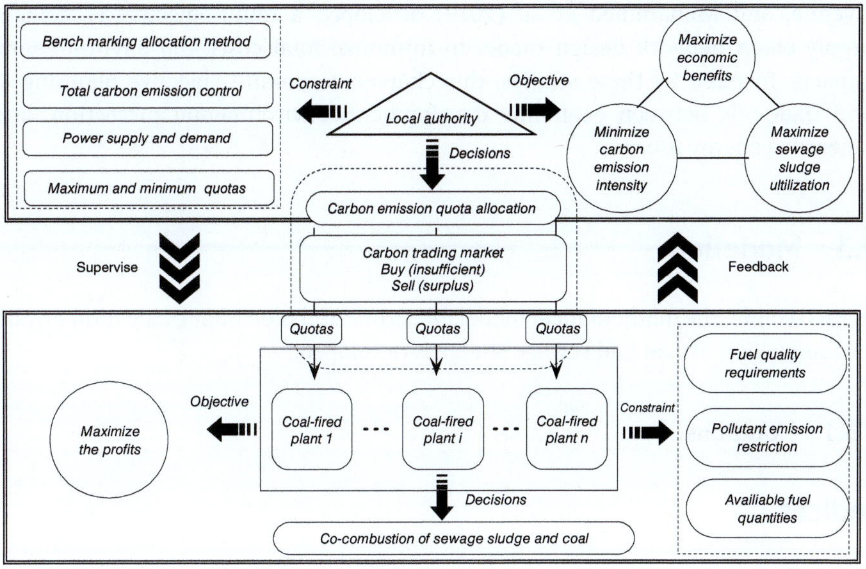

Figure 7.3 Flowchart of the bi-level structure.

the upper and lower decision-makers is similar to the Stackelberg game, as shown in Figure 7.3. Bi-level planning is an effective method for finding Stackelberg equilibrium and has been widely used in many fields (Lu et al. 2006). For example, Talari et al. (2019) proposed a bi-level model for stochastic day-ahead dispatch between independent system operators and different demand response aggregators, and Zolfaghari and Akbari (2018) developed a multi-stage bi-level model for transmission expansion planning with wind investments. Inspired by these studies, this chapter applies bi-level planning to determine the optimal relationship between authorities and coal-fired power plants.

The optimization of the allocation of carbon emission quotas for coal–sludge blending takes into account several conflicting objectives such as economic development, environmental protection, and renewable energy use. In this chapter, the level of economic development is expressed in terms of the total profits of coal-fired power plants. In their pursuit of maximum generation and maximum profit, coal-fired power plants emit large amounts of carbon emissions that are difficult and costly to remove. Therefore, carbon intensity represents environmental protection and conflicts with total profit. Furthermore, sewage sludge incineration is considered more sustainable than other traditional disposal methods such as landfill, composting, and agricultural use, as it reduces the sludge volume and recovers the calorific value contained (Woch et al. 2018). Therefore, increasing the utilization of sewage sludge in the coal–sludge co-combustion was identified as the third objective of this problem. Multi-objective planning has been shown to be a powerful tool for finding trade-offs between multiple objectives. For example, Mahmoudimehr and Sebghati (2019) developed a multi-objective model for solar power plant management to maximize daily electricity production and daily revenue, and Mohammed et al. (2019) developed a multi-objective G-resilient supply chain network design model to minimize total costs and environmental impacts. Inspired by these studies, this chapter uses multi-objective planning to find trade-offs between economic development, environmental protection, and renewable energy use.

7.3 Modeling

In this section, the mathematical model for carbon emission quota allocation toward co-combustion of coal and sewage sludge is developed.

7.3.1 Notations

Indices

i	Index of coal-fired power plant.
j	Index of fuels, $j = 1, 2, \dots, \theta, \dots, J$, where θ–J represent sewage sludge and others represent coal.
k	Index of pollution, $k = 1, 2, \dots, K$.
w	Index of quality, $w = 1, 2, \dots, W$.

Decision variables

y_i	Allocated threshold value for emission k in coal-fired power plant i.
Q_{ij}	The amount of fuel j used at coal-fired power plant i, where $1 - \theta$ represent coal and $\theta - J$ represent sewage sludge.

Certain parameters

R	Tax rate that coal-fired power plant j pays for each unit of sold power.
G	Trading price per unit of carbon emission quota.
L_j	Price of electricity generated from a unit of fuel j.
T_c	Baseline of carbon emission intensity from the last production period.
$B_i^{\max}$	The maximum value of carbon emission quotas at coal-fired power plant i.
$B_i^{\min}$	The minimum value of carbon emission quotas at coal-fired power plant i.
E	Regional power demand.
n_i	The operating costs of the coal-fired power plant i.
K	The supply of carbon emission quotas in the trading market.
RR_{ir}	The reduction rate of pollutant r at coal-fired power plant i.
EI_r	The concentration limits of pollutant r.
$u_{it}^{\max}$	The upper bound of fuel quality t at coal-fired power plant i.

$u_{it}^{\min}$ The lower bound of fuel quality t at coal-fired power plant i.
$H_{ij}^{\max}$ The maximum available amount of fuel j at coal-fired power plant i.
M_c The actual carbon emissions from the last production period.

Uncertain parameters

$\widetilde{P_{ij}}$ Power generated by per unit of fuel j at coal-fired power plant i.
$\widetilde{W_{ij}}$ The carbon emissions of each fuel j at coal-fired power plant i.
$\widetilde{C_{ij}^P}$ The procurement costs of each unit fuel j at coal-fired power plant i.
$\widetilde{DS_{ijr}}$ Mass flow rate of pollutant r of each unit fuel j at coal-fired power plant i.
$\widetilde{VF_{ij}}$ Volumetric flow rate of each unit fuel j at coal-fired power plant i.
$\widetilde{A_{ijt}}$ Quality t of each unit fuel j at coal-fired power plant i.

Policy control parameters

α The parameter of total carbon emission reduction level.
β The parameter of carbon emission intensity reduction target.

7.3.2 Allocation Scheme for the Authority

In this section, the economic benefits objective, the emission controlling objective, and the sludge utilization objective, and related constraints considered by the authority will be described in detail.

7.3.2.1 Maximizing Economic Benefits

The economic benefits consist of all the profits of the coal-fired plants in the area (Hong et al. 2017). Let f_i represent the profit of the ith coal-fired plant. The objective of maximizing the total economic benefits can be mathematically described as Eq. (7.1).

$$\max F_1 = \sum_{i=1}^{I} f_i \tag{7.1}$$

7.3.2.2 Minimizing Carbon Emission Intensity

This chapter regards negative externalities as carbon emissions when generating electricity. Therefore, in addition to maximizing total economic benefits, minimizing carbon emission intensity is also considered one of the objectives of the authorities. In order to cope with the increasingly serious global warming, it is necessary to reduce the intensity of carbon emissions, which is the carbon emissions per unit of physical product (Marco et al. 2013; Zhou and Wang 2016). Let carbon emissions from a unit of fuel i be represented by $\widetilde{W_{ij}}$ and handled as a trapezoidal fuzzy number. $\widetilde{W_{ij}}$ ranges from the minimum value d_1 to the maximum value d_4, and the most likely value is between d_2 and d_3 ($d_1 < d_2 \leq d_3 < d_4$). In order to determine the exact value of the trapezoidal fuzzy number, the expected value method proposed by Xu et al is used (Xu and Zhou 2011), and $\widetilde{W}_{ij}$ can be written as: $\widetilde{W}_{ij} \rightarrow E[\widetilde{W}_{ij}] = \frac{1-\delta}{2}(d_1 + d_2) + \frac{\delta}{2}(d_3 + d_4)$, where $0 \leq \delta \leq 1$. δ represents the optimistic–pessimistic parameter that determine the combined attitude of the decision-makers. $\widetilde{P}_{ij}$ represents power generation from a unit of fuel i, and thus $\sum_{i=1}^{I}\sum_{j=1}^{J} E[\widetilde{W}_{ij}]t_{ij}$ and $\sum_{i=1}^{I}\sum_{j=1}^{J} E[\widetilde{P}_{ij}]t_{ij}$, respectively, indicate the total carbon emissions and the total power. Therefore, the objective of minimizing carbon emission intensity can be written as Eq. (7.2).

$$\min F_2 = \frac{\sum_{i=1}^{I}\sum_{j=1}^{J} E[\widetilde{W}_{ij}]t_{ij}}{\sum_{i=1}^{I}\sum_{j=1}^{J} E[\widetilde{P}_{ij}]t_{ij}} \tag{7.2}$$

7.3.2.3 Maximizing Sludge Utilization

Sludge is a type of renewable energy source. Its incineration can greatly reduce its volume, recover the calorific value, and partially replace coal (Werle and Wilk 2010; Magdziarz et al. 2016). Let t_{ij} represent fuel j at power plant i, where $\theta - J$ represent sewage sludge. Therefore, this objective can be expressed as Eq. (7.3).

$$\max F_3 = \sum_{i=1}^{I}\sum_{j=\theta}^{J} x_{ij} \tag{7.3}$$

7.3.2.4 Benchmark Allocation Method

In the power sector, the authorities allocate carbon emission quotas to target coal-fired power plants in accordance with benchmark allocation rules (Jiang et al. 2014; Tang et al. 2015). The benchmarking method is adopted to allocate quotas according to the industry's carbon emission intensity level (Vespermann and Wald 2011; Zhou et al. 2009). T_c represents the sectoral-level carbon emission intensity baseline, and β represents the authority attitude toward carbon emission intensity reduction. When β equals 0.01, the authority aims to reduce the intensity by 1%. The higher the β value, the lower the carbon emission intensity. $\sum_{j=1}^{J} E[\widetilde{P}_{ij}]t_{ij}$ represents the power generated by power plant i; therefore, the carbon emission quota can be mathematically formulated as Eq. (7.4).

$$y_i = (1 - \beta)E[\widetilde{P}_{ij}]t_{ij}T_c \tag{7.4}$$

7.3.2.5 The Control of Carbon Emission

$\widetilde{W}_{ij}$ is the carbon emissions from a unit of fuel i, and $\sum_{i=1}^{I}\sum_{j=1}^{J} E[\widetilde{W}_{ij}]t_{ij}$ represents the total carbon emissions in the region. Let M_c be the carbon emissions of the last production period, and α be carbon emission reduction targets. When α is equal to 1, the aim of the authorities is to keep the total carbon emissions not exceeding the previous production period. The lower the α value, the less the total carbon emissions. Therefore, this limitation can be described as Eq. (7.5).

$$\sum_{i=1}^{I}\sum_{j=1}^{J} E[\widetilde{W}_{ij}]t_{ij} \leq (1 - \alpha)M_c \tag{7.5}$$

7.3.2.6 Power Supply and Demand Balance

The total electricity demand of the area must be met. Let $\widetilde{P_{ij}}$ be the conversion parameter from a unit of fuel i to power, and the total power supply can be expressed as $\sum_{i=1}^{I}\sum_{j=1}^{J} E[\widetilde{P_{ij}}]t_{ij}$. E represents the total power demand. Therefore, this limitation can be described as Eq. (7.6).

$$\sum_{i=1}^{I}\sum_{j=1}^{J} E[\widetilde{P_{ij}}]t_{ij} \geq E \tag{7.6}$$

7.3.2.7 Bounds of Quotas

The basic rights of coal-fired power plants need to be protected by the authorities. $B_i^{\min}$ indicates the carbon emissions when expenses balance receipts. When the carbon emission quota is below $B_i^{\min}$, coal-fired power plants will struggle to maintain daily operations. $B_i^{\max}$ indicates the carbon emissions when coal-fired power plant i is under full-load production. When the authority allocates more carbon emission quota than $B_i^{\max}$, it is unfair to other coal-fired power plants. Therefore, this limitation is formulated as Eq. (7.7).

$$B_i^{\min} \leq y_i \leq B_i^{\max} \tag{7.7}$$

7.3.3 Strategy for Coal-Fired Plants

In this section, the profits maximizing objective, the fuel quality requirements, the pollutant emission restrictions, and the available fuel quantities considered by each CPP will be described in detail.

7.3.3.1 Maximizing Profits

Coal-fired power plants are independent of each other and make their own decisions in order to maximize profits. This chapter divides profits into after-tax revenue, fuel costs, operating costs, and carbon-related costs. Since the feed-in tariff is always agreed upon in advance with a specific value and will not change within a period of time, the feed-in tariff is treated as a specific parameter L_j. Let $\widetilde{P_{ij}}$ be the conversion parameter from a unit fuel i to power, and R be added-value tax rate. Therefore, $\sum_{j=1}^{J}(1-R)L_jE[\widetilde{P_{ij}}]t_{ij}$ represents after-tax income. Because procurement costs are constantly changing and are affected by many factors, $\widetilde{Q_{ij}}$ representing the unit procurement cost is considered as an uncertain parameter. Then $\sum_{j=1}^{J} E[\widetilde{Q_{ij}}]t_{ij}$ is the total fuel costs. n_i represents fixed operating costs in a production period. Since the pricing of tradable carbon emission allowances is to internalize negative externalities, power plants will incur carbon-related costs. $\sum_{j=1}^{J} E[\widetilde{W_{ij}}]t_{ij} - y_i$ represents the trading quantity of carbon emission quota. When the value is positive, it means that carbon emission allowances are insufficient and more allowances need to be purchased; when the value is negative, it means that coal-fired power plants can sell excess carbon emission allowances (Xu et al. 2015). The transaction price of G changes due to the relationship between supply and demand, and there is uncertainty: when there are sufficient or surplus tradable allowances in the market, the trading price of carbon emission allowances will decrease; when there are insufficient tradable allowances in the market, the trading price of carbon emission allowances will increase. Based on the oligopolistic competition model, G can be expressed as: $a - b[K - \sum_{i=1}^{I}(y_i - \sum_{j=1}^{J} E[\widetilde{W}_{ij}]t_{ij})]$, where K represents the supply (ChaoQun et al. 2018; Tu et al. 2015). Combining these, this objective can be described as Eq. (7.8).

$$\max f_i = \sum_{j}^{J}(1-T)L_j E[\widetilde{P}_{ij}]t_{ij} - \sum_{j}^{J} E[\widetilde{Q}_{ij}]t_{ij} - n_i - \left(\sum_{j}^{J} E[\widetilde{W}_{ij}]T_{ij} - y_i\right)G \quad (7.8)$$

7.3.3.2 Quality Requirements on Fuel

When sewage sludge and coal are burned together, the quality of the mixed fuel must be acceptable within the capacity of the boiler designed for coal burning (China National Institute of Standardization 2010; Shanmugam et al. 2022). $\widetilde{A}_{ijt}$ denotes quality t per unit of fuel i. When $t = 1$, it represents volatile matter content; when $t = 2$, it represents heat rate; when $t = 3$, it represents ash content; when $t = 4$, it represents moisture content; when $t = 5$, it represents sulfur content; when $t = 6$, it represents ash-fusion softening temperature; and when $t = 7$, it represents grindability (China National Institute of Standardization 2010). Let $\sum_j E[\widetilde{A}_{ijt}]t_{ij}/\sum_j t_{ij}$ be quality t per unit of fuel mix, and $u_{it}^{\min}$ and $u_{it}^{\max}$, respectively, represent the lower and upper bound for quality t. Quality t per unit of fuel mix should range from the lower bound to the upper bound to avoid equipment failures. Therefore, the mathematical form is described as Eq. (7.9).

$$u_{it}^{\min} \leq \frac{\sum_j E[\widetilde{A}_{ijt}]t_{ij}}{\sum_j t_{ij}} \leq u_{it}^{\max}, \quad \forall i \in \psi \quad (7.9)$$

7.3.3.3 Restrictions on Pollutant Emission

Coal-fired power plants must comply with relevant pollutant emission regulations to discharge harmful pollutants such as sulfur dioxide, nitrogen oxides, and particulate matter. Based on Cristóbal et al.'s research, emission concentration can be calculated as mass flow rate of pollutant r divided by volumetric flow rate (Cristóbal et al. 2012). Let $\widetilde{DS}_{ijr}$ be mass flow rate of pollutant r: $r = 1$ represents SO_2, $r = 2$ represents NO_x, and $r = 3$ represents PM. Let $\widetilde{VF}_{ij}$ be volumetric flow rate of fuel i, then $\sum_j E[\widetilde{DS}_{ij}]/\sum_j t_{ij}$ and $\sum_j E[\widetilde{VF}_{ij}]t_{ij}/\sum_j t_{ij}$, respectively, represent the total mass flow rate of pollutant r and the total volumetric flow rate. Coal-fired power plants always take measures to control pollutants such as desulphurization and denitrification technologies; therefore, there is pollutant emission reduction RR_{ir}. As a result, $E[\widetilde{DS}_{ijr}](1 - RR_{ir})t_{ij}/\sum_j t_{ij}$ represents mass flow rate of pollutant r after disposal. Let EI_r represent emission limit value for pollutant r. Therefore, this constraint is written as Eq. (7.10).

$$\frac{\sum_j E[\widetilde{DS}_{ijr}](1 - RR_{ir})\frac{t_{ij}}{\sum_j t_{ij}}}{\sum_j E[\widetilde{VF}_{ij}]\frac{t_{ij}}{\sum_j t_{ij}}} \leq EI_r \quad (7.10)$$

7.3.3.4 Available Quantities of Fuel

The quantity of fuel used by each power plant cannot be negative, then $t_{ij} \geq 0$. Besides, the quantity cannot exceed its maximum available quantity. Let $H_{ij}^{\max}$ represent maximum available quantity, then $t_{ij} \leq H_{ij}^{\max}$. Therefore, the constraint is formed as Eq. (7.11).

$$0 \leq t_{ij} \leq H_{ij}^{\max} \tag{7.11}$$

7.3.4 Global Model

A bi-level multi-objective model for carbon emission quota allocation toward co-combustion of coal and sewage sludge under uncertainty is developed in this chapter. The mutual reactions between the authority and coal-fired power plants are like leader–follower hierarchical relationship. The authority develops a carbon emission quota allocation strategy with the aim of optimizing its economic benefit maximization, carbon emission intensity minimization, and sewage sludge utilization maximization objectives, as described in Eqs. (7.1)–(7.3), by considering its allocation method, total carbon emission control, power supply and demand, and maximum and minimum quotas, as outlined in Eqs. (7.4)–(7.7). Under the allocated carbon emission quotas, the coal-fired power plants then choose coal/sewage sludge co-combustion scheme to maximize their own profits, as shown in Eq. (7.8), under the constraints of fuel quality requirements, pollutant emission restrictions, and fuel quantity in Eqs. (7.9)–(7.11). Therefore, the global mathematical model is formulated as Eq. (7.12).

$$
\begin{aligned}
&\max F_1 = \sum_i f_i \\
&\min F_2 = \frac{\sum_i \sum_j E[\widetilde{Q}_{ij}] x_{ij}}{\sum_i \sum_j E[\widetilde{C}_{ij}] x_{ij}} \\
&\max F_3 = \sum_i \sum_{j=\theta}^{J} x_{ij} \\
&\text{s.t.} \begin{cases}
y_i = (1-\beta) \sum_j E[\widetilde{C}_{ij}] x_{ij} G_c, \quad \forall i \\
\sum_i \sum_j E[\widetilde{Q}_{ij}] x_{ij} \leq (1-\alpha) L_c \\
\sum_i \sum_j E[\widetilde{C}_{ij}] x_{ij} \geq D \\
V_i^{\min} \leq y_i \leq V_i^{\max}, \quad \forall i \\
\max f_i = \sum_j (1-T) P_j E[\widetilde{C}_{ij}] x_{ij} - \sum_j E[\widetilde{W}_{ij}] x_{ij} - z_i + (y_i - \sum_j E[\widetilde{Q}_{ij}] x_{ij}) p(r) \\
\text{s.t.} \begin{cases}
b_{it}^{\min} \leq \frac{\sum_j E[\widetilde{N}_{ijt}] x_{ij}}{\sum_j x_{ij}} \leq b_{it}^{\max}, \quad \forall i, \quad \forall t \\
\frac{\sum_j E[\widetilde{MF}_{ijr}](1 - RR_{ir}) \frac{x_{ij}}{\sum_j x_{ij}}}{\sum_j E[\widetilde{VF}_{ij}] \frac{x_{ij}}{\sum_j x_{ij}}} \leq EL_r, \quad \forall i, \quad \forall r \\
0 \leq x_{ij} \leq O_{ij}^{\max}, \quad \forall i, \quad \forall j
\end{cases}
\end{cases}
\end{aligned}
\tag{7.12}
$$

7.3.5 Model Solving

In this chapter, a multi-objective model is proposed to address the trade-off between economic development, environmental protection, and renewable energy utilization. The weighted sum method is employed to integrate the three objective functions of economic development, environmental protection, and renewable energy utilization into one objective problem. The weighted sum method is a basic and easy-to-use approach for multi-objective optimization (Tu et al. 2015). To ensure that the weighted sum method is applied appropriately, it is necessary to standardize all objective functions before integrating them into one objective problem. This is important because the objective functions may have different units and scales, which can cause a scale effect that would make it difficult to compare their relative importance. By standardizing the objective functions, they will have similar ranges and units, making it easier to compare and combine them in a meaningful way (Xu et al. 2019; Holik et al. 2019). Therefore, the integrated single-objective function can be formulated as Eq. (7.13).

$$\min F^e = \left\{ -w_1 \left(\frac{F_1 - \underline{F_1}}{\overline{F_1} - \underline{F_1}} \right) + w_2 \left(\frac{F_2 - \underline{F_2}}{\overline{F_2} - \underline{F_2}} \right) - w_3 \left(\frac{F_3 - \underline{F_3}}{\overline{F_3} - \underline{F_3}} \right) \right\} \tag{7.13}$$

where F^e is the equality objective function for the authority; w_1, w_2, and w_3 are the weighting coefficients to represent the authority's preference to economic development, environmental protection, and renewable energy utilization, and they satisfy $w_1 + w_2 + w_3 = 1$. $\underline{F_1}$, $\underline{F_2}$, and $\underline{F_3}$ represent the minimum values of the objectives, and $\overline{F_1}$, $\overline{F_2}$, and $\overline{F_2}$ represent the maximum values.

Equation (7.12) represents a bi-level model that reflects the interaction between the authority and the coal-fired power plants. Several solution methods have been proposed for bi-level models, and among them, KKT optimality conditions have been demonstrated to be the most effective (Colson et al. 2007; Hanson 1981). The KKT optimality conditions are incorporated into the lower-level model and then appended to the leader's problem to create a single-level single-objective model. Using weighted sum method and KKT optimal conditions, the bi-level multi-objective model is transformed into a single-level single-objective model in Eq. (7.14), which can be calculated using the matrix laboratory toolbox.

$$\begin{aligned}
&\min F^e \\
&\text{s.t.} \begin{cases}
\text{Eqs. (7.4)–(7.7)} \\
\text{Eqs. (7.9)–(7.11)} \\
f_i'(x_{ij}) = \sum_{v=1}^{7} u_{iv} g_1'(x_{ij}) + \sum_{v=8}^{14} u_{iv} g_2'(x_{ij}) + \sum_{v=15}^{17} u_{iv} g_3'(x_{ij}) + \sum_{v=18}^{19} u_{iv} g_4'(x_{ij}) \\
\quad + \sum_{v=20}^{21} u_{iv} g_5'(x_{ij}), \quad \forall i \\
\sum_{v=1}^{7} u_{iv} g_1 + \sum_{v=8}^{14} u_{iv} g_2 + \sum_{v=15}^{17} u_{iv} g_3 + \sum_{v=18}^{19} u_{iv} g_4 + \sum_{v=20}^{21} u_{iv} g_5 = 0, \quad \forall i \\
g_1 = b_{it}^{\min} \sum_{j=1}^{J} x_{ij} - \sum_{j=1}^{J} E[\tilde{N}_{ijt}] x_{ij} \le 0, \quad \forall i, \quad \forall t \\
g_2 = \sum_{j=1}^{J} E[\widetilde{N_{ijt}}] x_{ij} - b_{it}^{\max} \sum_{j=1}^{J} x_{ij} \le 0, \quad \forall i, \quad \forall t \\
g_3 = \sum_{j=1}^{J} E[\widetilde{MF}_{ijr}](1 - RR_{ir}) \frac{x_{ij}}{\sum_{j=1}^{J} x_{ij}} - EL_r \sum_{j=1}^{J} E[\widetilde{VF_{ij}}] \frac{x_{ij}}{\sum_{j=1}^{J} x_{ij}} \le 0, \quad \forall i, \quad \forall r \\
g_4 = -x_{ij} \le 0, \quad \forall i, \quad \forall j \\
g_5 = x_{ij} - O_{ij}^{\max} \le 0, \quad \forall i, \quad \forall j
\end{cases}
\end{aligned} \tag{7.14}$$

7.4 Case Study

In this section, a case study from Zhejiang Province, China, is given to demonstrate the model's practicality and efficiency.

7.4.1 Case Description

China is responsible for nearly 50% of the worldwide coal demand and almost a third of the global carbon emissions (Ye et al. 2016; Jeff 2015). Zhejiang Province, which is located in the southern part of the Yangtze River Delta and faces the East China Sea, is one of the most prosperous regions in China. In 2017, Zhejiang Province ranked fourth in China with a gross regional domestic product of 5.177 trillion (National Bureau of Statistics of China 2019). The total electricity consumption in Zhejiang Province reached 419.263 billion kilowatt hour in 2017 due to sustainable economic development and population growth; however, only 325.9 billion kilowatt hour can be locally provided (Zhejiang Provincial Bureau of Statistics 2019). Zhejiang Province heavily relies on coal for power generation, resulting in significant carbon emissions. Consequently, Zhejiang Province has been actively exploring measures to increase power production and reduce carbon emissions to address the electricity supply shortage and environmental pressure.

Zhejiang Province, a highly developed region in China, is endowed with ample reserves of sewage sludge. In 2013, the amount of sewage sludge generated in this province reached 456,000 tonnes, which constituted 7.3% of the total national production (Yang et al. 2015). In support of sustainable power generation, the National Energy Administration has promoted the co-combustion of coal and sewage sludge for power generation through the "Notice on Carrying Out Pilot Work of Technical Transformation of Coal-fired Coupled Biomass Power Generation" (National Energy Administration 2019). In addition, Zhejiang Province has implemented policies aimed at promoting the utilization of sewage sludge. As a result of these policies, many coal-fired power plants have begun to adopt the method of co-combustion of coal and sewage sludge. Due to the complexity of the computation, only three power plants were selected for analysis in this case (Figure 7.4). Jiaxing coal-fired power plant (Jiaxing CPP), Shaoxing coal-fired power plant (Shaoxing CPP), and Ningbo coal-fired power plant (Ningbo CPP) are chosen as the representative samples.

7.4.2 Data Collection

The data needed for the proposed model were divided into certain data in Tables 7.1 and 7.2 and uncertain data in Table 7.3.

Table 7.1 presents the power generation price of sewage sludge at 0.65 CNY/kWh, sourced from the policy document "Improving the Policy on Waste Incineration for Power Generation." The carbon emission quotas, quality boundaries, pollutant reduction rates, and operating costs for the coal-fired power plants were obtained from the official websites, the "Specification of Coal Used for Pulverized Coal-fired

Figure 7.4 Description of the case study.

Table 7.1 Certain data for the three coal-fired power plants.

Power plants	Units	Ningbo CPP		Jiaxing CPP		Shaoxing CPP	
Fuel		Sludge	Coal	Sludge	Coal	Sludge	Coal
Power price	CNY	0.65	0.45	0.65	0.45	0.65	0.45
Carbon emission allowance	$\times 10^6$ tonnes	B_{it}^{min}	B_{it}^{max}	B_{it}^{min}	B_{it}^{max}	B_{it}^{min}	B_{it}^{max}
		4.447	6.647	2.541	5.7541	1.37	2.29
Operating cost	$\times 10^8$ CNY	2.11		1.56		0.82	
Maximum quantity: O_{ij}^{max}	$\times 10^6$ tonnes	2.57	1.22	2.29	1.03	1.99	0.92
$r = 1$ SO_2		95.1		94.7		94.3	
$r = 2$ NO_x		85		84.4		83.8	
$r = 3$ *PM*		93.6		93.5		93.1	
Boundaries of quality		u_{it}^{min}	u_{it}^{max}	u_{it}^{min}	u_{it}^{max}	u_{it}^{min}	u_{it}^{max}
$t = 1$ Volatile matter	%	6.9	29.2	7.5	31.6	8.3	34.5
$t = 2$ Heat rate	Gj/tonne	24.5	—	23	—	21.7	—
$t = 3$ Ash	%	—	21	—	22.1	—	23
$t = 4$ Moisture	%	—	8.3	—	9.4	—	10.1
$t = 5$ Sulfur	%	—	1.2	—	1.3	—	1.5
$t = 6$ Ash-fusion softening temperature	°C	1250	—	1270	—	1290	—
$t = 7$ Grindability	%	45	—	52	—	57	—

Table 7.2 Other certain parameters used in the model.

Parameters	Units	Value	Parameters	Units	Value
Added-value tax	—	0.17	Power demand	$\times 10^{1}0$ kWh	1.50474
Actual CO_2 in the last production period	$\times 10^7$ tonnes	1.28	Pollutant discharge limit ($r = 1, 2, 3$)	mg/m^3	(50, 100, 20)
Emission allowance supply	tonnes	400,000	Carbon emission intensity baseline	%	0.85

Boiler for Power Generation," and other relevant documents. The pollution emissions limits presented in Table 7.2 were obtained from two sources: the Emission Standard of Air Pollutions for Thermal Coal-fired Power Plant, and the Best Feasible Technical Guide for Pollution Prevention and Control of Sludge Treatment and Disposal in Urban Sewage Treatment Plants.

Uncertain data were collected in a trapezoidal fuzzy form. Table 7.3 presents the conversion parameters that were used to estimate power generation, fuel qualities,

Table 7.3 Uncertain data for the three coal-fired power plants in a trapezoidal fuzzy form.

Parameters			Units	Ningbo CPP		Jiaxing CPP		Shaoxing CPP	
				Sludge	Coal	Sludge	Coal	Sludge	Coal
Fuel to power: $\widetilde{C}_{ij}(e)$			kWh/tonne	(2150,2250,2345,2450)	(1175,1290,1410,1520)	(1975,2070,2180,2290)	(1010,1120,1230,1360)	(1850,1935,2020,2115)	(1975,2070,2180,2295)
Procurement cost: $\widetilde{W}_{ij}$			CNY/tonne	(418,445,476,501)	(425,480,520,575)	(632,667,700,737)	(348,380,425,467)	(545,580,620,675)	(588,634,660,686)
Fuel to carbon: $\widetilde{Q}_{ij}$			tonne/tonne	(1.82,2.03,2.2,2.35)	(0.97,0.99,1.03,1.05)	(1.74,1.87,2.1,2.23)	(0.87,0.9,0.94,0.97)	(1.75,1.82,1.93,2.06)	(0.78,0.81,0.84,0.87)
Volumetric flow rate: $\widetilde{VF}_{ij}$			Nm^3/h	(8.76,9.55,10.35,11.14)	(7.5,8.44,9.43,10.43)	(8.61,9.48,10.3,11.21)	(7.1,8,8.7,9)	(8.76,9.54,10.16,10.9)	(7,7.7,8.5,9.2)
Mass	SO_2:	$r = 1$	kg/h	(1585,1680,1770,1853)	(450,515,588,663)	(1480,1540,1600,1668)	(425,480,532,579)	(1347,1415,1490,1556)	(380,435,490,551)
flow	NO_x:	$r = 2$	kg/h	(1595,1680,1762,1835)	(690,785,900,1001)	(1430,1540,1705,1885)	(675,760,840,925)	(1395,1510,1630,1745)	(620,715,805,920)
rate: $\widetilde{MF}_{ijr}$	*PM*:	$r = 3$	kg/h	(170,270,365,467)	(145,178,220,265)	(155,260,355,466)	(137,170,215,262)	(165,250,335,410)	(126,165,204,253)
Quality: $\widetilde{N}_{ijt}$	Volatile matter	$t = 1$	%	(23.6,25,26.5,28.1)	(31,35.4,40.1,44.7)	(24.3,25.8,27.1,28.4)	(40.5,44.6,48.9,51.2)	(25.2,26.3,27.5,28.7)	(43.7,46.8,50.2,54)
	Heat rate	$t = 2$	Gj/tonne	(23.3,24.6,26.2,27.5)	(16.5,17.3,18.1,18.9)	(22.4,23.5,24.4,25.3)	(14.5,15.2,16.1,17)	(21.3,22.2,22.9,23.6)	(13.2,14.1,15,16.1)
	Ash	$t = 3$	%	(12.4,13.7,14.5,15.4)	(6.8,7.6,8.5,9.5)	(13.4,14.4,15.5,16.7)	(7.1,7.9,9,10)	(16.7,17.5,18.4,19.4)	(7.6,8.8,9.5,10.1)
	Moisture	$t = 4$	%	(3.2,4,4.9,5.9)	(7,7.7,8.6,9.5)	(4.5,5.2,6.1,7)	(8.1,9,9.8,10.7)	(5.4,6.2,6.8,7.6)	(9.2,9.7,10.3,10.8)
	Sulfur	$t = 5$	%	(0.48,0.56,0.64,0.72)	(0.07,0.14,0.2,0.27)	(0.52,0.58,0.62,0.68)	(0.42,0.54,0.66,0.74)	(0.53,0.56,0.6,0.65)	(0.55,0.63,0.71,0.79)
	Ash-fusion softening temperature	$t = 6$	°C	(1275,1305,1334,1360)	(940,1010,1090,1160)	(1370,1400,1435,1475)	(1045,1130,1210,1290)	(1278,1348,1430,1504)	(1145,1190,1235,1270)
	Grindability	$t = 7$	%	(41,50,59,66)	(18,25,31,38)	(45,54,62,71)	(21,28,34,41)	(50,58,66,74)	(22,28,39,47)

fuel procurement costs, and the conversion parameters from a unit of sewage sludge or coal to power and carbon emissions, as well as pollution flow rates at every coal-fired power plant. These parameters were collected from various sources, including the National Bureau of Statistics, the Energy Research Institute, and the official websites of the power plants.

7.4.3 Calculation Results

In this section, results under different scenarios are calculated and presented.

7.4.3.1 Analysis Under Different Objective Weights

Thirty-six possible sets of objective weights were tested in the proposed model for the three weights w_1, w_2, and w_3. All three were valued at 0.1, 0.2, 0.3, 0.4, 0.5, 0.6, 0.7, and 0.8 on the condition that $w_1 + w_2 + w_3 = 1$. (0.1,0.8,0.1) means $w_1 = 0.1$, $w_2 = 0.8$, and $w_3 = 0.1$. The results obtained from these 36 possible sets of objective weights are shown in Figure 7.5.

Trade-Off Among the Conflict Objectives Analysis In the upper part of Figure 7.5, it is evident that the range of economic benefits is around 7.92×10^8 to around 8.13×10^8 CNY, and six sets of objective weights have achieved the highest values. From the central part of Figure 7.5, carbon emission intensity changes from the minimum value 0.838 to the maximum value 0.8412. Compared with carbons emission intensity of the last production period (0.85), it has the potential to reduce carbon emission intensity by 1.4%. The lower part of Figure 7.5 shows about 8.2×10^5 tonnes sewage sludge that can be handled by the method of coal and sewage sludge co-combustion. Therefore, it can be seen that the bi-level multi-objective model proposed in this study for carbon emission quota allocation toward co-combustion of coal and sewage sludge has the potential to contribute significantly to sustainable development by achieving coordination among the economy, the environment, and the society. By considering various economic, environmental, and social factors, the model provides a comprehensive approach to decision-making that promotes the efficient use of resources and reduces the negative impacts of power generation on the environment and human health.

Carbon Emission Quota Allocation Strategies Analysis The carbon emission quota allocation strategies for coal-fired power plants under various sets of objective weights are presented in Figure 7.6. Ningbo CPP typically receives the highest carbon emission quota, with Jiaxing CPP and Shaoxing CPP following closely. The relatively lower carbon emission intensity of 0.817 at Ningbo CPP gives it an advantage over the other two power plants. These power plants can be classified as insensitive or sensitive based on the weight of each objective function. The carbon emission quotas allocated to Shaoxing CPP fluctuated around 2×10^6 tonnes, which indicated that Shaoxing CPP was insensitive to the objective weights. However, Ningbo CPP and Jiaxing CPP were found to be sensitive to the objective weights. Jiaxing CPP had the highest variance in carbon emission quota under different objective weights,

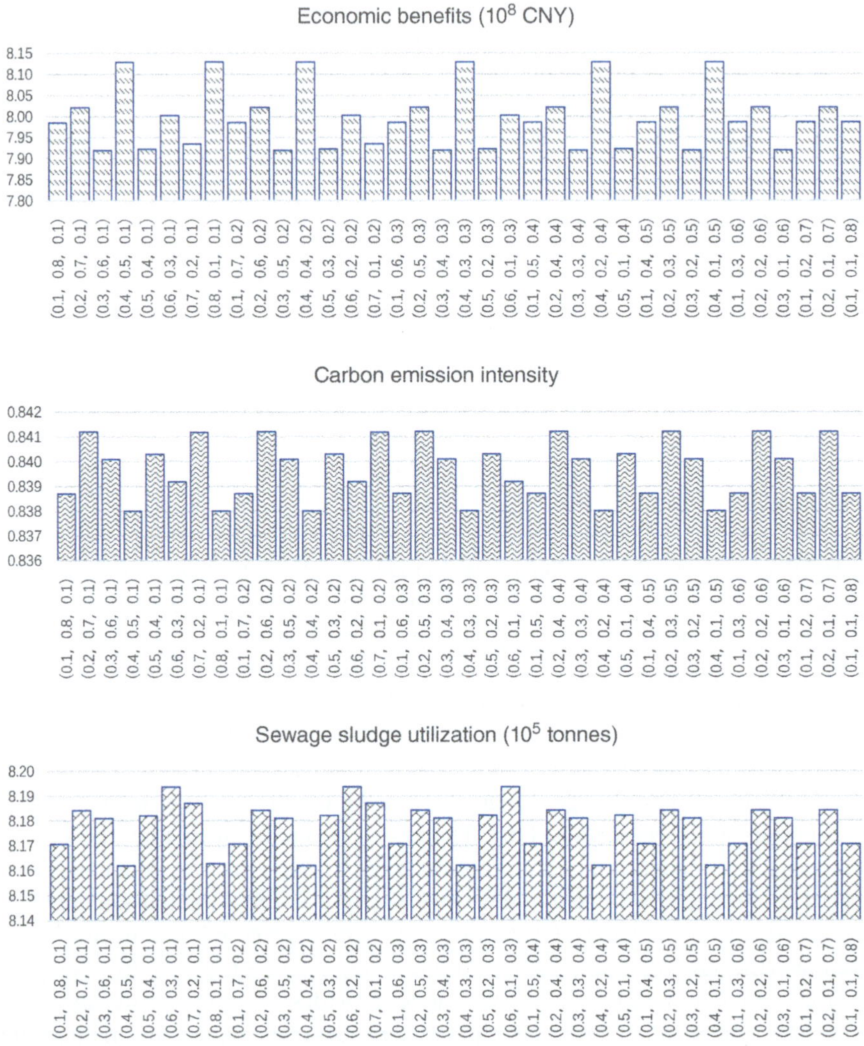

Figure 7.5 Economic benefits, carbon emission intensity, and sewage sludge utilization under different sets of objective weights.

which was 2.031×10^6 tonnes. This suggests that Jiaxing CPP was more sensitive to changes in the objective weights compared to the other two coal-fired power plants. Therefore, the carbon emission quota allocation strategies depend on the authority's preferences toward economic benefits, carbon emission intensity, and sewage sludge utilization.

7.4.4 Scenario Analysis

As previously mentioned, this scenario analysis employs balanced weights (1/3,1/3,1/3). The proposed model uses α and β to represent carbon emission

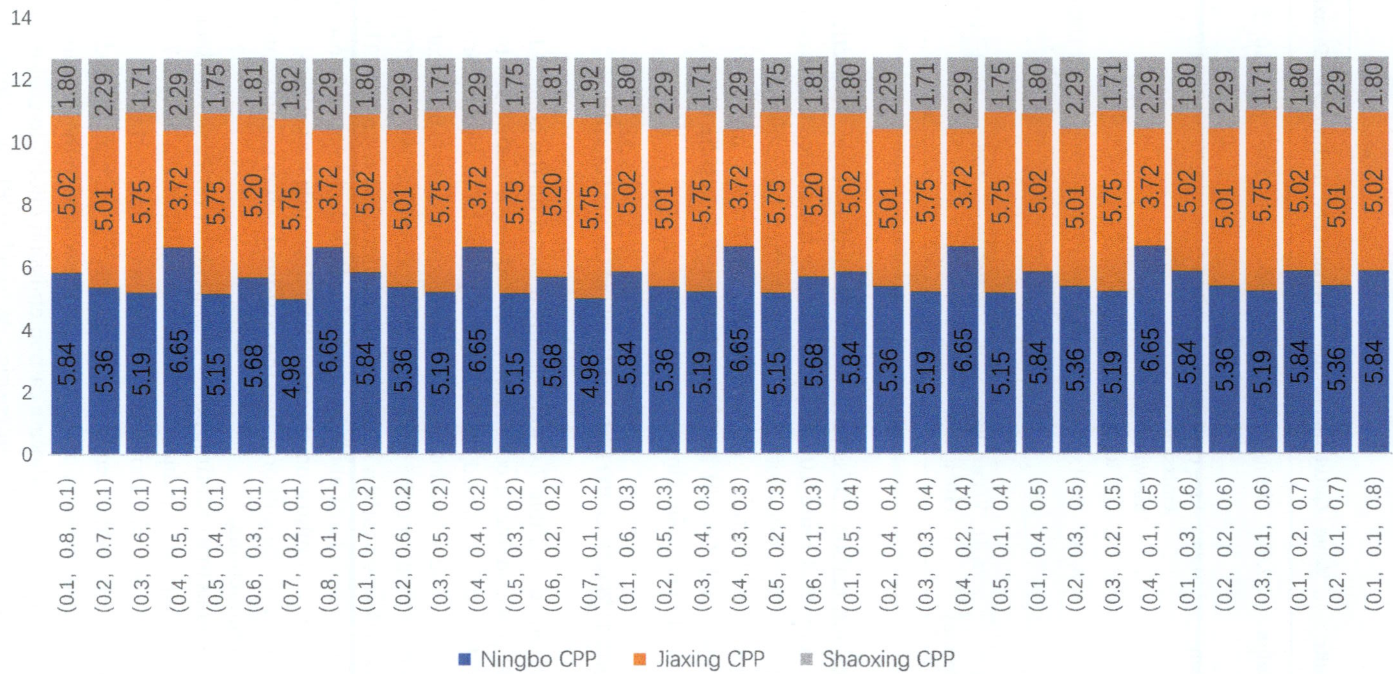

Figure 7.6 Carbon emission quotas under different sets of objective weights (unit: 10^6 tonnes).

Table 7.4 Results under balanced weights.

			Power plants		
Calculation results	**Value**	**Calculation results**	**Ningbo CPP**	**Jiaxing CPP**	**Shaoxing CPP**
Economic benefits F_1 (10^8 CNY)	7.93	Profits f (10^8 CNY)	3.06	3.76	10.73
Carbon emission intensity F_2	0.84	Free carbon emissions allowance y_i (10^6 tonnes)	5.14	5.75	1.76
Water sludge quantity F_3 (10^5 tonnes)	8.18	Coal x_{i2} (10^6 tonnes)	2.49	3.04	10.09
Allowance trade price r (CNY/tonne)	76.01	Water sludge x_{i1} (10^5 tonnes)	3.29	3.75	1.14

reduction level and carbon emission intensity reduction target, respectively. Initially, α and β were set at 1 and 0.01, respectively, which indicated that the authority wanted the total carbon emissions to be no more than the emissions in the last production period, and it aimed to reduce the carbon emission intensity by 1%. The results when $\alpha = 1$ and $\beta = 0.01$ are shown in Table 7.4. Results under different α and β are analyzed in the following.

7.4.4.1 Scenario 1: Results Under Different Levels of Carbon Emission Reductions

The results presented in Table 7.4 were obtained based on the collected data and under the total power demand constraint while varying α from 0.985 to 1. These results suggest that the proposed model has the potential to reduce carbon emissions by up to 1.5%. The allocation of carbon emission quotas is a critical factor in achieving this reduction, and the authority must trade-off between multiple objectives to develop effective carbon emission quota allocation strategies. The subordinate coal-fired power plants abide by the authority's decisions and develop coal and sewage sludge co-combustion schemes. Therefore, this hierarchical optimization structure between the authority and the coal-fired power plants contributes to carbon emission reductions.

Figure 7.7 illustrates that coal-fired power plants exhibit varied sensitivities toward changes in carbon emission reduction levels. The results from Tables 7.5 and 7.6 show that as α decreases, Ningbo CPP is granted more free carbon emission quotas, while Jiaxing CPP's carbon emission quotas remain unchanged, and Shaoxing CPP is given fewer carbon emission quotas. This is because in the decision-making process, the authority tends to allocate carbon emission quotas to coal-fired power plants with lower carbon emission intensities. As Ningbo CPP has a carbon emission intensity (0.8171) below the carbon emission intensity baseline (0.85), it is allocated

Table 7.5 Results for different α under balanced weights (1/2).

Changing α	Economic benefits F_1 (10^8 CNY)	Carbon emission intensity F_2	Water sludge quantity F_3 (10^5 tonnes)	Allowance trade price r (CNY)
−0.005	7.92	0.8403	8.1818	76.01
−0.010	7.91	0.8401	8.1811	76.05
−0.015	7.89	0.8383	8.1835	76.48

Table 7.6 Results for different α under balanced weights (2/2).

Changing α	Power plant	Profits f (10^8 CNY)	Free carbon emissions allowance y_i (10^6 tonnes)	Coal x_{i2} (10^6 tonnes)	Water sludge x_{i1} (10^5 tonnes)
−0.005	Ningbo CPP	3.09	5.16	2.47	3.27
	Jiaxing CPP	3.76	5.75	3.04	3.75
	Shaoxing CPP	1.07	1.75	1.03	1.16
−0.010	Ningbo CPP	3.11	5.18	2.48	3.29
	Jiaxing CPP	3.76	5.75	3.04	3.75
	Shaoxing CPP	1.04	1.72	1.01	1.14
−0.015	Ningbo CPP	3.47	5.53	2.65	3.52
	Jiaxing CPP	3.76	5.75	3.04	3.75
	Shaoxing CPP	0.66	1.37	0.81	0.91

greater carbon emission quotas. However, as Shaoxing CPP has the highest carbon emission intensity (0.8818), it is allocated less carbon emission quotas. Therefore, coal-fired power plants with lower carbon emission intensities are more competitive, which could encourage coal-fired power plants to focus on cleaner production.

7.4.4.2 Scenario 2: Results Under Different Carbon Emission Intensity Reduction Targets

This scenario analysis examines the impact of changing the carbon emission intensity reduction target, β. The results were obtained for values ranging from a low of 0.01 to a high of 0.04. Tables 7.7 and 7.8 provide the results for this range of values. When the emission intensity reduction target is set at 1%, the total carbon emission intensity is 0.8409, while setting the target at 2% reduces the total intensity to 0.8407. Therefore, as β increases, the total intensity finally reduces to 0.8377, clearly indicating that the bi-level multi-objective model for carbon emission quota allocation toward co-combustion of coal and sewage sludge has the ability to reduce carbon emission intensity and improve the environment.

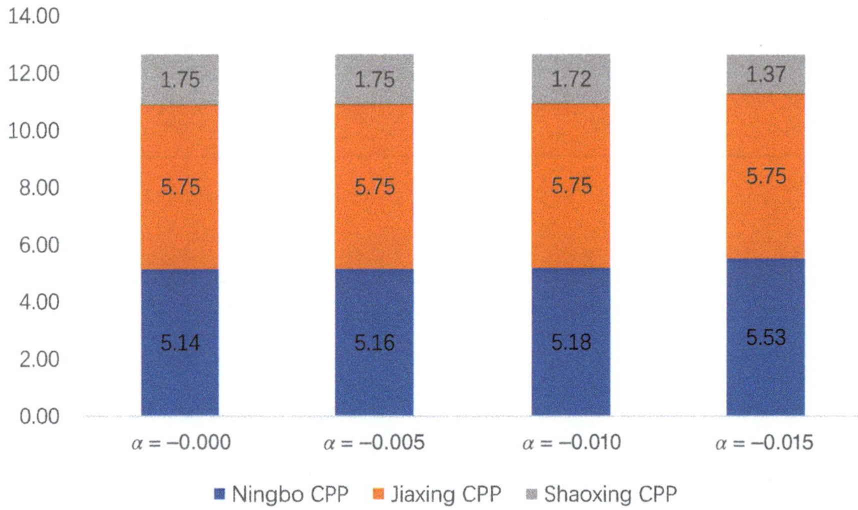

Figure 7.7 Carbon emission quotas of every coal-fired power plants with α reducing from inside to outside under balanced objective weights (unit: 10^6 tonnes).

Table 7.7 Results for different β under balanced weights (1/2).

Changing β	Economic benefits F_1 (10^8 CNY)	Carbon emission intensity F_2	Water sludge quantity F_3 (10^5 tonnes)	Allowance trade price r (CNY)
$\beta = +0.01$	7.94	0.8409	8.182	73.67
$\beta = +0.02$	7.80	0.8399	8.178	71.75
$\beta = +0.03$	7.88	0.8377	8.159	70.14

As shown in Table 7.7, β has an impact on the carbon emission quota trading price. A tighter emission intensity reduction target results in a higher quota trading price. According to Table 7.7, when β is fixed at 0.01, the quota trading price is 73.67 CNY/tonne. However, if β is increased to 0.02, the quota trading price decreases to 71.75 CNY/tonne. When the authority adopts a much stricter target ($\beta = 0.03$), the quota trading price decreases further to 70.14 CNY/tonne. As the carbon quota trading price is sensitive to β fluctuations and is also influenced by supply and demand, different carbon emission quota allocation strategies result in supply-demand changes, further affecting the trading price. Therefore, the authority can indirectly adjust the quota trading prices through its emission intensity reduction policies.

7.5 Comprehensive Discussion

In this section, a model comparison is conducted and policy implementations are proposed.

Table 7.8 Results for different β under balanced weights (2/2).

Changing β	Power plant	Profits f (10^8 CNY)	Free carbon emissions allowance y_i (10^6 tonnes)	Coal x_{i2} (10^6 tonnes)	Water sludge x_{i1} (10^5 tonnes)
$\beta = +0.01$	Ningbo CPP	3.40	5.46	2.64	3.50
	Jiaxing CPP	2.88	4.78	2.55	3.15
	Shaoxing CPP	1.67	2.29	1.36	1.53
$\beta = +0.02$	Ningbo CPP	3.38	5.42	2.65	3.51
	Jiaxing CPP	3.15	5.06	2.73	3.37
	Shaoxing CPP	1.28	1.93	1.16	1.30
$\beta = +0.03$	Ningbo CPP	4.64	6.65	3.29	4.35
	Jiaxing CPP	1.55	3.34	1.82	2.24
	Shaoxing CPP	1.69	2.29	1.39	1.57

7.5.1 Model Comparison

To compare the bi-level model and the single-level model, we formulated a single-level model that only considered the objectives of the authority, as shown in Eq. (7.15).

$$
\begin{aligned}
&\max F_1 = \sum_{i=1}^{I}\left\{\sum_{j=1}^{J}(1-T)P_jE[\widetilde{C_{ij}}]x_{ij} - \sum_{j=1}^{J}E[\widetilde{W_{ij}}]x_{ij} - z_i + (y_i - \sum_{j=1}^{J}E[\widetilde{Q_{ij}}]x_{ij})p(r)\right\}\\
&\min F_2\\
&\max F_3\\
&\text{s.t.}\begin{cases}\text{Eqs. (7.4)–(7.7)}\\ \text{Eqs. (7.9)–(7.11)}\end{cases}
\end{aligned}
\tag{7.15}
$$

Tables 7.9–7.10 provide comparative results for the bi-level model and the single-level model under balanced objective weights. Although the single-level model can provide a carbon emission quota allocation strategy that has higher economic benefits and lower carbon emission intensity, there are unfair allocations and unbalanced development. In the single-level model, Ningbo CPP and Jiaxing CPP receive higher quotas, while Shaoxing CPP receives the lowest quota. This indicates that the single-level model ignores the profits at the coal-fired power plants. However, in the bi-level model, the authority considers not only its own objectives but also those of the power plants when developing carbon emission quota allocation strategies. This means that more reasonable and acceptable carbon emission quota allocation strategies are developed when using the proposed bi-level model.

Table 7.9 Comparison between the bi-level model and the single-level model (1/2).

Models	Economic benefits F_1 (10^8 CNY)	Carbon emission intensity F_2	Water sludge quantity F_3 (10^5 tonnes)	Allowance trade price r (CNY)
Bi-level	8	0.8403	8	76.01
Single-level	8	0.8370	98	76.87

Table 7.10 Comparison between the bi-level model and the single-level model (2/2).

Changing α	Power plant	Profits f (10^8 CNY)	Free carbon emissions allowance y_i (10^6 tonnes)	Coal x_{i2} (10^6 tonnes)	Water sludge x_{i1} (10^5 tonnes)
Bi-level	Ningbo CPP	3.06	5.14	2.49	3.29
	Jiaxing CPP	3.76	5.75	3.04	3.75
	Shaoxing CPP	1.07	1.75	1.01	1.14
Single-level	Ningbo CPP	4.19	6.28	3.02	3.78
	Jiaxing CPP	3.25	5.21	2.76	3.28
	Shaoxing CPP	0.66	1.37	0.81	0.91

7.5.2 Policy Implications

Based on the above analysis and discussion, some policy implications are given in this section.

First, establishing a competition mechanism for carbon emission quota allocation is necessary. As the authority prefers to allocate a higher free carbon emission quota to coal-fired power plants that have lower carbon emission intensity, each coal-fired power plant needs to improve its carbon emission intensity performances to gain an advantage over other coal-fired power plants when competing for carbon emission quota. In this competitive process, the coal-fired power plants would compete to generate more power in carbon-constrained situations and gradually move toward cleaner production. This finding provides an important support for the Chinese government to introduce the Administrative Measures for Carbon Emission Trading (Trial) (The Ministry of Ecology and Environment of China 2020).

Second, setting flexible environmental protection policies is of great importance. To ensure sustainable development, the authority must continuously adapt environmental protection policies in response to the prevailing economic conditions. During periods of economic hardship, the authority may opt to relax carbon emission reduction requirements and intensity targets, thus promoting economic recovery. Conversely, during periods of economic stability, the authority may tighten environmental protection policies to ensure long-term sustainability.

Finally, considering the profits of coal-fired power plants is of significant importance. The authority and the coal-fired power plants are involved in carbon emission quota allocations. In the decision-making system, carbon emission quota allocation influences the relationship between the authority and the coal-fired power plants. Therefore, if profits at each coal-fired power plant are not considered, there would be unbalanced development and a poor relationship.

References

Aien, M., Hajebrahimi, A., and Fotuhi-Firuzabad, M. (2016). A comprehensive review on uncertainty modeling techniques in power system studies. *Renewable and Sustainable Energy Reviews* 57: 1077–1089.

Akdağ, A.S., Atak, O., Atimtay, A.T., and Sanin, F.D. (2018). Co-combustion of sewage sludge from different treatment processes and a lignite coal in a laboratory scale combustor. *Energy* 158: 417–426.

Çalı, S. and Balaman, Ş.Y. (2019). A novel outranking based multi criteria group decision making methodology integrating ELECTRE and VIKOR under intuitionistic fuzzy environment. *Expert Systems with Applications* 119: 36–50.

ChaoQun, M., YiShuai, R., YueJun, Z., and Basil, S. (2018). The allocation of carbon emission quotas to five major power generation corporations in China. *Journal of Cleaner Production* 189: 1–12.

China National Institute of Standardization (2010). Specification of coal used for pulverized coal-fired boiler for power generation (in Chinese).

Colson, B., Marcotte, P., and Savard, G. (2007). An overview of bilevel optimization. *Annals of Operations Research* 153(1): 235–256.

Cristóbal, J., Guillén-Gosálbez, G., Jiménez, L., and Irabien, A. (2012). Multi-objective optimization of coal-fired electricity production with CO_2 capture. *Applied Energy* 98: 266–272.

Folgueras, M.B., Díaz, R.M., and Xiberta, J. (2005). Pyrolysis of blends of different types of sewage sludge with one bituminous coal. *Energy* 30 (7): 1079–1091.

Hanson, M.A. (1981). On sufficiency of the Kuhn-Tucker conditions. *Journal of Mathematical Analysis & Applications* 80 (2): 545–550.

Holik, M., živić, M., Virag, Z., and Barac, A. (2019). Optimization of an organic rankine cycle constrained by the application of compact heat exchangers. *Energy Conversion and Management* 188: 333–345.

Hong, J., Xu, C., Hong, J. et al. (2013). Life cycle assessment of sewage sludge co-incineration in a coal-based power station. *Waste Management* 33 (9): 1843–1852.

Hong, Z., Chu, C., Zhang, L.L., and Yu, Y. (2017). Optimizing an emission trading scheme for local governments: a Stackelberg game model and hybrid algorithm. *International Journal of Production Economics* 193: 172–182.

Jeff, T. (2015). Scientists step in to assess carbon-emissions pledges. *Nature* 521 (7553): 404–405.

Jiang, J.J., Ye, B., and Ma, X.M. (2014). The construction of Shenzhen's carbon emission trading scheme. *Energy Policy* 75: 17–21.

Khaloie, H., Abdollahi, A., Rashidinejad, M., and Siano, P. (2019). Risk-based probabilistic-possibilistic self-scheduling considering high-impact low-probability events uncertainty. *International Journal of Electrical Power & Energy Systems* 110: 598–612.

Lei, K., Zhang, R., Ye, B. et al. (2020). Combustion of single particles from sewage sludge/pine sawdust and sewage sludge/bituminous coal under oxy-fuel conditions with steam addition. *Waste Management* 101: 1–8.

Lu, J., Shi, C., and Zhang, G. (2006). On bilevel multi-follower decision making: general framework and solutions. *Information Sciences* 176 (11): 1607–1627.

Magdziarz, A., Wilk, M., Gajek, M. et al. (2016). Properties of ash generated during sewage sludge combustion: a multifaceted analysis. *Energy* 113: 85–94.

Mahmoudimehr, J. and Sebghati, P. (2019). A novel multi-objective dynamic programming optimization method: performance management of a solar thermal power plant as a case study. *Energy* 168: 796–814.

Marco, S., Fortunat, J., and Stocker, T.F. (2013). Allowable carbon emissions lowered by multiple climate targets. *Nature* 499 (7457): 197–201.

Mohammed, A., Harris, I., Soroka, A., and Nujoom, R. (2019). A hybrid MCDM-fuzzy multi-objective programming approach for a G-resilient supply chain network design. *Computers & Industrial Engineering* 127: 297–312.

National Bureau of Statistics of China (2019). Provincial annual data. http://data.stats.gov.cn/easyquery.htm?cn=E0103 (accessed 26 March 2019) (in Chinese).

National Energy Administration (2019). Notice on carrying out pilot work of technical transformation of coal-fired coupled biomass power generation. http://zfxxgk.nea.gov.cn/auto84/201712/t20171204_3065.htm (accessed 17 October 2019).

Shanmugam, K., Gadhamshetty, V., Tysklind, M. et al. (2022). A sustainable performance assessment framework for circular management of municipal wastewater treatment plants. *Journal of Cleaner Production* 339: 130657.

Talari, S., Shafie-Khah, M., Wang, F. et al. (2019). Optimal scheduling of demand response in pre-emptive markets based on stochastic bilevel programming method. *IEEE Transactions on Industrial Electronics* 66 (2): 1453–1464.

Tang, L., Wu, J., Yu, L., and Bao, Q. (2015). Carbon emissions trading scheme exploration in China: a multi-agent-based model. *Energy Policy* 81: 152–169.

The Ministry of Ecology and Environment of China (2020). Administrative Measures for Carbon Emission Trading (Trial). https://www.gov.cn/zhengce/zhengceku/2021-01/06/content_5577360.htm (accessed 20 October 2023).

The United Nations (2019). The sustainable development goals. https://www.un.org/sustainabledevelopment/sustainable-development-goals/ (accessed 11 October 2019).

Tu, Y., Zhou, X., Gang, J. et al. (2015). Administrative and market-based allocation mechanism for regional water resources planning. *Resources, Conservation and Recycling* 95: 156–173.

Tyagi, V.K. and Lo, S.L. (2013). Sludge: a waste or renewable source for energy and resources recovery? *Renewable and Sustainable Energy Reviews* 25: 708–728.

Vespermann, J. and Wald, A. (2011). Much Ado about Nothing?–An analysis of economic impacts and ecologic effects of the EU-emission trading scheme in the aviation industry. *Transportation Research Part A: Policy and Practice* 45 (10): 1066–1076.

Wang, Z., Hong, C., Xing, Y. et al. (2018). Combustion behaviors and kinetics of sewage sludge blended with pulverized coal: with and without catalysts. *Waste Management* 74: 288–296.

Werle, S. and Wilk, R.K. (2010). A review of methods for the thermal utilization of sewage sludge: the polish perspective. *Renewable Energy* 35 (9): 1914–1919.

Woch, M.W., Radwańska, M., Stanek, M. et al. (2018). Relationships between waste physicochemical properties, microbial activity and vegetation at coal ash and sludge disposal sites. *Science of the Total Environment* 642: 264–275.

Xu, J. and Zhou, X. (2011). *Fuzzy-Like Multiple Objective Decision Making*. Springer.

Xu, J., Xin, Y., and Tao, Z. (2015). A tripartite equilibrium for carbon emission allowance allocation in the power-supply industry. *Energy Policy* 82 (1): 62–80.

Xu, J., Huang, Q., Lv, C. et al. (2018). Carbon emissions reductions oriented dynamic equilibrium strategy using biomass-coal co-firing. *Energy Policy* 123: 184–197.

Xu, J., Lv, C., Yao, L., and Hou, S. (2019). Intergenerational equity based optimal water allocation for sustainable development: a case study on the upper reaches of Minjiang River. *China. Journal of Hydrology* 568: 835–848.

Yang, G., Zhang, G., and Wang, H. (2015). Current state of sludge production, management, treatment and disposal in China. *Water Research* 78: 60–73.

Ye, Q., Stern, N., Tong, W. et al. (2016). China's post-coal growth. *Naturegeoscience* 9 (8): 564–566.

Yu, Y., Hong, Y., Fang, T. et al. (2018). Environmental assessment by dynamic full-field synergy model in co-combustion of coal and sludge with high water content. *Journal of Cleaner Production* 198: 54–70.

Zadeh, L.A. (1978). Fuzzy sets as a basis for a theory of possibility. *Fuzzy Sets and Systems* 1 (1): 3–28.

Zhang, Y.-J., Wang, A.-D., and Da, Y.B. (2014). Regional allocation of carbon emission quotas in China: evidence from the Shapley value method. *Energy Policy* 74: 454–464.

Zhejiang Provincial Bureau of Statistics (2019). Zhejiang statistical yearbook. http://tjj.zj.gov.cn/col/col1525563/index.html (accessed 26 March 2019) (in Chinese).

Zhou, P. and Wang, M. (2016). Carbon dioxide emissions allocation: a review. *Ecological Economics* 125: 47–59.

Zhou, X., James, G., Liebman, A. et al. (2009). Partial carbon permits allocation of potential emission trading scheme in Australian electricity market. *IEEE Transactions on Power Systems* 25 (1): 543–553.

Zolfaghari, S. and Akbari, T. (2018). Bilevel transmission expansion planning using second-order cone programming considering wind investment. *Energy* 154: 455–465.

8

Reliable–Economical Scheduling of Hybrid Solar–Hydro System

We are like tenant farmers chopping down the fence around our house for fuel when we should be using Nature's inexhaustible sources of energy - sun, wind and tide. ... I'd put my money on the sun and solar energy. What a source of power! I hope we don't have to wait until oil and coal run out before we tackle that.

Thomas A. Edison
Feb. 11, 1847 – Oct.18, 1931

There have been major difficulties in guaranteeing steady and consistent energy supplies due to the inherent uncertainty associated with renewable energy sources. As a result, developing hybrid generating systems to make use of various compensation resources has garnered more study interest. This chapter suggests a short-term optimization technique for the hybrid hydro-solar power supply to enable logical scheduling decisions based on natural limitations and policy impacts. Multi-objective optimization, which takes into account both power shortages/excesses and renewable electricity subsidy policy, is used to create short-term operation decisions that take reliable power supply and a beneficial power-generating equilibrium into consideration. Nine discrete scheduling scenarios are created to address the seasonal uncertainties of hydro and solar electricity, and trapezoidal fuzzy sets are used to characterize the uncertainty of daily water inflows and solar irradiation. The viability and effectiveness of the optimization method are demonstrated through a case study of the hybrid hydro-solar power system at Longyang Gorge in China. According to the calculations from the nine scheduling scenarios under various water inflows and solar irradiation, this multi-objective approach was successfully able to resolve conflicts between reliable and economical power supplies. Furthermore, renewable electricity subsidy policies can effectively help to adjust energy supply structure, especially in hybrid energy supply systems. To more thoroughly investigate equilibrium scheduling for hybrid power production systems, management recommendations are made for the power supplier, the electricity sector, and the administration department.

Sustainable Hybrid Energy Systems: Carbon Neutral Approaches, Modeling, and Case Studies,
First Edition. Jiuping Xu and Fengjuan Wang.

8.1 Background Review

Future global energy resources must be secured due to quick socio-economic development, ongoing population increase, and rising living standards (Chen et al. 2016). According to the "BP Statistical Review of World Energy (2018)," the world's remaining coal, natural gas, and oil reserves would only last for 134, 52.5, and 50.2 years, respectively, if fossil fuels were extracted at the rate of 2017 (BP Global 2018). As a result, it is now a global priority to develop alternative energy sources such as nuclear, biomass, wind, and solar (Drechsler et al. 2017). However, fossil fuels continue to be a major part of the world's energy supply. According to the "Key World Energy Statistics (2018)" report by the International Energy Agency (IEA), coal, natural gas, and oil collectively contribute to 86.7% of the world's total primary energy (International Energy Agency 2018). Additionally, the widespread use of fossil fuels has had detrimental repercussions such as climate change, acid rain, and photochemical smog, all of which are aggravating existing environmental issues worldwide and bolstering the case for the development of renewable energy sources (Xu et al. 2018a).

Solar energy, which is available everywhere, fossil fuel-free, and ecologically beneficial, has the largest market share of all renewable energy sources and is regarded as an excellent integration source (Huang et al. 2015). The four main solar energy technologies – solar heating, solar photovoltaic power, solar thermal electricity, and solar architecture – are being employed to address the world's most urgent energy issues (Singh 2013). The data made available by the IEA showed that the total world solar photovoltaic (PV) electricity production was 328 TWh in 2016, which was a 32.79% increase over that in 2015 (International Energy Agency 2017). According to a recent report from the International Renewable Energy Agency (IRENA), the total global solar photovoltaic net installed capacity had reached a 400 GW milestone by the end of 2017 (International Renewable Energy Agency 2018). However, there are still several obstacles to ensuring dependable, continuous solar electricity supply because solar energy is intermittent, has high volatility, and has low solar irradiation predictability (Cervantes and Choobineh 2018).

Modern solar irradiation forecast system has been largely developed over the past decades with numerous studies addressing the uncertainty of solar PV. For example, using a statistical time series method, Yona et al. (2013) developed PV output forecasting based on 1-day ahead insolation predictions. Almonacid et al. (2014) proposed a methodology that predicted 1-h ahead power PV generator output. Applying a physical method, Fernandez-Jimenez et al. (2012) developed a short-term prediction system for hourly electrical energy generation in real grid-connected photovoltaic (GCPV) plant, and Alonso-Montesinos et al. (2015) developed short- and mid-term solar irradiation forecasting approaches for all sky conditions. In general, accurate information on solar source forecasts can remove uncertainty in solar PV output, increase system stability, and save costs associated with ancillary device maintenance, all of which help grid operators address the issue of solar power shortages. However, more thorough investigation and development are still needed to effectively predict solar irradiation (Lan et al. 2018).

Hybrid solar-storage generation systems have therefore been playing a more important role in providing efficient, reliable power through an adjustment of the charge and discharge times and rates to increase the correlation between PV supply and electricity demand (Sun et al. 2019). For example, Hemmati et al. (2017) presented an optimal planning and scheduling method for hybrid wind–solar power system energy storage systems. Reddy (2017) solved an optimal scheduling problem for hybrid thermal–wind–solar power systems with storage. Sabooria et al. (2017) reviewed state-of-the-art energy storage planning for electric power distribution networks. The technical maturity and economic effectiveness of energy storage systems have grown rapidly over the past decades as they help provide reliable generation systems. But large-scale applications are still hampered by drawn-out construction and expensive investment prices, particularly in developing nations like China and India.

Numerous studies have been conducted on integrating solar energy with existing power sources. For instance, hydropower has been discovered to be the perfect complement to solar photovoltaic since it is simple to start and stop power generation and can swiftly adapt to load changes (Chang et al. 2017). Dujardin et al. (2017) proved that hydropower was able to effectively deal with intermittent energy sources. François et al. (2015) conducted a sensitivity analysis on the climate characteristics of a 100% renewable energy mix and found that a large hydropower share could reduce energy balance variability at hourly scales. Ogueke et al. (2016) conducted a potential evaluation study of small hydro-photovoltaic hybrid systems in Nigeria and found that PV systems were a practical hydropower supplement to fill hydropower supply gaps. While there has been significant progress in the creation of workable hybrid hydro-solar generation systems, most of these developments have been in the realm of small-scale or off-grid power systems, where the demands for power generation quality are less stringent than in large-scale grid-connected systems (Li and Qiu 2016).

The modeling and optimization of large-scale grid-connected hydro-solar production systems has therefore received more attention in research due to the quick technological advancement in utility-scale photovoltaic energy sources. For instance, Yang et al. (2016) assessed the dispatch flexibility of coordinated large-scale solar and hydro generation. Ming et al. (2017) optimized the size of a utility-scale photovoltaic plant to facilitate the integration of photovoltaic in hydro-solar generation systems. Fang et al. (2017) optimized the size of a utility-scale photovoltaic generation to enhance the quality of the photovoltaic power. Li and Qiu (2016) developed a long-term operation model that took annual power generation as well as the regularity of power output into account. Yang et al. (2018) developed long-term operating rules for large-scale hydro–photovoltaic power systems to increase operational effectiveness. All of these studies have significantly aided the integration of large-scale photovoltaic energy into current hydropower systems, resulting in increased power generation with reduced power fluctuation and greater operational flexibility. Few studies, meanwhile, considered both reliable power supply and beneficial power generation while taking into account natural limitations and policy impacts.

Therefore, this chapter tries to suggest a short-term scheduling strategy for large-scale solar–hydro power generation systems under natural limitations and policy impacts. This strategy balances reliable power supply and beneficial power generation while also taking into account natural solar irradiation uncertainty, seasonal water flow limitations, and policy impacts on renewable electricity.

8.2 Key Problem Statement

Compensations of the hybrid solar–hydro power generation system, trade-off between reliable and economical power supply, and methods to handle renewable energy uncertainties will be described here.

8.2.1 System Description

In the hybrid solar–hydro power generation system considered in this chapter, the generation sources to be dispatched by the power supplier are fully constructed, geographically close and managed by the same firm (Beluco et al. 2012). As an illustration, consider the Longyang Gorge hybrid solar–hydro project in Qinghai, China, where the hydro and solar power plants are dispatched simultaneously by the Huanghe Hydropower Development Co., Ltd. The hydropower plant serves as a backup for the photovoltaic plant. Hydropower plants can reduce the intermittent nature of solar power by altering their production in response to changes in the photovoltaic power output caused by the weather or time of day. This ensures a steady and reliable supply of electricity. Additionally, hydropower's quick response time helps make up for variations in projected solar irradiation (Yang et al. 2016); see some details in Figure 8.1. Hydropower generation, however, is higher during the summer flood season and lower during dry seasons because of its strong seasonality and significant annual variations; in contrast, solar power output has weaker annual variability, with lower power output in summer and higher output in winter. Therefore, hydropower can make up for the hourly and daily variations in

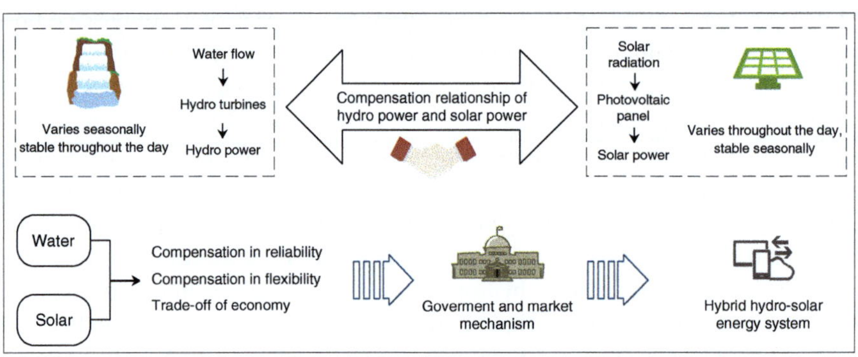

Figure 8.1 Compensations of the hybrid solar–hydro power generation system.

solar power production, and solar power can make up for the seasonal and annual variations in hydropower output (Zhang et al. 2019).

8.2.2 Trade-Off Between Reliable and Economical Power Supply

Determining reliable and economical power supply equilibrium under natural limitations and policy impacts remains a complex problem, as shown in Figure 8.2. Solar power plants can only produce electricity during the daytime because of the nature-driven intermittency; as a result, there is a limited capacity for load following, which leads to dispatch inflexibility. Fast start and stop hydro turbines, on the other hand, have the potential to ramp up quickly, with water reservoir storage offering solid backup for fluctuating power supply. Hydropower can therefore be offered as a quick backup method of compensating for solar power loss. If the company seeks more flexible and reliable operation schedules, more hydropower access would be necessary. However, economics is also an important goal that companies must pursue (Kok et al. 2018). As stated in the *Solar Power on-grid Price Policy ([2017] No.2196)*, the on-grid price for solar PV power is 0.55–0.75 CNY/kWh (Chinese National Development and Reform Commission 2017). But the on-grid price for hydropower is only 0.25–0.40 CNY/kWh, far less than that of solar power. Therefore, if a company seeks higher electricity sales revenue, more solar power access would be better. It has been discovered that multi-objective programming is particularly effective in resolving trade-offs between objectives. For instance, Chen et al. (2017) developed a multi-objective model to deal with a short-term hydro-thermal-wind economic emissions dispatch problem. Barakat and Sharma (2018) proposed a multi-objective optimization for bulldozers and their blades for soil cutting. Hafezalkotob et al. (2017) developed a multi-resource trade-off between project contractors in a cooperative environment. Encouraged by these successful applications, the multi-objective method is used to optimize the scheduling of hybrid solar–hydro power under reliable and economical objectives.

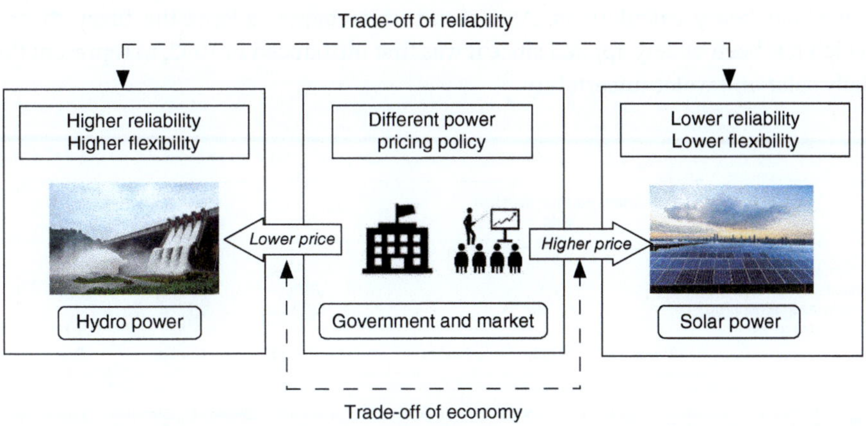

Figure 8.2 Trade-off between reliability and economy in the hybrid solar–hydro power system. Source: Thaiprayboy/Adobe Stock Photos (Hydro power); 安琦 王/Adobe Stock (Solar power).

8.2.3 Handling Renewable Energy Uncertainties

Representing and handling the climate-driven hydro and solar power uncertainties is also a key problem (Han et al. 2019). Because of geographical location and climate change influences, hydropower has strong seasonality and annual variability, on the other side, although solar energy is abundantly available, seasonal variations and weather conditions make it highly unpredictable in terms of consistency (Snaith and Hacke 2018). Therefore, it is unreasonable to have a single predetermined scheduling scenario since the strong natural properties of solar and hydropower both make generation scheduling challenging. Using weather forecasts and historical weather data, Séguin et al. (2017) developed a scenario tree to represent hydro inflow uncertainties, and Yang et al. (2016) considered the stochastic nature of solar irradiance using forecasting error distribution and the Monte Carlo Method. Although these research projects significantly reduced hydro and solar power uncertainties by using stochastic approaches or robust optimization, they were burdened by high computing burdens (Wang et al. 2008), or sometimes at the expense of the economy (Chen et al. 2012). Therefore, to overcome these problems, this chapter proposes nine typical operating schedule scenarios for hybrid solar–hydro power plants (see Figure 8.3), where three water inflow conditions (flood, normal, and dry) and three weather conditions (sunny, cloudy, and rainy) are considered.

However, even in determined scenarios, there are solar irradiation and water inflow uncertainties. For example, in natural conditions, minute-to-minute variations in solar irradiation can cause significant variations in solar power output, and even cloud shadowing can abruptly alter solar power production. These factors all pose serious challenges for complementary hybrid system scheduling. Reddy (2017) used bi-model distribution to model daily solar irradiance uncertainties, while (Zubo et al. 2018) employed scenario-tree techniques. Their efforts are very important in scientifically modeling the solar irradiation uncertainty, but require somewhat heavy calculations. As a result, this chapter adopts the fuzzy theory, which has been widely applied since it was first introduced in 1993, to represent the daily solar and water uncertainty.

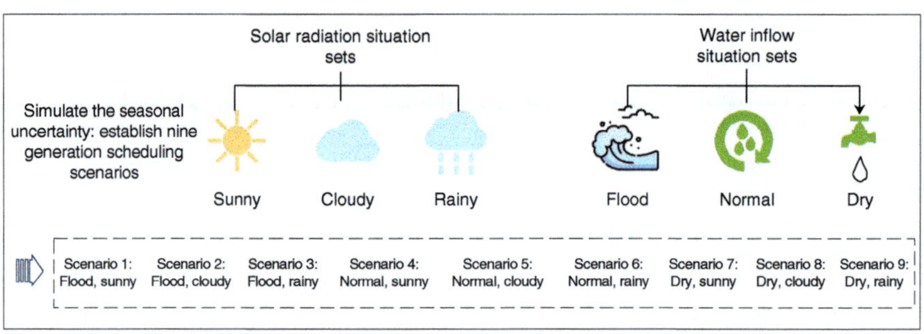

Figure 8.3 Method to handle renewable energy associated uncertainties.

8.3 Modeling

Nine typical scheduling schemes and fuzzy theory are applied to provide a method to deal with the seasonal uncertainty related to the weather-driven characteristics of hydropower and solar power generation. However, in each case, reliable economic trade-offs need to be properly managed to determine reasonable scheduling decisions. Therefore, this section establishes the mathematical model for short-term dispatching of hydropower–photovoltaic hybrid system.

8.3.1 Notations

(1) Indices

i Index for photovoltaic power plant, $j = 1, 2, \ldots, I$.
k Index for water turbines, $i = 1, 2, \ldots, K$.
t Index for time period, $n = 1, 2, \ldots, 24$.

(2) Decision variables

h_{kt} Water discharged through hydro turbine k during time t.
s_{it} Photovoltaic arrays to be operated at solar park i during time t.
O_{kt} On/off status of water turbine k at time t, binary variable.

(3) Certain parameters

Q_t Quantity of water left in the reservoir at time period t.
$\overline{S}_i$ Maximum available photovoltaic arrays at photovoltaic park i.
$\underline{Te}_i$ Minimum technical power output at photovoltaic power plant i.
$\overline{Te}_i$ Maximum technical power output at photovoltaic power plant i.
D_t Power demand at time t.
$\underline{F}_k$ Minimum discharging water flowing into the hydro turbine k.
$\overline{F}_k$ Maximum discharging water flowing into the hydro turbine k.
S_0 The solar irradiation in the standard test conditions.
H Water head of hydropower station.
ϕ'_t Total power not supplied at time t.
a^- Lower transmission variation bound of the transmission line.
a^+ Upper transmission variation bound of the transmission line.
$\overline{a}$ Themaximum transmission capacity of the transmission line.
c_k^{rh} Rated power output of hydro turbine k.
c_i^{rs} Rated output of photovoltaic arrays.
p_{kt} Power output of hydro turbine k at time t.
p_{it} Output of photovoltaic power plant i at time t.
ϕ_t Total power supplied at time t.

$A_{\min}$ Minimum available generation water in the hydropower plant.
$A_{\max}$ Maximum available generation water in the hydropower plant.
T_s Temperature under standard test conditions, set as 25°C.
p^h Electricity online price of hydropower.
p^s Electricity online price of solar power.
e Penalty for each megawatt power not supplied.
α_s Temperature coefficients of solar cells, which is −0.35%/°C.
ρ The specific weight of water, 9.81 kN/m^3.
Δt Time conversion factor from hour to second, equals to 3600.
λ_i Comprehensive output coefficient for the hydro turbine k.

(4) Uncertain parameters

$\widetilde{W}_t$ Water flow of hydropower plant at time t.
$\widetilde{R}_{it}$ Irradiation observed in photovoltaic power plant i at time t.
$\widetilde{T}_{it}$ Actual air temperature at the photovoltaic power plant.

8.3.2 Hybrid System's Reliability and Economy Equilibrium

As previously mentioned, a bigger proportion of hydropower access would be required if the power plant pursued schedules with flexible and stable operation, but a higher proportion of solar power access would be more effective if the plant sought higher energy sales revenue. For power suppliers, however, earnings from power sales and reliability of supply are essential. Consequently, it is essential to take these two goals into account simultaneously.

8.3.2.1 Maximize Power Supply Reliability

A steady supply of electricity can guarantee resource preservation and a healthy economy. The operation reliability of hybrid power generation systems must therefore be guaranteed, especially for those systems that use uncertain renewable energy. As stated in the main problem statement, PV quickly transforms radiation into electric energy, which causes a significant level of output uncertainty. Even if hydropower provides reasonable peak shaving capacity on the dispatching day, there are uncertainties in rainy and dry seasons. In literature, Ming et al. (2017) achieved system reliability by maximizing the total number of times when the system power output was more than the necessary output, and Yang et al. (2018) defined system output reliability in the same way. These approaches define power supply reliability by scheduling more power than required, but new operation rules are crucial to enhancing operational performance. Therefore, this chapter takes into account both power shortage and surplus, and it holds that the the closer the supply and demand are balanced, the better. However, if there is more power generated than necessary, it is also a waste of resources and energy (Lv et al. 2016).

Let p_{kt} and p_{it} represent the output of hydro station k and PV park i at time t. and D_t represent the power demand at time t Maximizing operational reliability means reducing the amount of power that is exceeded or not provided. Therefore, the objective of minimizing power supply demand risk can be written as Eq. (8.1).

$$\min f_1 = \sum_t \left(\sum_k p_{kt} + \sum_i p_{it} - D_t \right)^2 \tag{8.1}$$

8.3.2.2 Maximize Electricity Sales Revenue

The pursuit of greater and more steady profitability is the continual aim of power producers. However, the optimization objective has changed from maximizing total power generation to maximizing net power supply income as a result of changes in the power supply industry and power market conditions (Lohmann and Rebennack 2017). For example, Li and Qiu (2016) tried to maximize the total power generation of hydropower–photovoltaic systems, but soon, Fang et al. (2017) sought to maximize the life-cycle profit of solar–hydro hybrid power generation system. These studies lay the groundwork for modeling the hybrid power generation system's economic objective. However, the feed-in tariff policy of hydropower and solar power generation becomes more complicated when taking into account varying levels of technological maturity and application scope. As discussed before, under the renewable energy subsidy policy scheme, the on-grid price of PV power generation is higher than hydropower's. Therefore, a higher percentage of solar energy connected to the grid will undoubtedly result in better power sales revenue for hybrid generators with both hydropower and PV power plants. Let p_{kt} be the power output of hydro turbine k at time t, and p^h be the electricity online price of hydropower. Therefore, the financial returns from the hydropower generation is $\sum_i \sum_n p_{kt} p^h$. Let p_{it} be the power output of photovoltaic park i at time t, and p^s be the electricity online price of solar power. Therefore, the financial returns from solar power generation is $\sum_j \sum_n p_{it} p^s$. Let ϕ_t' be the quantity of power that is not supplied by the hybrid system, which equals to the difference of power demand and supply when $D_n \geq \left(\sum_i p_{kt} + \sum_j p_{it} \right)$ and equals to zero when $D_n \leq \left(\sum_i p_{kt} + \sum_j p_{it} \right)$. Let e be the penalty; therefore, the total penalties is $\sum_t \phi_t' e$. Therefore, the objective to maximize power supply financial returns is Eq. (8.2).

$$\max f_2 = \sum_k \sum_t p_{kt} p^h + \sum_i \sum_t p_{it} p^s - \sum_t \phi_t' \cdot e \tag{8.2}$$

8.3.3 Constraints of the Hybrid System

In this section, the constraints for the solar power plant, the hydropower plant, and the power transmission will be described in detail.

8.3.3.1 Photovoltaic Power Plant's Output

Solar radiation and temperature determine the PV power plant's actual output. Let $\widetilde{R}_{it}$ represent the solar irradiation observed in photovoltaic power plant i at time

t; $\widetilde{T_{it}}$ be the actual air temperature at the photovoltaic power plant where the PV modules are operating. c_i^{rs} represent the rated power output of solar arrays; S_0 represent the solar irradiation in standard test conditions, set as 1 kw/m^2 and α_s be the temperature coefficients of power output of solar cells, which is −0.35%/°C in this chapter; T_s is the temperature under standard test conditions, set as 25°C; and s_{it} be the photovoltaic arrays to be operated at solar park i at time t, which is the decision variable. Using the model developed by the National Renewable Energy Laboratory (NREL) (Li and Qiu 2016), the output of a photovoltaic power plant can be written as Eq. (8.3).

$$p_{it} = s_{it} c_i^{rs} \left(\frac{E\left(\widetilde{R}_{it}\right)}{S_0} \right) \left[1 + \alpha_s \left(E\left(\widetilde{T}_{it}\right) - T_s \right) \right] \tag{8.3}$$

The parameters $E\left(\widetilde{R}_{it}\right)$ and $E\left(\widetilde{T}_{it}\right)$ are expected values of $\widetilde{R}_{it}$ and $\widetilde{T}_{it}$. As described in Section 8.2.3, trapezoidal fuzzy numbers can be used to represent solar radiation and air temperature. In order to achieve the estimation of fuzzy parameters, academics have put forth numerous efficient approaches for calculating the average of these quantities. Of which, the expected value operator proposed by Xu and Zhou Xu and Zhou (2011) has been widely applied in energy-and resourcerelated areas and obtained satisfactory results. In Xu's method, a trapezoidal fuzzy number can be represented by its corresponding expected value using the following operator: $\widetilde{H} \to E\left[\widetilde{H}\right] = \frac{1-\varphi}{2}(a+b) + \frac{\varphi}{2}(c+d)$, in which $0 \le \varphi \le 1$ is confidence level. By applying the operator, the expected value of fuzzy number can be obtained.

$$\widetilde{R}_{it} \to E\left[\widetilde{R}_{it}\right] = \frac{1-\varphi}{2}\left(r_{it1} + r_{it2}\right) + \frac{\varphi}{2}\left(r_{it3} + r_{it4}\right) \tag{8.4}$$

$$\widetilde{T_{it}} \to E\left[\widetilde{T}_{it}\right] = \frac{1-\sigma}{2}\left(t_{it1} + t_{it2}\right) + \frac{\sigma}{2}\left(t_{it3} + t_{it4}\right) \tag{8.5}$$

In which $\widetilde{R}_{it} = \left(r_{it1}, r_{it2}, r_{it3}, r_{it4}\right)$ and $r_{it1} \le r_{it2} \le r_{it3} \le r_{it4}$. Meaning that the maximum value of $\widetilde{R}_{it}$ can be g_{it4}, the minimum value is g_{it1}, and the most likely value is between g_{it2} and g_{it3}. Similarly, the parameter $\widetilde{T}_{it} = \left(t_{it1}, t_{it2}, t_{it3}, t_{it4}\right)$, in which $t_{it1} \le t_{it2} \le t_{it3} \le t_{it4}$. t_{it4} is the maximum available value of $\widetilde{T_{it}}$, t_{it1} is the minimum available value, and the most likely value of $\widetilde{T_{it}}$ is between t_{it2} and t_{it3}.

8.3.3.2 Accessible Photovoltaic Arrays

For solar parks, there are a certain number of PV arrays that can be used. Let $\overline{S_i}$ represent the installed PV arrays; this constraint can be written as Eq. (8.6).

$$s_{it} \le \overline{S_i} \tag{8.6}$$

8.3.3.3 Solar Power Output Limitation

Technical power output for the PV power plant is constrained by real-world factors. Therefore, let $\underline{Te_i}$ and $\overline{Te_i}$ represent the minimum and maximum value of technical power output at photovoltaic power plant i, this constraint is Eq. (8.7).

$$\underline{Te_i} \le p_{it} \le \overline{Te_i} \tag{8.7}$$

8.3.3.4 Hydro Turbine Output

Let λ_i represent the comprehensive coefficient of power output for the hydro turbine k; c_k^{rh} represent the rated power output of hydro turbine k; H represent the water head (m); and $\underline{F}_k$ and $\overline{F}_k$ represent the minimum and maximum discharging water flowing into the hydro turbine k, respectively, which means that the water flow thoughthe hydro turbine k must be larger/lower than this parameter. Decision variables, including h_{kt} and O_{kt}, represent the water discharged through the hydro turbine k during time t $\left(\mathrm{m}^3/\mathrm{s}\right)$ and its on/off status $O_{kt} = 1$ means turn on, otherwise $O_{kt} = 0$. The specific weight of water ρ is 9.81 kN/m^3, then p_{kt}, which represents the power to be generated by hydro turbine k at time t, can be written as Eq. (8.8).

$$p_{kt} = \begin{cases} 0, h_{kt} \leq \underline{F}_k O_{kt} \\ \eta_i \rho h_{kt} H O_{kt}, \underline{F}_k O_{kt} \leq h_{kt} \leq \overline{F}_k O_{kt} \\ c_k^{rh} O_{kt}, \overline{F}_k O_{kt} \leq h_{kt} \end{cases} \tag{8.8}$$

8.3.3.5 Limitation on Available Water

Since the water flow is depicted in seconds, and the time conversion factor of hour to second $\Delta t = 3600$, then the total amount of water used for power generation in the observed time period can be expressed as $\sum_i \sum_n O_{kt} h_{kt} \Delta t$. However, total amount of water available for power generation has upper and lower limits. Let $A_{\min}$ and $A_{\max}$ represent the minimum and maximum amount of water available for power generation, then these constraints can be written as Eq. (8.9).

$$A_{\min} \leq \sum_i \sum_n O_{kt} h_{kt} \Delta t \leq A_{\max} \tag{8.9}$$

8.3.3.6 Dynamic Water Inventory

Let $\widetilde{W}_t$ be the water flow into water reservoir at time t $\left(\mathrm{m}^3/\mathrm{s}\right)$, to handle the climate- and rainfall-caused uncertainty of water inflows, this uncertain parameter is represented as a trapezoidal fuzzy number $\widetilde{W}_t = \left(w_{t1}, w_{t2}, w_{t3}, w_{t4}\right)$, where $w_{t1} \leq w_{t2} \leq w_{t3} \leq w_{t4}$. w_{t4} is the maximum available value of $\widetilde{W}_t$, w_{t1} is the minimum available value, and the most likely water flow is between w_{t2} and w_{t3}. Based on the calculation method proposed by Xu and Zhou (2011), a corresponding expected value $E\left[\widetilde{W}_t\right]$ can be calculated as follows, and $0 \leq \tau \leq 1$.

$$\widetilde{W}_t \rightarrow E\left[\widetilde{W}_t\right] = \frac{1-\tau}{2}\left(w_{t1} + w_{t2}\right) + \frac{\tau}{2}\left(w_{t3} + w_{t4}\right) \tag{8.10}$$

Therefore, the total water flow into the reservoir can be written as $E\left[\widetilde{W}_t\right] \Delta t$, where the Δt is time conversion parameter for hour to second, which equals to 3600. During time period t, the water quality flow out of the reservoir to generate electricity in total can be written as $\sum_i h_{kt} O_{kt} \Delta t$. Let Q_t represent the water that left in reservoir at time t, and $\overline{a}$ is the reservoir water storage at the beginning of the scheduling period, then the dynamic water inventory constraint can be written as Eq. (8.11).

$$Q_t = Q_{t-1} + E\left[\widetilde{W_t}\right]\Delta t - \sum_i h_{kt} O_{kt} \Delta t$$
$$Q_1 = \overline{a} + E\left[\widetilde{w_1}\right]\Delta t - \sum_i h_{kt} O_{kt} \Delta t \tag{8.11}$$

8.3.3.7 Limit on the Ability of Power Transmission

There are upper and lower transmission capacity restrictions for every transmission line. The power transmitted at time t cannot exceed the maximum bound in order to ensure a constant and steady power supply. Let $\overline{a}$ represent the maximum transmission capacity of the transmission line. The constraint can then be written as Eq. (8.12).

$$\phi_t \leq \overline{a} \tag{8.12}$$

8.3.3.8 Limit on the Stability of Power Transmission

Because too much fluctuation from renewable energy sources could jeopardize the stability of the power grid, the range of power output variation must be restricted and system resilience must be increased (Ju et al. 2016). Let a^- represent the lower transmission variation bound of the transmission line, and a^+ represent the upper variation bound; then this constraint can be written as Eq. (8.13).

$$a^- \leq C_t - C_{t-1} \leq a^+ \tag{8.13}$$

8.3.4 Global Model

For hybrid power suppliers, it is challenging to come up with a feasible schedule that can achieve both a stable power supply and cost-effective power generation. This is due to the unpredictability of hydropower and solar energy resources as well as the regulation of government regulations. This chapter uses the nonlinear multi-objective model along with integer variables and piecewise functions to optimize the supplier's plan in order to resolve the issue. The power excess or shortage and renewable energy subsidy policies are taken into consideration, and the trade-off between reliable and economical generation scheduling is resolved using the multi-objective approach. To address the seasonal uncertainty associated with hydropower and the weather-dependent characteristics of solar power generation, nine typical dispatching scenarios are proposed. Fuzzy numbers and the expected value operator are used to deal with the daily uncertainty in solar and hydropower generation. By integrating the objectives and corresponding constraints of the model, a global optimization model is established to determine the discharge of the turbine, the on/off state of the turbine, and the operation of the photovoltaic array of the photovoltaic power station, so as to provide the best operation decision for the hybrid power supplier.

$$
\begin{aligned}
&\min f_1 = \textstyle\sum_t\left(\sum_k p_{kt} + \sum_i p_{it} - D_t\right)^2 \\
&\max f_2 = \textstyle\sum_k\sum_t p_{kt}p^h + \sum_i\sum_t p_{it}p^s - \sum_t \phi_t' \cdot e \\
&\text{s.t.}\begin{cases}
\phi_t = \sum_k p_{kt} + \sum_i p_{it} \\
p_{it} = s_{it}c_i^{rs}\left(\frac{E(\widetilde{R}_{it})}{S_0}\right)\left[1 + \alpha_s\left(E\left(\widetilde{T}_{it}\right) - T_s\right)\right] \\
\widetilde{R}_{it} \to E\left[\widetilde{R}_{it}\right] = \frac{1-\varphi}{2}\left(r_{it1} + r_{it2}\right) + \frac{\varphi}{2}\left(r_{it3} + r_{it4}\right) \\
\widetilde{T}_{it} \to E\left[\widetilde{T}_{it}\right] = \frac{1-\sigma}{2}\left(t_{it1} + t_{it2}\right) + \frac{\sigma}{2}\left(t_{it3} + t_{it4}\right) \\
s_{it} \le \overline{S}_i \\
\underline{Te}_i \le p_{it} \le \overline{Te}_i \\
p_{kt} = \begin{cases} 0, h_{kt} \le \underline{F}_k O_{kt} \\ \eta_i \rho h_{kt} H O_{kt}, \underline{F}_k O_{kt} \le h_{kt} \le \overline{F}_k O_{kt} \\ c_k^{rh} O_{kt}, \overline{F}_k O_{kt} \le h_{kt} \end{cases} \\
A_{\min} \le \sum_i\sum_n O_{kt}h_{kt}\Delta t \le A_{\max} \\
W_t \to E\left[\widetilde{W}_t\right] = \frac{1-\tau}{2}\left(w_{t1} + w_{t2}\right) + \frac{\tau}{2}\left(w_{t3} + w_{t4}\right) \\
Q_t = Q_{n-1} + E\left[\widetilde{W}_t\right]\Delta t - \sum_i h_{kt}O_{kt}\Delta t \\
Q_1 = \overline{a} + E\left[\widetilde{w_1}\right]\Delta t - \sum_i h_{kt}O_{kt}\Delta t \\
\phi_t \le \overline{a} \\
a^- \le C_t - C_{t-1} \le a^+
\end{cases}
\end{aligned}
\tag{8.14}
$$

This model has a more comprehensive, organized structure than earlier optimal hybrid generation system approaches. First, in order to minimize the drawbacks of single generation, two climate-related power plants–solar and hydropower plantsare coordinated. Second, in order to achieve flexible power dispatch, a dependable power supply, and advantageous power generation, a multi-objective optimization method is further applied to the hybrid solar–hydro power generation system. Third, interactive water flow scenarios and weather condition scenario sets are constructed for nine alternative operating schedules to deal with the natural water flows and solar irradiation uncertainties and limits. The mathematical form of global model Eq. (8.14) is suitable for every discrete, determined scheduling scenario we proposed one only needs to adjust the corresponding calculation data when solving the model in other scenarios. Fourth, fuzzy numbers are used to characterize the seasonal water flows and daily solar irradiation for each scenario in order to ensure more scientific and accurate judgments. A particular Chinese pricing policy condition for renewable energy is also taken into consideration.

8.3.5 Model Solving

It is very difficult to find an optimal solution that satisfies all objectives, as these objectives are usually conflicting with each other (Lv et al. 2016). Many previous works have discussed multi-objective programming solution approaches, from classical optimization algorithms to intelligent optimization algorithms (Cui et al. 2017). Of which, the *ε-constraint method*, which was first proposed by Haimes et al. (1971) and enhanced by Chankong and Haimes (1983), has been proven to be a very valuable method with wide range of successful applications. This method is based on the maximization of one objective function, and considering other objectives as constraints bounded by some allowable level ε. Then the level ε may be altered to generate the entire Pareto-optimal set (Guillén et al. 2005).

As previously stated, the pursue of higher revenue and steadier economic development are unchanged goals for power suppliers; therefore, it is obligated for power suppliers to increase financial revenue to ensure steady enterprise survival and normal social development. Consequently, some power suppliers may transform their intention of minimizing power shortage/excess into an acceptable range. Let ε be the upper bound of power suppliers' reliability purpose, then the multi-objective model in Eq. (8.14) can also be transformed into Eq. (8.15).

$$\begin{aligned} &\max f_2 \\ &\text{s.t.} \begin{cases} \text{Eqs. (8.3)–(8.13)} \\ f_1 \leq \varepsilon \end{cases} \end{aligned} \tag{8.15}$$

In this problem, the real-world meaning of parameter ε can be further extended to $\beta \times L$, as shown in Eq. (8.16), where β ensures that the square of the unsupplied power is less than or equal to β times parameter L, and L is the square of total unsupplied power in the previous scheduling cycle.

$$\begin{aligned} &\max f = \textstyle\sum_k \sum_t P_{it} w_h + \sum_i \sum_t P_{jt} w_s - \sum_t L_t w \\ &\text{s.t.} \begin{cases} \sum_t \left(\sum_k P_{it} + \sum_i P_{jt} - P_{dt}\right)^2 \leq \beta L \\ \text{Eqs. (8.3)–(8.13)} \end{cases} \end{aligned} \tag{8.16}$$

By introducing the allowable level parameters ε_1 and ε_2, the global model can be transformed into its equivalent single form with integer variables, piecewise functions, and dynamic water inventory. Lingo software, which can easily analyze, modify, and solve mathematical optimization problems, has been proven very effective in solving the single-objective model. For example, Xu et al. (2018b) used Lingo solver to solve a bi-level multi-objective waste load allocation model for river water pollution control problem. Ware et al. (2014) used Lingo software to solve a dynamic supplier selection problem modeled as a mixed-integer non-linear program. Therefore, this software can be applied to solve the transformed model Eq. (8.16).

In summary, the model solving approach can be described as follows:

- Select one objective $f_i(x)$ as the optimized objective and the other $f_j(x)$ as constraints of the optimization problem, and transform the model in Eq. (8.14) into its equivalent single-objective form, as shown in Eq. (8.16).

- Set estimated bound value ε of the objective function $f_j(x)$. The advantage of ε-constraint method lies in that it guarantees that the objective $f_i(x)$ get the ideal value and gives certain consideration to the other $f_j(x)$ objectives. But if the upper bounds are too small, the algorithm may not find the feasible solution; or if the upper bounds are too large, the other objectives as constraints may have excessive loss. Therefore a detailed survey with decision-makers in advance will help to set a reasonable upper bound value (Cui et al. 2017).
- Solve the transformed single-objective model Eq. (8.16) with Lingo software according to the real-world situation and obtain the corresponding solutions.

8.4 Case Study

In twybrid solar–hydro generation scheduling example from Longyang Gorge, China, is given to demonstrate the practicality and efficiency of the proposed optimization method.

8.4.1 Case Description

Qinghai is located in northwest China and is rich in solar resources, with the annual solar radiance being as high as 5800–7400 MJ/m^2 (Li and Qiu 2016). A total of 7.91 million kilowatts of solar PV power plant capacity has been installed in Qinghai Province as of the end of 2017, placing it second in China, while 11.3 billion kilowatt hours had been generated, placing it first. But Qinghai is experiencing significant issues with solar energy curtailment because of the flexible units' limited capacity and the lack of extensive energy storage (Zhang et al. 2016). Although Qinghai Province took several steps to reduce solar power curtailment in 2016, it still experienced 750 million kilowatt hours of curtailment in 2017; therefore, it has become urgent to increase the solar power online rate (Northwest China Energy Regulatory Bureau of National Energy Administration of the People's Republic of China 2018).

At an average elevation of over 3000 m, Qinghai is also rich in hydro energy sources. To relieve the pressures associated with solar power integration and pursue a cleaner future, Qinghai Province built the largest hybrid solar–hydro power plant in the world, the Longyang Gorge Project, with installed hydro and solar power capacities of 128 and 85 thousand megawatts, respectively. However, as a new demonstration project, the use of traditional scheduling techniques was declining, necessitating the development of new scheduling techniques for more practical circumstances. The location of the hydro and solar power plants and some related transmission works are shown in Figure 8.4.

Based on solar energy resources and engineering construction conditions, China's solar PV resources were categorized into three categories (*Benchmark on-grid Price for PV Power ([2013] No.1638)* published by the Chinese National Development and Reform Commission (2013)). The on-grid PV power plant project cost in Qinghai, China, was 0.9 CNY/kWh as a result of the region being categorized as a first

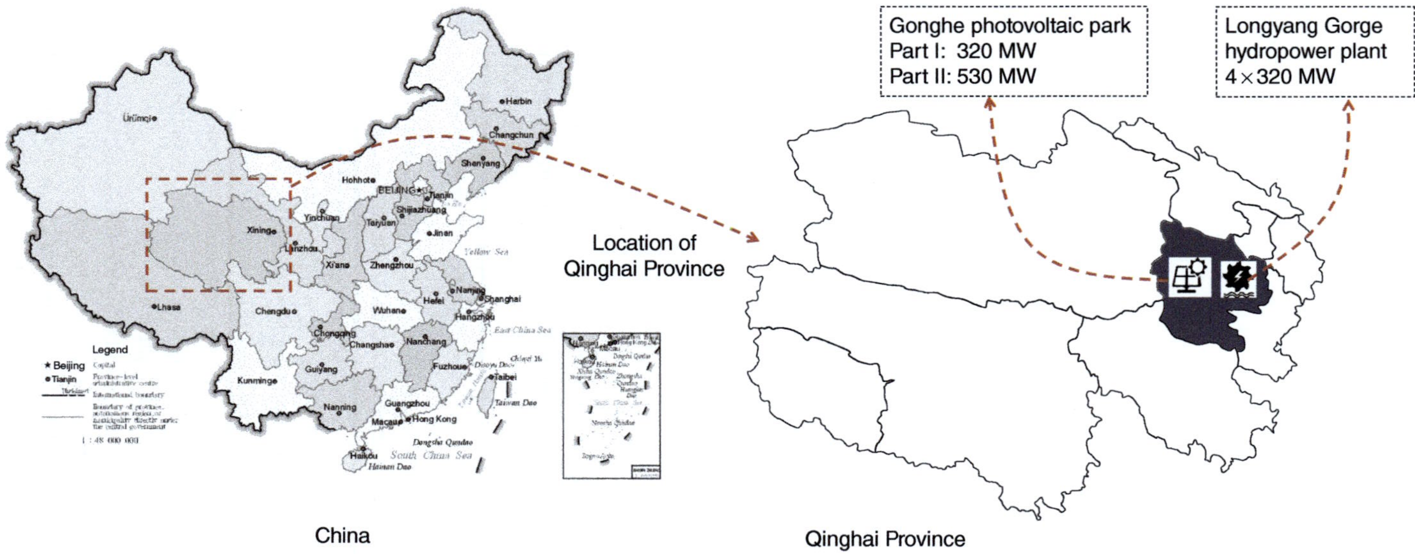

Figure 8.4 Location of Longyang Gorge Project. Source: Adapted from (Pang and Zhang 2017).

solar energy resource area type. The equivalent coal thermal power benchmark price (0.3247 CNY/kWh in 2015) was borne by the power grid, with the National Renewable Energy Development Fund covering the remaining costs (Qinghai Provincial Development and Reform Commission 2016). To ensure a reasonable use of renewable energy, the National Development and Reform Commission of China (CNDRC) issued *Management Measures for Full Supportability Purchase of Renewable Energy Generation ([2016] No.625)*, in which the annual solar power generation was split into a supported part and a market transaction part, with the supported part receiving generation priority and being fully purchased (Chinese National Development and Reform Commission 2016).

8.4.2 Data Collection

In this section, some relevant, certain and uncertain data were collected separately. The hydropower station parameters were collected from An (2016) and are presented in Table 8.1.

According to Kougias et al. (2016) research, the water flow typically exhibits little variation on an hourly scale, then the water flow is therefore assume to be comparatively constant within the 24-hour scheduling scale. Take the typical daily water flow data of dry season in Table 8.1 as an example. The data $\widetilde{W}_t = (370, 400, 420, 450)$ means that for each time step (1 h) in typical dry season day, 450 m^3/s is the highest possible water flow value, 370 m^3/s is the lowest possible water flow value, and 400 to 420 m^3/s is the range in which water flow will most commonly occur. The Longyang Gorge hydropower plant's basic load and load reserve were set at 200 and 80 MW, respectively. From November to March of the following year, individual turbine inspections are scheduled. As a result, when a turbine check is taking place, the hydro station's maximum and adjustable power outputs are both 680 MW. However, when there is no turbine inspection, the adjustable output is 800 MW, and the maximum power output is 1000 MW, providing 200 MW additional to assure the security of the power supply.

Table 8.1 Parameters of Longyang Gorge hydropower station.

Water flow scenarios		Dry season	Normal season	Flood season
Typical daily water flow $\widetilde{W}_t = (b_{t1}, b_{t2}, b_{t3}, b_{t4})$	m^3/s	(370, 400, 420, 450)	(700, 720, 780, 830)	(1150, 1200, 1210, 1250)
Load reserve capacity	MW	80	80	80
Accident spare capacity	MW	0	200	200
Inspection capacity	MW	320	0	0
Adjustable output	MW	680	800	800
Base load	MW	200	200	200
Maximum output	MW	880	1000	1000

Table 8.2 Parameters of hydro turbines.

Parameters	Value	Units	Parameters	Value	Units
Turbine highest water head	150	m	Turbine-designed water flow	298	m^3/s
Turbine average water head	130	m	Turbine rated power output	326.5	MW
Turbine-designed water head	122	m	Turbine maximum power output	356.1	MW
Turbine minimum water head	99.5	m	Maximum turbine efficiency	93.50%	None

Certain hydro turbine information, including designed water flow, rated power output, maximum turbine efficiency, and some water head parameters, was gathered from published papers (Xu and He 1987) and are presented in Table 8.2.

The uncertain parameter data, including the hourly solar irradiation data and the actual temperature data, were gathered and separately displayed in Tables 8.3 and 8.4. Hourly solar irradiation data for different weather conditions were was first gathered from public reports (Northwest Engineering Corporation Limited 2014) and paper (Xie et al. 2017) and then improved based on discussions with specialists and engineers. Corresponding actual temperature data were collected from Word Weather Online (World Weather Online 2018) and China Central Meteorological Station (China Central Meteorological Station 2018), then displayed in the fuzzy form due to their highly hourly uncertainty. Solar irradiation, actual temperature, and power consumption statistics vary depending on whether it is dry, normal, or flood season in reality, however due to space restrictions, only the data from normal seasons were shown in the following.

Typical daily power demand data supplied by the Longyang Gorge hybrid solar–hydro generation systems were collected form the *Feasibility study report of Longyang Gorge hybrid hydro-solar generation project-the second phase* (Northwest Engineering Corporation Limited 2014). This report stated that in addition to producing electricity, Longyang Gorge Reservoir also needs to do numerous extra duties, including irrigation, flood control, and anti-flooding. Therefore, during the flood season (July–September), under the premise of ensuring flood control and irrigation the power station can fully generate electricity (Liu et al. 2015); However, during the dry season (December–January–February), there is proportionately little water available for power generation due of the low upstream precipitation (Wang 2010). Therefore, different power generation responsibilities during the flood, dry, and regular seasons were allocated to the hybrid generation system, as shown in Table 8.5.

8.4.3 Calculation Results

In this section, three analyses; peak load capacity analysis, power output analysis, and economic benefits analysis; for the nine calculated results are conducted.

Table 8.3 Solar irradiation data in fuzzy form (Normal Season).

Time	1 h	2 h	3 h	4 h	5 h	6 h
Sunny days (w/m^2)	(0, 0, 0, 0)	(0, 0, 0, 0)	(0, 0, 0, 0)	(0, 0, 0, 0)	(0, 0, 0, 0)	(0, 0, 0, 0)
Cloudy days (w/m^2)	(0, 0, 0, 0)	(0, 0, 0, 0)	(0, 0, 0, 0)	(0, 0, 0, 0)	(0, 0, 0, 0)	(0, 0, 0, 0)
Rainy days (w/m^2)	(0, 0, 0, 0)	(0, 0, 0, 0)	(0, 0, 0, 0)	(0, 0, 0, 0)	(0, 0, 0, 0)	(0, 0, 0, 0)
Time	**7 h**	**8 h**	**9 h**	**10 h**	**11 h**	**12 h**
Sunny days (w/m^2)	(120, 130, 140, 150)	(235, 250, 260, 280)	(413, 420, 435, 460)	(547, 556, 560, 578)	(652, 660, 666, 678)	(765, 777, 789, 799)
Cloudy days (w/m^2)	(70, 75, 80, 90)	(164, 170, 175, 182)	(375, 380, 390, 400)	(422, 430, 440, 450)	(471, 480, 488, 492)	(609, 612, 620, 630)
Rainy days (w/m^2)	(60, 70, 77, 85)	(141, 153, 162, 173)	(375, 382, 394, 400)	(312, 323, 334, 345)	(290, 300, 309, 315)	(156, 162, 173, 179)
Time	**13 h**	**14 h**	**15 h**	**16 h**	**17 h**	**18h**
Sunny days (w/m^2)	(850, 860, 865, 870)	(805, 810, 815, 821)	(765, 770, 780, 788)	(703, 710, 720, 723)	(632, 640, 644, 750)	(438, 448, 467, 475)
Cloudy days (w/m^2)	(688, 693, 700, 712)	(569, 573, 582, 589)	(646, 652, 660, 671)	(544, 552, 560, 568)	(557, 567, 576, 582)	(292, 300, 310, 320)
Rainy days (w/m^2)	(425, 435, 442, 451)	(162, 173, 184, 192)	(280, 288, 293, 301)	(396, 403, 412, 423)	(464, 473, 482, 488)	(97, 103, 112, 124)
Time	**19 h**	**20 h**	**21 h**	**22 h**	**23 h**	**24 h**
Sunny days (w/m^2)	(273, 284, 299, 302)	(0, 0, 0, 0)	(0, 0, 0, 0)	(0, 0, 0, 0)	(0, 0, 0, 0)	(0, 0, 0, 0)
Cloudy days (w/m^2)	(164, 170, 180, 185)	(0, 0, 0, 0)	(0, 0, 0, 0)	(0, 0, 0, 0)	(0, 0, 0, 0)	(0, 0, 0, 0)
Rainy days (w/m^2)	(87, 93, 103, 121)	(0, 0, 0, 0)	(0, 0, 0, 0)	(0, 0, 0, 0)	(0, 0, 0, 0)	(0, 0, 0, 0)

Table 8.4 Temperature data in fuzzy form (Normal Season,°C).

Time	1 h	2 h	3 h	4 h	5 h	6 h
Sunny days (°C)	(2, 5, 6, 8)	(3, 6, 7, 8)	(4, 5, 6, 8)	(5, 6, 7, 9)	(6, 7, 9, 10)	(5, 8, 9, 11)
Cloudy days (°C)	(3, 4, 5, 6)	(4, 5, 6, 7)	(4, 5, 6, 7)	(5, 6, 7, 8)	(5, 6, 7, 8)	(5, 7, 8, 9)
Rainy days (°C)	(2, 3, 4, 5)	(3, 4, 5, 6)	(3, 4, 5, 7)	(4, 5, 6, 7)	(4, 5, 6, 8)	(5, 7, 8, 9)
Time	**7 h**	**8 h**	**9 h**	**10 h**	**11 h**	**12 h**
Sunny days (°C)	(6, 9, 10, 11)	(8, 10, 12, 13)	(9, 11, 13, 16)	(10, 12, 14, 16)	(9, 12, 15, 17)	(10, 13, 15, 18)
Cloudy days (°C)	(6, 8, 9, 10)	(7, 9, 11, 12)	(8, 10, 12, 13)	(8, 10, 12, 14)	(9, 11, 13, 14)	(10, 12, 14, 16)
Rainy days (°C)	(5, 7, 8, 10)	(6, 8, 9, 11)	(6, 8, 9, 12)	(6, 8, 9, 13)	(8, 10, 12, 13)	(9, 10, 12, 14)
Time	**13 h**	**14 h**	**15 h**	**16 h**	**17 h**	**18 h**
Sunny days (°C)	(10, 13, 16, 19)	(10, 13, 17, 20)	(12, 14, 16, 18)	(9, 11, 13, 16)	(8, 10, 12, 14)	(7, 9, 11, 12)
Cloudy days (°C)	(10, 12, 15, 17)	(9, 11, 13, 16)	(8, 10, 13, 15)	(8, 10, 12, 13)	(7, 9, 10, 12)	(7, 8, 9, 11)
Rainy days (°C)	(9, 10, 12, 15)	(8, 11, 13, 16)	(7, 8, 9, 12)	(7, 8, 9, 11)	(6, 7, 8, 10)	(6, 7, 8, 9)
Time	**19 h**	**20 h**	**21 h**	**22 h**	**23 h**	**24 h**
Sunny days (°C)	(6, 8, 9, 11)	(6, 7, 8, 10)	(5, 7, 8, 9)	(5, 6, 7, 8)	(4, 5, 6, 7)	(3, 4, 5, 6)
Cloudy days (°C)	(6, 8, 9, 10)	(5, 7, 8, 9)	(5, 6, 7, 8)	(4, 5, 6, 8)	(3, 4, 5, 6)	(3, 4, 5, 6)
Rainy days (°C)	(4, 5, 6, 8)	(3, 5, 6, 7)	(2, 3, 4, 6)	(2, 3, 4, 6)	(2, 3, 4, 6)	(2, 3, 4, 6)

Table 8.5 Typical daily power demand data of Longyang Gorge hybrid solar–hydro generation systems (MW) (Northwest Engineering Corporation Limited 2014).

Flood season	Time	1 h	2 h	3 h	4 h	5 h	6 h	7 h	8 h	9 h	10 h	11 h	12 h
	Demand	1280	1280	1280	1280	1280	1280	1292	1360	1428	1496	1578	1620
	Time	13 h	14 h	15 h	16 h	17 h	18 h	19 h	20 h	21 h	22 h	23 h	24 h
	Demand	1580	1500	1365	1300	1400	1500	1600	1650	1700	1670	1500	1280
Dry season	Time	1 h	2 h	3 h	4 h	5 h	6 h	7 h	8 h	9 h	10 h	11 h	12 h
	Demand	760	760	760	760	780	790	800	880	980	1080	1160	1200
	Time	13 h	14 h	15 h	16 h	17 h	18 h	19 h	20 h	21 h	22 h	23 h	24 h
	Demand	1180	1120	1020	940	920	920	960	920	840	760	740	720
Normal season	Time	1 h	2 h	3 h	4 h	5 h	6 h	7 h	8 h	9 h	10 h	11 h	12 h
	Demand	912	912	923	923	933	933	1000	1060	1160	1280	1380	1380
	Time	13 h	14 h	15 h	16 h	17 h	18 h	19 h	20 h	21 h	22 h	23 h	24 h
	Demand	1350	1280	1230	1160	1100	1060	1080	1100	1050	1000	950	912

8.4.3.1 Technical Output Analysis

Figure 8.5 displays the Longyang Gorge hybrid hydro-solar system's maximum technical output under various water input and weather conditions. The highest technical output during the flood season and during normal times prior to the incorporation of solar electricity is 1280 MW. Since the repair work on the turbines is scheduled for the dry season, the hydropower plant's maximum technological output is lowered to 960 MW. Changes in the technical output of the power generation system after PV access is shown in Table 8.6. From the data in Figure 8.5 and Table 8.6, two main findings can be obtained. First, from the "Rate of technical output changes" in Table 8.6, it is clear that the maximum technical production greatly increased in each of the nine scenarios, ranging from 10.08% in a normal

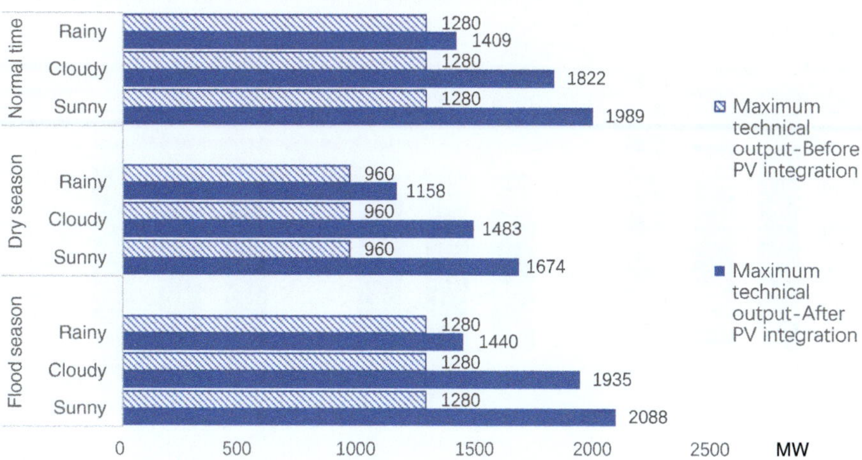

Figure 8.5 Maximum technical output under different scenarios.

Table 8.6 Changes of technical output under different scenarios (MW).

Scenarios		Maximum technical output-after	Maximum technical output-before	Maximum technical output-difference	Rate of technical output changes (%)
Flood season	Sunny	2088	1280	808	63.09
	Cloudy	1935	1280	655	51.13
	Rainy	1440	1280	160	12.50
Dry season	Sunny	1674	960	714	74.38
	Cloudy	1483	960	523	54.48
	Rainy	1158	960	198	20.63
Normal time	Sunny	1989	1280	709	55.39
	Cloudy	1822	1280	542	42.34
	Rainy	1409	1280	129	10.08

time when it is rainy to 74.38% in the dry season when it is sunny. Second, the output of a hybrid generation system at its highest technical level is directly correlated with the weather. For example, even if the hydropower plant's 1280 MW maximum technical output is the same during the flood season, the maximum technical output after establishing the hybrid generation system is,respectively, 2088, 1935, and 1440 MW. As can be observed, the rise in technical production on sunny days is more than that on cloudy days, which in turn is greater than that on days when it is raining. Therefore, photovoltaic output was proven to be directly influenced by the weather conditions.

8.4.3.2 Power Output Ratio Analysis

The quantities and ratios for the total hydroand solar power output in the different scheduling scenarios are presented in Figure 8.6. Itclearly shows that the

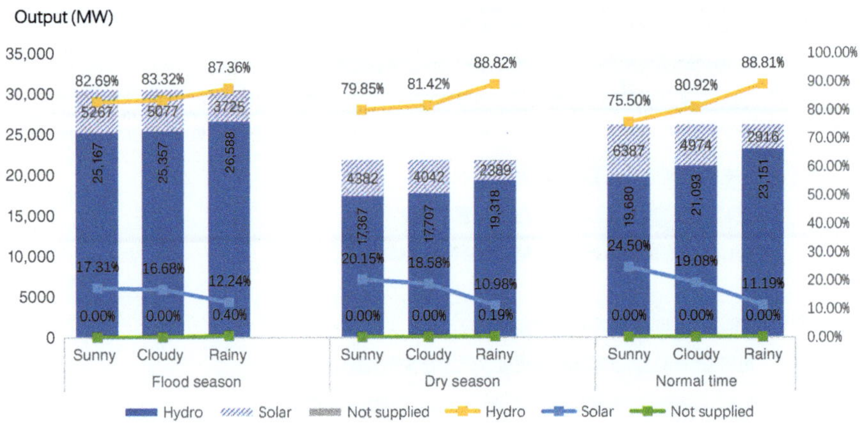

Figure 8.6 Quantities and ratios of total hydroand solar power output in different scheduling scenarios.

hydropower output is larger during the flood season than it is during normal times, which is higher than it is during the dry seasons. For instance, when it is sunny, the hydropower output is 25,167, 19,680, and 17,367 MW in the flood season, at normal time, and in the dry season, respectively. Further, as the seasons change from sunny to cloudy and from cloudy to rainy, it is clear that the solar power output gradually decreases while the hydropower output rises in response. This is a good example of how solar power and hydropower complement one another. It can also be seen that the highest solar power output was in the normal time on sunny days at 6387 MW, and the lowest in the dry season on rainy days at 2389 MW. This was due to the fact that the hydro power output, when running normally, has the most flexibility to account for variations in solar power because it is not required to operate at full capacity during the flood season or perform maintenance on the hydraulic turbines during the dry season.

8.4.3.3 Hourly Power Output Analysis

In this part, the hydro and solar power hourly output is plotted to observe the hourly output characteristics in the different scheduling scenarios. As can be seen from Figure 8.7, the first power supply peak appears at 12 noon, and the second peak appears between 7 and 8 p.m. in the evening, at which time hydropower takes the basic load, especially during the first demand peak. The hourly hydropower output variations on sunny days were very smooth (for example, see Scenarios 1, 4 and 7 in Figure 8.7). However, this smoothness was progressively replaced by glaring fluctuations as the weather changed from being sunny to cloudy and from cloudy to rainy (for example, see Scenarios 7, 8, and 9 in Figure 8.7). Further, on cloudy days, the hydropower output variations in the normal time (Scenario 8) are larger than in the dry season (Scenario 5), and in the dry season were larger than those in the flood season (Scenario 2), which indicated that the hydropower plant's flexibility is maximum during normal times, followed by the dry season and the flood season. Additionally, it can be seen that the total power supply of the hybrid system is higher during the flood season (Scenario 3) than during normal times (Scenario 9), and higher than during the dry season (Scenario 6). The solar power supply is highest during normal times on sunny days (Scenario 7) and lowest during the dry season on rainy days (Scenario 6). Similarly, the hydropower supply is highest on rainy days during the flood season (Scenario 3), and lowest on sunny days during the dry season (Scenario 4).

8.4.3.4 Economic Benefits Analysis

In this section, we present the power supply income under different scenarios are presented in Table 8.7. From "Total power supply,'column' it can be observed that the overall revenue steadily declines when the weather shifts from sunny to cloudy and from cloudy to rainy, regardless of whether it is the flood season, normal time, or the dry season. Taking the normal time income as an example, the income for sunny, cloudy, and rainy weather conditions is, respectively, 8317, 7417, and 6106 ($\times 10^3$ CNY); however, the total power demand in all three sub-scenarios is the same. As can be observed, the variations in solar power output were the main reason of

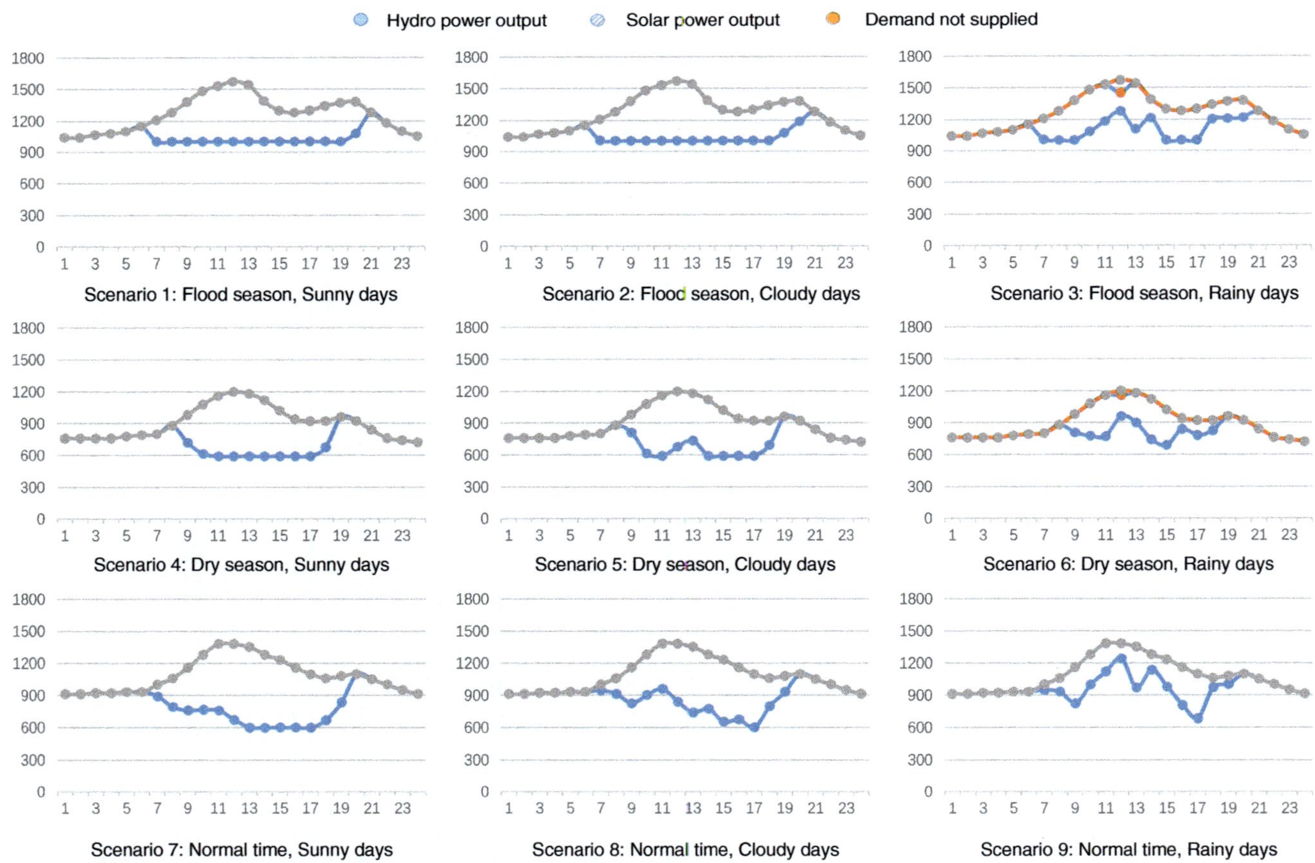

Figure 8.7 Hourly power supply and demand of different scenarios.

Table 8.7 Income of power supply of different scenarios (10^3 CNY).

Scenarios		Hydropower output (MW)	Solar power output (MW)	Demand not supplied $MW)	Income of hydro power supply ($\times 10^3$ CNY)	Income of solar power supply ($\times 10^3$ CNY)	Penalty of power not supplied ($\times 10^3$ CNY)	Total income of power supply ($\times 10^3$ CNY)
Flood season	Sunny	25167	5267	0	4102	4214	0	8316
	Cloudy	25357	5077	0	4133	4062	0	8195
	Rainy	26588	3846	0	4334	3077	0	7411
Dry season	Sunny	17367	4382	0	2831	3506	0	6336
	Cloudy	17707	4042	0	2886	3234	0	6120
	Rainy	19318	2431	0	3149	1945	0	5094
Normal time	Sunny	19680	6387	0	3208	5110	0	8317
	Cloudy	21093	4974	0	3438	3979	0	7417
	Rainy	23151	2916	0	3774	2333	0	6106

these revenue discrepancies. The solar power output, for instance, in this situation (normal time) is 6387 MW on bright days, 4974 MW on cloudy days, and 2916 MW on rainy days. Consequently, the income from the solar power production reduces proportionally from 5110 to 3979 to 2333 ($\times 10^3$ CNY). These differences are because of the on-grid pricing for the Longyang Gorge hybrid hydro-solar system. In Qinghai Province, the on-grid hydropower power price is 163 CNY/MW and the on-grid solar power price is 800 CNY/MW, nearly five times what on-grid hydropower costs. Therefore, greater solar power output increases sales revenue. Further, although the electricity supply during the flood season and during normal times is different, the sales revenue is nearly the same. This is largely due to the hydro turbine flexibility during normal times, which results in more photovoltaic power being used on the main grid and higher income than when using only hydropower.

8.5 Discussion

In this section, a comprehensive discussion from the calculation results of the nine scheduling scenarios is given, including comparative study, related propositions, and policy implications.

8.5.1 Comparative Study

Many attempts have been made to integrate uncertain renewable energy sources by establishing hybrid generation systems. In this section, a comparative discussion is given.

Comparison of the intention of applying the multi-objective optimization method. To balance the trade-offs between power output process smoothness and total annual power generation, Li and Qiu (2016) created a multi-objective optimization model for hydro-photovoltaic power systems. Fang et al. (2017) suggested a long-term operation model for the hydro-PV plant with the goal of increasing energy production and a guaranteed rate of power generation over the entire operational duration. These efforts made significant contributions to the efficient modeling of the long-term renewable power integration. However, few of these efforts concentrate on the short-term trade-offs between a reliable and profitable hybrid power supply, especially in light of the unique power on-grid price policy for various renewable electricity sources. Therefore, in this chapter, the reason for applying the multi-objective method was power supply reliability and energy generation economy trade-offs. The approach in this chapter not only maintains the stability of the power supply but also achieves a certain level of energy conservation by concurrently taking into account both power shortages and power excesses. As for profitable electricity dispatch for power suppliers, the Benchmark on-grid Price for PV Power ([2013] No.1638) and the Management Measures for Full Supportability Purchase of Renewable Energy Generation ([2016] No.625); which have a significant impact on the scheduling plans, are given full consideration, therefore, making this model more reasonable and practical.

Comparison of the method to handle renewable energy sourcesrelated uncertainties. In 2018, Yang et al. (2018) developed the operating rules for a sizable hydro-photovoltaic power system. To address the uncertainties in reservoir inflow and PV power, implicit stochastic optimization was used. Initially, a model was created to maximize total energy production and the assurance rate. This problem was then solved using dynamic programming. The parameters were then refined using a parameterization-simulation-optimization based framework, and the operating rules were then constructed from an optimal trajectory using the linear fitting approach. These initiatives were beneficial because they provided instructions for the efficient operation of hybrid hydro/PV power systems and were successful in increasing the system benefits when compared to traditional operating regulations. However, more consideration should have been given to the hourly, daily, monthly, and annual uncertainties in the hydro water flow and solar irradiation, which are crucial factors for scheduling hydro-solar compensatory operations. For example, Séguin et al. (2017) built a scenario tree for hydro inflow uncertainties, and Yang et al. (2016) considered the stochastic nature of solar irradiance using the Monte Carlo Method. However, these attempts were sometimes excessively conservative and suffered from the heavy computing loads brought on by the vast number of scenarios. In order to simultaneously account for seasonal uncertainties in water flow and daily uncertainties in solar irradiation, nine discrete scheduling scenarios were created using the method described in this chapter. Hourly solar irradiation and actual temperatures were described using trapezoidal fuzzy sets from field research interviews with experts and engineers as well as historical data. Despite the fact that our method is based on historical data and the system operators' experience, it not only dealt with climate-related solar and hydro power uncertainties successfully but also freed the system operators from the laborious and cautious calculations associated with numerous uncertain scenarios, resulting in a more practical approach for system operators.

8.5.2 Related Propositions

Three main propositions are proposed here to clearly outline the core conclusions drawn from the calculation results.

Firstly, for a hybrid solar–hydro generation system, an equilibrium-based scheduling technique can improve power supply economy without compromising the reliability of electricity production. Prior to adopting the equilibrium scheduling method, the solar power output was limited to no more than 10% of the total amount of power supplied; however, with the help of the suggested technique, the solar power output ratio reached as high as 24.5% (see normal time, sunny days in Figure 8.4), and the lowest solar power output ratio was 10.98% (see dry season, rainy days in Figure 8.4). The power supply benefits increased in each scenario because the on-grid price of solar energy is around five times higher than that of hydropower. It is important to note that to fully guarantee a certain range of power supply in each scenario, the first priority was to ensure the basic requirements of economic development and social progress. Therefore, the power supply economy was obtained under a reliability premise; that is, to ensure that the power supply demand was fully satisfied and that there was neither an inadequate nor an excessive amount of power supply, the attitude parameter *beta* was set at a minimal value when the optimization model was solved.

Secondly, when several power supplies are merged into a hybrid system, the power generation system's scheduling flexibility increases. In a hybrid system, which combines different power sources, each source complements the other in terms of output curve and other qualities, particularly for climate-related solar and hydropower. From the calculations, it was demonstrated that once the photovoltaic power station was connected to the hydropower station as a hybrid system, the scheduling flexibility of the combined power supply was greatly improved. From the "Rate of technical output changes"column in Table 8.6, in all nine scenarios, it is demonstrated that the maximum technical production increased dramatically. The technical output ranged from 10.08% during normal time when it rained to 74.38% during the dry season when it was sunny. As a result, the climate-driven uncertainty of the energy sources affects the hybrid power–generating system's scheduling flexibility, with the technical output increases on sunny days being larger than on cloudy days (for example, in Table 8.6, the change on sunny days was 55.39% and on cloudy days it is 42.24%), and the output being larger on cloudy days than on rainy days (42.24% on cloudy days and 10.08% on rainy days – Table 8.6 at normal times), because the weather directly affects the maximum photovoltaic output, which affects the hybrid generating system's technical output.

Further, hydropower can make up for solar energy's hourly and daily output variability, and solar energy can make up for hydropower's seasonal and yearly output variability. Although climate-related hydro and solar power have individual uncertainties, they compliment one another. First, hydropower can make up for the hourly and daily variations in solar output. Even though the solar power output on cloudy or rainy days changed across different time periods, as seen in Scenarios 2, 3, 5, 6, 8, and 9, these fluctuations were successfully reduced by utilizing hydropower. Further, as shown in Scenario 8 (normal time, cloudy), when the photovoltaic power output was unavailable at night, the hydropower station was able to adjust the output accordingly. Second, solar power is able to compensate for the hydropower output seasonal and annual fluctuations. For example, as shown in Figure 8.5, there was seasonal complementarity, as the solar power output grew as the hydropower output declined from the flood season to the dry season. This was also true for annual output since, despite variations in annual water flow from year to year, there are not many variations in annual solar power generation.

Finally, policies that subsidize renewable electricity can effectively aid in adjusting the structure of the energy supply, particularly in hybrid energy supply systems. Because the demand curve in a hybrid energy supply system is allocated as a full profile, the system operator must plan reasonable unit commitments to ensure the supply of demand while also maximizing profit. Take the Longyang Gorge hybrid system as an example, the power demand curve is firstly assigned to Huanghe Hydropower Development Co., Ltd. dispatch center, and then scheduled to the hydro turbines and solar power plants. However, because the policy supports the growth of solar energy, the National Renewable Energy Development Fund will offer a 0.53 CNY/kWh subsidy for solar power access to the grid, indicating that the on-grid price of electricity for solar power is substantially greater than that for hydropower. Therefore, the power supplier would access more solar power to the main grid. In other instances, though, more hydropower might be accessed if the online price was greater. Therefore, various power subsidy programs can aid in adjusting the structure of the energy supply.

8.5.3 Management Recommendations

Based on the comprehensive comparison and related proposition analyses, some management recommendations are given in this section to better explore hybrid solar–hydro power system applications.

For the hybrid system power supplier, the suggested reliable–economical equilibrium-based scheduling method should be taken into consideration as a feasible alternative to encourage climate-related renewable source integration within the constraints of the environment and policy implications. Climate-related renewable energy sources like wind, solar, and hydropower all have unpredictable weather patterns, making it difficult to provide reliable and stable power supplies. However, this problem can be partly relieved by combining complementary sources into hybrid systems and applying the proposed reliable–economical scheduling strategy. As was shown in the Longyang Gorge Project case study, electricity generation flexibility and economy were increased and power supply reliability (including power shortages and power excesses) guaranteed. Therefore, for a more reliable, economical, and energy-saving future, this method should be considered by system operators as a more reasonable and practical production plan.

For the electricity generation industry, the accuracy of associated climate uncertain factor forecast needs to be increased for the electricity generation sector to improve generating complementarity. Although the seasonal water flow uncertainty and daily solar irradiation uncertainty can be handled to some extent by referring to the nine typical scheduling scenarios, this balance can be disrupted when the solar irradiation or water flow unpredictably changes on rainy days during the flood or dry seasons. However, as discussed earlier, the most effective strategy to establish compensatory operating scheduling for hybrid generating systems is to increase system stability and decrease maintenance costs by precisely estimate uncertain climate related elements to avoid the influence of output uncertainty. Therefore, to better manage climate-related uncertainty, the timeliness and accuracy of projections need to be enhanced for a cleaner future.

For the administration departments, policy support should be offered to assist the development of hybrid generation systems and greater generation complementarity. For example, the Chinese government started to promote multi-energy complementary integration and optimization demonstration projects in 2016 (National Development and Reform Commission of China, National Energy Administration of China 2016), and further issued guidance to explore high-quality development paths for multi-energy complementary in 2021 (National Development and Reform

Commission of China, National Energy Administration of China 2021), and the findings in this paper make an important contribution to such policies. Cooperation between multiple power supply sources is still inefficient and insufficient, despite the fact that many academics and engineers have been researching the viability of developing and deploying hybrid energy systems. To explore wider applications, governments and administrative agencies must support policies and regulations that reduce carbon emissions, increase the use of clean energy, increase energy system efficiencies, and improve generation complementarity, particularly for the production of electricity.

References

Almonacid, F., Perez-Higueras, P.J., Fernández, E.F., and Hontoria, L. (2014). A methodology based on dynamic artificial neural network for short-term forecasting of the power output of a PV generator. *Energy Conversion & Management* 85(9): 389–398.

Alonso-Montesinos, J., Batlles, F.J., and Portillo, C. (2015). Solar irradiance forecasting at one-minute intervals for different sky conditions using sky camera images. *Energy Conversion & Management* 105: 1166–1177.

An, Y. (2016). The study on the theories and methodology of complementary hydrophotovoltaic operation. PhD thesis. Xi'an: Xi'an University of Technology.

Barakat, N. and Sharma, D. (2018). Evolutionary multi-objective optimization for bulldozer and its blade in soil cutting. *International Journal of Management Science and Engineering Management*. https://doi.org/10.1080/17509653.2018.1500953.

Beluco, A., de Souza, P.K., and Krenzinger, A. (2012). A method to evaluate the effect of complementarity in time between hydro and solar energy on the performance of hybrid hydro PV generating plants. *Renewable Energy* 45(3): 24–30.

BP Global (2018). Statistical review of world energy. https://www.bp.com/en/global/corporate/energy-economics/statistical-review-of-world-energy.html (accessed 20 October 2023).

Cervantes, J. and Choobineh, F. (2018). Optimal sizing of a nonutility-scale solar power system and its battery storage. *Applied Energy* 216: 105–115.

Chang, J., Li, Y., Yuan, M., and Wang, Y. (2017). Efficiency evaluation of hydropower station operation: a case study of Longyangxia station in the Yellow River, China. *Energy* 135: 23–31.

Chankong, V. and Haimes, Y.Y. (1983). *Multiobjective Decision Making: Theory and Methodology*. Elsevier.

Chen, C., Li, Y., Huang, G.H., and Li, Y. (2012). A robust optimization method for planning regional-scale electric power systems and managing carbon dioxide. *International Journal of Electrical Power & Energy Systems* 40(1): 70–84.

Chen, Y., He, L., Li, J. et al. (2016). An inexact bi-level simulation-optimization model for conjunctive regional renewable energy planning and air pollution control for electric power generation systems. *Applied Energy* 183: 969–983.

Chen, F., Zhou, J., Wang, C., and Li, C. (2017). A modified gravitational search algorithm based on a non-dominated sorting genetic approach for hydro-thermal-wind economic emission dispatching. *Energy* 121: 276–291.

China Central Meteorological Station (2018). Weather forecast for Gonghe county. http://www.nmc.cn/publish/forecast/AQH/gonghe.html (accessed 20 October 2023).

Chinese National Development and Reform Commission (2013). Notice of CNDRC on exploiting price leverage to promote healthy development of PV industry. https://www.ndrc.gov.cn/xxgk/zcfb/tz/201308/t20130830_963934.html (accessed 14 December 2023).

Chinese National Development and Reform Commission (2016). Notice of the national development and reform commission on management measures for full supportability purchase of renewable energy generation. http://www.ndrc.gov.cn/zcfb/zcfbtz/201603/t20160328_796404.html.

Chinese National Development and Reform Commission (2017). Solar power on-grid price policy. http://zfxxgk.ndrc.gov.cn/web/iteminfo.jsp?id=2947 (accessed 20 October 2023).

Cui, Y., Geng, Z., Zhu, Q., and Han, Y. (2017). Multi-objective optimization methods and application in energy saving. *Energy* 125: 681–704.

Drechsler, M., Egerer, J., Lange, M., and Masurowski, F. (2017). Efficient and equitable spatial allocation of renewable power plants at the country scale. *Nature Energy* 6(9): 17124.

Dujardin, J., Kahl, A., Kruyt, B. et al. (2017). Interplay between photovoltaic, wind energy and storage hydropower in a fully renewable Switzerland. *Energy* 135: 513–525.

Fang, W., Huang, Q., Huang, S. et al. (2017). Optimal sizing of utility-scale photovoltaic power generation complementarily operating with hydropower: a case study of the world's largest hydro-photovoltaic plant. *Energy Conversion & Management* 136: 161–172.

Fernandez-Jimenez, L.A., Jiménez, A.M., Falces, A., and Mendoza-Villena, M. (2012). Short-term power forecasting system for photovoltaic plants. *Renewable Energy* 44(4): 311–317.

François, B., Borga, M., Creutin, J.D. et al. (2015). Complementarity between solar and hydro power: sensitivity study to climate characteristics in Northern-Italy. *Renewable Energy* 86: 543–553.

Guillén, G., Mele, F., Bagajewicz, M.J., and Espuna, A. (2005). Multiobjective supply chain design under uncertainty. *Chemical Engineering Science* 60(6): 1535–1553.

Hafezalkotob, A., Hosseinpour, E., Moradi, M., and Khalili-Damghani, K. (2017). Multi-resource trade-off problem of the project contractors in a cooperative environment: highway construction case study. *International Journal of Management Science and Engineering Management* 13(2): 129–138.

Haimes, Y.Y., Lasdon, L.S., and Wismer, D.A. (1971). On a bicriterion formulation of the problems of integrated system identification and system optimization. *IEEE Transactions on Systems Man & Cybernetics* SMC-1(3): 296–297.

Han, S., Zhang, L., Liu, Y. et al. (2019). Quantitative evaluation method for the complementarity of wind–solar–hydro power and optimization of wind–solar ratio. *Applied Energy* 236: 973–984.
Hemmati, R., Saboori, H., and Jirdehi, M.A. (2017). Stochastic planning and scheduling of energy storage systems for congestion management in electric power systems including renewable energy resources. *Energy* 133: 380–3870.
Huang, C., Li, F., and Jin, Z. (2015). Maximum power point tracking strategy for large-scale wind generation systems considering wind turbine dynamics. *IEEE Transactions on Industrial Electronics* 62(4): 2530–2539.
International Energy Agency (2017). Key world energy statistics. https://www.oecd-ilibrary.org/energy/key-world-energy-statistics-2017_key_energ_stat-2017-en (accessed 14 December 2023).
International Energy Agency (2018). Key world energy statistics. https://webstore.iea.org/download/direct/2291?fileName=Key_World_2018.pdf.
International Renewable Energy Agency (2018). Renewable Energy Statistics. http://www.irena.org/publications/2018/Jul/Renewable-Energy-Statistics-2018 (accessed 20 October 2023).
Ju, L., Li, H., Yu, X., and Fugui, D. (2016). A combined external delivery optimization model for wind power and thermal power considering carbon trading and tradable green certificates based on Credibility theory. *International Journal of Electrical Power & Energy Systems* 78: 51–60.
Kok, A.G., Shang, K., and Yucel, S. (2018). Impact of electricity pricing policy on renewable energy investments and carbon emissions. *Management Science* 64(1): 131–148.
Kougias, I., Szabó, S., Monforti, F., and Huld, T. (2016). A methodology for optimization of the complementarity between small-hydropower plants and solar PV systems. *Renewable Energy* 87(2): 1023–1030.
Lan, H., Yin, H., Hong, Y.-Y. et al. (2018). Day-ahead spatio-temporal forecasting of solar irradiation along a navigation route. *Applied Energy* 211: 15–27.
Li, F. and Qiu, J. (2016). Multi-objective optimization for integrated hydro-photovoltaic power system. *Applied Energy* 167: 377–384.
Liu, J., Wang, S., and Wang, M. (2015). Influence analysis of hydro-photovoltaic power complementary system of Longyangxia hydro-power station. *Power System and Clean Energy* 31(9): 83–87.
Lohmann, T. and Rebennack, S. (2017). Tailored benders decomposition for a long-term power expansion model with short-term demand response. *Management Science* 63(6): 2017–2048.
Lv, C., Xu, J., Xie, H. et al. (2016). Equilibrium strategy based coal blending method for combined carbon and PM10 emissions reductions. *Applied Energy* 183: 1035–1052.
Ming, B., Liu, P., Guo, S. et al. (2017). Optimizing utility-scale photovoltaic power generation for integration into a hydropower reservoir by incorporating long- and short-term operational decisions. *Applied Energy* 204: 432–445.
National Development and Reform Commission of China, National Energy Administration of China (2016). Implementing opinions on promoting the construction of demonstration projects of multi-energy implementation opinions on

promoting the construction of multi-energy complementary integration and optimization demonstration projects. https://www.gov.cn/gongbao/content/2017/content_5174532.htm (accessed 20 October 2023).

National Development and Reform Commission of China, National Energy Administration of China (2021). Guiding Opinions on Promoting the Integration of Electricity Source, Grid, Demand and Storage and the Development of Multi-energy Complementarity. https://www.gov.cn/zhengce/zhengceku/2021-03/06/content_5590895.htm (accessed 20 October 2023).

Northwest China Energy Regulatory Bureau of National Energy Administration of the People's Republic of China (2018). Grid connected operation condition of new energy in northwest China in 2017. http://xbj.nea.gov.cn/website/Aastatic/news-186169.html (accessed 20 October 2023).

Northwest Engineering Corporation Limited (2014). Feasibility study report of Longyang Gorge hybrid hydro-solar generation project. https://jz.docin.com/p-1882233248.html (accessed 20 October 2023).

Ogueke, N.V., Ikpamezie, I.I., and Anyanwu, E.E. (2016). The potential of a small hydro/photovoltaic hybrid system for electricity generation in FUTO. *Nigeria. International Journal of Ambient Energy* 37(3): 256–265.

Pang, X. and Zhang, W. (2017). Hydro-photovoltaic complementary technology research and application. *Journal of Hydroelectric Engineering* 36(7): 1–13.

Qinghai Provincial Development and Reform Commission (2016). Qinghai provincial development and reform commission forwarded the notice of the national development and reform commission on reducing the on-grid price of coal-fired power generation and the price of electricity for general industrial and commercial use. http://www.qhfgw.gov.cn/zfxxgk/fgwwj/qfgwj/201608/t20160804_657610.shtml (accessed 20 October 2023).

Reddy, S.S. (2017). Optimal scheduling of thermal-wind-solar power system with storage. *Renewable Energy* 101: 1357–1368.

Sabooria, H., Hemmati, R., Ghiasi, S.M.S., and Dehghan, S. (2017). Energy storage planning in electric power distribution networks-A state-of-the-art review. *Renewable and Sustainable Energy Reviews* 79: 1108–1121.

Séguin, S., Fleten, S.E., Côté, P. et al. (2017). Stochastic short-term hydropower planning with inflow scenario trees. *European Journal of Operational Research* 259(3): 1156–1168.

Singh, G.K. (2013). Solar power generation by PV (photovoltaic) technology: a review. *Energy* 53(5): 1–13.

Snaith, H.J. and Hacke, P. (2018). Enabling reliability assessments of pre-commercial perovskite photovoltaics with lessons learned from industrial standards. *Nature Energy* 3(6): 459.

Sun, K., Li, K., Pan, J. et al. (2019). An optimal combined operation scheme for pumped storage and hybrid wind-photovoltaic complementary power generation system. *Applied Energy* 242: 1155–1163.

Wang, Z. (2010). Analysis of the run off series representativeness of the upper Yellow River above Longyangxia hydropower station. *Northwest Hydropower* 2: 1–5+8.

Wang, J., Shahidehpour, M., and Li, Z. (2008). Security-constrained unit commitment with volatile wind power generation. *IEEE Transactions on Power Systems* 23(3): 1319–1327.

Ware, N.R., Singh, S.P., and Banwet, D.K. (2014). A mixed-integer non-linear program to model dynamic supplier selection problem. *Expert Systems with Applications* 41(2): 671–678.

World Weather Online (2018). Gonghe historical weather. https://www.worldweatheronline.com/gonghe-weather-history/qinghai/cn.aspx (accessed 14 December 2023).

Xie, X., Wei, X., Pang, X. et al. (2017). Research and application of key technologies for large-scale hydro and solar complementation. *China Science and Technology Achievements* 18(16): 70–71. https://doi.org/10.3772/j.issn.1009-5659.2017.16.033.

Xu, L. and He, X. (1987). Design and manufacture of 320MW hydroelectric generating set for Longyangxia hydropower station. *Dongfang Electric Review* 02: 11–20+62.

Xu, J. and Zhou, X. (2011). *Fuzzy-Like Multiple Objective Decision Making*. Berlin Heidelberg: Springer-Verlag.

Xu, J., Wang, F., Lv, C. et al. (2018a). Economic-environmental equilibrium based optimal scheduling strategy towards wind-solar-thermal power generation system under limited resources. *Applied Energy* 231: 355–371.

Xu, J., Wang, F., Lv, C., and Xie, H. (2018b). Carbon emission reduction and reliable power supply equilibrium based daily scheduling towards hydro-thermal-wind generation system: a perspective from China. *Energy Conversion and Management* 164: 1–14.

Yang, J., Zhang, N., Kang, C., and Xia, Q. (2016). Assessing the dispatch flexibility of coordinated solar and hydro generation. *Power and Energy Society General Meeting*, 1357–1368.

Yang, Z., Liu, P., Cheng, L. et al. (2018). Deriving operating rules for a large-scale hydro-photovoltaic power system using implicit stochastic optimization. *Journal of Cleaner Production* 195: 562–572.

Yona, A., Senjyu, T., Funabashi, T., and Kim, C.H. (2013). Determination method of insolation prediction with fuzzy and applying neural network for long-term ahead PV power output correction. *IEEE Transactions on Sustainable Energy* 4(2): 527–533.

Zhang, Y., Tang, N., Niu, Y., and Du, X. (2016). Wind energy rejection in China: current status, reasons and perspectives. *Renewable & Sustainable Energy Reviews* 66: 322–344.

Zhang, H., Lu, Z., Hu, W. et al. (2019). Coordinated optimal operation of hydro-wind-solar integrated systems. *Applied Energy* 242: 883–896.

Zubo, R.H.A., Mokryani, G., and Abd-Alhameed, R. (2018). Optimal operation of distribution networks with high penetration of wind and solar power within a joint active and reactive distribution market environment. *Applied Energy* 220: 713–722.

9

Reliable–Economical Equilibrium-Based Short-Term Scheduling of Hybrid Solar–Wind–Gas System

I predict the electricity generated by water power is the only thing that is going to keep future generations from freezing. Now we use coal whenever we produce electric power by steam engine, but there will be a time when there'll be no more coal to use. That time is not in the very distant future. ... Oil is too insignificant in its available supply to come into much consideration.

Charles Proteus Steinmetz
Apr. 9, 1865 – Oct. 26, 1923

There has been an increased interest in hybrid power generation systems that integrate renewable energy and reliable fossil fuels, which can ensure reliable–economic benefits and power supply. However, establishing a balance between reliability and economic benefits in hybrid power generation that uses solar, wind, and natural gas while taking into account renewable uncertainties has also been a significant problem and has not been thoroughly investigated. This chapter suggests a multi-objective optimization that simultaneously takes into account reliability and economic benefits in order to improve the short-term scheduling of solar, wind, and natural gas power generation under natural constraints. Twelve scheduling scenarios based on seasonal weather conditions were established, in which the hourly renewable resource data were represented using trapezoidal fuzzy numbers. This was done to deal with the solar irradiation, wind speed, and air temperature uncertainties that adversely affect renewable power output. The best possible solution was identified using a fuzzy satisfactory strategy after the suggested multi-objective model was transformed using the ε-constraint method and associated attitude parameters. The viability and effectiveness of the suggested scheduling strategies were tested using a representative real-world scenario from Golmud, China. The outcomes amply proved that the cooperative complementarity in this hybrid solar–wind–gas power system could successfully generate sufficient revenue and guarantee generation reliability. There are also management suggestions for government agencies, system operations, and the energy sector.

Sustainable Hybrid Energy Systems: Carbon Neutral Approaches, Modeling, and Case Studies,
First Edition. Jiuping Xu and Fengjuan Wang.

9.1 Background Review

Global electricity demand has significantly increased as a result of socioeconomic trends (Hu et al. 2021). Coal is still the main energy source, according to the BP Statistical Review of World Energy (BP Globe 2019). Even though coal has long been the primary energy source, particularly in developing nations, the excessive carbon emissions and environmental harm that are associated with coal have prompted many nations to look for clean energy sources to replace conventional fossil fuels (Gu et al. 2020; Wang et al. 2018; Luo et al. 2016).

As carbon-free energy sources, renewable energies such as solar and wind power are becoming increasingly popular (Johlas et al. 2020). According to Renewable Energy Statistics (IRENA 2019), by the end of 2018, the world's renewable energy production stood at over 2356 GW, which accounted for more than 40% of all energy growth worldwide. Without adaptable auxiliary energy sources, renewable energy is often unreliable and unable to constantly meet load demand because of its inherent variability (Hussain et al. 2020). As a result, natural gas (NG), which is eco-friendly and provides a flexible scheduling option, has been steadily included in power systems (Wang and Yuan 2020). According to *Nature* (Tollefson 2008), natural gas has been replacing coal as the primary fuel for generating energy in the United States. According to BPS 2019, by the end of 2018, natural gas had largely replaced coal as the primary fuel for generating electricity in North America (about 3%), the Commonwealth of Independent States (about 51%), the Middle East (about 70%), and Africa (about 40%) (BP Globe 2019). This presents significant opportunities for the development of hybrid solar–wind–gas power generation systems.

Recent studies have examined how natural gas and renewable energy can be integrated into hybrid solar–wind–gas systems, combined cooling, heating, and power (CCHP) simultaneously. Paolo et al. (2020) proposed a circulatory system to efficiently combine natural gas power and renewable energy. For this system, a solid oxide electrolysis system was used to store excess solar power by converting it to hydrogen, which was then converted back to electric power as needed using natural gas turbines. The viability of this system was demonstrated on the University of California microgrid. Zhang et al. (2019) created a hybrid CCHP system that simultaneously provided cooling, heating, and electric power using natural gas, biogas, and geothermal energy sources. The system's findings demonstrated its energy efficiency and economic advantages. Shima and Nima (2019) suggested a multi-carrier energy system with three subsystems: electricity networks, natural gas networks, and energy hubs. This system had a higher efficiency of energy usage and connected solar, wind, and natural gas power sources to meet the demands for electricity. Even though all of these earlier experiments contributed to the development of hybrid natural gas and renewable energy generation systems, the amount of natural gas used is still quite minimal. It is vital to discover whether a hybrid solar–wind–gas system can successfully meet large-scale power demands because there have not been many research studies exploring how natural gas and renewable energy complement each other in the production of electric power.

The feasibility, system planning, and interaction complementarity between natural gas and various renewable power sources have been looked at. For example, Lei et al. (2020) created a multi-objective expansion planning model for a regional energy system that integrates solar, wind, and natural gas. This model was proven to be effective in a set of multi-dimensional correlation scenarios. Rashid et al. (2019) examined the viability of a synergistic hybrid power system, in which concentrated solar power was used to generate steam in a Rankine cycle and natural gas was used as a backup source of steam when there was less solar irradiation to fully exploit the complementarity of generation between concentrated solar power and natural gas. Xi et al. (2020) built an integrated multi-energy power system that harnesses the power of solar, wind, and natural gas, for which a nonlinear complementary equilibrium model was developed to analyze the operational interactions between the subsystems. Thus, while the complementary nature of natural gas, solar, and wind energy has been demonstrated, there has been less discussion of the optimal dispatch strategy for effective, practical operation.

The majority of research on power scheduling using solar, wind, and natural gas has been geared toward increasing system efficiency (Shams et al. 2018), lowering operating costs(Jadidbonab et al. 2019), or balancing variations in electrical and thermal loads (Chamandoust et al. 2020). However, few studies have concurrently considered the economic benefits and reliability of hybrid solar–wind–gas power generation systems under renewable uncertainties. Due to their inherent intermittency, solar, and wind power offer a cheap but unstable source of electricity; as a result, natural gas can make up for changes in solar and wind power, however doing so would result in a large cost rise. Therefore, a balance between economic effectiveness and reliability exists that has not been taken into account in recent studies. At the same time, it is well known that the unpredictable nature of the production of renewable energy presents a problem for the dispatching strategy, which urgently requires effective resolution.

This chapter suggests a day-ahead scheduling optimization for a hybrid solar–wind–gas power generation system that fully addresses the reliable-economical equilibrium and renewable energy restrictions in an effort to partially fill this research gap. The multi-objective model is constructed for twelve common scheduling scenarios, in which the complementarities of solar, wind, and natural gas power are examined under a framework of economically sound equilibrium, and the uncertainties associated with renewable energy sources are successfully addressed. Utilizing a case from Golmud, China, the effectiveness and applicability of the created model and scheduling strategies for hybrid power generation are confirmed.

9.2 Key Problem Statement

The primary goal of a hybrid solar–wind–gas power system is to produce and provide daily power while simultaneously balancing power supply and demand and generating economic benefits. Prior to constructing the multi-objective model and

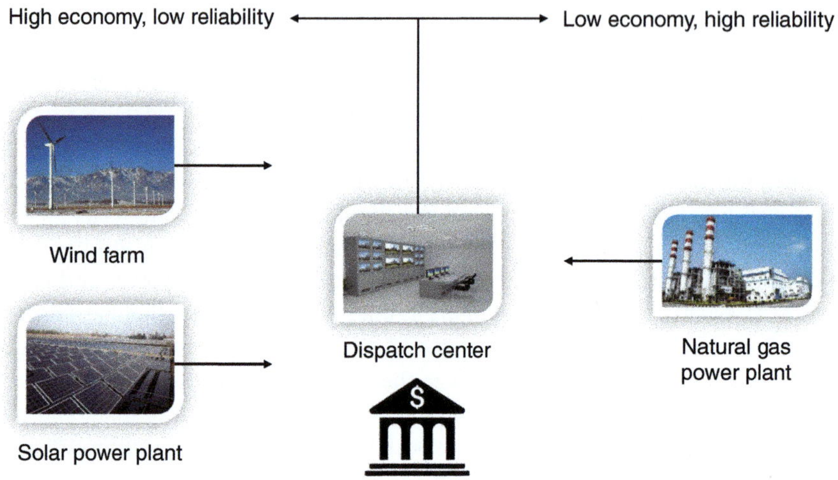

Figure 9.1 Structure and characteristics of the proposed hybrid solar–wind–gas power generation system.

creating the scheduling plans, certain background information and descriptions are initially provided.

9.2.1 System Description

As depicted in Figure 9.1, the hybrid system consists of three sub-systems in nearby locations: a solar power plant, a wind farm, and a natural gas power plant. These sub-systems are managed by one dispatch center, which receive power from photovoltaic array generating units, wind turbines, and natural gas-fired turbines. The solar and wind energy in this hybrid generation system is complemented at different times, meaning that there is enough solar energy during the day and in the summer and enough wind energy at night and in the winter. However, as system reliability is challenged by these changeable natural conditions, the natural gas power plant is utilized to compensate for these variations. On the other hand, natural gas-fired turbines require raw materials to convert the heat from the natural gas combustion, which lowers the revenue, but solar and wind power can offer large economic benefits due to their low input costs (Yin et al. 2021) and government subsidies (Wang et al. 2020b). Therefore, compared to conventional fossil energy, the hybrid system has higher conversion efficiencies, less pollutant emissions (Yang et al. 2019), and more reliable operations (Zhao et al. 2016). Under this circumstance, coordinated solar–wind–gas power generation is encouraged to be developed.

9.2.2 Resolving Renewable Energy Uncertainties

The integration of weather-driven renewable energy sources, such as solar and wind, into power-generating systems introduces uncertainty because their variations are unrelated to the features of power demand and are also outside the scope

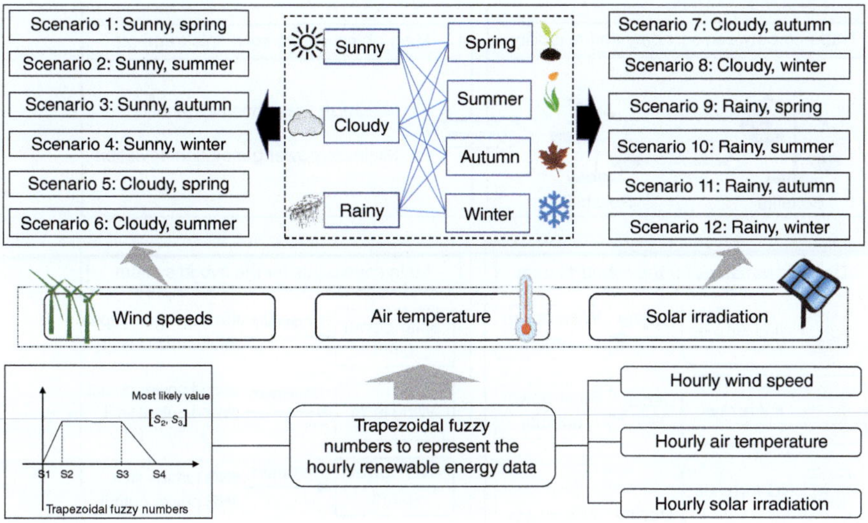

Figure 9.2 Methods to solve renewable energy uncertainties.

of conventional scheduling planning (Reddy 2017). In order to depict wind speed and solar irradiation, Wang and Gooi (2011) used probability density functions. The solar and wind resource uncertainties were determined by Reddy (2017) using a genetic algorithm, a two-point estimate method, and Monte Carlo simulations; however, while these studies made substantial contributions, they came at the cost of extremely high computer data processing loads. Therefore, to overcome these burdens, based on Xu et al. (2018) research and the known renewable energy variability features, this chapter established twelve dispatch scenarios to handle the seasonal and daily uncertainties of renewable sources, as shown in Figure 9.2. These scenarios are created by combining three different weather conditions (sunny, cloudy, and rainy) with each of the four seasons (spring, summer, autumn, and winter).

However, even with a determined scenario, there can be unpredictable sudden changes in natural conditions, making day-ahead power generation scheduling difficult (Xu et al. 2018). However, real-time data on renewable energy is crucial, and finding a solution has been the focus of extensive study (Wang et al. 2020a, Zhang et al. 2020b). Wang et al. (2019) have found these data are changing within a certain range, of which the most probable value can be estimated on a relatively small scale. Using historical data from Golmud as an example, at 12 am on a sunny summer day, the air temperature ranges from 14 to 24°C, with 17 to 23°C being the most likely range (WWO 2019). This changing feature is similar to that of the trapezoidal fuzzy numbers, which were originally defined as a map from the possibility space to the real value (Gupta 1992), and then have been widely applied to represent vague values (Sui et al. 2020; Huang and Xu 2020). In order to calculate the hourly data of wind speeds, sun irradiation, and air temperatures, we used trapezoidal fuzzy numbers in this work.

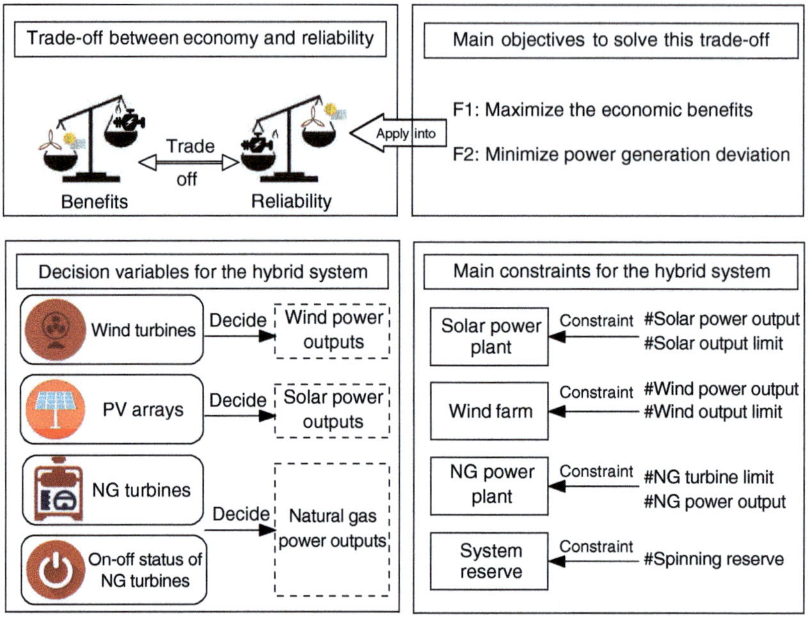

Figure 9.3 Methods to solve renewable energy uncertainties.

9.2.3 Achieving Reliable-Economical Equilibrium

In each scenario, the power system operator needs to trade off the power generation income and reliability, which remains another key problem, as shown in Figure 9.3. Higher economic benefits come from power sales and lower generation costs. While renewable energy has been found to be more economically efficient and has lower operational costs (Yin et al. 2021), in both developed and developing countries (Wang et al. 2020b). However, the high penetration of renewable energy can also bring unavoidable limitations to damage the continuity of power supply. To guarantee system reliability, natural gas with its controllability, flexibility, and shorter start-up time is taken as the complementary power source in this chapter. However, natural gas power generation is comparatively more expensive than solar and wind power, which conflicts with economic objectives. Therefore, scheduling decisions comprise a trade-off between economic benefits and generation reliability. Multi-objective models have been proven to be very effective tools for resolving contradictory objectives. For example, Zhong et al. (Salkuti 2019) implemented multi-objective programming to resolve the trade-offs between reliability maximization and cost minimization. Zhang et al. (2020a) proposed a multi-objective model to simultaneously smooth the net load curve and economy toward a coordinated electricity and natural gas system. Sheng et al. (2015) proposed multi-objective scheduling strategy for a combined system with cooling, heating, and power microgrid, which determined the trade-off between running costs and pollutant emission minimization. Given the above successful analysis, we built a multi-objective model in this chapter to implement the economy and reliability equilibrium for power generation.

9.3 Modeling

In this section, considering the reliable-economical equilibrium a mathematical model for the hybrid solar–wind–gas power generation system is built, and an efficient model transformation solution is also provided.

9.3.1 Notations

Indices

i : Index for solar power plant.

j : Index for wind farm.

l : Index for natural gas power plant.

t : Index for time period.

Decision variables

s_{it} : Photovoltaic arrays to be operated at solar power plant i time t.

w_{jt} : Wind turbines to be operated at wind farm j at time t.

g_{lt} : Quantity of natural gas combusted at gas turbines of natural gas power plant l at time t.

O_{lt} : Turn-on or turn-off status of gas turbines of natural gas power plant l at time t. $O_{lt} = 0$ means turn-off, $O_{lt} = 1$ means turn-on.

Certain parameters

ρ_n : Natural gas price.

p_{it} : Total power output of solar power plant i at time t.

p_{jt} : Total power output of wind farm j at time t.

p_{lt} : Total power output of natural gas power plant l at time t.

$\overline{p}_{lt}$: Maximum power output of natural gas power plant l at time t.

$\underline{p}_{lt}$: Minimum power output of natural gas power plant l at time t.

$\overline{p}_{jt}$: Maximum power output of wind farm j at time t.

$\underline{p}_{jt}$: Minimum power output of wind farm j at time t.

b_l^+ : Upper climbing limits of gas turbines at gas power plant l.

b_l^- : Lower climbing limits of gas turbines at gas power plant l.

$\overline{p}_{it}$: Maximum power output of solar power station i at time t.

$\underline{p}_{it}$: Minimum power output of solar power plant i at time t.

D_t : Power demand at time t.

α_s : Temperature coefficients of photovoltaic arrays, $-0.35\%/°C$.

T_s : Standard temperature, set as $25°C$.

c_i^{rs} : Rated power of photovoltaic arrays of solar power plant l.
v_{in} : Cut-in wind speed of wind turbines.
v_{out} : Cut-out wind speed of wind turbines.
v_r : Rated wind speed of wind turbines.
c_j^{rw} : Rated power of wind turbines of wind farm j.
p_j^w : On-grid price of wind power.
η_l : Energy conversion efficiency of gas turbines at plant l.
p_j^g : On-grid price of natural gas power.
ϕ^+ : Deviation of power supply and demand when supply is less than or equal to demand.
S_0 : Solar irradiation in standard test conditions (set as 1 kw/m^2).
p_i^s : On-grid price of solar power.
m_l^{on} : Minimum startup time of turbines at natural gas power plant l.
m_l^{of} : Minimum shutdown time of turbines at power plant l.
O_{lt}^{on} : Continuous operation time of gas turbines at natural gas power plant l at time t.
O_{lt}^{of} : Continuous shutdown time of gas turbines at natural gas power plant l at time t.
ϕ^- : Deviation of power supply and demand when supply is more than or equal to demand.
s : System load spinning reserve rate.

Uncertain parameters

$\widetilde{R}_{it}$: Solar irradiation obtained at solar power plant i at time t.
$\widetilde{T}_{it}$: Air temperature at solar power plant at time t.
$\widetilde{V}_{it}$: Wind speed at wind farm j at time t.

9.3.2 To Guarantee Economic Benefits and Reliability

The hybrid solar–wind–gas power system operator is responsible for guaranteeing both the economic benefits and the power generation reliability. Therefore, maximizing income and minimizing power supply and demand deviations are the main objectives in this chapter.

9.3.2.1 To Maximize Total Income

The income from hybrid power generation needs to be high enough to ensure adequate economic development. As stated in Hadayeghparast et al. (2019), sales of solar-, wind-, and natural gas–powered energy generation generate direct income, with natural gas purchases accounting for the majority of expenditures. Therefore, let p_i^s and p_j^w be the on-grid price for solar and wind power; p_{it} and p_{jt} denote the total solar and wind output at time t; p_j^w denote the natural gas electricity price; p_{lt} represent the natural gas plant; l denote power output at time t; and ρ_l denote the natural gas purchase costs for the natural gas power plant l. Therefore, the objective of maximizing total income can be formulated as Eq. (9.1).

$$\max f_1 = \sum_t \sum_i p_i^s p_{it} + \sum_t \sum_j p_j^w p_{jt} + \sum_t \sum_l p_j^g p_{lt} - \sum_t \sum_l \rho_l g_{lt} \tag{9.1}$$

where p_{it}, p_{jt}, and p_{lt} are decided by the decision variables s_{it}, w_{jt}, and g_{lt}, which represent the numbers of solar panels, wind turbines, and natural gas consumed, respectively, as shown in Eqs. (9.6), (9.8), and (9.10).

9.3.2.2 To Minimize the Deviation of Power Supply and Demand

To ensure hybrid power generation reliability, a feasible operating schedule to meet the hourly power demands over one day and the specific objective of minimizing supply and demand power deviations is designed. As stated by Garcia-Gonzalez and Castro (2001), let D_t be the power demand at time t; ϕ_t denote the deviation of power supply and demand at time t, then this objective can be formulated as Eq. (9.2). The function represented in Eq. (9.2) can be transformed linearly into a normal form with two auxiliary continuous variables (Garcia-Gonzalez and Castro 2001).

$$\phi_t = \left| \sum_t D_t - \left(\sum_t \sum_i p_{it} + \sum_t \sum_j p_{jt} + \sum_t \sum_l p_{lt} \right) \right| \tag{9.2}$$

When the value is positive, define the deviation of power supply and demand as ϕ_t^+, which can be written as Eq. (9.3).

$$\phi_t^+ = \sum_t D_t - \left(\sum_t \sum_i p_{it} + \sum_t \sum_j p_{jt} + \sum_t \sum_l p_{lt} \right) \tag{9.3}$$

When the value is negative, the deviation indicated as ϕ_t^- can be formulated as Eq. (9.4).

$$\phi_t^- = \left(\sum_t \sum_i p_{it} + \sum_t \sum_j p_{jt} + \sum_t \sum_l p_{lt} \right) - \sum_t D_t \tag{9.4}$$

So, the power supply and demand deviation minimization objective can be substituted as Eq. (9.5).

$$\min f_2 = \sum_t \left(\phi_t^+ + \phi_t^- \right) \tag{9.5}$$

9.3.3 Constraints of System Components

The following solar power plant, wind farm, natural gas power plant, and reserve spinning limits were taken into account to ensure a workable hybrid power generation scheduling optimization calculation:

9.3.3.1 Output of Solar Power Plants

Since the power output of each photovoltaic array at time t relies on the corresponding solar irradiation and air temperature, trapezoidal fuzzy numbers are employed to account for seasonal and weather uncertainty. Let $\widetilde{R}_{it}$ be the solar irradiation at solar power plant i at time t; $\widetilde{T}_{it}$ be the air temperature at solar power plant i at time t when the photovoltaic arrays are working; c_i^{rs} be the rated photovoltaic array power; S_0 be the irradiation in standard test conditions (set as 1 kw/m^2); T_s be the standard temperature set at 25°C; α_s be the photovoltaic array temperature coefficients, set at −0.35%/°C; and s_{it} be the decision variable for the photovoltaic arrays to be operated at solar power plant i at time t. It is assumed that the stationary solar power plant's photovoltaic arrays have the same technical specifications and are free of transmission disorders. Based on the model developed by the National Renewable Energy Laboratory (NREL) (Li and Qiu 2016), the solar power plant i output at time t can be formulated as Eq. (9.6).

$$p_{it} = \sum_t \sum_i \left\{ s_{it} \times c_s^r \times \frac{E\left[\widetilde{R}_{it}\right]}{S_0} \times \left[1 + \alpha_s \left(E\left[\widetilde{T}_{it}\right] - T_s\right)\right] \right\} \tag{9.6}$$

9.3.3.2 Solar Power Output Limitation

As each photovoltaic array has upper and lower power output bounds (Villalva et al. 2009), let $p_{it}^{\max}$ and $p_{it}^{\min}$ be the maximum and the minimum solar power plant power i output at time t. Solar power output limitation can be represented by Eq. (9.7).

$$\underline{p}_{it} \le p_{it} \le \overline{p}_{it} \tag{9.7}$$

9.3.3.3 Power Output of Wind Farm

The wind speed and the wind turbine power rate, both of which are unpredictable, determine the hourly wind power production at time t; therefore, trapezoidal fuzzy numbers are used to represent the wind speed at time t using parameter $\widetilde{V}_{it}$. Let w_{jt} be the wind turbines to be operated at wind farm j at time t; c_j^{rw} be the wind turbine rated power; v_{in} v_{out}, and v_r be the wind turbine cut-in, cut-out, and rated wind speed of the wind turbine. Then, based on the model proposed by (Sun et al. 2017), the power output of wind farm j at time t can be written as Eq. (9.8).

$$p_{jt} = \begin{cases} 0, \ E\left[\widetilde{V}_{it}\right] < v_{in} \text{ and } E\left[\widetilde{V}_{it}\right] > v_{out} \\ \sum_t \sum_j \left(w_{jt} \times c_j^{rw} \frac{\left(E\left[\widetilde{V}_{it}\right] - v_{in}\right)}{(v_r - v_{in})} \right), \quad v_{in} \le E\left[\widetilde{V}_{it}\right] \le v_r \\ \sum_t \sum_j \left(w_{jt} \times c_j^{rw} \right), \ v_r \le E\left[\widetilde{V}_{it}\right] \le v_{out} \end{cases} \tag{9.8}$$

9.3.3.4 Wind Power Output Limitation

Each wind turbine has upper and lower power output bounds (Villalva et al. 2009). Let $p_{jt}^{\max}$ and $p_{jt}^{\min}$ be the maximum and the minimum power outputs at wind farm j at time t. This constraint can be written as Eq. (9.9).

$$\underline{p}_{jt} \le p_{jt} \le \overline{p}_{jt} \tag{9.9}$$

9.3.3.5 Output of Natural Gas Power Plants

The natural gas power output at time t depends on the energy conversion rate (η_l) (Rashid et al. 2019), as shown in Eq. (9.10), where O_{lt} is the gas turbine status, and $O_{lt} = 0$ be turn-off and $O_{lt} = 1$ be turn-on.

$$p_{lt} = \sum_t \sum_l \eta_l g_{lt} O_{lt} \tag{9.10}$$

9.3.3.6 Operation Limitations of Natural Gas Turbines

For conventional gas-fired power turbines, there are climbing, startup–shutdown, and power output constraints (Ju et al. 2019). Let b_{lt}^+ and b_{lt}^- denote the upper and lower power turbine climbing limits at natural gas power plant l at time t; O_{lt}^{on} and O_{lt}^{of} represent the gas power turbine continuous operation and shutdown times at natural gas power plant l at time t; m_l^{on} and m_l^{of} represent the minimum gas power turbine startup and shutdown times at natural gas power plant l; and $\overline{p}_{lt}$ and $\underline{p}_{lt}$ denote the maximum and the minimum power output from natural gas power plant l at time t. These constraints are respectively shown in Eqs. (9.11)–(9.13).

$$O_{lt} b_{lt}^- \leq p_{lt} - p_{t-1}^l \leq O_{lt} b_{lt}^+ \tag{9.11}$$

$$\begin{cases} \left(t_{l(t-1)}^{on} - m_l^{on}\right)\left(e_{t-1}^l - O_{lt}\right) \geq 0 \\ \left(t_{l(t-1)}^{of} - m_l^{of}\right)\left(O_{lt} - e_{t-1}^l\right) \geq 0 \end{cases} \tag{9.12}$$

$$\underline{p}_{lt} \leq p_{lt} \leq \overline{p}_{lt} \tag{9.13}$$

9.3.3.7 System Spinning Reserve

According to the report published by Lin et al. (2020), renewable energy is mostly connected to the system using electronic inverters. Because of their low inertia, these inverter-based resources could result in high change frequencies. Therefore, to ensure power system stability, enough untapped capacity must be reserved. We let s be the system load spinning reserve rate (Xu et al. 2018). Then this constraint can be formulated as Eq. (9.14).

$$\begin{cases} \overline{p}_{lt} + p_{it} + p_{jt} \geq \sum_t D_t (1 + s\%) \\ \underline{p}_{lt} + p_{it} + p_{jt} \leq \sum_t D_t \end{cases} \tag{9.14}$$

9.3.4 Global Model

Based on the previous descriptions, a multi-objective model with reliable-economical equilibrium is developed for hybrid solar–wind–gas power generation in Eq. (9.15).

$$\max f_1 = \sum_t \sum_i p_i^s p_{it} + \sum_t \sum_j p_j^w p_{jt} + \sum_t \sum_l p_j^g p_{lt} - \sum_t \sum_l \rho_l g_{lt}$$

$$\min f_2 = \sum_t \left(\phi_t^+ + \phi_t^-\right)$$

$$
\text{s.t.} \begin{cases}
\phi_t^+ = \sum_t D_t - \left(\sum_t \sum_i p_{it} + \sum_t \sum_j p_{jt} + \sum_t \sum_l p_{lt} \right) \\
\phi_t^- = \left(\sum_t \sum_i p_{it} + \sum_t \sum_j p_{jt} + \sum_t \sum_l p_{lt} \right) - \sum_t D_t \\
p_{it} = \sum_t \sum_i \left\{ s_{it} \times c_s^r \times \frac{E[\tilde{R}_{it}]}{S_0} \times \left[1 + \alpha_s \left(E\left[\widetilde{T}_{it}\right] - T_s \right) \right] \right\} \\
p_{jt} = \begin{cases} 0, \ E\left(\widetilde{V}_{it}\right) < v_{in} \text{ and } E\left(\widetilde{V}_{it}\right) > v_{out} \\ \sum_t \sum_j \left(w_{jt} \times c_j^{rw} \frac{\left(E\left[\widetilde{V}_{it}\right] - v_{in}\right)}{(v_r - v_{in})} \right), \quad v_{in} \leq E\left[\widetilde{V}_{it}\right] \leq v_r \\ \sum_t \sum_j \left(w_{jt} \times c_j^{rw} \right), \quad v_r \leq E\left[\widetilde{V}_{it}\right] \leq v_{out} \end{cases} \\
p_{lt} = \sum_t \sum_l \eta_l g_{lt} O_{lt} \\
\underline{p}_{it} \leq p_{it} \leq \overline{p}_{it} \\
\underline{p}_{jt} \leq p_{jt} \leq \overline{p}_{jt} \\
O_{lt} b_{lt}^- \leq p_{lt} - p_{t-1}^l \leq O_{lt} b_{lt}^+ \\
\begin{cases} \left(t_{l(t-1)}^{on} - m_l^{on} \right) \left(e_{t-1}^l - O_{lt} \right) \geq 0 \\ \left(t_{l(t-1)}^{of} - m_l^{of} \right) \left(O_{lt} - e_{t-1}^l \right) \geq 0 \end{cases} \\
\underline{p}_{lt} \leq p_{lt} \leq \overline{p}_{lt} \\
\begin{cases} \overline{p}_{lt} + p_{it} + p_{jt} \geq \sum_t D_t (1 + s\%) \\ \underline{p}_{lt} + p_{it} + p_{jt} \leq \sum_t D_t \end{cases}
\end{cases}
\tag{9.15}
$$

The reliability goal is to reduce power supply and demand variations, whereas the economic goal is to increase overall profits. For each power source, decision variables are set explicitly in order to find the best hybrid power generation scheduling: the numbers of photovoltaic arrays (s_{it}), wind turbine (w_t^i), natural gas consumption (g_t^i), and the natural gas turbines status (O_{lt}). Based on the solar–wind–gas power system characteristics, corresponding constraints are considered in Eqs. (9.6)–(9.14). Therefore, this global model can be formulated in Eq. (9.15).

The different power sources have concurrent economic and reliability characteristics therefore, the multi-objective model suggested above can help system operators identify an ideal reliable-economical equilibrium solution. In other words, natural gas power ensures dependability by reducing the variations in solar and wind power while still generating significant income.

The global model proposed in this chapter is general and adjustable. The formulations suggested can be used by other power systems with different generating equipment and conditions by adjusting the solar, wind, and natural gas power output parameters, and system operators can change, add, or remove objectives and constraints depending on their unique goals and environments.

9.3.5 Mathematical Solving

Because of the difficulties associated with concurrently satisfying the conflicting objectives, the ε-constraint method is applied to solve the proposed global model (Eq. (9.15)) as it has been proven to effectively transform complex mathematical models into equivalent single forms (Nojavan et al. 2017), and the fuzzy satisfying approach, which was successfully applied in Mohseni-Bonab et al. (2016), is employed to determine the best compromise solution.

9.3.5.1 Transforming the Multi-Objective Model Using ε-Constraint Method

In the ε-constraint method, the main objective is optimized, and the other is treated as an inequality constraint (Mohseni-Bonab et al. 2016). The system operator in this chapter is the electrical company, whose primary goal is to increase profits; therefore, the economic objective is optimized and the reliability objective is viewed as the constraint.

$$\max f_1 = \sum_t \sum_i p_i^s p_{it} + \sum_t \sum_j p_j^w p_{jt} + \sum_t \sum_l p_j^g p_{lt} - \sum_t \sum_l \rho_l g_{lt}$$

$$\text{s.t.} \begin{cases} \sum_t \left(\phi_t^+ + \phi_t^-\right) \leq \vartheta\mu \\ \text{Eqs. (9.7)–(9.15)} \end{cases} \tag{9.16}$$

Further, two parameters are utilized to define the control parameter ε. The first one is ϑ, which is the system operator's attitude toward system reliability; the second one is μ, which is the power supply and demand deviation. The multi-objective model is transformed as in Eq. (9.16).

9.3.5.2 Select the Optimal Solution Using Fuzzy Satisfying Method

To select the best solution from the Pareto optimal set described above, a fuzzy satisfying approach proposed by Chen (1994) is utilized to solve the multi-objective optimization problem, in which the fuzzy membership is in the interval [0, 1], which is indicated as κ_n^i, and indicates the degree of optimality for the nth objective function in the ith solution. The higher the optimality degree, the greater the satisfaction with the solution. Specifically, the decision-maker is fully satisfied when $\kappa_n^i = 1$, and dissatisfied when $\kappa_n^i = 0$. Let f_n^i be the value of the nth objective function in the ith Pareto optimal solution and $f_n^{\min}$ and $f_n^{\max}$ be the minimum and maximum objective function values. Therefore, for the objective function to be maximized, a larger objective function is better and vice versa (Esmaeili et al. 2016), which means that the fuzzy membership functions in this chapter can be developed based on the original equation form (Balaman 2016), as shown in Eqs. (9.17)–(9.18), where κ_1^i represents the fuzzy membership function for objective maximization while κ_2^i represents the fuzzy membership function for objective minimization, the optimal κ^opt is obtained, as shown in Eq. (9.19).

$$\kappa_1^i = \begin{cases} 1 & f_1^i > f_1^{\max} \\ \dfrac{f_1^i - f_1^{\min}}{f_1^{\max} - f_1^{\min}} & f_1^{\min} < f_1^i \leq f_1^{\max} \\ 0 & f_1^i \leq f_1^{\min} \end{cases} \tag{9.17}$$

$$\kappa_2^i = \begin{cases} 1 & f_2^i \leq f_2^{\min} \\ \dfrac{f_2^{\max} - f_2^i}{f_2^{\max} - f_2^{\min}} & f_2^{\min} < f_2^i \leq f_2^{\max} \\ 0 & f_2^i > f_2^{\max} \end{cases} \tag{9.18}$$

$$\kappa^{opt} = \max\left[\min\left(\kappa_1^n, \kappa_2^i, \ldots, \kappa_N^n\right)\right] \tag{9.19}$$

Detailed model-solving process is clarified as having eight steps, which are as follows:

(1) Collecting the data for model solving, including hourly solar irradiation, hourly wind speed, hourly air temperature, and typical power demand;
(2) Transform the global model with f_1 as the only objective function and f_2 as the constraint;
(3) Set the decision-maker's attitude parameters (ϑ and μ) toward objective f_2;
(4) Maximize f_1 subject to all of the constraints, and calculate the maximum value of objective f_2;
(5) Minimize f_2 subject to all of the constraints, and calculate the minimum value of objective f_1;
(6) Change the value of f_2 in discrete steps with the maximum and minimum value, and a Pareto front is obtained.
(7) The fuzzy satisfying approach is utilized to set optimal solutions from the Pareto front, and the objective functions are transformed with Eqs. (9.17)–(9.18);
(8) The best possible solution is obtained using Eq. (9.19), which has the maximum weakest membership function.

9.4 Case Study

In this section, the practicality and efficiency of the proposed optimization model is demonstrated on the hybrid solar–wind–gas power generation system in Golmud City, Qinghai Province, China.

9.4.1 Case Description

Because natural gas-powered generation is highly reliable, efficient, and environmentally friendly, the 13th Five-Year National Energy Plan ([2016] No. 2743) promoted its inclusion in hybrid power generation systems (NDRC 2016) as a supplement to renewable energy. Golmud, also known as "Photovoltaic City," is located in the hinterland of the Qinghai-Xizang plateau in the middle of Qinghai Province. It boasts the highest annual solar radiation, up to 7400 MJ/m^2, and the second-highest average number of sunshine hours, between 2550 and 3400 h, after Tibet (Yang et al. 2017). Golmud also possesses the largest untapped wind energy potential in China with the lowest operating and maintenance expenses due to its proximity to the Qinghai Tibet Plateau (Shi et al. 2017; Yang et al. 2018).

Golmud, however, has experienced severe solar power curtailment issues due to insufficient grid connection technology, which has impeded the development of solar energy (Guangfu 2016). According to the National Energy Administration, the amount of solar electricity curtailed in the first half of 2019 was 6.3%, or 5.2 billion kilowatt hours. This was an increase of 3% from the same period in 2018 (NEA 2019). The total renewable generating capacity in Golmud by the middle of 2019 was 47.3 GW, of which solar photovoltaic power contributed around 29.3 GW and wind power only 2.5 GW, showing that the wind power developments were substantially less robust than the solar photovoltaic power (Ma 2017). The local electricity provider has shown little enthusiasm to expand wind power because of the requirement to invest in wind energy storage. In order to reduce curtailment and increase the online rate, which will encourage the further development of solar and wind power, Golmud's solar, wind, and natural gas resources were combined to create a physical "solar–wind–gas power generation bundling mode", as shown in Figure 9.4. As shown in the figure, the eastern Golmud solar power station, the Dongtai wind farm, and the Golmud natural gas power station are the three energy sources for the hybrid power generation system, all of which are connected to the main grid. The eastern Golmud power plant, which has a cumulative installed capacity of 800MW and is connected to the main grid with a 110 kV line from Golmud to Nomuhong, was selected as the solar power scheduling object. By the end of 2019, Golmud had invested 3.75 billion CNY in 300 MW of installed wind power at the Dongtai wind farm. The Golmud natural gas power plant, which is located on the highest altitude in northwest China, has a total installed capacity of 300 MW and plays an important role in resource and economic development in Golmud.

9.4.2 Data Collection

Before applying the modeling technique, the certain and uncertain parameter data were collected. The certain data for the hybrid solar–wind–gas power generation system are given in Table 9.1. Based on the benchmark on-grid price for solar, wind, and natural gas power published in China, the price for solar power is 0.88 CNY/kWh, for wind power is 0.57 CNY/kWh, and for natural gas power is 0.36 CNY/kWh, with the natural gas purchase price being 1.15 CNY/m^3. The relevant data on natural gas power were collected from (Poullikkas 2005), the corresponding technical data on solar and wind power were mainly obtained from the chapter (Khaloie et al. 2020); and the power output bounds, available generating units, and operation

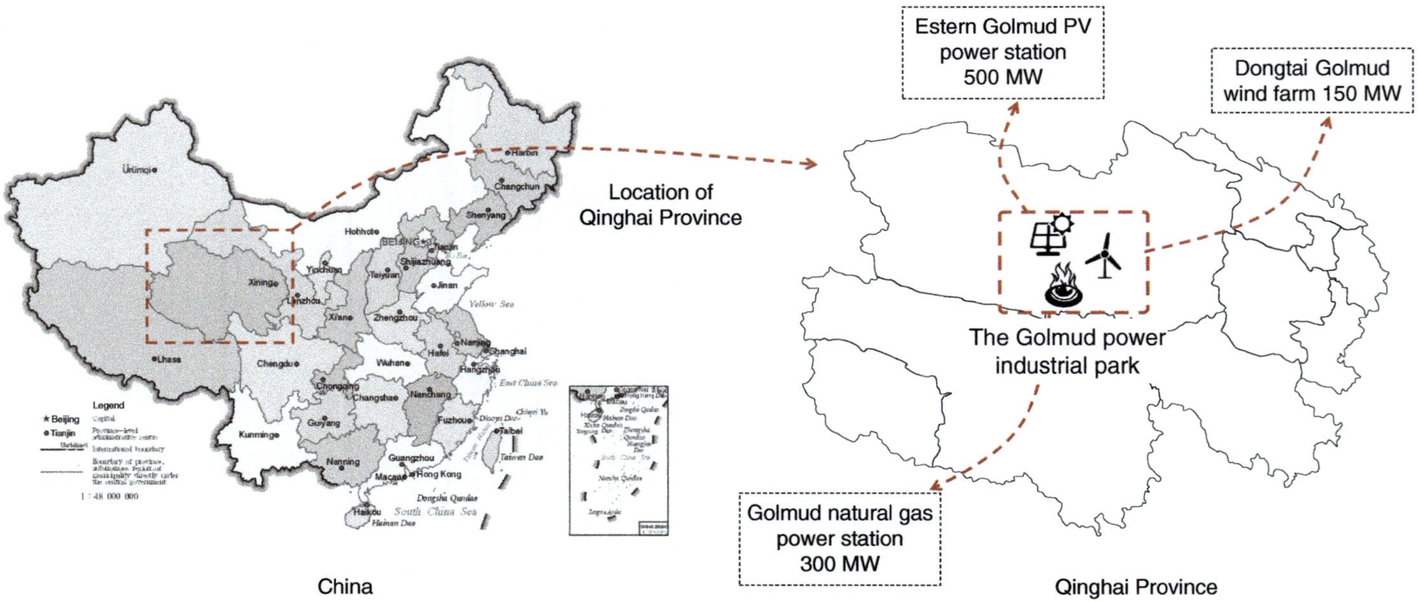

Figure 9.4 Physical model for hybrid power generation in solar–wind–gas bundling mode in Golmud.

Table 9.1 Certain parameters for the hybrid solar–wind–gas power system.

Parameters	Values	Units
Conversion rate between NG consumption and electric power	0.0046	MWh/m^3
Minimum startup time of natural gas turbine	1	h
Minimum shutdown time of natural gas turbine	1	h
Minimum shutdown time of natural gas turbine	1.15	CNY/m^3
Minimum shutdown time of natural gas turbine	0	MW
Maximum output of wind farm	300	MW
Maximum output of wind farm	800	MW
Maximum output of natural gas power plant	300	MW
Minimum output of natural gas power plant	100	MW
Lower climbing limits of natural gas turbines	−140	MW/h
Upper climbing limits of natural gas turbines	140	MW/h
On-grid price of natural gas in Qinghai Province	360	CNY/MW
On-grid price of wind in Qinghai Province	570	CNY/MW
On-grid price of solar in Qinghai Province	880	CNY/MW

parameters were accessed from news reports on the Golmud government website (Geermu 2020).

Based on historical data and Golmud power demand reports, Figure 9.5a shows the forecasted power demand in the one hour time step within a normal day. The data for the uncertain parameters, which include the hourly solar irradiation, the real-time air temperature, and the actual wind speed for the different seasons and climatic conditions, were collected from the Golmud weather bureau (Qinghai Meteorological Bureau 2019) and the World Weather Online (WWO 2019) websites, as shown in Figure 9.5b–d. However, due to the limitations on space, only data for solar irradiation in spring, while air temperature and wind speed in sunny days are shown below. The data for these climate indicators were processed with trapezoidal fuzzy numbers in this chapter due to the uncertainty associated with highly renewable energy sources. The trapezoidal fuzzy numbers for current air temperature, wind speed, and solar irradiation were gathered through field research interviews with local officials, experts, and engineers involved in this hybrid power system. Four values, ranging from minimum to maximum, were obtained from these individuals, with the most likely values falling between the two center figures. After analysis, the initial data were chosen, and the extreme values were eliminated.

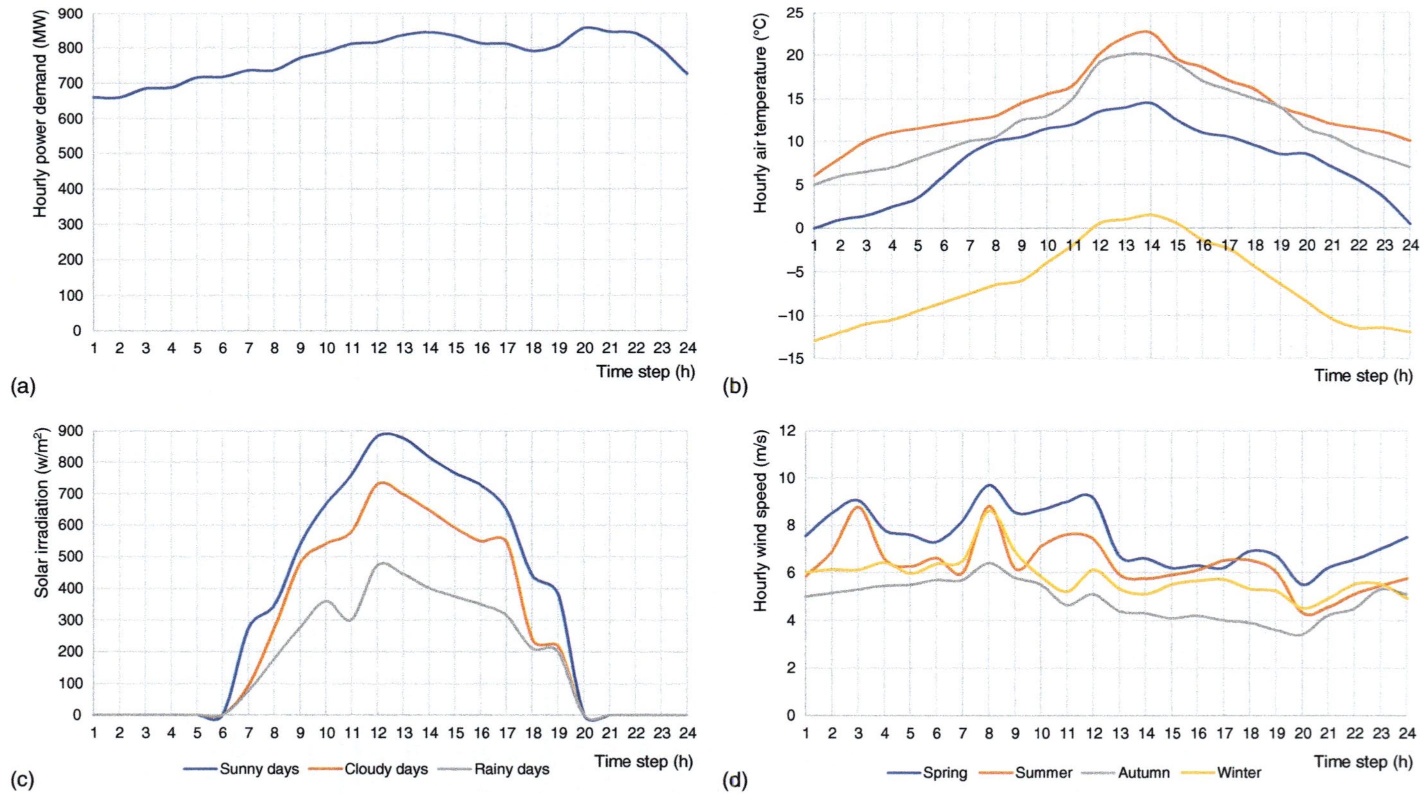

Figure 9.5 Typical uncertain data. (a) Typical power demand data, (b) typical temperature in sunny days, (c) typical solar irradiation in spring, and (d) typical wind speed in sunny days.

9.5 Calculation Results and Analysis

To elucidate the effectiveness of the proposed solar–wind–gas power generation system under the 12 scheduling scenarios, two cases are analyzed in the following.

9.5.1 Optimal Solutions

The Pareto fronts for the scheduling scenarios are shown as three-dimensional diagrams in Figure 9.6, from which the trade-offs between the economic benefits and system reliability can be seen; that is, as the total benefits (f_1) decrease, the system reliability increases (f_2), which indicates the inherent contradictions in the hybrid power generation system.

Because each of the maximum weakest membership functions under the 12 scenarios has its own maximum limits, there is only one best compromise solution. For example, from sunny weather to rainy weather in the spring season, it can be seen that the optimal solutions respectively have maximum weakest memberships of 0.76, 0.71, and 0.75, economic benefits of 1023, 910, and 881×10^4 CNY, and power deviations of 1013, 716.7, and 1070 MW. The same procedure can be easily adapted to obtain the multi-objective optimization Pareto optimal solutions for the other scenarios.

9.5.2 Economic Benefits Analysis

The economic benefits changing under the different scheduling scenarios are summarized in Table 9.2, which shows that the total benefits vary with the season, peaking in summer and bottoming out in autumn. For instance, on sunny days in spring, summer, autumn, and winter, the total benefits are, respectively, 28,144, 31,176, 25,984, and 26,660 ($\times 10^3$ CNY), and regardless of the season, the total benefits slowly decline from the maximum on sunny days to the minimum on rainy days. Therefore, from the combined results in Table 9.2, it was concluded that the differences in benefits are primarily influenced by solar power generation and natural gas consumption as the on-grid solar power price is the highest at 880 CNY/MW, which is almost 1.4 times that of natural gas (360 CNY/MW). However, the natural gas purchase expenses are also the highest at 1.15 CNY/m^3, which demonstrates the complementary influences of solar and natural gas power on the economic benefits; that is, the economic benefits increase as solar energy supply increases and natural gas consumption decreases. For example, in summer, the solar power output is 23,094.915 MW and the natural gas consumption is 31,176($\times 10^3$ CNY), while in spring, the solar power output is 15,523 MW and the natural gas consumption is 32,131($\times 10^3$ CNY).

However, comparing the results in winter and autumn, the lower solar output (10,887 MW in winter and 15,531 MW in autumn) with the higher wind power output income (15,718 $\times 10^3$ CNY in winter and 10,787$\times 10^3$ CNY in autumn) caused the lower costs of natural gas purchase (3094$\times 10^3$ CNY in winter and 3476 $\times 10^3$ CNY in autumn). The hybrid solar–wind–gas power system's complementary nature is

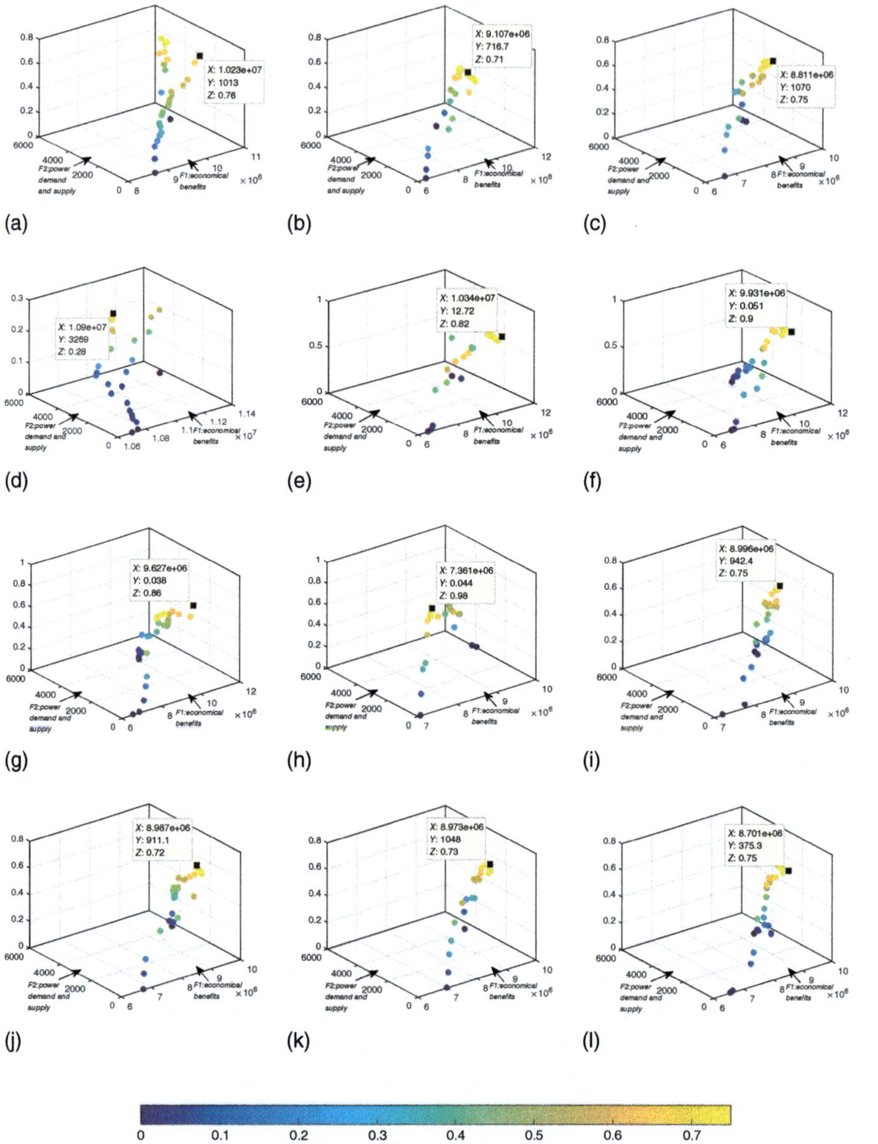

Figure 9.6 Optimal solutions for different scheduling scenarios. (a) Spring-sunny, (b) spring-cloudy, (c) spring-rainy, (d) summer-sunny, (e) summer-cloudy, (f) summer-rainy, (g) autumn-sunny, (h) autumn-cloudy, (i) autumn-rainy, (j) winter-sunny, (k) winter-cloudy, and (l) winter-rainy.

amply illustrated by the way their respective power outputs and financial benefits complement one another.

The output ratios were compared in order to better assess the solar, wind, and natural gas power outputs under the various scheduling situations, as shown in Figure 9.7. When the weather is constant, the ratios and power outputs vary according to the seasons, with summer contributing the most solar power, followed by spring, autumn, and winter. According to Golmud's wind resource patterns, the

Table 9.2 Economic benefits in optimal solution under 12 scheduling scenarios.[a)]

	Spring				Summer			
Scenarios	**Sunny**	**Cloudy**	**Rainy**	**Total**	**Sunny**	**Cloudy**	**Rainy**	**Total**
Solar outputs	7571	4798	3153	15,523	8417	7723	6953	23,094
Wind outputs	5450	7726	9754	22,930	4964	5685	6081	16,731
NG outputs	4160	4360	4331	12,852	6055	2772	3133	11,961
Solar benefits	6662	4223	2774	13,660	7407	6796	6119	20,323
Wind benefits	3106	4403	5560	13,070	2829	3240	3466	9537
NG benefits	1497	1569	1559	4626	2180	998	1128	4306
NG costs	1040	1090	1082	3213	1513	693	783	2990
Total benefits	10,227	9106	8811	28,144	10,903	10,342	9930	31,176

	Autumn				Winter			
Scenarios	**Sunny**	**Cloudy**	**Rainy**	**Total**	**Sunny**	**Cloudy**	**Rainy**	**Total**
Solar outputs	6975	4541	4013	15,531	4105	3857	2924	10,887
Wind outputs	5383	4794	8747	18,925	8580	8931	10,063	27,575
NG outputs	3809	5748	4350	13,905	4394	4427	3556	12,378
Solar benefits	6138	3996	3532	13,667	3612	3395	2573	9581
Wind benefits	3068	2732	4985	10,787	4890	5091	5736	15,718
NG benefits	1371	2068	1566	5005	1581	1593	1280	4456
NG costs	952	1436	1087	3476	1098	1106	889	3094
Total benefits	9626	7361	8996	25,984	8986	8973	8700	26,600

a) The unit for power outputs of solar, wind, and natural gas (NG) is MW, and that for power benefits is $\times 10^3$ CNY.

winter months contribute the most wind power, while the fall months contribute the least.

As a result, in the fall, when there is the least amount of solar and wind energy, there is a greater requirement for natural gas electricity. Due to the rapid conversion of solar energy to electricity when the season is fixed, the solar power output steadily falls from sunny weather to rainy weather (Reddy and Momoh 2015). In order to maintain a consistent supply of electricity when solar output falls, the contributions of wind and natural gas steadily rise, with wind power taking precedence because it earns more money. For example, on sunny and cloudy days in spring, when the solar power output falls to about 15.65%, the wind power increases to 14.04% and the natural gas power increases to only 1.61%.

9.5.3 System Reliability Analysis

Figure 9.8 shows the maximum technical power generation output with/without natural gas, and Table 9.3 shows the specific reliability improvement rates. As can

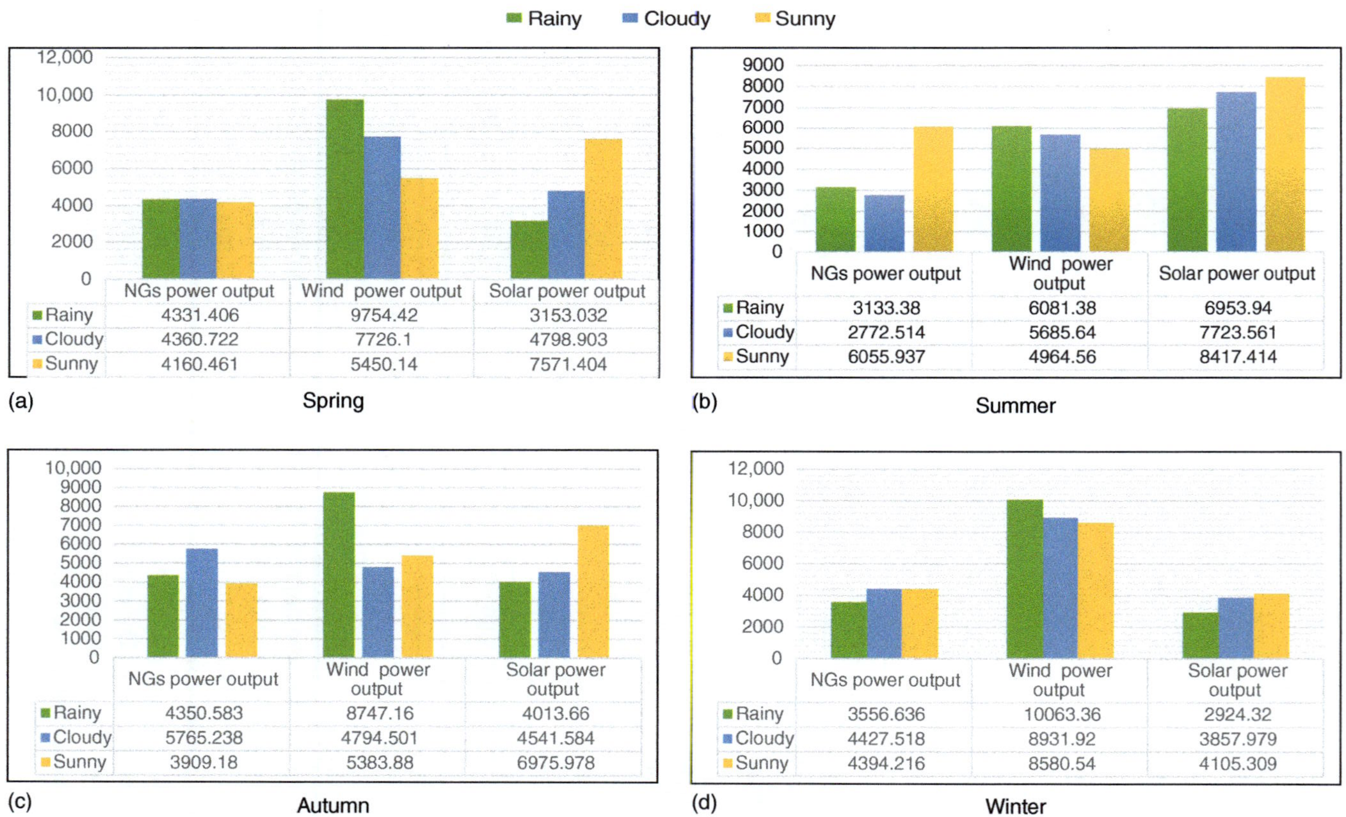

Figure 9.7 Power output ratios of solar, wind, and natural gas in optimal solutions under 12 scheduling scenarios (MW). (a) Spring, (b) summer, (c) autumn, and (d) winter.

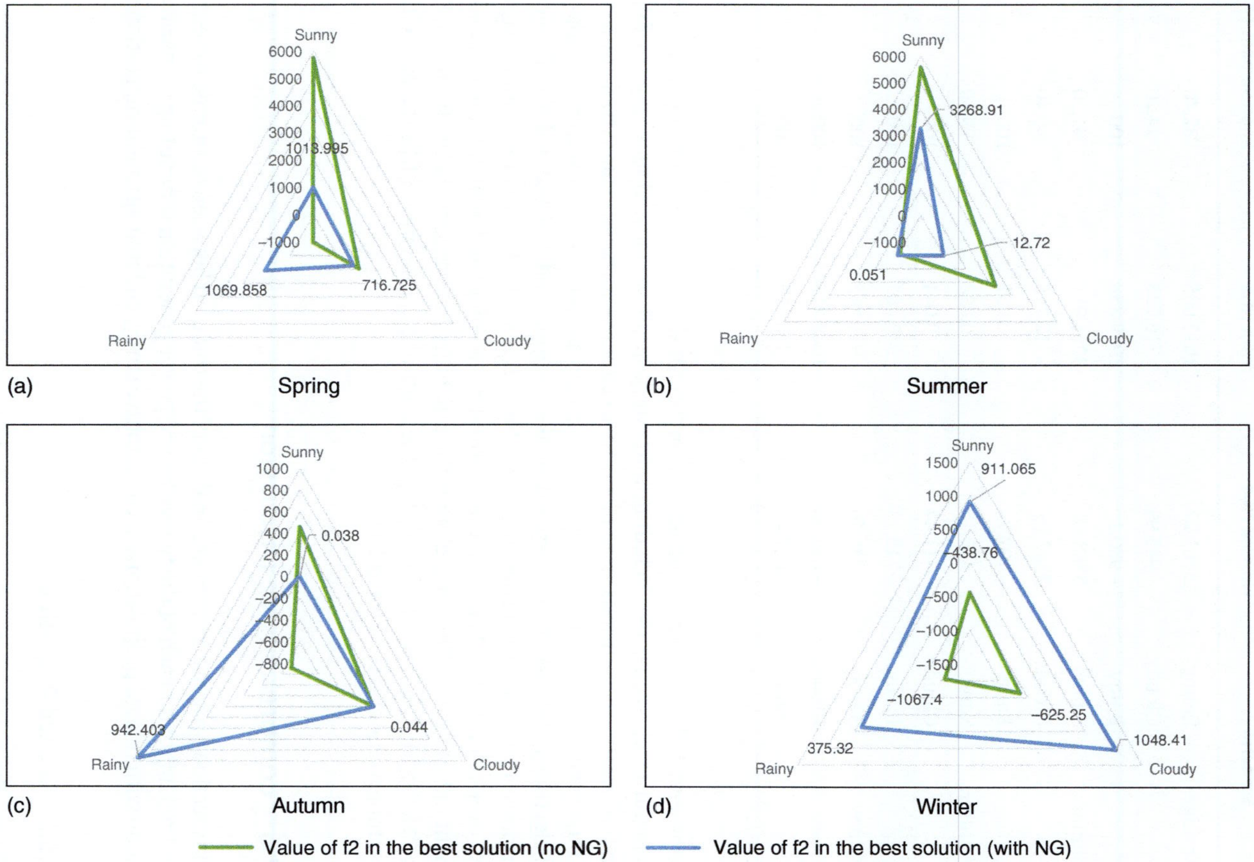

Figure 9.8 Maximized economic benefits under 12 scenarios (MW). (a) Spring, (b) summer, (c) autumn, and (d) winter.

Table 9.3 Improvement on system reliability under income maximum with natural gas integration.

Scenario		Value of f2 without NGs	Value of f2 with NGs	Rate of improvement (%)
Spring	Sunny	5761.22	1013.005	82.40
	Cloudy	945.99	716.725	24.20
	Rainy	−980.1	1069.858	100
Summer	Sunny	5588.68	3268.911	41.50
	Cloudy	2287.15	12.715	99.40
	Rainy	−102.74	0.051	100
Autumn	Sunny	456.17	0.038	99.90
	Cloudy	−11.44	0.044	100
	Rainy	−715.13	942.403	100
Winter	Sunny	−438.76	911.065	100
	Cloudy	−652.25	1048.417	100
	Rainy	−1067.4	375.316	100

be observed, the solar and wind maximum technical outputs cannot meet the power demand in all scenarios, including rainy days in all seasons, cloudy days in fall, and all weather conditions in winter, due to either insufficient solar irradiation or wind speed. However, the incorporation of natural gas ensures a sufficient supply of electricity and lessens the disparity between supply and demand (value of f_2). For instance, during sunny spring days, the value of f_2 is 5761.22 MW without natural gas but drops to 1013.005 MW when the natural gas is integrated, which is an improvement rate of 82.4%. When it is cloudy, the reliability increases by 24.2%, from 945.99 to 716.725 MW, and when it israining, the natural gas integration fills the gap with an improvement rate of up to 100%.

9.6 Comprehensive Discussion

This section provides a thorough analysis of the trade-off between economic benefits and reliability for each of the twelve hybrid power generation scheduling scenarios, along with related propositions, a comparative study, and management suggestions.

9.6.1 Related Propositions

Firstly, the solar–wind–gas hybrid power generation system can be equipped to increase the economic benefits and the system reliability. Because hybrid power system operators can flexibly schedule unit commitments based on the characteristics of each power source under the demand curve, they are better equipped to achieve an optimal multi-objective scheduling strategy. For instance, in order to maximize profits, the operator supplies as much solar energy as feasible given that the policy in Qinghai province prioritizes solar electricity, which has the highest price of 0.88 CNY/kWh and is almost 2.5 times more expensive than natural gas, as shown in Table 9.2 and Figure 9.7. Therefore, the operator would be more likely to limit wind and natural gas generation when the overall power output is sufficient to allow for solar power generation to have the biggest proportion, which would then create higher income. In summer, for example, when the total system output satisfies the power demand, solar power provides 25.54%, 35.14%, and 37.61% of the power, and natural gas provides 31.16%, 17.13%, and 19.38% of the power, respectively returning income of 10,903, 10,342, and 9930 ($\times 10^3$ CNY). However, a higher natural gas component is required to guarantee a steady supply of electricity when the output from renewable sources is insufficient. For example, as shown in Table 9.3 and Figure 9.8, on rainy days, the power supply and demand mismatch was −980.1, −1022.74, −715.13, and −1067.4 MW before the natural gas integration. The output of solar and wind cannot meet the demand.

Secondly, in the general case, the economic benefits decrease from sunny to rainy days, except in autumn. Solar and wind power yield higher returns than natural gas power in China as a result of the government's policies on renewable energy subsidies. For instance, in the hybrid power system planned in Golmud, the on-grid prices of solar and wind energy (including the on-grid price and government subsidies) are 0.88 and 0.57 CNY/kWh, respectively, while the cost of natural gas (which is not supported by a subsidy policy) is 0.36 CNY/kWh. As a result, bigger sale returns are achieved with increased solar power output, which often translates to more income from sunny conditions than from wet ones. However, due to the climatic conditions in Golmud, the solar production is only marginally greater than on rainy days and the wind output is significantly lower on rainy autumn days. As a result, the maximum natural gas input is required, which results in lower returns and higher expenses, as shown in Figure 9.7 and Table 9.2.

Further, natural gas has a supplementary advantage over unstable power sources when the output of renewable energy is insufficient due to its scheduling flexibility. As shown in Figure 9.7, when the weather changes from sunny to cloudy in the autumn, the solar output ratio falls from 43.14% to 30.11%, the wind output rises from 33.3% to 31.79%, and the natural gas output falls from 23.56% to 38.1%. This change rate of 14.54%, or nearly 10 times the change in wind power, shows that natural gas has a higher complementarity than wind power because it can be more flexibly scheduled. In this case also, the wind power output nearly reaches its maximum, which means that natural gas contributes more power when the solar output is insufficient. To sum up, because of its flexibility and controllability, natural gas has a higher complementary advantage than the renewable energy in the hybrid power system.

Finally, since the output of renewable energy sources is constrained by environmental factors, solar and wind power can make up for their corresponding climatic and seasonal output fluctuations. Renewable energy's output is unpredictable because it depends on natural factors. For instance, solar energy output depends on air temperature and irradiation intensity, which fluctuate with the seasons, the weather, and specific events, with the maximum value occurring on sunny days in the summer and the minimum value occurring on rainy days in the winter. Solar irradiation increases at 7 a.m., peaks around 1 pm, and decreases gradually until sunset. In a similar manner, the air temperature increases during the night, peaks at 12 a.m., and then slowly drops. As a result, depending on the season, solar power generation is low in the morning and evening and concentrated between 7 a.m. and 7 p.m. However, wind speed power production is at its highest during rainy winter days, varies little, and increases slightly at dawn and dusk; therefore, solar and wind power can compensate for each other. To summarize, choosing solar and wind as renewable energy sources proved to be reasonable and valid in this hybrid power system.

9.6.2 Comparative Study

A comparative study is presented in this section, with specific attention paid to the cooperation complementarity of hybrid power systems, equilibrium optimization, and techniques for resolving uncertainty in renewable energy sources.

Comparison of cooperation complementarity analysis in hybrid solar–wind–gas power generation. As is reviewed earlier Paolo et al. (2020) developed a campus microgrid integrating solar power and natural gas, in which natural gas worked as a backup to compensate for solar power variations. Zhang et al. (2019) designed a hybrid CCHP system, integrating natural gas with biogas and geothermal energy sources, aiming at providing cooling, heating, and electric power simultaneously. Shima and Nima (2019) proposed a multi-carrier energy system, which included electricity networks, natural gas networks, and energy hubs. All of the aforementioned studies examined multi-energy systems, in which natural gas was only used on a very modest scale and primarily as a backup supply. To assess if natural gas can successfully replace conventional fossil fuels in the future, particularly in large-scale operations, it is crucial to analyze the specific collaboration complementarity of renewable energy and natural gas in the generation of electric power. As a result, this chapter created an optimization for the generation of hybrid electric power that took into account the complementary nature of the cooperative solar, wind, and natural gas resources.

Comparison of equilibrium optimization in hybrid power generation with trade-off in economic benefits and reliability. Jadidbonab et al. (2019) proposed a risk-constrained scheduling strategy with renewable energy and natural gas, aiming at improving the system efficiency. Shams et al. (2018) developed the stochastic day-ahead scheduling toward a multi-energy system with renewable energy and natural gas, with objective to minimize operation costs. Few research have focused on the trade-off between economical and dependable power generation, despite the fact that these studies made substantial contributions to the scheduling optimization of natural gas and renewable energy hybrid power systems. In order to increase revenue and provide a steady supply of energy, a multi-objective scheduling model for a hybrid solar–wind–gas power system was designed in this chapter. This model is more reasonable and feasible for system operators and can guarantee a reliable supply of energy.

Comparison of the handling of renewable power generation uncertainty. Based on the literature reviews above, the available methods to solve renewable energy uncertainty, are mainly probability density functions (Wang and Gooi 2011), Monte Carlo simulations (Reddy 2017), deep learning technology (Wang et al. 2020a), and stochastic simulation scenarios (Zhang et al. 2020b), all of which proved relatively effective. However, no prior research has been able to anticipate the power resource hourly data reliably without requiring a great deal of mathematics. This chapter created 12 scenarios based on various solar irradiation, air temperature, and wind speed situations using effective historical data analysis and expressed the hourly data using trapezoidal fuzzy numbers, which is more practical and convenient.

9.6.3 Management Recommendations

The following includes a few managerial suggestions based on the above analysis and discussion.

Firstly, to deliver consistent, affordable power generation, the technology for integrating natural gas into high-penetration renewable energy systems needs to be improved further. For example, the Chinese government has proposed to build an intelligent and efficient dispatch and operation system to explore a joint dispatch mechanism for electricity, heat and natural gas (National Development and Reform Commission of China, National Energy Administration of China 2022), and the findings in this chapter have provided important support for such policies. Although there has been a lot of research done on the viability of using natural gas in hybrid power systems, most of it has been theoretical. The technologies for energy conversion, transposition, storage, and hybrid scheduling need to be improved in practice. Second, in order to develop effective day-ahead power scheduling systems, technology for anticipating fluctuations in renewable power generation must also be enhanced. Although this issue can be addressed using techniques like building scenarios and solar and wind probability density estimates, the unpredictable nature of solar radiation, wind speed, and air temperature can cause schedule conflicts, financial losses, and/or unstable power supply. Therefore, to promote effective, efficient energy industry development, natural gas power integration and advances in renewable energy forecasting technologies are needed.

Secondly, the proposed hybrid solar–wind–gas power generation could be a workable solution for the system operators to get over the risks associated with renewable energy while also generating reliable returns. Natural gas can be incorporated into hybrid power generation systems to make up for the limitations of renewable energy, ensure a stable power supply, and achieve an equilibrium between economic benefits and reliability. Renewable energies, such as solar, wind, and hydro, generate higher returns due to their low operational costs and the current preferential government policies, but generation uncertainties affect the reliability of the power supply. A hybrid solar–wind–gas power system can increase power supply reliability and energy sales returns, as was demonstrated in the Golmud case discussed in this chapter. Therefore, it is important to develop complementary solar–wind–gas power generation systems to ensure reliable power supplies that give good economic returns.

Finally, enhancing the share of hybrid power generation systems and further promoting natural gas power as a competitive alternative to coal power are two management proposals for the government's development of the electricity industry. While numerous studies have been conducted on hybrid power generation systems (including site selection, scale, establishment, unit commitment, and scheduling strategy) and natural gas to power, have been carried out the majority of these studies have been theoretical, and the few practical operations are still largely ineffective. As a result, more technical guidance as well as corresponding policy support is required. In Golmud, the natural gas electricity price is relatively low and the cost of raw material purchase is relatively high, which could dissuade investment. In order to develop useful, dependable, and financially feasible power production systems, it is also necessary to establish a hybrid power generation system infrastructure and centralized operations. Therefore, supportive policies are needed to encourage the development of hybrid power and natural gas power systems.

References

Balaman, Ş.Y. (2016). Investment planning and strategic management of sustainable systems for clean power generation: an ε-constraint based multi objective modelling approach. *Journal of Cleaner Production* 137: 1179–1190.

BP Globe (2019). Energy developments on BP Statistical Review of World Energy (2019). https://www.bp.com/en/glob-al/corporate/energy-economics/statistical-review-of-world-energy.html (accessed 20 October 2023).

Chamandoust, H., Derakhshan, G., and Hakimi, S.M. (2020). Multi-objectives optimal scheduling in smart energy hub system with electrical and thermal responsive loads. *Environmental and Climate Technologies* 24: 209–232.

Chen, Y. (1994). Interactive fuzzy satisfying method for optimal multi-objective VAR planning in power systems. *IEE Proceedings-Generation, Transmission and Distribution* 141 (6): 554–560.

Esmaeili, M., Sedighizadeh, M., and Esmaili, M. (2016). Multi-objective optimal reconfiguration and DG (Distributed Generation) power allocation in distribution networks using Big Bang-Big Crunch algorithm considering load uncertainty. *Energy* 103: 86–99.

Garcia-Gonzalez, J. and Castro, G.A. (2001). Short-term hydro scheduling with cascaded and head-dependent reservoirs based on mixed-integer linear programming. *2001 IEEE Porto Power Tech Proceedings (Cat. No. 01EX502)*, Volume 3, 6–pp. IEEE.

Geermu (2020). Corresponding energy reports in Golmud. http://www.geermu.gov.cn/newsListSearch (accessed 20 October 2023).

Gu, Y., Chen, Q., Xue, J. et al. (2020). Comparative techno-economic study of solar energy integrated hydrogen supply pathways for hydrogen refueling stations in China. *Energy Conversion and Management* 223: 113240.

Guangfu (2016). The dilemma of Photovoltaic city: a survey on the construction of Golmud solar farm. http://guangfu.bjx.com.cn/news/20160114/701132.shtml (accessed 20 October 2023).

Gupta, M.M. (1992). Fuzzy set theory and its applications. *Fuzzy Sets and Systems* 47 (3): 396–397.

Hadayeghparast, S., Farsangi, S.N., and Shayanfar, H. (2019). Day-ahead stochastic multi-objective economic/emission operational scheduling of a large scale virtual power plant. *Energy* 172: 630–646.

Hu, M., Xiao, F., and Wang, S. (2021). Neighborhood-level coordination and negotiation techniques for managing demand-side flexibility in residential microgrids. *Renewable and Sustainable Energy Reviews* 135: 110248.

Huang, Q. and Xu, J. (2020). Bi-level multi-objective programming approach for carbon emission quota allocation towards co-combustion of coal and sewage sludge. *Energy* 211: 118729.

Hussain, C.M.I., Norton, B., and Duffy, A. (2020). Comparison of hybridizing options for solar heat, biomass and heat storage for electricity generation in Spain. *Energy Conversion and Management* 222: 113231.

IRENA (2019). Renewable Energy Statistics 2019. https://cisp.cachefly.net/assets/articles/attachments/78833_irena_renewable_energy_statistics_2019.pdf (accessed 20 October 2023).

Jadidbonab, M., Mousavi-Sarabi, H., and Mohammadi-Ivatloo, B. (2019). Risk-constrained scheduling of solar-based three state compressed air energy storage with waste thermal recovery unit in the thermal energy market environment. *IET Renewable Power Generation* 13: 920–929.

Johlas, H., Witherby, S., and Doyle, J.R. (2020). Storage requirements for high grid penetration of wind and solar power for the MISO region of North America: a case study. *Renewable Energy* 146: 1315–1324.

Ju, L., Tan, Q., Zhao, R. et al. (2019). Multi-objective electro-thermal coupling scheduling model for a hybrid energy system comprising wind power plant, conventional gas turbine, and regenerative electric boiler, considering uncertainty and demand response. *Journal of Cleaner Production* 237: 117774.

Khaloie, H., Abdollahi, A., Shafie-Khah, M. et al. (2020). Co-optimized bidding strategy of an integrated wind-thermal-photovoltaic system in deregulated electricity market under uncertainties. *Journal of Cleaner Production* 242: 118434.

Lei, Y., Wang, D., Jia, H. et al. (2020). Multi-objective stochastic expansion planning based on multi-dimensional correlation scenario generation method for regional integrated energy system integrated renewable energy. *Applied Energy* 276: 115395.

Li, F. and Qiu, J. (2016). Multi-objective optimization for integrated hydro-photovoltaic power system. *Applied Energy* 167: 377–384.

Lin, Y., Eto, J.H., Johnson, B.B. et al. (2020). Research Roadmap on Grid-Forming Inverters. https://www.nrel.gov/docs/fy21osti/73476.pdf (accessed 14 December 2023).

Luo, G., Li, Y., Tang, W., and Wei, X. (2016). Wind curtailment of China's wind power operation: evolution, causes and solutions. *Renewable and Sustainable Energy Reviews* 53: 1190–1201.

Ma, J. (2017). The yearbooks of Golmud in 2014–2016. *Local Chronicles Committee in Golmud*, 1–344.

Mohseni-Bonab, S.M., Rabiee, A., and Mohammadi-Ivatloo, B. (2016). Voltage stability constrained multi-objective optimal reactive power dispatch under load and wind power uncertainties: a stochastic approach. *Renewable Energy* 85: 598–609.

National Development and Reform Commission of China, National Energy Administration of China (2022). Plan for developing a modern energy system during the 14th Five-Year Plan period. https://www.gov.cn/zhengce/zhengceku/2022-03/23/content_5680759.htm.

NDRC (2016). The 13th Five-Year National Energy Plan. https://www.ndrc.gov.cn/fggz/fzzlgh/gjjzxgh/201705/t20170517_1196766.html (accessed 14 December 2023).

NEA (2019). Photovoltaic power generation construction and operation in the first half of 2019, 2019. http://www.nea.gov.cn/2019-08/23/c_138330885.htm (accessed 14 December 2023).

Nojavan, S., Majidi, M., Najafi-Ghalelou, A. et al. (2017). A cost-emission model for fuel cell/PV/battery hybrid energy system in the presence of demand response program: ε-constraint method and fuzzy satisfying approach. *Energy Conversion and Management* 138: 383–392.

Paolo, C., Alireza, S., Massimo, S., and Jack, B. (2020). Dynamic dispatch of solid oxide electrolysis system for high renewable energy penetration in a microgrid. *Energy Conversion and Management* 204: 112322.

Poullikkas, A. (2005). An overview of current and future sustainable gas turbine technologies. *Renewable and Sustainable Energy Reviews* 9 (5): 409–443.

Qinghai Meteorological Bureau (2019). Report on the change in weather characteristics. http://qh.cma.gov.cn/bmgk/sqxj/201812/t20181228_107277.html (accessed 20 October 2023).

Rashid, K., Safdarnejad, S.M., and Powell, K.M. (2019). Dynamic simulation, control, and performance evaluation of a synergistic solar and natural gas hybrid power plant. *Energy Conversion and Management* 179: 270–285.

Reddy, S.S. (2017). Optimal scheduling of thermal-wind-solar power system with storage. *Renewable Energy* 101: 1357–1368.

Reddy, S.S. and Momoh, J.A. (2015). Realistic and transparent optimum scheduling strategy for hybrid power system. *IEEE Transactions on Smart Grid* 6 (6): 3114–3125.

Salkuti, S.R. (2019). Day-ahead thermal and renewable power generation scheduling considering uncertainty. *Renewable Energy* 131: 956–965.

Shams, M.H., Shahabi, M., and Khodayar, M.E. (2018). Stochastic day-ahead scheduling of multiple energy carrier microgrids with demand response. *Energy* 155: 326–338.

Sheng, W., Peng, S., Tang, Y. et al. (2015). Stochastic multi-objective scheduling of a combined cooling, heating and power microgrid containing a fuel cell. *Journal of Renewable and Sustainable Energy* 7 (6): 063123.

Shi, R., Fan, X., and He, Y. (2017). Comprehensive evaluation index system for wind power utilization levels in wind farms in China. *Renewable and Sustainable Energy Reviews* 69: 461–471.

Shima, R. and Nima, A. (2019). Optimal operation strategy for multi-carrier energy systems including various energy converters by multi-objective information gap decision theory and enhanced directed search domain method. *Energy Conversion and Management* 198: 111804.

Sui, Y., Ding, R., and Wang, H. (2020). A novel approach for occupational health and safety and environment risk assessment for nuclear power plant construction project. *Journal of Cleaner Production* 258: 120945.

Sun, Y., Dong, J., and Ding, L. (2017). Optimal day-ahead wind-thermal unit commitment considering statistical and predicted features of wind speeds. *Energy Conversion and Management* 142: 347–356.

Tollefson, J. (2008). Natural gas back in favour with US power companies. *Nature* 451 (7180): 753.

Villalva, M.G., Gazoli, J.R., and Filho, E.R. (2009). Comprehensive approach to modeling and simulation of photovoltaic arrays. *IEEE Transactions on Power Electronics* 24 (5): 1198–1208.

Wang, M. and Gooi, H.B. (2011). Spinning reserve estimation in microgrids. *IEEE Transactions on Power Systems* 26 (3): 1164–1174.

Wang, S. and Yuan, S. (2020). Interval optimization for integrated electrical and natural-gas systems with power to gas considering uncertainties. *International Journal of Electrical Power & Energy Systems* 119: 105906.

Wang, X., Chang, J., Meng, X., and Wang, Y. (2018). Short-term hydro-thermal-wind-photovoltaic complementary operation of interconnected power systems. *Applied Energy* 229: 945–962.

Wang, F., Xie, Y., and Xu, J. (2019). Reliable-economical equilibrium based short-term scheduling towards hybrid hydro-photovoltaic generation systems: case study from China. *Applied Energy* 253: 113559.

Wang, F., Xuan, Z., Zhen, Z. et al. (2020a). A day-ahead PV power forecasting method based on LSTM-RNN model and time correlation modification under partial daily pattern prediction framework. *Energy Conversion and Management* 212: 112766.

Wang, K., Su, C., Lobon, O.R., and Moldovan, N.C. (2020b). Chinese renewable energy industries' boom and recession: evidence from bubble detection procedure. *Energy Policy* 138: 111200.

WWO (2019). World Weather Online for Golmud Weather Forecast and Current Report. https://www.worldweatheronline.com/golmud-weather/qinghai/cn.aspx (accessed 14 December 2023).

Xi, Y., Zeng, Q., Chen, Z. et al. (2020). A market equilibrium model for electricity, gas and district heating operations. *Energy* 206: 117934.

Xu, J., Wang, F., Lv, C. et al. (2018). Economic-environmental equilibrium based optimal scheduling strategy towards wind-solar-thermal power generation system under limited resources. *Applied Energy* 231: 355–371.

Yang, L., Gao, X., Lv, F. et al. (2017). Study on the local climatic effects of large photovoltaic solar farms in desert areas. *Solar Energy* 144: 244–253.

Yang, S., Chen, Q., Liu, Z. et al. (2018). Performance analysis of the wind energy integrated with a natural-gas-to-methanol process. *Energy Conversion and Management* 173: 735–742.

Yang, L., Zhao, X., Li, X. et al. (2019). An MILP-Based optimal power and gas flow in electricity-gas coupled networks. *Energy Procedia* 158: 6399–6404.

Yin, J.N., Huang, G.H., Xie, Y.L. et al. (2021). Carbon-subsidized inter-regional electric power system planning under cost-risk tradeoff and uncertainty: a case study of Inner Mongolia, China. *Renewable and Sustainable Energy Reviews* 135: 110439.

Zhang, X., Zeng, R., Mu, K. et al. (2019). Exergetic and exergoeconomic evaluation of co-firing biomass gas with natural gas in CCHP system integrated with ground source heat pump. *Energy Conversion and Management* 180: 622–640.

Zhang, B., Hu, W., Li, J. et al. (2020a). Dynamic energy conversion and management strategy for an integrated electricity and natural gas system with renewable energy: deep reinforcement learning approach. *Energy Conversion and Management* 220: 113063.

Zhang, Z., Qin, H., Li, J. et al. (2020b). Short-term optimal operation of wind-solar-hydro hybrid system considering uncertainties. *Energy Conversion and Management* 205: 112405.

Zhao, B., Conejo, A.J., and Sioshansi, R. (2016). Unit commitment under gas-supply uncertainty and gas-price variability. *IEEE Transactions on Power Systems* 32 (3): 2394–2405.

10

Reliable–Economical–Social Equilibrium-Based Scheduling of Hybrid Solar–Wind–Hydro System

As yet, the wind is an untamed, and unharnessed force; and quite possibly one of the greatest discoveries hereafter to be made, will be the taming, and harnessing of it.

Abraham Lincoln
Feb. 2, 1809 – Apr. 15, 1865

Many studies have been conducted on the optimal size or operation strategy of the hybrid solar–hydro, hydro–wind, or solar–wind–hydro system, but few studies have addressed the sizing and scheduling issues for large-scale hybrid wind–PV–hydro systems equipped with long-distance transmission lines with respect to resource complementarity. In order to assess the output complementarity of wind and solar electricity, this chapter suggests a mathematical metric and incorporates it into the multi-objective sizing and operating model of a hybrid solar–wind–hydro generation system. In the hybrid system, the trade-offs of resource complementarity, system reliability, generation profits, and power curtailments are simultaneously taken into account. Additionally, the decision-maker's attitude parameters are incorporated into the model-solving process. A case study in China found that adding wind power to a hybrid solar–hydro system can increase the non-dispatchable renewable energy exported to the demand side by 5% and increase the load that the system can handle by 4.11% by relieving the system's nighttime hydropower compensatory burden. Furthermore, it is discovered that if a power curtailment of 2% is allowed, 1.53% more load may be served, and 14.9% more profits can be made.

10.1 Background Review

A total of 136 nations have made the commitment to achieve net-zero emissions by the middle of the century in order to address the global climate crisis (Net Zero Tracker 2022). The race to zero has been accelerated in the the electrical business by the development of ever-cheaper renewable energy technology (International

Sustainable Hybrid Energy Systems: Carbon Neutral Approaches, Modeling, and Case Studies,
First Edition. Jiuping Xu and Fengjuan Wang.

Energy Agency 2021). According to the International Energy Agency (IEA) (International Energy Agency 2021), if humans want to reach net zero by 2050 and keep the increase in global temperature to 1.5 °C, solar PV and wind energy must supply half of the world's electricity demand. Large-scale integration of wind and solar power into the current grid, however, poses risks to the stability, reliability, and flexibility of the power system because of the inherent instability and uncertainty (Ding et al. 2021). As a result, creating hybrid energy systems that utilize the capabilities of many energy sources is gaining popularity (Arent et al. 2021).

For instance, China has proposed developing a modern energy system in its newly issued "14th Five-Year Plan and 2035 Vision" in order to meet its pledge to reach the carbon peak by 2035 (Central People's Government of the People's Republic of China 2021). Building hybrid clean energy bases will likely take the lead in China's future power system growth among all energy-related undertakings (Mallapaty 2020). According to the published document, nine wind/PV/storage energy bases, wind/PV/thermal/storage energy bases, and wind/PV/hydro/storage energy bases will be created in the near future However, there is a high risk of experiencing serious power abandonment issues if irrational and disorderly construction is carried out without adequate consideration (O'Shaughnessy et al. 2020), which will inevitably lower enterprise profits, decrease investor confidence, and result in more than expected carbon dioxide emissions (Cui et al. 2020). Therefore, careful assessment of resource endowments and full utilization of consumption channels are required prior to permitting the development of numerous diverse projects (Parker et al. 2019).

The already constructed long-distance transmission lines will continue to play an indispensable role in transferring electricity from the north and west of China to the south and east due to the unequal geographic distribution of China's energy resources and energy demands, as noted in Lin and Wu (2017). These cables serve as "backbones" to send "industry blood," tying supply and demand together (Ming et al. 2016). The "bundled delivery" mode was developed to combine wind and solar power with dispatchable hydro and coal-based electricity so that they could be transmitted to consumers simultaneously through long-distance transmission lines. This not only mitigates the consumption problem of wind and solar electricity but also boosts the green ratio at the consumer side (Shu and Chen 2018). China's 22 UHV lines produced 531.8 billion kilowatt hour of power in 2020 (National Energy Administration 2021), of which 45.9% came from renewable sources. However, additional work needs to be done to integrate a higher ratio of non-dispatchable new energy at the supply side in order to achieve the carbon neutrality goal.

Because of its benefits of flexible start and stop, hydropower has traditionally been recognized as a significant resource to offset the uncertainty, intermittency, and fluctuation of wind and solar power (Sterl et al. 2020; International Renewable Energy Agency 2018). Additionally, the reservoir can store excess, non-dispatchable new energy and then regenerate it when necessary, smoothing output variations and boosting the amount of available power (Siemonsmeier et al. 2018). In addition, the storage of water resources reduces the financial outlay for installing large-scale energy storage systems and prevent environmental contamination from nonbiodegradable battery components (Ralon et al. 2017). As a result, numerous studies have been conducted on the creation of hybrid wind–hydro, PV–hydro, and wind–PV–hydro systems (Xiong et al. 2021). The frequent start-and-stop operation of the turbine and frequent slope climbing used to achieve this compensation, however, do not fully take into account the spatial and temporal complementarity of local PV and wind resources in the sizing and operating problems of hybrid generation systems.

In an ideal situation, photovoltaic and wind power capacities should be wisely deployed, and their output ratios should be optimized in order to achieve the best complementary effect on their own (Zhang et al. 2020a). After reducing the fluctuations of wind and solar power through time and space complementary effects (Parker et al. 2019), hydropower can be used as a backup resource to fill in any remaining gaps in the power supply. Such a mode can release the hydropower's compensatory pressure, harness the complementary potential of nearby wind and solar resources, and ultimately enhance the quantity of electricity that can be delivered to consumers (Kang et al. 2020; Wang et al. 2021b). Few studies have, however, thoroughly integrated these dimensions while also taking into account the long-distance power transmission infrastructure.

Therefore, this optimizes wind and solar power capacities to be integrated by hydropower as well as the outputs of the three generation resources in meeting the demand through long-distance transmission at the same time. Based on hourly production fluctuation, the complementary effect of wind and solar electricity is statistically assessed. For the coordinated operation of the hybrid solar–wind–hydro system, a multi-objective model is proposed that weighs the trade-off of resource complementarity, power curtailments, system reliability, and economic profits. The goal is to maximize the wind and solar power capabilities as well as the outputs of the power sources. The Hainan hybrid solar–wind–hydro generation system, which is located in the Chinese province of Qinghia, will be examined for the complementary effects of its wind and solar resources.

10.2 Key Problem Statement

The hybrid solar–wind–hydro generation system is detailed in this section, along with the trade-offs that must be balanced during the coordinated generation process of the hybrid system and the innate uncertainties of supply and demand.

10.2.1 System Description

The hybrid solar–wind–hydro generation system considered in this chapter comprises a solar PV power plant, a wind power plant, and a hydropower plant with a reservoir. They all are situated adjacent to one another within a clean energy base and are run by the same firm (Zhou et al. 2019). For instance, the Huanghe Hydropower Development Co. Ltd. is the owner of the Gonghe solar PV power plant (850 MW), Mohe wind power plant (400 MW), and Longyang Gorge hydropower plant (1280 MW), all of which are situated in the Hainan Clean Energy Base in Qinghai Province (XinhuaNet 2020). They all act as supply points for a long-distance

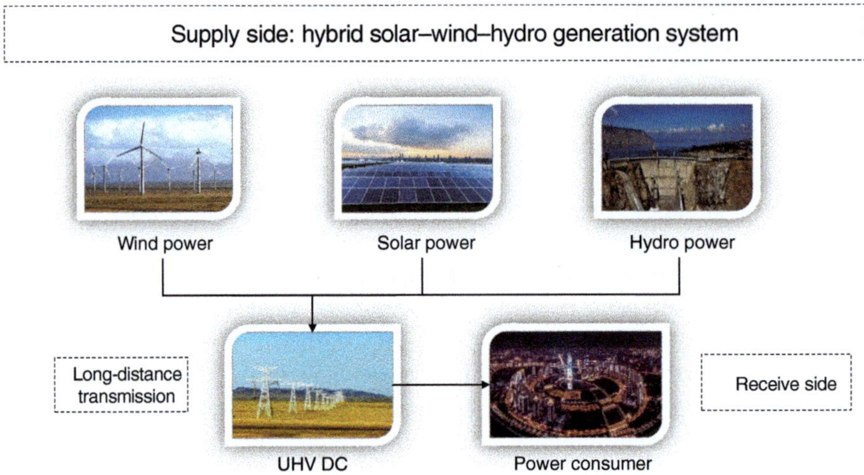

Figure 10.1 Framework of the hybrid solar–wind–hydro system. Source: 安琦 王/Adobe Stock (Solar power).

transmission network that connects to distant electricity users. Such a large-scale hybrid power production system must be built in stages due to time constraints. To ensure that the hybrid system can export as much clean and reliable power as possible, it is vital to assess the optimal size of each new generation source and modify the operation schedule accordingly. The diagrammatic sketch of the hybrid solar–wind–hydro generation system is represented in Figure 10.1.

10.2.2 Multi-objective Decision-Making Problem

The operation of a hybrid solar–wind–hydro system is a multi-objective problem because conflicting, multidimensional factors must be taken into account (Xie et al. 2021). It is recommended to export as much solar PV and wind power as possible because, for instance, the on-grid prices of solar and wind power are relatively higher than those of hydropower. However, such a schedule will inevitably put the system at a high risk of intermittent power supply, which is contrary to the objective of reliable power supply (Tong et al. 2021). However, because hydropower is flexible and wind and solar PV cannot be dispatched, it is preferable to integrate and export as much stable hydropower as possible if the system operator wants to achieve the highest level of power supply reliability. However, this schedule not only reduces power supply income but also puts the hybrid generation system in the difficult situation of having to curtail wind and solar power (Huang et al. 2021). Furthermore, it is yet unclear how wind and solar resources might reach maximum complementarity and how this complementarity will affect other goal-related characteristics. Multi-objective programming is required to define this coordinated operation challenge due to the complex relationship that needs to be balanced (Zhang et al. 2020b). The optimal capacities of wind and solar PV to be integrated, as well as the power production of wind, solar, and hydro resources during the operating time will be determined by this method while taking into account the aforementioned

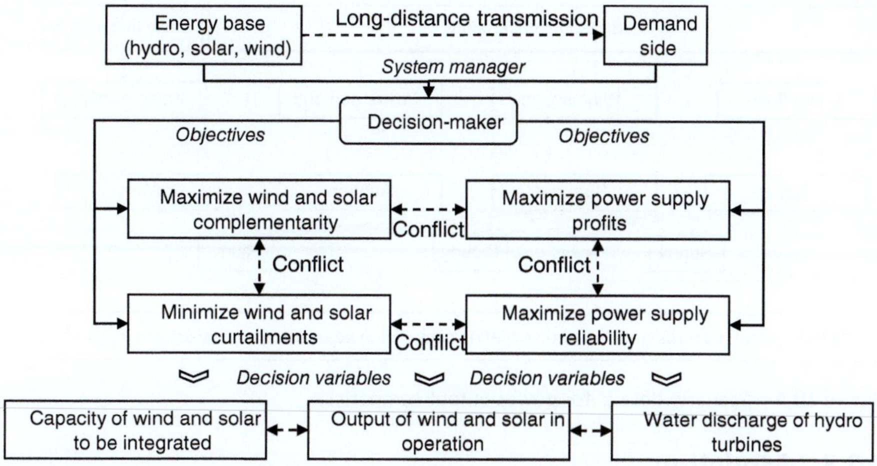

Figure 10.2 Optimization scheme of hybrid solar–wind–hydro system's coordinated operation.

conflict goals in full (Wang et al. 2021a). The detailed optimization scheme of hybrid solar–wind–hydro system coordinated operation is shown in Figure 10.2.

10.2.3 Seasonal and Daily Uncertainties

Unavoidable issues with a hybrid solar–wind–hydro generation system include the inherent uncertainties of wind, solar, and hydro resources as well as the demand (Sunar and Birge 2019). The average wind speed, solar irradiation, and water flows vary significantly across different seasons due to geographic location, climate condition, and rainfall irregularities (Bett and Thornton 2016). Additionally, there is a clear seasonal pattern to how much energy is consumed due to the greater need for cooling in the summer and heating in the winter (Conevska and Urpelainen 2020). All these features constitute the inherent seasonal uncertainty of the hybrid solar–wind–hydro generation system (Iris and Lam 2021). Meanwhile, the solar power is also quite variable under different weather conditions, which contributes to the daily uncertainty of solar PV power (Carannante et al. 2009). These characteristics made wind, solar and hydropower contribute differently in different scenarios. For system operators, comparing power output in various common conditions is a crucial reference point for daily and seasonal scheduling strategies (Yi et al. 2019). Therefore, twelve representative scenarios are proposed to deal with the aforementioned seasonal and daily uncertainties of energy resources and electricity demand in order to compare the output and complementarity of wind, solar, and hydropower. Scenarios for water flow, wind speed, power demand, and solar irradiation are defined for spring, summer, autumn, and winter. As for the weather-driven character of solar irradiation, three weather conditions (sunny, cloudy, and rainy) are established for each season. Each of the twelve scheduling scenarios involves the interaction of one of the four season conditions and one of the three weather conditions, as shown in Figure 10.3.

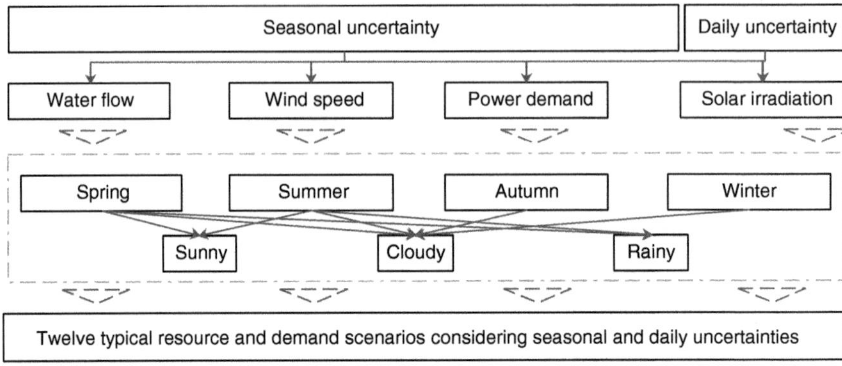

Figure 10.3 Scenario details for power output comparison.

10.3 Modeling

Mathematical modeling of the above-described problem is processed here, including a detailed description of objectives, constraints, and the model-solving method.

10.3.1 Notations

Indices

k	Index of seasons.
l	Index of weather.
t	Index of time period.

Decision variables

S	Solar capacity to be integrated.
W	Wind capacity to be integrated.
h_{klt}	Hydropower output in season k, weather condition l, time period t.
s_{klt}	Solar power output in season k, weather condition l, time period t.
w_{klt}	Wind power output in season k, weather condition l, time period t.
q_{klrt}	Water flows through hydro turbine.

Parameters

s_0	Rated capacity of wind turbine.
a_{kl}	Penalty of fails supplying power loads.
C_i^s	Seasonally complementary rate.
D_{kt}	Power demand in the receiving side.
Q_k	Water of the reservoir at season k.
$\underline{T}$	Lower bound of transmission capacity.
$\overline{T}$	Upper bound of transmission capacity.
N_{kl}	Number of l weather day in season k.
p	Unit penalty failed to or exceed supply.

w_0 On-grid price of wind power.
s_0 On-grid price of solar power.
h_0 On-grid price of hydropower.
$\underline{H_k}$ Lower bound of hydropower.
$\overline{\overline{H_k}}$ Upper bound of hydropower.
R_{klt} Solar irradiation of solar power plant.
S_{ijt} Technical solar power output in season k, weather condition l, time period t.
T_{klt} Temperature of solar power plant.
Q_0 Reservoir water inventory value at the beginning.
v^{in} Cut-in wind speed of wind turbines.
v^r Rated wind speed of wind turbines.
v^{out} Cut-out wind speed of wind turbines.
v_{kt} Wind speed in season k, time period t.
w^0_{kt} Wind turbine power output in season k, time period t.
G^D_{klt} Gaps of power supply and demand.
V^s_{klt} Hourly output variation of PV array.
V^w_{klt} Hourly output variation of wind turbine.
μ Transmission loss.
e_r Efficiency of hydro turbines.
$\overline{q_r}$ Technical maximum water flow limitation of hydro turbines.
$\underline{q_k}$ Lower bound of water flow limitation.
$\overline{\overline{q_k}}$ Upper bound of water flow limitation.
q^{in}_k Average hourly water flows into the reservoir.

Attitude parameters

α Attitude parameter of complementarity rate.
β Attitude parameter of reliability rate.
λ Attitude parameter of curtailment rate.

10.3.2 Four Main Goals Considered for the Hybrid System

In this subsection, the goals of hybrid solar–wind–hydro system coordinated operation are described in detail.

10.3.2.1 Maximizing Complementary Rate

Due to fluctuations in resources, the hourly output variation values of solar ($\Delta V^s_{klt} = V^s_{klt} - V^s_{kl(t-1)}$) and wind power ($\Delta V^w_{klt} = V^w_{klt} - V^w_{kl(t-1)}$) can be either negative or positive. This also applies to the sum of these two variations. The sum of hourly wind and solar power output variation can be classified into six scenarios. First (scenarios 1 and 2), both wind and solar variation values are positive (or negative), and the absolute value $\left|\Delta V^s_{klt} + \Delta V^w_{klt}\right| = \left|\Delta V^s_{klt}\right| + \left|\Delta V^w_{klt}\right|$. These two scenarios are defined as "no complementarity." Second (scenarios 3 and 4), one of the variation values is positive and the other is negative, so that $\left|\Delta V^s_{klt} + \Delta V^w_{klt}\right| < \left|\Delta V^s_{klt}\right| + \left|\Delta V^w_{klt}\right|$, which means that the wind and solar resources variations are "partly complemented." Lastly (scenarios 5 and 6), the ideal situation is that the sum of ΔV^s_{klt} and ΔV^w_{klt} can reach zero, which means the variation of the two is totally complemented. The hourly "complementary rate" can be expressed as $1 - \left|\Delta V^s_{klt} + \Delta V^w_{klt}\right| / \left|\Delta V^s_{klt}\right| + \left|\Delta V^w_{klt}\right|$.

The objective of maximizing yearly complementary effects of wind and solar can be represented as Eq. (10.1), where W and S are wind and solar power capacities to be integrated in the hybrid generation system, respectively, and both are decision variables. V^w_{klt} and V^s_{klt} are hourly output variation of a unit of wind turbine or PV array. N_{kl} is the number of l weather day in season k. The value of complementarity rate is within 0 and 1. A value of 0 means wind and solar power output fluctuations are not complemented, and value of 1 means they are fully complemented. The closer the value is to 1, the higher the complementary effects.

$$\max f_1 = 1 - \frac{\sum_k \sum_l N_{kl} \cdot \sum_t \left|W\Delta V^w_{klt} + S\Delta V^s_{klt}\right|}{\sum_k \sum_l N_{kl} \sum_t \left|W\Delta V^w_{klt}\right| + \sum_k \sum_l N_{kl} \sum_t \left|S\Delta V^s_{klt}\right|} \tag{10.1}$$

A unit of a wind turbine or solar module's hourly power output variation in season k and weather condition l can be expressed as $\Delta V^w_{klt} = w^0_{kt} - w^0_{k(t-1)}$ and $V^s_{klt} = s_{ijt} - s_{ij(t-1)}$, respectively. Equation (10.2) can be used to define the technical output of a wind turbine unit (w^0_{kt}) for a certain season and time period, where v_{kt} is the real wind speed in season k during time period t (m/s), s_0 is the rated capacity of a unit of wind turbine (MW), and v^{in}, v^r, and v^{out} are technical characteristics of wind turbines, indicating cut-in, rated, and cut-out wind speed. Δt is equal to one hour.

$$w^0_{kt} = \begin{cases} 0, & v_{kt} < v^{in} \\ a_0 \frac{v_{kt} - v^{in}}{v^r - v^{in}} \Delta t, & v^{in} \le v_{kt} < v^r \\ s_0\, \Delta t, & v^r \le v_{kt} < v^{out} \\ 0, & v_{kt} \ge v^{out} \end{cases} \tag{10.2}$$

Equations (10.3) can be used to express the technical output of a solar module (s^0_{ijt}) in season k, weather condition l, and time period t (Reddy et al. 2014), where b_0 stands for the rated capacity of a solar array (MW), R_{klt} and T_{klt} are the actual solar irradiation and atmosphere temperature of the solar power plant in season k, weather condition l, and time period t. R_0 and T_0 are standard test conditions of solar PV modules, which are 1 kW/m^2 and 25°C, respectively. a is PV module's output efficiency related to temperature, which is equal to −0.35%/°C. Δt is equal to one hour.

$$s^0_{ijt} = b_0 \left(\frac{R_{klt}}{R_0}\right) \left[1 + a\left(T_{klt} - T_0\right)\right] \Delta t \tag{10.3}$$

10.3.2.2 Maximizing Power Supply Reliability

This chapter uses the power supply **Reliability Rate**, as given in Eq. (10.4), to assess the reliability of the power supply of the hybrid generating system via long-distance transmission. G^D_{klt} is the difference of power supply and demand in season k, weather condition l, and time period t, which can be both negative or positive, and D_{kt} is power demand in the receiving side in season k time period t.

$$\max f_2 = 1 - \frac{\sum_k \sum_l N_{kl} \cdot \sum_t \left|G^D_{klt}\right|}{\sum_k \sum_l N_{kl} \sum_t D_{kt}} \tag{10.4}$$

G^D_{klt} can be further expressed by Eq. (10.5), where w_{klt}, s_{klt}, and h_{klt} represent the production of wind, solar, and hydropower plants in season k, weather condition l, and time period t; b stands for transmission loss.

$$G^D_{klt} = (w_{klt} + s_{klt} + h_{klt})\,(1 - b) - D_{kt} \tag{10.5}$$

10.3.2.3 Minimizing New Energy Curtailments

China has major issues with wind and solar power curtailments. Therefore, the hybrid generating system must take these issues into account with the compensating operation of hydro units (Aigner et al. 2021). This chapter uses the **Curtailment Rate** to express this objective, as shown in Eq. (10.6), where $\sum_l N_{kl} \sum_t \left(W \times w^0_{kt} - w_{klt}\right)$ and $\sum_l N_{kl} \sum_t \left(S \times s^0_{ijt} - s_{klt}\right)$ are wind and solar power to be curtailed in season k during the operation. The expressions $\sum_l N_{kl} \sum_t W \times w^0_{kt}$ and $\sum_l N_{kl} \sum_t S \times s^0_{ijt}$ are technical output of wind and solar power, respectively.

$$\min f_3 = \frac{\sum_k \left(\sum_l N_{kl} \sum_t \left(W w^0_{kt} - w_{klt}\right) + \sum_l N_{kl} \sum_t \left(S \times s^0_{ijt} - s_{klt}\right)\right)}{\sum_k \left(\sum_l N_{kl} \sum_t W \times w^0_{kt} + \sum_l N_{kl} \sum_t S \times s^0_{ijt}\right)} \tag{10.6}$$

10.3.2.4 Maximizing Yearly Power Supply Profits

The energy bases responsible for power transmission should ensure not only the steady development of society but also the economic success of businesses (Kök et al. 2018). Let w_{klt}, s_{klt}, and h_{klt} represent output of wind, solar, and hydropower plants in season k, weather condition l, and time period t, respectively. w_0, s_0, and h_0 represent the on-grid price of wind, solar, and hydropower separately. p is the unit penalty per kilowatt-hour when electricity energy base fails to or exceeds supply. Then the yearly profits of power supply can be represented by Eq. (10.7).

$$\max f_4 = \sum_k \sum_l N_{kl} \left(\sum_t w_{klt} w_0 + \sum_t s_{klt} s_0 + \sum_t h_{klt} h_0 - \sum_t p \left|G^D_{klt}\right| \right) \tag{10.7}$$

10.3.3 Constraints of the Hybrid System

In this section, the constraints for the wind power plant, the solar power plant, the hydropower plant, and the power transmission will be described in detail.

10.3.3.1 New Energy Output Limitation

In certain conditions, the combined hourly output of wind and solar energy cannot be greater than its maximum technical output, as shown in Eqs. (10.8) and (10.9), where $W \times w^0_{kt}$ is the maximum technical wind power output in season k and time period t, and $S \times s_{ijt}$ is the maximum technical solar power output in season k, weather condition l, and time period t.

$$w_{klt} \leq W \times w_{kt}^{0} \tag{10.8}$$

$$s_{klt} \leq S \times s_{ijt} \tag{10.9}$$

10.3.3.2 Hydropower Output Limitation

There are upper and lower limits to the amount of hydropower that can be produced due to installed capacity, turbine maintenance requirements, and power plant maintenance. Let $\underline{H_k}$ and $\overline{H_k}$ represent the minimum guaranteed and maximum outputs for hydropower, respectively. The hydropower output limitation can be shown in Eq. (10.10), where e_r is the comprehensive efficiency of hydro turbines, q_{klrt} (m/s^3) is the water flow through the hydro turbine, h (m) is the water head, k (9.81 kN/m^3) is the specific gravity of water.

$$\underline{H_k}\Delta t \leq \sum_{l} \eta_l q_{klrt} hk\Delta t \leq \overline{H_k}\Delta t \tag{10.10}$$

10.3.3.3 Water Flow Limitation

Water flows through hydro turbines are limited by their technical maximum water flow capacity $(\overline{q_r})$, as shown in Eq. (10.11). The discharge flow of the reservoir has upper and lower limits because of the need for irrigation, flood management, and snow prevention, as shown in Eq. (10.12), where $\underline{q_k}$ represents the minimum water flow the reservoir needs to discharge, and $\overline{q_k}$ represents the maximum water flow the reservoir can discharge.

$$q_{klrt} \leq \overline{q_r} \tag{10.11}$$

$$\underline{q_k} \leq \sum_{l} q_{klrt} \leq \overline{q_k} \tag{10.12}$$

10.3.3.4 Water Volume Limitation

The water volume in reservoir changes dynamically, which can be expressed as Eqs. (10.13) and (10.14), where Q_k and Q_{k-1} represent the water volume of the reservoir at season k and season $k-1$, q_k^{in} represents average hourly water flows into the reservoir, $\sum_l q_{klrt}$ is the hourly water flow out of the reservoir, and $\underline{\Delta t}$ is equal to 3600 seconds. $\left(q_k^{in} - \sum_l q_{klrt}\right)\underline{\Delta t}$ is the hourly water inventory variation, $\sum_t \left(q_k^{in} - \sum_l q_{klrt}\right)\underline{\Delta t}$ is the daily water inventory variation, and $\sum_l N_{kl} \cdot \sum_t \left(q_k^{in} - \sum_l q_{klrt}\right)\underline{\Delta t}$ is the seasonal water inventory variation. The initial value of the reservoir water inventory is Q_0. The reservoir's remaining water volume needs to be within the allowed range. The reservoir's minimum and maximum technical permissible water inventories are represented by the letters $\underline{g}$ and $\overline{g}$, respectively. Therefore, the seasonal water inventory should meet the conditions shown in Eq. (10.15).

$$Q_k = Q_{k-1} + \sum_{l} N_{kl} \cdot \sum_{t} \left(q_k^{in} - \sum_{l} q_{klrt} \right) \underline{\Delta t} \tag{10.13}$$

$$Q_1 = Q_0 \tag{10.14}$$

$$\underline{g} \leq Q_k \leq \overline{g} \tag{10.15}$$

10.3.3.5 Transmission Capacity Limitation

The hourly exported power must fall within the lower and upper limits of authorized transmission capacity due to the long-distance transmission line's technical maximum and minimum transmission capacities, as shown in Eq. (10.16).

$$\underline{M} \le w_{klt} + s_{klt} + h_{klt} \le \overline{M} \tag{10.16}$$

10.3.4 Global Model

The multi-objective sizing and operation model for the hybrid solar–wind–hydro system can be developed by combining the above-mentioned conflicting aims and technological restrictions, as stated in Eq. (10.17). The complementarity of wind and solar resources is fully taken into account in this model, and it also suggests mathematical metrics to measure this complementarity. The proposed method has the advantage of relieving the hydropower plants' compensation burden and enhancing the hybrid system's load-serving capacity by determining the capacities of wind and solar power plants to be integrated into the existing power grid and letting hydropower serve as a backup of the two resources. Additionally, the model closely considers the renewable energy curtailment issue because it is more effective to pursue an optimal rather than a minimal amount of curtailment (O'Shaughnessy et al. 2020). The reliability rate that ensures the hybrid system supplied electricity will not exceed or fail to meet the demand is used to model the reliability of power supply and demand through long-distance transmission lines, ensuring that society's development and the state of resource utilization can both be maintained. To secure the long-term development of power plants and to ensure investors' returns, companies that meet demand for electricity also maximize their economic profitability while taking into account the price variations among different energy sources. The hybrid system's coordinated operation and built-in uncertainties are both taken into account.

$$\begin{aligned}
&\max f_1 = 1 - \frac{\sum_k \sum_l N_{kl} \cdot \sum_t \left| WV_{klt}^w + SV_{klt}^s \right|}{\sum_k \sum_l N_{kl} \sum_t \left| WV_{klt}^w \right| + \sum_k \sum_l N_{kl} \sum_t \left| SV_{klt}^s \right|} \\
&\max f_2 = 1 - \frac{\sum_k \sum_l N_{kl} \cdot \sum_t \left| G_{klt}^D \right|}{\sum_k \sum_l N_{kl} \sum_t D_{kt}} \\
&\min f_3 = \frac{\sum_k \left(\sum_l N_{kl} \sum_t (W \times w_{kt}^0 - w_{klt}) + \sum_l N_{kl} \sum_t (Ss_{ijt} - s_{klt}) \right)}{\sum_k \left(\sum_l N_{kl} \sum_t W \times w_{kt}^0 + \sum_l N_{kl} \sum_t Ss_{ijt} \right)} \\
&\max f_4 = \sum_k \sum_l N_{kl} \left(\sum_t w_{klt} w_0 + \sum_t s_{klt} s_0 + \sum_t h_{klt} h_0 - \sum_t p \left| G_{klt}^D \right| \right) \\
&\text{s.t.} \begin{cases}
w_{klt} \le W \times w_{kt}^0 \\
s_{klt} \le Ss_{ijt} \\
\underline{H_k} \Delta t \le \sum_l \eta_l q_{klrt} hk \Delta t \le \overline{H_k} \Delta t \\
q_{klrt} \le \overline{q_r} \\
\underline{q_k} \le \sum_l q_{klrt} \le \overline{q_k} \\
Q_k = g_{i-1} + \sum_l N_{kl} \cdot \sum_t \left(q_k^{in} - \sum_l q_{klrt} \right) \underline{\Delta t} \\
Q_1 = Q_0 \\
\underline{g} \le Q_k \le \overline{g} \\
\underline{M} \le w_{klt} + s_{klt} + h_{klt} \le \overline{M}
\end{cases}
\end{aligned} \tag{10.17}$$

With the aforementioned model, the system operator, acting as the decision-maker, can ascertain the optimal wind and solar power capacities that can maximize the complementarity of local non-dispatchable resources and will relieve as much of the burden of hydropower compensation as possible. The operator can make cost-effective and detailed seasonal, daily, and hourly dispatch plans for typical scheduling scenarios that consist of various weather conditions for hybrid solar–wind–hydro generation in each season. The operator can also create lucrative, comprehensive dispatch plans for typical scheduling scenarios that include various weather conditions for seasonal hybrid solar–wind–hydro generation.

10.3.5 Model Solving

The above-proposed non-linear multi-objective model has well characterized the features of hybrid solar–wind–hydro generation; however, it is challenging to find the optimal value. Dasdemir et al. (2021). Therefore, in this section, the proposed model is converted into its equivalent single objective form using *varepsilon* and decision-maker's attitude factors. Previous research article (Wang et al. 2019) provides a thorough description and process.

In order to guarantee that wind and solar power can have at least an α of yearly complementary impacts, a decision-maker's attitude parameter (α) is introduced. With this parameter, Eq. (10.1) can be transformed as a related constraint, as shown in Eq. (10.18).

$$\begin{aligned} \max f_1 \Rightarrow (1-\alpha)\sum_l N_{kl}\sum_t \left(\left|WV_{klt}^w\right| + \left|SV_{klt}^s\right|\right) \\ \geq \sum_l N_{kl}\cdot\sum_t \left|WV_{klt}^w + SV_{klt}^s\right| \end{aligned} \tag{10.18}$$

The decision-maker's attitude parameter is used to ensure that at least a β ratio of annual power supply reliability must be maintained, which means that the annual supply and demand differential should be limited within $(1-\beta)$. It is further strict as hourly supply reliability need to be higher than β, as shown in Eq. (10.19).

$$\max f_2 \Rightarrow (1-\beta)\sum_t D_{kt} \geq G_{klt}^D \geq -(1-\beta)\sum_t D_{kt} \tag{10.19}$$

The decision-maker's attitude parameter λ is introduced here to ensure that yearly wind and solar power curtailments will not exceed λ. Therefore, Eq. (10.6) can be transformed as Eq. (10.20).

$$\begin{aligned} \max f_3 \Rightarrow \sum_k\left(\sum_l N_{kl}\sum_t (W\times w_{kt}^0 - w_{klt}) + \sum_l N_{kl}\sum_t (S\times s_{ijt} - s_{klt})\right) \\ \leq \lambda\sum_k\left(\sum_l N_{kl}\sum_t W\times w_{kt}^0 + \sum_l N_{kl}\sum_t S\times s_{ijt}\right) \end{aligned} \tag{10.20}$$

With the above attitude parameters introduced and integrated, the multi-objective model can be transformed as its equivalent single-objective form, as shown

Eq. (10.21), which can be directly solved using LINGO software.

$$\max f = \sum_k \sum_l N_{kl} \left(g_{ij}^w + g_{ij}^s + g_{ij}^h - a_{kl} \right)$$

$$\text{s.t.} \begin{cases} (1-\alpha) \sum_l N_{kl} \sum_t \left(\left| WV_{klt}^w \right| + \left| SV_{klt}^s \right| \right) \\ \geq \sum_l N_{kl} \cdot \sum_t \left| WV_{klt}^w + SV_{klt}^s \right| \\ (1-\beta) \sum_t D_{kt} \geq G_{klt}^D \geq -(1-\beta) \sum_t D_{kt} \\ \sum_k \left(\sum_l N_{kl} \sum_t \left(W \times w_{kt}^0 - w_{klt} \right) + \sum_l N_{kl} \sum_t \left(S \times s_{ijt} - s_{klt} \right) \right) \\ \leq \lambda \sum_k \left(\sum_l N_{kl} \sum_t W \times w_{kt}^0 + \sum_l N_{kl} \sum_t S \times s_{ijt} \right) \\ \text{Eqs. (10.5) and (10.8)–(10.17)} \end{cases} \tag{10.21}$$

10.4 Case Study

In this section, the efficacy and viability of the suggested model are assessed using a number of wind, solar, and hydropower plants owned by Huanghe Hydropower Development Co. Ltd. and situated in the Qinghai Province's Hainan Clean Energy Base.

10.4.1 Case Description

Qinghai, which is in the northwest of China, has several advantages and great potential for the growth of renewable energy (The People's Government of Qinghai Province 2021a). It is ranked second in China for solar energy's technical development potential and sixth for hydropower resources (The People's Government of Qinghai Province 2021b). By 2020, the province has installed a total of 36.7 GW of clean energy, or 90.2% of its installed capacity. This capacity generated 94.8 TWh of renewable electricity, or 89.3% of the province's total power generation (China Electric Council 2021). Hainan Clean Energy Base is one of the tens of millions of kilowatt-level clean energy bases built in Qinghai; it is an important supporting power supply for Qingyu UHV DC transmission lines (The People's Government of Qinghai Province 2021a). Among the supporting power sources of the Qingyu DC UHV transmission line, the Huanghe Hydropower Development Co. Ltd. has completed 81% of the building work on the Qingyu DC UHV transmission line's auxiliary power sources, comprising 14.6 GW of hydropower, 30 GW of photovoltaic, and 20 GW of wind power (Inengyuan 2020).

The Longyangxia solar–hydro system, which is part of the Hainan Clean Energy Base and supplies the Qingyu UHV DC transmission lines, is the largest hybrid hydro–solar production power plant in the world and has been in operation since 2013 (State-owned Assets Supervision and Administration Commission of the State Council 2014). However, the company's Mohe Wind Power Plants, which are situated in the Hainan Clean Energy Base and are a source of supply for the Qingyu UHV DC transmission lines, were finished in 2019 (Beijixing 2019). In order to relieve the reserve weight of compensation on hydropower, it is desirable

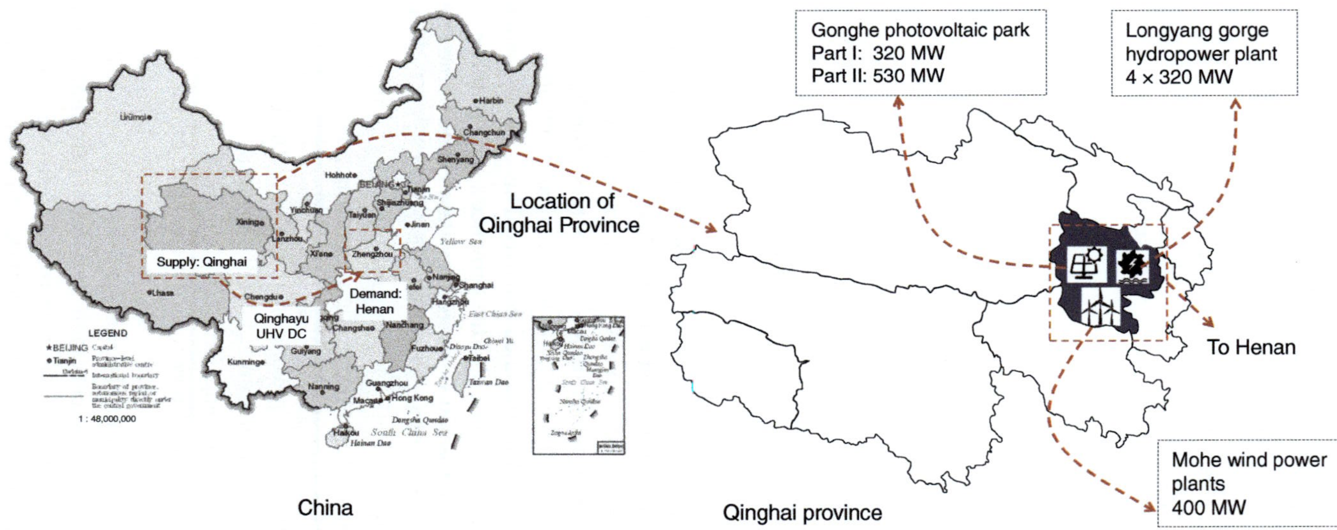

Figure 10.4 Location of hybrid solar–wind–hydro generation system.

to evaluate whether wind and solar sources can complement one another to some extent. This allows the the hybrid solar–wind–hydro system to transfer more electricity and make the best use of the transmission channels. In order to explore the appropriate ratio of wind and solar PV capacities to be integrated, maximize wind and solar power complementarity, relieve pressure on hydropower compensation, and maximize power-supporting benefits, this chapter uses the data from the aforementioned solar–wind–hydro system as an example.

10.4.2 Data Collection

The Climate Center of Qinghai Province has recently updated its assessment of the wind energy resources in the province, and it finds that the diurnal variations in wind speed in southern Qinghai often exhibit a "single-peak" change that shifts to the afternoon (Wang et al. 2020). The typical daily wind speed curves of the four seasons were sorted out using the hourly wind speed data from the Gonghe Shazhuyu Wind Measuring Tower (36.17°N, 100.23°E) at a height of 70 m. Wind turbines have a 3 MW installed capacity, and their individual cut-in, rated, and cur-out wind speeds are 2.5, 9, and 18 m/s, respectively (Goldwind Technology 2021). The receiving side of Qingyu UHV DC is Henan province. (Liu et al. 2017) conducted a detailed analysis of the power load characteristics. The Henan electricity grid's normal daily load curves for the spring, summer, autumn, and winter seasons can be derived by sorting the data in the article.

The average photovoltaic intensity and temperature data of Tara Beach (where Longyang Gorge solar projects are located) in various seasons and weather conditions are obtained from Solargis, as shown in Tables 10.1 and 10.2. Days with various weather conditions are gathered from the website (http://lishi.tianqi.com/gonghe/202001.html) for each season. The Longyang solar PV power plants consists of 850 solar arrays, each consist of forty 255 Wp polysilicon battery modules, with a rated capacity of 1 MW (Li and Qiu 2016). Technical parameters of hydro turbines collected from Xu and He (1987) are shown in Tables 10.3 and 10.4. Each year, during the dry season, the four hydro turbines, each of which has a rated capacity of 320 MW, are overhauled separately. Additionally, 80 and 200 MW will act as independent load reserves and emergency reserves (An 2016).

The operating parameters of Longyang Gorge reservoir are shown in Table 10.5. The feasibility study report by (Yang et al. 2014) has information on water flow for each season, which is 190, 343, 1003, and 727 m^3/s in the spring, summer, autumn, and winter, respectively. A hydropower plant's operation must also take into account the needs of irrigation, flood control, and anti-freezing. According to the data from the literature (Liu et al. 2015), the hydropower plant's average daily discharge flow is not less than 290 m^3/s from April to June each year.

The on-grid price of hydropower plants is 157.4 CNY/MWh (Qinghai Provincial Development and Reform Commission 2019). The PV power plant was separated as project I and II: project I (320 MW) has an on-grid price of 1000 CNY/MWh (National Development and Reform Commission 2011) and project II (530 MW) has an on-grid price of 950 CNY/MWh (National Development and Reform

Table 10.1 Solar irradiation data in different weather conditions of each season at Talatan (W/m^2).

Time period	1	2	3	4	5	6	7	8	9	10	11	12
Spring_sunny	0	0	0	0	0	0	120	236	413	548	653	766
Spring_cloudy	0	0	0	0	0	0	60	165	375	422	471	609
Spring_rainy	0	0	0	0	0	0	60	141	375	313	290	156
Summer_sunny	0	0	0	0	0	0	441	591	693	793	891	949
Summer_cloudy	0	0	0	0	0	0	230	414	630	612	643	755
Summer_rainy	0	0	0	0	0	0	230	355	630	453	396	194
Autumn_sunny	0	0	0	0	0	0	0	127	203	271	386	490
Autumn_cloudy	0	0	0	0	0	0	0	89	185	209	278	390
Autumn_rainy	0	0	0	0	0	0	0	76	185	155	171	100
Winter_sunny	0	0	0	0	0	0	0	0	200	340	450	500
Winter_cloudy	0	0	0	0	0	0	0	0	182	262	325	398
Winter_rainy	0	0	0	0	0	0	0	0	182	194	200	102

Time period	13	14	15	16	17	18	19	20	21	22	23	24
Spring_sunny	850	805	766	703	632	438	274	0	0	0	0	0
Spring_cloudy	689	569	647	545	558	292	164	0	0	0	0	0
Spring_rainy	425	163	281	396	465	97	88	0	0	0	0	0
Summer_sunny	994	1000	999	960	783	686	566	467	0	0	0	0
Summer_cloudy	805	707	844	744	691	458	340	220	0	0	0	0
Summer_rainy	497	202	366	540	576	153	181	180	0	0	0	0
Autumn_sunny	600	575	531	440	348	211	0	0	0	0	0	0
Autumn_cloudy	486	406	448	341	307	141	0	0	0	0	0	0
Autumn_rainy	300	116	195	247	256	121	0	0	0	0	0	0
Winter_sunny	590	515	445	350	200	0	0	0	0	0	0	0
Winter_cloudy	478	364	376	271	176	0	0	0	0	0	0	0
Winter_rainy	295	104	163	197	147	0	0	0	0	0	0	0

Commission 2013). Wind power on-grid price is 520 CNY/MWh (Beijixing 2019; National Development and Reform Commission 2019a). The Loss rate of Qingyu DC is set as $b = 6.5\%$, according to National Development and Reform Commission (2019b).

10.5 Results

In this section, complementary effects of local wind and solar resources are calculated, results under changing reliability rates, complementarity rates, and

Table 10.2 Temperature data for different weather conditions of each season in Gonghe County (°C).

Time period	1	2	3	4	5	6	7	8	9	10	11	12
Spring_sunny	5	6	5	6	7	8	9	10	11	12	12	13
Spring_cloudy	4	5	5	6	6	7	8	9	10	10	11	12
Spring_rainy	4	5	5	6	6	7	7	8	9	9	10	11
Summer_sunny	12	11	10	10	11	11	12	14	15	17	19	20
Summer_cloudy	13	13	12	12	11	11	12	13	12	14	16	18
Summer_rainy	13	12	13	12	13	13	14	15	16	16	17	18
Autumn_sunny	0	1	1	2	4	5	6	7	8	9	10	11
Autumn_cloudy	1	2	2	3	4	6	7	8	9	9	10	11
Autumn_rainy	2	3	4	4	5	5	6	6	7	7	8	8
Winter_sunny	−8	−9	−9	−8	−7	−6	−6	−5	−4	−2	−2	−2
Winter_cloudy	−11	−10	−10	−9	−8	−7	−7	−6	−5	−4	−3	−3
Winter_rainy	−13	−12	−12	−11	−11	−10	−10	−9	−9	−8	−8	−7

Time period	13	14	15	16	17	18	19	20	21	22	23	24
Spring_sunny	13	12	12	11	10	9	8	8	7	6	5	5
Spring_cloudy	12	11	10	10	9	8	8	7	6	5	4	4
Spring_rainy	10	10	9	9	8	7	6	5	4	4	4	4
Summer_sunny	21	22	23	22	22	23	22	20	19	16	14	13
Summer_cloudy	19	20	21	21	21	20	19	18	17	16	15	14
Summer_rainy	18	17	16	14	12	10	8	7	6	8	10	14
Autumn_sunny	11	10	10	9	8	7	6	5	4	3	2	0
Autumn_cloudy	10	10	10	9	8	6	5	4	3	2	2	1
Autumn_rainy	8	7	7	6	6	5	4	3	2	3	2	2
Winter_sunny	−4	−3	−3	−4	−5	−6	−6	−7	−8	−9	−10	−10
Winter_cloudy	−3	−4	−4	−5	−6	−7	−8	−9	−10	−9	−9	−10
Winter_rainy	−7	−8	−8	−9	−9	−10	−10	−11	−12	−12	−13	−13

Table 10.3 Technical parameters of hydro turbines.

Parameter	Units	Value	Parameter	Units	Value
Highest water head	m	150	Designed water flow	m^3/s	298
Average water head	m	130	Rated capacity	MW	326.5
Designed water head	m	122	Maximum capacity	MW	356
Lowest water head	m	99.5	Highest efficiency	%	93.5

Table 10.4 Lower and upper bounds of hydropower output in each season.

Seasons	Spring-normal	Summer-flood	Autumn-normal	Winter-dry
Load reserve	80	80	80	80
Emergency reserve	200	200	200	0
Overhaul capacity	0	0	0	320
Base load ($\underline{H_k}$)	200	200	200	200
Adjustable output	800	800	800	680
Maximum output ($\overline{H_k}$)	1280	1280	1280	960

Table 10.5 Reservoir storage capacity.

Parameter	Units	Value	Parameter	Units	Value
Highest storage level	m	2607	Highest storage capacity	Billion cubic meter	274
Normal storage level	m	2600	Normal storage capacity	Billion cubic meter	247
Normal dead water level	m	2560	Normal dead water level storage capacity	Billion cubic meter	117.8
Extreme dead water level	m	2530	Extreme dead water level storage capacity	Billion cubic meter	53.4

curtailment rates are obtained, and comparisons of different systems and typical days are conducted.

10.5.1 Complementary Rates of New Energies

The complementary rate of the output of a 1 MW wind turbine and 1 MW PV module on typical days is calculated and illustrated in Figure 10.5 to examine the complementary effects of nearby wind and solar resources.

During a typical spring day, solar and wind resources complement each other from 7 a.m. to 19 p.m., particularly between 7 a.m. and 9 a.m., and between 14 and 19 p.m. If there is complementarity, the complementary rate in spring fluctuates from 0.01 to 1, with the maximum rate occurring at 8 a.m. on sunny days and the lowest occurring at 10 a.m. on sunny days. The complementary rates are 0.30 on sunny days, 0.33 on cloudy day, and 0.25 on rainy days. Seasonally, the complementary rate in spring is 0.30.

The complementarity of wind and solar resources is generally lower in summer than it is in spring. The complementary rate in summer (if there is complementarity) varies from 0.13 to 0.89, with 0.89 occurring at 14 p.m. and 0.13 occurring at 16 p.m. on cloudy days. The complementation typically happens from

Typical days		Spring			Summer			Autumn			Winter		
		Sunny	Cloudy	Rainy	Sunny	Cloudy	Rainy	Sunny	Cloudy	Rainy	Sunny	Cloudy	Rainy
Time Period	1	0.00	0.00	0.00	0.00	0.00	0.00	0.00	0.00	0.00	0.00	0.00	0.00
	2	0.00	0.00	0.00	0.00	0.00	0.00	0.00	0.00	0.00	0.00	0.00	0.00
	3	0.00	0.00	0.00	0.00	0.00	0.00	0.00	0.00	0.00	0.00	0.00	0.00
	4	0.00	0.00	0.00	0.00	0.00	0.00	0.00	0.00	0.00	0.00	0.00	0.00
	5	0.00	0.00	0.00	0.00	0.00	0.00	0.00	0.00	0.00	0.00	0.00	0.00
	6	0.00	0.00	0.00	0.00	0.00	0.00	0.00	0.00	0.00	0.00	0.00	0.00
	7	0.68	0.98	0.98	0.27	0.46	0.46	0.00	0.00	0.00	0.00	0.00	0.00
	8	1.00	0.96	0.84	0.55	0.47	0.63	0.34	0.45	0.51	0.00	0.00	0.00
	9	0.27	0.23	0.21	0.00	0.00	0.00	0.66	0.56	0.51	0.00	0.00	0.00
	10	0.01	0.03	0.00	0.00	0.80	0.32	0.00	0.00	0.72	0.58	0.82	0.35
	11	0.00	0.00	0.49	0.25	0.64	0.00	0.00	0.00	0.00	0.00	0.00	0.00
	12	0.00	0.00	0.63	0.26	0.14	0.00	0.00	0.00	0.52	0.00	0.00	0.44
	13	0.00	0.00	0.00	0.00	0.00	0.00	0.00	0.00	0.00	0.00	0.00	0.00
	14	0.58	0.93	0.56	0.00	0.89	0.43	0.45	1.00	0.60	0.80	0.63	0.42
	15	0.84	0.00	0.00	0.41	0.00	0.00	0.70	0.00	0.00	0.94	0.00	0.00
	16	0.75	0.99	0.00	0.33	0.13	0.00	0.62	0.55	0.00	0.75	0.80	0.00
	17	0.62	0.00	0.00	0.77	0.65	0.00	0.94	0.58	0.00	1.00	0.77	0.50
	18	0.34	0.26	0.19	0.84	0.47	0.28	0.53	0.45	0.53	0.07	0.08	0.10
	19	0.31	0.38	0.48	0.86	0.87	0.00	0.00	0.00	0.00	0.00	0.00	0.00
	20	0.00	0.00	0.00	0.54	0.45	0.03	0.00	0.00	0.00	0.00	0.00	0.00
	21	0.00	0.00	0.00	0.00	0.00	0.00	0.00	0.00	0.00	0.00	0.00	0.00
	22	0.00	0.00	0.00	0.00	0.00	0.00	0.00	0.00	0.00	0.00	0.00	0.00
	23	0.00	0.00	0.00	0.00	0.00	0.00	0.00	0.00	0.00	0.00	0.00	0.00
	24	0.00	0.00	0.00	0.00	0.00	0.00	0.00	0.00	0.00	0.00	0.00	0.00
Complementary rate	Cijd	0.30	0.33	0.25	0.31	0.34	0.17	0.25	0.23	0.22	0.37	0.31	0.15
	Cis		0.30			0.26			0.31			0.33	
	f1						0.30						

Figure 10.5 Complementary effects of local wind and solar resources. Cijd means daily complementary rate, Cis means seasonal complementary rate, and f1 means yearly complementary rate.

7 a.m. to 19 p.m., especially at 7 a.m. and 8 a.m. in the morning, and between 17 and 19 p.m. The corresponding rates are 0.31 on sunny days, 0.34 on cloudy days, and 0.17 on rainy days. The supplemental rate for the summer is 0.26 seasonally.

The complementary rate in autumn (if there is complementarity) varies from 0.34 to 1, with 0.34 occurring at 8 a.m. on sunny days and 1 occurring at 14 p.m. on cloudy days. In a typical autumn day, the complementation of wind and solar resources primarily occurs at 8 a.m. and at 9 a.m. in the morning, as well as 14 between 14 p.m. and 18 p.m. in the afternoon. The complementary rates are 0.25 on sunny days, 0.23 on cloudy days, and 0.22 on rainy days. The supplementary rate in spring is 0.31 seasonally.

The complementary rate in winter (if there is complementarity) varies from 0.35 to 1, with 0.35 occurring at 10 a.m. on rainy days and 1 occurring at 17 p.m. on sunny days. In a typical winter day, the complementation of wind and solar resources primarily occurs at 10 a.m., 14 p.m., and 17 p.m. The corresponding rates are 0.37 on sunny days, 0.31 on cloudy days, and 0.15 on rainy days. The supplementary rate of winter is 0.33 seasonally.

10.5.2 Results Under Different Reliability and Complementarity Rates

In this section, results under three reliability rate scenarios are compared, where β equals to 100%, 99.99%, and 99.9%. In each scenario, renewable wind and solar power curtailments are not permitted, which means λ is set to 0, and complementary rate parameter (α) is set to vary from 5% to 30% with a step of 5%. It is important to note that when β is equal to 100% and α is 20%, the complementary rate cannot reach 20%. Therefore, we adjust α by 1% and discover that the highest complementary rate is 19%. Similarly, a highest complementary rate of 29% can be reached when β = 99.99% and β = 99.9%. Detailed calculation results can be seen in Tables 10.6–10.8.

The capacity of renewable resources to be integrated in each scenario is shown in Figure 10.6. The integrated capacity of solar power decreases and the integrated capacity of wind power increases as the complementarity rate varies when the

Table 10.6 Calculation results when $\lambda = 0$, $\beta = 100\%$, and α is changing.

Detailed results items	Units	Changing complementary rate (α)							
		5%	10%	15%	19%	20%	25%	29%	30%
Integrated wind	MW	39	81	122	162	0	0	0	0
Integrated PV	MW	411	393	375	358	0	0	0	0
Wind power output ratio	%	0.02	0.03	0.05	0.07	0.07	0.10	0.13	0.00
Solar power output ratio	%	0.11	0.11	0.10	0.10	0.10	0.09	0.07	0.00
Hydropower output ratio	%	0.87	0.86	0.85	0.83	0.83	0.81	0.80	1.00
Power supply profits	10^5 CNY	18,134	18,059	17,980	17,910	11,368	11,368	11,368	11,368

Table 10.7 Calculation results when $\lambda = 0$, $\beta = 99.99\%$, and α is changing.

Detailed results items	Units	Changing complementary rate (α)							
		5%	10%	15%	19%	20%	25%	29%	30%
Integrated wind	MW	39	79	122	162	173	239	310	0
Integrated PV	MW	411	394	375	358	353	325	270	0
Wind power output ratio	%	0.02	0.03	0.05	0.07	0.07	0.10	0.13	0.00
Solar power output ratio	%	0.11	0.11	0.10	0.10	0.10	0.09	0.07	0.00
Hydropower output ratio	%	0.87	0.86	0.85	0.83	0.83	0.81	0.80	1.00
Power supply profits	10^5 CNY	18,134	18,065	17,980	17,910	17,886	17,768	17,254	11,368

Table 10.8 Calculation results when $\lambda = 0$, $\beta = 99.9\%$, and α is changing.

Detailed results		Changing complementary rate (α)							
items	Units	5%	10%	15%	19%	20%	25%	29%	30%
Integrated wind	MW	39	80	122	161	173	238	309	0
Integrated PV	MW	412	394	376	358	354	325	271	0
Wind power output ratio	%	0.02	0.03	0.05	0.07	0.00	0.00	0.00	0.00
Solar power output ratio	%	0.11	0.11	0.10	0.10	0.00	0.00	0.00	0.00
Hydropower output ratio	%	0.87	0.86	0.85	0.83	1.00	1.00	1.00	1.00
Power supply profits	10^5 CNY	18,150	18,070	17,995	17,921	17,902	17,773	17,265	11,334

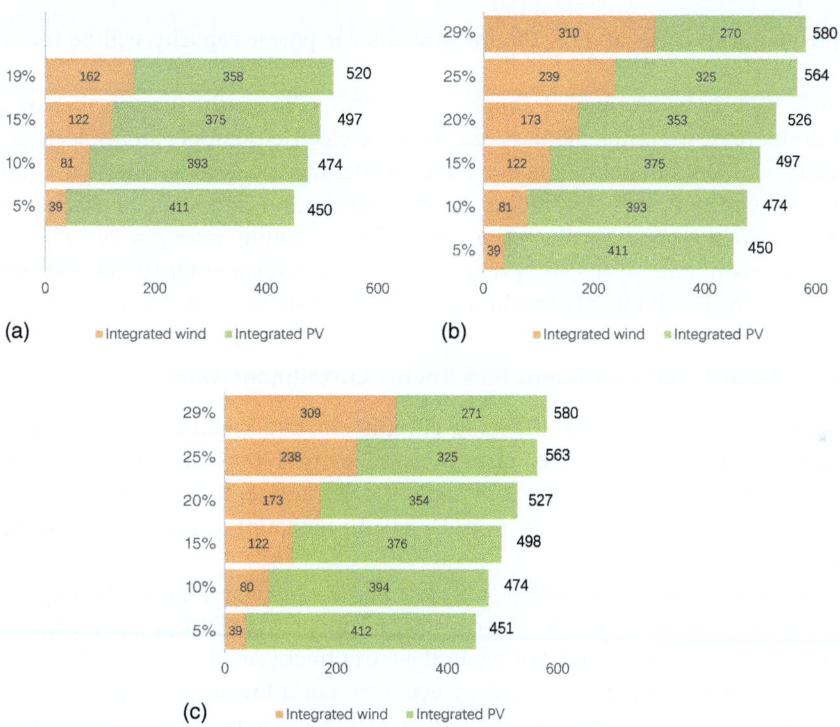

Figure 10.6 Integrated capacity results under different system reliability and complementary rates. (a) $\lambda = 0\%$, $\beta = 100\%$, and α is changing. (b) $\lambda = 0\%$, $\beta = 99.99\%$, and α is changing. (c) $\lambda = 0\%$, $\beta = 99.9\%$, and α is changing.

reliability rate is set to 100%. At the maximum complementarity rate of $\alpha = 19\%$, a total of 520 MW of wind and solar power can be integrated, of which solar power is 358 MW and wind power is 162 MW. The same pattern can be observed when the reliability rate is set at 99.99% and 99.9%; in these two cases, a total of 580 MW of non-dispatchable renewable energy can be integrated. However, when

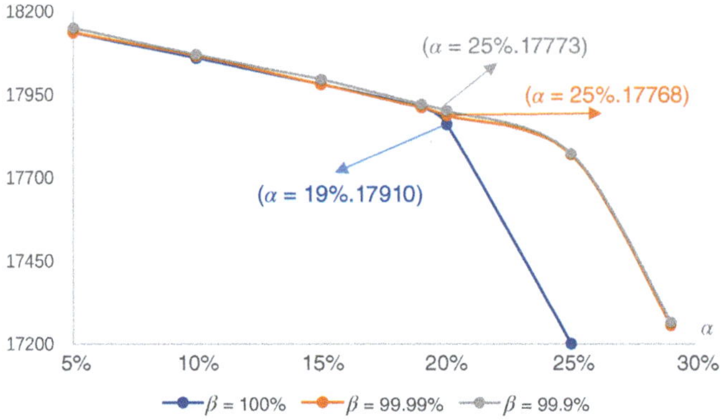

Figure 10.7 Power supply profits under different system reliability and complementary rates (10^5 CNY).

complementarity is set at 29%, the integrated solar power capacity will be lower than wind power.

Figure 10.7 shows the power supply profit (10^5 CNY) results in each scenario. With the increase in complementary rate, the total electricity supply profits decrease accordingly. When reliability rate $\beta = 100\%$, profits sharply decline when the complementarity rate exceeds 19%, indicating that a maximum complementarity rate of 19% is reasonable economically in this case. When reliability rates $\beta = 99.99\%$ and $\beta = 99.9\%$, economic profits sharply decline when complementarity rate exceeds 25%, despite the possibility of a maximum complementarity rate of 29%, respectively.

10.5.3 Results Under Different New Energy Curtailment Rates

In this section, results under changing curtailment rate scenarios are analyzed. In each scenario, the reliability rate (β) is equal to 100%, the complementary rate parameter (α) is equal to 0.19, and the curtailment rate parameter (λ) ranges from 0% to 16% with a 2% step. Detailed calculation results are shown in Table 10.9.

The integrated wind and solar power capacity is shown in Figure 10.8 for various curtailment rate scenarios. The capacity of integrated wind power increased as curtailment rates increased, but when the curtailment rate is less than 10%, the increase in installed capacity caused by every 2% curtailment gradually declines. For instance, if 2% curtailment is allowed, 60 MW more wind power capacity can be integrated, but when the curtailment rate is allowed to increase by 2–4%, this increase can only integrate an additional 46 MW of capacity. There is a highest rise of 109 MW integrated capacity when the curtailment rate exceeds 10%. The integrated installed capacity for solar electricity gradually rises and eventually stabilizes at the maximum value of 850 MW as the permissible curtailment rate rises. Similar to this, the installed capacity gain for every 2% curtailment rate is decreasing. The installed capacity of solar power is always greater than that of wind power in every situation.

Figure 10.9 shows curtailed wind and solar power under different curtailment rate scenarios. Both the quantity and the rate of wind power curtailment increased with

Table 10.9 Calculation results when $\alpha = 0$, $\beta = 100\%$, and λ is changing.

Detailed results (%)	Integrated wind (MW)	Integrated PV (MW)	Complementarity rate	Wind power output ratio	Solar power output ratio	Hydropower output ratio
0	162	358	0.19	0.07	0.10	0.83
2	222	494	0.19	0.09	0.13	0.78
4	268	593	0.19	0.10	0.16	0.74
6	309	687	0.19	0.12	0.18	0.70
8	346	761	0.19	0.13	0.20	0.68
10	373	828	0.19	0.13	0.21	0.66
12	482	849	0.22	0.17	0.22	0.62
14	588	850	0.24	0.20	0.22	0.59
16	673	850	0.26	0.22	0.22	0.57

Detailed results (%)	Curtailed wind power (GWh)	Curtailed solar power (GWh)	Wind curtailment rate	Solar curtailment rate	Total curtailed power (GWh)	Power supply profits (10^5) CNY	Maximum load (%)
0	0	0	0.00	0.00	0	17,910	104.11
2	29	3	0.04	0.00	33	20,311	105.64
4	59	20	0.07	0.02	79	21,942	106.12
6	91	46	0.10	0.03	137	23,387	106.20
8	130	73	0.12	0.05	204	24,470	106.29
10	170	106	0.15	0.06	276	25,350	106.34
12	258	118	0.18	0.07	376	25,990	106.57
14	365	119	0.20	0.07	484	26,359	106.80
16	475	119	0.23	0.07	593	26,606	106.97

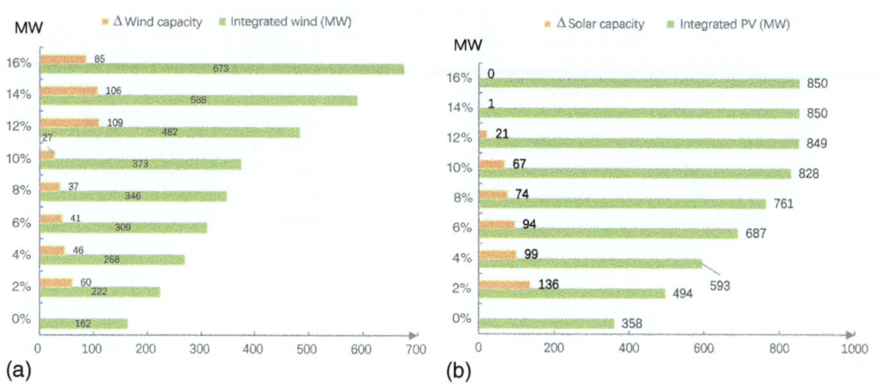

Figure 10.8 Integrated wind (a) and solar (b) power capacities under different curtailment rate scenarios.

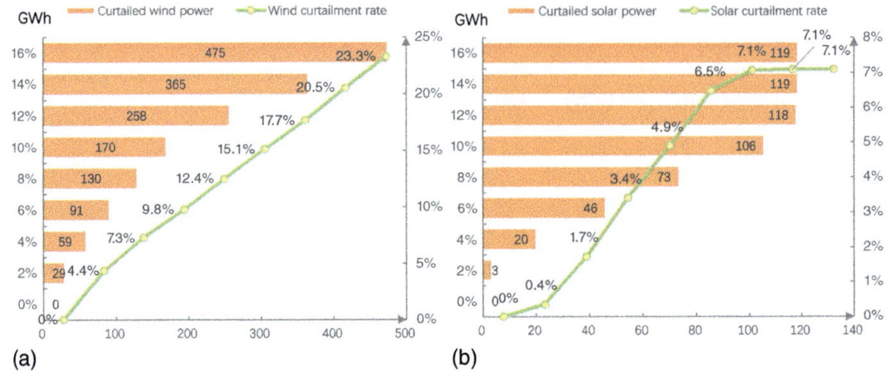

Figure 10.9 Curtailed wind (a) and solar (b) power under different curtailment rate scenarios.

the increase in curtailment rates; solar power curtailment and curtailment rate first increased and then stabilized; when the curtailment rate exceeds 12%, 118–119 GWh of photovoltaic energy cannot be used, accounting for 7.1% of the technology's capacity to generate electricity. The amount of solar electricity that must be restricted is always less than the amount of wind power.

The power output ratio, power supply profits, and maximum load results under various curtailment rate scenarios are displayed in Figure 10.10. Profits from the

Figure 10.10 Power output ratios (a), power supply profits (b), and maximum load (c) results under different curtailment rate scenarios.

power supply also increased as a result of the rise in curtailment rates, which also affected the output ratios of wind and solar electricity. It is important to note that the 2% curtailment rate will result in an increase in income of 2401×10^5 CNY. The amount of revenue growth will decrease when the curtailment rate continues to rise by 2%; for example, 1631×10^5 CNY for $\lambda = 4\%$, 1445×10^5 CNY for $\lambda = 6\%$. The maximum load that can be supported by a hybrid solar–wind–hydro system increases by 1.5% when λ increases from 0% to 2% but only by 0.5% when λ increases from 2% to 4%. As λ continues to grow, the increments get smaller and smaller until they finally stabilize at 0.1–0.2%.

10.5.4 Comparison of Different Systems

Results of the hydro-alone system, solar–hydro system, and solar–wind–hydro system are calculated and compared in Table 10.10 to assess the output features and benefits of the hybrid solar–wind–hydro system, where two groups of results (λ = 0%, λ = 2%) are presented.

From Table 10.10, as can be seen, a hybrid solar–hydro system could substitute 12% of the overall load supply responsibility by integrating 428 MW solar arrays, increasing power supply profits from 11,368 to 18,208 × 105 CNY. A total of 520 MW of non-dispatchable renewable energy, including 162 MW of wind and 358 MW of solar, can be integrated into the hybrid solar–wind–hydro system to supply the same load, burdening 17% of the total load task. However, total power supply profits will fall to 17,910 × 105 CNY (compared to solar–hydro system). In hybrid solar–hydro and solar–wind–hydro systems, a 2% power curtailment can both boost the capacities and contributions of wind and solar generation. It is important to keep in mind that, despite the fact that non-dispatchable renewable energy can support a larger share of the load burden in a hybrid solar–wind–hydro system, its power supply profit is always smaller than that of a solar–hydro system, regardless of whether there is curtailment or if the curtailment is limited to 2%.

The investigation into how much excess load can be supplied by the hybrid solar–wind–hydro system was further detailed in Table 10.11. Setting $\alpha = 0.19$, $\beta = 100\%$, and $\lambda = 0\%$ and increasing the load in each typical day with a step of 1%, it was discovered that a total load of 4% can be increased for the hybrid system. The load was then increased with a step of 0.1% at the base of 4%, and it was discovered that a total load of 4.1% can be served while 4.2% is impossible. Finally, it was discovered that a solar–wind–hydro system is capable of meeting a maximum of 104.11% of the load without reducing its use of solar or wind energy. Similar to this, a solar, wind, and hydro system can handle up to 105.64% of the load if a 2% curtailment is allowed. When comparing $\lambda = 2\%$ and $\lambda = 0\%$ situations, it can be seen that with 2% of curtailment allowed, 1.53% more load can be served, resulting in an increase of 14.9% in power supply profits.

10.6 Discussion

Based on the results described above, some important features and core findings of the coordinately operated solar–wind–hydro system can be obtained.

Table 10.10 Calculation results of hydro-only system, solar–hydro system, and solar–wind–hydro system.

Scenarios	Systems	Integrated wind capacity	Integrated solar capacity	Wind power output ratio (%)	Solar power output ratio (%)	Hydropower output ratio (%)	Power supply profits
$\alpha = 0.19$	Hydro	0	0	0	0	100	11,368
$\beta = 100\%$	Solar–hydro	0	428	0	12	88	18,208
$\lambda = 0\%$	Solar–wind–hydro	162	358	7	10	83	17,910
$\alpha = 0.19$	Hydro	0	0	0	0	100	11,368
$\beta = 100\%$	Solar–hydro	0	611	0	16	84	20,938
$\lambda = 2\%$	Solar–wind–hydro	222	494	9	13	78	20,463

Table 10.11 Maximum load results of different generation systems.

Scenarios	Systems	Maximum load	Maximum power supply profits
$\alpha = 0.19$, $\beta = 100\%$, $\lambda = 0\%$	Hydro	100% of base case demand	11,368
	Solar–hydro	100% of base case demand	18,208
	Solar–wind–hydro	104.11% of base case demand	18,845
$\alpha = 0.19$, $\beta = 100\%$, $\lambda = 2\%$	Hydro	100% of base case demand	11,368
	Solar–hydro	100% of base case demand	20,938
	Solar–wind–hydro	105.64% of base case demand	21,657

10.6.1 Core Findings

When it comes to the complementary impacts of local sources, it can be noted that winter (0.33) and summer (0.26) have the highest complementarity for local wind and solar resources, respectively, while autumn (0.31) and spring (0.30) rank second and third, respectively. However, in autumn and winter, the complementary effects of sunny days are the highest, cloudy days rank second, and rainy days are the lowest. In spring and summer, the complementary benefits of overcast days are higher than sunny days, and sunny days are higher than rainy days.

Results under varying dependability and complementarity rates show that a hybrid solar–wind–hydro system can achieve a maximum complementarity rate of 0.19 of wind and solar output with a 100% reliability of the power supply. Renewable wind and solar resources can meet 10% and 7% of the total loads in this scenario, respectively, and 520 MW of wind and solar energy can be integrated as a supplier of the long-distance transmission line. If the power supply reliability rate is increased to 99.99%, 580 MW of wind and solar power can be integrated. However, from an economic perspective, a complementarity rate of 0.25 would be more reasonable given that complementarity rates above this value result in sharply lower power supply profits.

Results under varying curtailment rates show that as the allowable curtailment rate rises, the amount of wind and solar power that the hybrid system can integrate rises in lockstep. When the curtailment rate is less than 10%, the increase in solar power is always greater than that of wind power, while the situation will be the opposite when the curtailment rate exceeds 10%. Additionally, the rate of solar power curtailment will initially rise and then stabilize at 7.1%, while the rate of wind power curtailment will continue to expand linearly. Most importantly, limiting the curtailment rate to 2% is appropriate in terms of raising power supply income and power transmission volume since 2% curtailment results in the biggest rise in power supply profits and the maximum load that can be provided. This demonstrates that seeking an optimal amount of curtailment rather than a minimal one becomes more effective.

With the integration of wind resources, non-dispatchable renewable energy can carry a higher proportion of the load burden in the hybrid solar–wind–hydro system than in the hybrid solar–hydro system. However, the power supply profits of the former system are always lower than those of the latter system due to the cost advantage of solar power. However, the integration of wind resources released the load burden of hydro at night, allowing the hybrid system to provide 104.11% of the solar–hydro system's load and generate larger supply profits. Additionally, it can be seen from comparing the $\lambda = 2\%$ and $\lambda = 0\%$ scenarios that 1.53% more load can be provided with the 2% of allowable curtailment, which will raise 14.9% of power supply profits.

10.6.2 Management Recommendations

The hybrid solar–wind–hydro sizing and scheduling model, which takes into account resource complementarity, can be utilized by system operators as a crucial reference tool for creating system operation plans. The difficulty of balancing multi-dimensional goals arises from the nature of multi-energy complementary systems. The multi-objective model put forth here has a more systematic and comprehensive viewpoint and considers the economics and dependability of the power supply as well as the rates of curtailment and complementarity of renewable energy sources. The model can assist system operators in not only choosing appropriate system sizes but also in optimizing short-term scheduling plans, which can enhance system operation effectiveness and economic benefits.

A strong lever for the power sector to create a diversified power supply system is to encourage the creation of multi-source complementary systems. It is necessary to take the proper steps to encourage power producers to integrate different types of power plants to operate as complementary systems in order to achieve the effect of one plus one more than two, which can not only build a diversified power supply structure and improve the flexibility of the power system but also reduce operating costs. The results of the previous calculations show that a three-source supply system can improve the scheduling flexibility and operating economy of the power system.

For the management division, using wind and solar electricity with a"reasonable abandonment" mentality rather than "full consumption" is more supportive of goals for net zero emissions and power restructuring. This finding provides an effective reference for the Chinese government's proposal to maintain the utilization rate of renewable electricity at a reasonable level in the Plan for Renewable Energy Development during the 14th Five-Year Plan Period (The Chinese Government 2022). In order to ensure that wind and solar energy are fully utilized, alternative controllable power sources, such as flexible hydropower and energy storage technologies, are required. This is due to the high dependency of wind and solar energy on environmental conditions. Because energy storage technology is still expensive, it is not yet the right time to fully implement it. Flexible power sources like hydropower are also crucial for controlling fluctuations in wind and PV, but if "full consumption" is the operating principle, the integrated wind and PV power generation capacity is more conservative. As a result, a reasonable calculation of the abandonment threshold is a more suitable solution.

References

Aigner, K.-M., Clarner, J.-P., Liers, F., and Martin, A. (2021). Robust approximation of chance constrained DC optimal power flow under decision-dependent uncertainty. *European Journal of Operational Research* https://doi.org/10.1016/j.ejor.2021.10.051.

An, Y. (2016). Research on theory and method of hydro-PV complementary coordinated operation. PhD thesis. Xi'an University of Technology.

Arent, D.J., Bragg-Sitton, S.M., Miller, D.C. et al. (2021). Multi-input, multi-output hybrid energy systems. *Joule* 5(1): 47–58.

Beijixing (2019). All wind turbines in Mohe Wind Farm of Huanghe Company are connected to the grid. https://news.bjx.com.cn/html/20190708/991295.shtml (accessed 19 October 2023).

Bett, P.E. and Thornton, H.E. (2016). The climatological relationships between wind and solar energy supply in Britain. *Renewable Energy* 87: 96–110.

Carannante, G., Fraddanno, C., Pagano, M., and Piegari, L. (2009). Experimental performance of MPPT algorithm for photovoltaic sources subject to inhomogeneous insolation. *IEEE Transactions on Industrial Electronics* 56(11): 4374–4380.

Central People's Government of the People's Republic of China (2021). 14th Five-Year Plan and 2035 Vision. http://www.gov.cn/xinwen/2021-03/13/content_5592681.htm (accessed 19 October 2023).

China Electric Council (2021). *China Electric Power Statistical Yearbook*. China Statistics Press.

Conevska, A. and Urpelainen, J. (2020). Weathering electricity demand? Seasonal variation in electricity consumption among off-grid households in rural India. *Energy Research & Social Science* 65: 101444.

Cui, Q., Liu, Y., Ali, T. et al. (2020). Economic and climate impacts of reducing China's renewable electricity curtailment: a comparison between CGE models with alternative nesting structures of electricity. *Energy Economics* 91: 104892.

Dasdemir, E., Testik, M.C., Öztürk, D.T. et al. (2021). A multi-objective open vehicle routing problem with overbooking: exact and heuristic solution approaches for an employee transportation problem. *Omega* 108: 102587.

Ding, Z., Wen, X., Tan, Q. et al. (2021). A forecast-driven decision-making model for long-term operation of a hydro-wind-photovoltaic hybrid system. *Applied Energy* 291: 116820.

Goldwind Technology (2021). Operating parameters of direct-drive permanent magnet smart wind turbine GW 2S [EB/OL]. https://www.goldwind.com/cn/windpower/product-gw2s/ (accessed 19 October 2023).

Huang, K., Liu, P., Ming, B. et al. (2021). Economic operation of a wind-solar-hydro complementary system considering risks of output shortage, power curtailment and spilled water. *Applied Energy* 290: 116805.

Inengyuan (2020). Qingyu UHV DC supporting power supply project put into operation [EB/OL]. http://www.inengyuan.com/kuaixun/4090.html (accessed 19 October 2023).

International Energy Agency (2021). Net zero by 2050: a roadmap for the global energy sector. https://www.iea.org/reports/net-zero-by-2050 (accessed 19 October 2023).

International Renewable Energy Agency (2018). Power System Flexibility for the Energy Transition: Part 1, Overview for Policy Makers. *Technical report*. https://www.irena.org/-/media/Files/IRENA/Agency/Publication/2018/Nov/IRENA_Power_system_flexibility_1_2018.pdf (accessed 19 October 2023).

Iris, Ç. and Lam, J.S.L. (2021). Optimal energy management and operations planning in seaports with smart grid while harnessing renewable energy under uncertainty. *Omega* 103: 102445.

Kang, J., Ng, T.S., and Su, B. (2020). Optimizing electricity mix for CO_2 emissions reduction: a robust input-output linear programming model. *European Journal of Operational Research* 287(1): 280–292.

Kök, A.G., Shang, K., and Yücel, Ş. (2018). Impact of electricity pricing policies on renewable energy investments and carbon emissions. *Management Science* 64(1): 131–148.

Li, F.F. and Qiu, J. (2016). Multi-objective optimization for integrated hydro–photovoltaic power system. *Applied Energy* 167: 377–384.

Lin, B. and Wu, W. (2017). Cost of long distance electricity transmission in China. *Energy Policy* 109: 132–140.

Liu, J., Wang, S., and Wang, M. (2015). Analysis of influence of hydro-solar complementary system on comprehensive operation of Longyangxia Hydropower Station. *Grid and Clean Energy* 9: 83–87.

Liu, J., Wang, J., Bai, H. et al. (2017). Analysis and prediction of electric power load characteristics in Henan province. *Management Informatization in China* 20(23): 140–141.

Mallapaty, S. (2020). How China could be carbon neutral by mid-century. *Nature* 586(7830): 482–484.

Ming, Z., Lilin, P., Qiannan, F., and Yingjie, Z. (2016). Trans-regional electricity transmission in China: status, issues and strategies. *Renewable and Sustainable Energy Reviews* 66: 572–583.

National Development and Reform Commission (2011). Notice on improving the on-grid tariff policy for solar photovoltaic power generation [EB/OL]. http://www.gov.cn/zwgk/2011-08/01/content_1917358.htm (accessed 19 October 2023).

National Development and Reform Commission (2013). Notice on using price leverage to promote the healthy development of the photovoltaic industry [EB/OL]. https://www.ndrc.gov.cn/xxgk/zcfb/tz/201308/t20130830_963934.html?code=&state=123 (accessed 19 October 2023).

National Development and Reform Commission (2019a). Notice on improving the on-grid tariff policy for wind power. https://www.ndrc.gov.cn/xxgk/zcfb/tz/201905/t20190524_962453.html?code=&state=123 (accessed 19 October 2023).

National Development and Reform Commission (2019b). Notice on reducing general industrial and commercial electricity prices [EB/OL]. https://www.ndrc.gov.cn/

xxgk/zcfb/tz/201905/t20190515_962447.html?code=&state=123 (accessed 19 October 2023).

National Energy Administration (2021). 2020 National renewable energy electric power development monitoring and evaluation report [EB/OL]. http://zfxxgk.nea.gov.cn/2021-06/20/c_1310039970.htm (accessed 19 October 2023).

Net Zero Tracker (2022). https://zerotracker.net/ (accessed 19 October 2023).

O'Shaughnessy, E., Cruce, J.R., and Xu, K. (2020). Too much of a good thing? Global trends in the curtailment of solar PV. *Solar Energy* 208: 1068–1077.

Parker, G.G., Tan, B., and Kazan, O. (2019). Electric power industry: operational and public policy challenges and opportunities. *Production and Operations Management* 28(11): 2738–2777.

Qinghai Provincial Development and Reform Commission (2019). Notice on reducing general industrial and commercial electricity prices [EL/OB]. http://fgw.qinghai.gov.cn/xwzx/tzgg/201906/t20190613_71168.html (accessed 19 October 2023).

Ralon, P., Taylor, M., Ilas, A. et al. (2017). Electricity Storage and Renewables: Costs and Markets to 2030. *Technical report*. International Renewable Energy Agency. https://www.irena.org/-/media/Files/IRENA/Agency/Publication/2017/Oct/IRENA_Electricity_Storage_Costs_2017_Summary.pdf (accessed 19 October 2023).

Reddy, S.S., Bijwe, P.R., and Abhyankar, A.R. (2014). Real-time economic dispatch considering renewable power generation variability and uncertainty over scheduling period. *IEEE Systems Journal* 9(4): 1440–1451.

Shu, Y. and Chen, W. (2018). Research and application of UHV power transmission in China. *High Voltage* 3(1): 1–13.

Siemonsmeier, M., Baumanns, P., Van Bracht, N. et al. (2018). Hydropower Providing Flexibility for a Renewable Energy System: Three European Energy Scenarios. *Technical report*. HydroFlex. https://www.h2020hydroflex.eu/wp-content/uploads/2019/01/HydroFlex-Report_Three-European-Energy-Scenarios.pdf (accessed 19 October 2023).

State-owned Assets Supervision and Administration Commission of the State Council (2014). The second phase of Longyangxia, the world's largest hydro-solar complementary project, starts construction [EB/OL]. http://www.sasac.gov.cn/n2588025/n2588124/c3781704/content.html (accessed 19 October 2023).

Sterl, S., Vanderkelen, I., Chawanda, C.J. et al. (2020). Smart renewable electricity portfolios in West Africa. *Nature sustainability* 3(9): 710–719.

Sunar, N. and Birge, J.R. (2019). Strategic commitment to a production schedule with uncertain supply and demand: renewable energy in day-ahead electricity markets. *Management Science* 65(2): 714–734.

The Chinese Government (2022). Plan for Renewable Energy Development during the 14th Five-Year Plan Period. https://www.ndrc.gov.cn/xxgk/zcfb/ghwb/202206/P020220601501054858882.pdf (accessed 20 October 2023).

The People's Government of Qinghai Province (2021a). Qinghai province's work plan for building a national clean energy demonstration province [EB/OL]. http://www.guoluo.gov.cn/html/33/303429.html (accessed 19 October 2023).

The People's Government of Qinghai Province (2021b). Interpretation of "Qinghai's action plan for building a national clean energy industry highland" [EB/OL]. http://

www.qinghai.gov.cn/zwgk/system/2021/09/29/010393797.shtml (accessed 19 October 2023).

Tong, D., Farnham, D.J., Duan, L. et al. (2021). Geophysical constraints on the reliability of solar and wind power worldwide. *Nature Communications* 12(1): 1–12.

Wang, F., Xie, Y., and Xu, J. (2019). Reliable-economical equilibrium based short-term scheduling towards hybrid hydro-photovoltaic generation systems: case study from China. *Applied Energy* 253: 113559.

Wang, Q., Wen, T., Yu, D., and Wang, Z. (2020). Reassessment of wind energy resources in Qinghai Province based on multi-source data. *Qinghai Technology* 27(04): 72–81+102.

Wang, F., Xu, J., Liu, L. et al. (2021a). Optimal design and operation of hybrid renewable energy system for drinking water treatment. *Energy* 219: 119673.

Wang, Z., Wen, X., Tan, Q. et al. (2021b). Potential assessment of large-scale hydro-photovoltaic-wind hybrid systems on a global scale. *Renewable and Sustainable Energy Reviews* 146: 111154.

Xie, J., Zheng, Y., Pan, X. et al. (2021). A short-term optimal scheduling model for wind-solar-hydro hybrid generation system with cascade hydropower considering regulation reserve and spinning reserve requirements. *IEEE Access* 9: 10765–10777.

XinhuaNet (2020). "Qingyu UHV DC Project" supporting power project put into operation on schedule [EB/OL]. http://www.xinhuanet.com/fortune/2020-09/30/c_1126564210.htm (accessed 19 October 2023) (in Chinese).

Xiong, H., Egusquiza, M., Østergaard, P.A. et al. (2021). Multi-objective optimization of a hydro-wind-photovoltaic power complementary plant with a vibration avoidance strategy. *Applied Energy* 301: 117459.

Xu, L. and He, X. (1987). Design and manufacture of the 320MW hydro-generator set of Longyangxia Hydropower Station. *Dongfang Electric Review* (002): 13–22+64.

Yang, T., Zhang, P., Lu, K., and Fan, X. (2014). Feasibility Study Report of Longyangxia Hydro-Solar Hybrid Phase II 530MWp Grid-Connected Photovoltaic Power Station Project. *Technical Report*. Northwest Survey and Design Institute.

Yi, Z., Xu, Y., Gu, W., and Wu, W. (2019). A multi-time-scale economic scheduling strategy for virtual power plant based on deferrable loads aggregation and disaggregation. *IEEE Transactions on Sustainable Energy* 11(3): 1332–1346.

Zhang, Y., Lian, J., Ma, C. et al. (2020a). Optimal sizing of the grid-connected hybrid system integrating hydropower, photovoltaic, and wind considering cascade reservoir connection and photovoltaic-wind complementarity. *Journal of Cleaner Production* 274: 123100.

Zhang, Z., Qin, H., Li, J. et al. (2020b). Short-term optimal operation of wind-solar-hydro hybrid system considering uncertainties. *Energy Conversion and Management* 205: 112405.

Zhou, Y., Scheller-Wolf, A., Secomandi, N., and Smith, S. (2019). Managing wind-based electricity generation in the presence of storage and transmission capacity. *Production and Operations Management* 28(4): 970–989.

11
Optimal RPS Implementation Strategy Considering Both Power Suppliers and Users

Limiting warming not only to 'just below' but to 'well below' 2 degrees Celsius or 1.5 degrees Celsius urgently requires policies and actions to bring about steep emission reductions this decade, aligned with mid-century global net-zero CO_2 emissions.

Malte Meinshausen et al. Published on Nature 2022, 604:304–309

Renewable portfolio standards (RPS) have been found to be an effective policy to promote the integration of renewable energy (RE). But in many countries, the implementation of RPS has not yet been fully developed. This chapter proposes a social, economic, and environmental equilibrium distribution strategy for a single province in China based on RPS from both the demand-side power user and the supply-side power plant to promote the creation of annual power plans. Three aspects, including policy acceptance, generation cost, and renewable power target, are taken into consideration at the same time. The multi-objective optimal model is reduced to a single form and solved with LINGO by using a fuzzy membership function. A case study from the Chinese province of Guangdong is offered to show the applicability and effectiveness of the model as well as comparison analysis with existing plans. Based on the decision-maker preferences, three scenarios – the non-hydro renewable targets scenario, the renewable development scenario, and the environmental impacts scenario –are proposed and analyzed. Compared with the current plans, the proposed optimal model could effectively maintain the power tariffs, decrease generation costs, and decrease Guangdong's electricity consumption emissions.

11.1 Background Review

The renewable portfolio standards (RPS) policy sets a minimum amount of renewable energy (RE) for power consumption to ensure the country's development of RE, promote its energy structure transition, and reduce emissions (Joshi 2021). The RPS policy in China was first conceptualized in 2011, and was legally enacted

Sustainable Hybrid Energy Systems: Carbon Neutral Approaches, Modeling, and Case Studies, First Edition. Jiuping Xu and Fengjuan Wang.

in 2019. The final document included advice on how to meet these RPS objectives and provided information on the amount of RE and non-hydro RE consumption for each province (Li et al. 2021). The China National Energy Administration releases an annual monitoring and evaluation report on the province's performance and the RPS goals for the upcoming year in order to encourage the implementation of the RPS policy. However, the outcomes were highly uneven, with some provinces meeting the RPS objectives while others did not (National Energy Administration 2019b). In line with the aims of the RPS targets set by the Chinese government to motivate the development of provincial RE, several provinces released their own RE consumption guarantee implementation plans in 2020, outlining the departmental responsibilities and RE consumption proportions for the various market stakeholders. While RPS policy mechanisms at the state and provincial levels are playing an increasingly significant role (Carley et al. 2018), the RPS guarantee implementation strategy in each of China's provinces has yet to have any tangible effects.

Previous studies have looked into the RPS policy's RE schedule at the national or regional level. For example, Bird et al. (2011) presented a regional energy deployment system model to assess how cap-and-trade and RPS policy choices will affect the US power industry. Similarly, Choi et al. (2015) used a bottom-up energy system model and scenario-based quantitative analysis to explore how an RPS regulation might affect the growth of the Korean power industry. Wang et al. (2019) proposed a cost-minimizing methodology that took into account energy, environment, and economy to investigate commercially viable and equitable RPS allocations in Chinese provinces. Wang et al. (2020b) built a mid-to-long-term optimization model for province growth for the China Southern Power Grid Corporation based on power plan limits and the RPS targets from 2016 to 2030. These studies offered viable RPS allocation models and drew significant conclusions for the establishment of national RE development plans. However, as the RPS policy must be implemented by the provinces (Chen et al. 2009), the design of the electricity schedule in each province under the RPS policy is crucial to the development of RE for the entire nation.

Consequently, the application of provincial RPS policies has drawn a lot of attention. For example, Ji et al. (2017) created a two-stage programming methodology with risk-explicit intervals to study electricity generating expansion under RPS in Guangdong. Rouhani et al. (2016) added demand functions to the RPS calculator model created by Energy and Environmental Economics (E3) Inc. to calculate the implications of the RPS targets on pollutants, fuel and generation cost distributions, and greenhouse gas emissions in California. Rountree (2019) used a thorough case study methodology to investigate how stakeholders felt about the motivations and strategies behind specific policy design elements in Nevada's RPS and how they affected the social and political dynamics in the state's power sector. Kagan (2019) conducted a Multiple Streams Framework to investigate the political process that resulted in Hawaii's 100% RPS. While the development of local or state RE under RPS policy has benefited greatly from these research works, they were all generation-focused. The power consumer is now an essential component of power plans thanks to the reform of the electricity markets. As power users in various industry sectors have diverse needs for RE, RPS policies must be consumer-centered.

Some research studies have been focused on the electricity demand side of RPS policies. For example, Barbose et al. (2016) presented the first national-level evaluation of state RPS program benefits in the United States, demonstrating that the RPS policy supported jobs and lowered wholesale electricity prices. Hao et al. (2020) compared the impact of feed-in-tariffs and RPS policies using market transactions to suggest central planning, bi-level regulations, and regulation under screening in three market structure scenarios. Ying et al. (2020) built benchmark market transaction models for on-grid prices using systems dynamics and analyzed the incentive impact of pertinent RPS system designs on the demand side of tradable green certificates using on-grid prices. Kim et al. (2012) employed contingent valuation to examine whether Korean families would be ready to pay extra for energy produced by wind, solar, and hydropower under an RPS policy. Despite the fact that these studies have acknowledged the crucial role power consumers play in RPS policy and made substantial recommendations, few studies have taken into account both generation and demand when creating realistic RPS allocation systems (Bu and Zhang 2020).

This chapter established an economic, social, and environmental optimization model for annual power plans under an RPS policy for a single province in an attempt to partially fill this research vacuum by taking into account both power consumers on the demand side and power generation on the supply side. From the perspectives of policy acceptance, generation costs, and RE proportion, the multi-objective equilibrium scheduling model put forth in this chapter offers an improvement over earlier work as a method for provincial governments to create annual power plans that incorporate a power distribution strategy for power plants and users. The case chosen to show the effectiveness of the suggested methodology is Guangdong, one of the first provinces to begin power market changes. Due to the many uncertainties, the optimal model is solved using a fuzzy membership function, and three scenarios are provided to assess the diverse preferences of the provincial governments and offer more useful advice for them.

11.2 Key Problem Statement

In this section, the detailed descriptions of the decision process, the classification of power users, and the multi-objective trade-offs will be presented.

11.2.1 Decision Process Description

When provincial governments try to comply with RPS, they first assess the local RE supply to determine whether to import RE from other provinces, then they create allocation plans for their related subordinate departments to ensure the stated targets can be accomplished (Bu and Zhang 2021). Due to high provincial demand, they are forced to buy RE and non-RE from other provinces, and the price, types, and generation costs vary depending on the particulars of the economic and energy markets in each province. On the demand side, as the power consumers are required to consume a particular percentage of RE, the allocation of RE to the various power users is the main concern on the demand side; as shown in Figure 11.1.

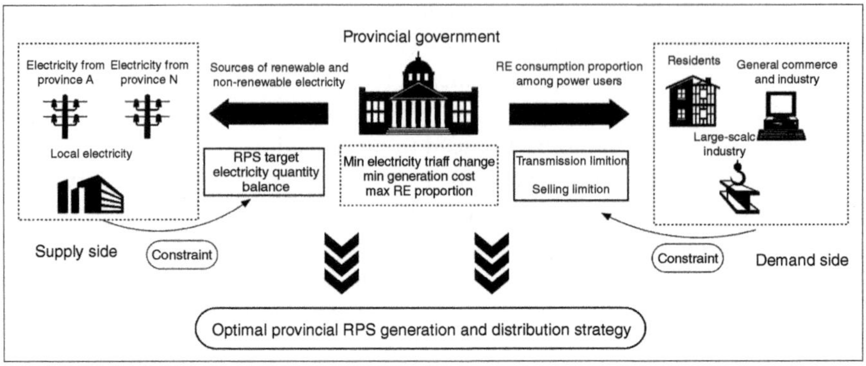

Figure 11.1 Electricity consumer's RPS allocation model.

11.2.2 Power User Classifications

Current policies indicate that all power users should be allocated the same RE proportion (National Energy Administration 2019a). However, as different types of power users have varied needs, structures, tariffs, and other features, identifying the various power users and setting diverse RE consumption proportions could lead to more affordable, effective RPS allocations (Wendling et al. 2020). For example, Wang et al. (2017) divided electricity users into eight categories: residential, non-residential, large-scale industrial, general industrial, non-industrial, commercial, wholesale, and agricultural, which made it challenging to implement. Power users can be categorized into four groups based on voltage levels and user types: residences, general industry and commerce, large industry, and others (Shin et al. 2014). However, the Chinese market electricity tariff has only three types: residents, general industry and commerce, and large-scale industry (Lin and Jiang 2012). RPS could play a part in the reform of the power user tariff mechanism because the power pricing mechanism has long had an inappropriate price structure and significant tariff cross-subsidization (China Electricity Council 2020). In order to ease the severe tariff cross-subsidization and lessen the strain on commercial and industrial power consumers, it is imperative that residents be included in the RPS policy. For these reasons, this chapter split the power users into residents, general industry and commerce, and large-scale industry in order to implement the provincial RPS.

11.2.3 Multi-Objectives of Demand and Supply Sides

Provincial governments must take into account the cost of power customers' purchases as well as the total cost of generation from the power plants to ensure the success of RPS. Previous research has shown that while higher tariffs are unavoidable when RPS standards are tightened (Considine and Manderson 2015), RPS is less accepted if the tariffs are too high (Lee et al. 2020). As a result, provincial governments must adopt a comprehensive approach when attempting to control the costs of the various users (Yao et al. 2021). Different RE consumption target allocation schemes can have different economic, environmental, and energy consequences (Barbose et al. 2015). It is important that provincial governments pay attention to both power generation and environmental protection, which means that the generation costs and the costs associated with pollutant emissions need to be closer to the lower level (Ang and Su 2016). Energy and resource planning has made extensive use of multi-objective programming techniques, which have been found to be effective in resolving trade-offs between several goals. For example, Jiao et al. (2018) suggested a multi-objective optimization for an integrated energy system based in a park, and Yao et al. (2020) developed a multi-resource trade-off between a reservoir in a pump-hydro system, which (Li et al. 2020) afterwards used. Based on the diverse needs of provincial governments and encouraged by these successful applications, it is also applied in this chapter.

11.3 Modeling

In this section, the mathematical model for the social–economic–environment equilibrium-based provincial annual power plan scheduling is built.

11.3.1 Assumptions

Encouraged by Bu and Zhang (2021), provinces were split into two groups: (i) those abundant in RE resources but with relatively lower electricity demand, and (ii) those located near the load center but with lower integration of renewables. Due to their stronger potential and motivation to develop RE, those in category 2 were selected as the research subjects for this study. The new provincial RPS policy has two kinds of targets (National Energy Administration 2019a): an RE (within hydro) target and a non-hydro RE target. The assumptions in this chapter are as follows:

(1) This is an annual power plan decision problem.
(2) This chapter divides power into hydropower, thermal power, and non-hydro RE, with wind, solar, and biomass all categorized as non-hydro RE.
(3) The decision province's local electricity supply is unable to satisfy demand and needs to import electricity from other provinces.

11.3.2 Notations

Indices

i Index of power users; $i = 1, 2, 3$ means residential power users, general industrial and commercial users, and large-scale industrial users.

j Index of different provinces; $j = 1, 2, 3$ means Guangdong Province, Guizhou Province, and Yunnan Province.

Decision variables

x_{ij} The ratio of hydroelectricity consumed by power user i from province j to total electricity.

y_{ij} The ratio of fossil electricity consumed by power user i from province j to total electricity.

z_{ij} The ratio of non-hydro RE (wind, solar., etc.) electricity consumed by power user i from province j to total electricity.

Parameters

Q_j^p The yearly electricity consumption of province j.
Q_i^u The yearly consumption of user i.
a_j^g The yearly generation of province j.
a_j^x The yearly hydroelectricity generation of province j.
a_j^y The yearly fossil electricity generation of province j.
a_j^z The yearly non-hydro RE generation of province j.
C_T The overall policy acceptance for all power users.
E_T The overall comprehensive generation cost.
b_j^x The average sales price of hydroelectricity in province j.
b_j^y The average sales price of fossil electricity in province j.
b_j^z The average sales price of non-hydro RE in province j.
P_i The average electricity tariff for power user i in local province.
C_j^T The transmission cost from province j to local province.
e_j^x The average feed-in-tariff of hydroelectricity in province j.
e_j^y The average feed-in-tariff of fossil electricity in province j.
e_j^z The average feed-in-tariff of non-hydro RE in province j.
c_j^x The generation cost per 1 kWh hydroelectricity in province j.
c_j^y The generation cost per 1 kWh fossil electricity in province j.
c_j^z The generation cost per 1 kWh non-hydro RE in province j.
C_j^e Emission cost per 1 kWh fossil electricity in province j.
g Government's attitude toward pollution, higher value means tougher attitude.
$T_j^{\max}$ Max transmission capacity from province j to local province.
RPS_j^1 Renewable portfolio standards for the province j.
RPS_j^2 Non-hydro renewable portfolio standards for the province j.

11.3.3 Objectives

The provincial government must create an annual power plan, choose how much energy will be produced locally and how much will need to be imported from other provinces, then coordinate the power users in order to meet the usual power demand and the RPS target. Obtaining different power inputs from province j and allocate these to power user i are two steps of this process. Decision variable x_{ij} is the ratio of hydroelectricity consumed by power user i from province j to total electricity, and

y_{ij}, z_{ij} refers to the thermal power and the non-hydro RE. The provincial government must take into account three factors in order to create a suitable annual power plan.

11.3.3.1 To Minimize the Electricity Tariff Variations

The first factor is how much the power users accept RPS policy (Stokes and Warshaw 2017), which is mostly reliant on the stability of the electricity tariff. An RPS strategy is less likely to be accepted, according to earlier studies, as power costs rise (Lee et al. 2020). The government must reduce tariff volatility and improve user acceptance when distributing the RPS to relieve strain on power users and encourage commercial and industrial growth (Kung 2012). The weighted average of a power user's tariff volatility can be used to determine the comprehensive power tariff volatility depending on the types of users, as shown in Eq. (11.1), where $i = 1, 2, 3$ refer to residents, general commerce and industry, and large-scale industry electricity users, respectively.

$$\text{Min}C_T = \frac{\sum_{i=1}^{4} Q_i C_i}{Q_0^T} \tag{11.1}$$

The tariff volatility (C_i) of power users is defined as the ratio between the change (N_i) and the previous tariff (P_i), as shown in Eq. (11.2).

$$C_i = \left(\frac{P_i - N_i}{P_i}\right)^2 \tag{11.2}$$

The new tariff (N_i) can be further expressed by Eq. (11.2), which is noted by x_{ij}, y_{ij}, z_{ij} and the power price in the different provinces.

$$N_i = \sum_{j=1}^{3} (x_{ij} b_j^x + y_{ij} b_j^y + z_{ij} b_j^z) \tag{11.3}$$

The power price in the different provinces (b_j^x, b_j^y, b_j^z) is calculated using the sum of the average feed-in-tariff and the transmission costs (National Development and Reform Commission 2019), as shown in Eqs. (11.4)–(11.6), where e_j^x, e_j^y, e_j^z is the average feed-in-tariff for hydropower, thermal power, and non-hydro RE in province j and C_j^T is the transmission costs from province j to the local province.

$$b_j^x = e_j^x + C_j^T \tag{11.4}$$

$$b_j^y = e_j^y + C_j^T \tag{11.5}$$

$$b_j^z = e_j^z + C_j^T \tag{11.6}$$

11.3.3.2 To Minimize Total Costs

The second factor is the comprehensive generation costs, which are primarily borne by the province responsible for scheduling. The RPS program aims to advance sustainability by encouraging the use of RE sources and the growth of associated generation businesses. Since different provinces have different resource endowments and different production technologies, the costs of non-hydro RE generation vary widely

(Shin et al. 2014), as do the costs of traditional power generation (China Electric Power Press 2019). The local provincial government must be concerned with reducing both the environmental costs and the generation costs when power is purchased from other provinces (Tan et al. 2021). Let c_j^x, c_j^y, c_j^z represent the generation cost per 1 kWh for hydropower, thermal power, and non-hydro RE in province j, which is measured using the levelized cost of power (LCOE) (Harvey 2020), C_j^e means the emission cost per 1 kWh of thermal power in province j (Basnet et al. 2015). g represents the government's attitude toward pollution; it is designed to evaluate the impact of the cost of emissions in a flexible manner. The higher the number, the stricter the stance. Then this objective can be written as Eq. (11.7)

$$\mathrm{Min}E_T = \sum_{i=1}^{3}\sum_{j=1}^{3} Q_1^p \left[x_{ij}c_j^x + y_{ij}(c_j^y + gC_j^e) + z_{ij}c_j^z\right] \tag{11.7}$$

11.3.3.3 To Maximize RPS

The third factor is the RPS target. Depending on the rules (Guangdong Province Energy Administration 2020), provinces have both restrictive and incentive RPS. For example, the restricted RPS for Guangdong in 2020 was 28.5%, while the incentive RPS was 32%. Governments are under pressure to increase the share of RE against the backdrop of energy saving and carbon reduction, as shown in Eq. (11.8).

$$\mathrm{Max}RPS = \sum_{i=1}^{3}\sum_{j=1}^{3}(x_{ij} + z_{ij}) \tag{11.8}$$

11.3.4 Provincial Power Constraints

In this section, the provincial power constraints will be described in detail, including the power generation and consumption balance, the power sales limitations, and the power transmission limitations, etc.

11.3.4.1 Power Generation and Consumption Balance

First, all provincial power users' power consumption should not exceed the combined power generated locally plus the input power sent into the province, which should equal the province's power demands (Munoz et al. 2017), as shown in Eq. (11.9).

$$\sum_{j=1}^{3} Q_0^T(x_{ij} + y_{ij} + z_{ij}) \geq Q_i \tag{11.9}$$

Let a_j^x, a_j^y, a_j^z denote the total hydro, thermal, and non-hydro RE power generation in province j, then the power generation and consumption balance can be written as Eqs. (11.10)–(11.12). The hydropower generation and consumption balance is shown in
Eq. (11.10).

$$\sum_{i=1}^{3} Q_1^p x_{i1} \leq a_1^x \tag{11.10}$$

The thermal power generation and consumption balance is shown in Eq. (11.11).

$$\sum_{i=1}^{3} Q_1^P y_{i1} \leq a_1^y \tag{11.11}$$

The non-hydro RE power generation and consumption balance is shown in Eq. (11.12).

$$\sum_{i=1}^{3} Q_1^P z_{i1} \leq a_1^z \tag{11.12}$$

11.3.4.2 Power Sale Limitations

The RE sold to the other provinces should be less than the excess RE generation, and the thermal power purchased from province j also needs to be less than the excess fuel power generation because each province must meet the power demand within its own borders, which means that the RE generated in province j must first fit the RPS target in its own province (Wang et al. 2020b). Therefore, the power sale limitations can be written as Eqs. (11.13)–(11.15).

$$\sum_{i=2}^{3} Q_0^T x_{ij} \leq a_j^x - Q_j^p RPS_j^1 \tag{11.13}$$

$$\sum_{i=2}^{3} Q_0^T z_{ij} \leq a_j^z - Q_j^p RPS_j^2 \tag{11.14}$$

$$\sum_{i=2}^{3} Q_0^T (x_{ij} + y_{ij} + z_{ij}) \leq a_j^g - Q_j^p \tag{11.15}$$

11.3.4.3 Power Transmission Limitations

The transmission limits for power through the grid should be adhered to. Let $P_j^{\max}$ be the maximum transmission capacity from local province to province j, then the total provincial power trading volume is less than the maximum transmission capacity, as Eq. (11.16).

$$\sum_{i=1}^{3} Q_1^P (x_{ij} + y_{ij} + z_{ij}) \leq T_j^{\max} \times 8760 \tag{11.16}$$

11.3.4.4 RPS and Non-hydro RPS Target Limitations

The province's energy consumers are obligated to contribute to the RPS goals issued to the local government by the federal government. Let RPS_j^1 be the RE consumption proportion for province j and RPS_j^2 be the non-hydro RE consumption proportion for province j, then this constraint can be represented as Eqs. (11.17) and (11.18).

$$\sum_{i=1}^{3} \sum_{j=1}^{3} (x_{ij} + z_{ij}) \geq RPS_j^1 \tag{11.17}$$

$$\sum_{i=1}^{3} \sum_{j=1}^{3} z_{ij} \geq RPS_j^2 \tag{11.18}$$

11.3.5 Global Model

Based on the unbalanced power generation capacity distribution among provinces and the various power user structures. A multi-objective model for power schedule in the local province and RE consumption proportion allocation is established, which takes into account both the power user on the demand side and the power plant on the supply side to achieve the provincial RPS target. The RPS target and the other related constraints are shown in Eqs. (11.17)–(11.18). The local provincial government considers the minimal user tariff volatility, the minimal comprehensive generation costs, and the maximum RPS, as shown in Eqs. (11.1), (11.7), (11.8). By integrating the model objectives and corresponding constraints, the global optimization model, as shown in Eq. (11.19), is effectively formed to determine power generation and importation into the local province, with the proportional distribution of RE and non-RE consumption among power consumers to acquire the best operational decisions for the local provincial government under the provincial RPS policy.

$$
\begin{aligned}
&\mathrm{Min}C_T = \frac{1}{Q_1^T}\sum_{i=1}^{3} Q_i\left[1-\frac{1}{P_i}\sum_{j=1}^{3}(x_{ij}b_j^z + y_{ij}b_j^y + z_{ij}b_j^z)\right]^2\\
&\mathrm{Min}E_T = \sum_{i=1}^{3}\sum_{j=1}^{3} Q_1^p\left[x_{ij}c_j^x + y_{ij}(c_j^y + gC_j^e) + z_{ij}c_j^z\right]\\
&\mathrm{Max}RPS = \sum_{i=1}^{3}\sum_{j=1}^{3}(x_{ij}+z_{ij})\\
&\text{s.t.}\begin{cases}
\sum_{j=1}^{3} Q_1^p(x_{ij}+y_{ij}+z_{ij}) \geq Q_i\\
\sum_{i=1}^{3} Q_1^p x_{i1} \leq a_1^x,\ \sum_{i=1}^{3} Q_1^p y_{i1} \leq a_1^y,\ \sum_{i=1}^{3} Q_1^p z_{i1} \leq a_1^z\\
\sum_{i=2}^{3} Q_1^p(x_{ij}+y_{ij}+z_{ij}) \leq T_j^{\max}\times 8760\\
\sum_{i=2}^{3} Q_1^p x_{ij} \leq a_j^x - Q_j^p RPS_j^1,\ \sum_{i=2}^{3} Q_1^p z_{ij} \leq a_j^z - Q_j^p RPS_j^2\\
\sum_{i=2}^{3} Q_1^p(x_{ij}+y_{ij}+z_{ij}) \leq a_j^g - Q_j^p\\
\sum_{i=1}^{3}\sum_{j=1}^{3}(x_{ij}+z_{ij}) \geq RPS_1^1\\
\sum_{i=1}^{3}\sum_{j=1}^{3} z_{ij} \geq RPS_1^2\\
\sum_{i=1}^{3}\sum_{j=1}^{3}(z_{ij}+y_{ij}+z_{ij}) = 1\\
0 \leq x_{ij}, y_{ij}, z_{ij} \leq 1
\end{cases}
\end{aligned}
\tag{11.19}
$$

The aforementioned multi-objective programming model was created to calculate the distribution of RE consumption among power users as well as power generation and imports. The suggested model can be effectively used in a province with a high demand for electricity and a low supply, but because supply and demand for electricity vary by province, it can also be an effective tool for implementing the RPS policy in a sizable region by taking into account the province's power sales. Additionally, by modifying the necessary parameters, this model can produce an accurate and scientific estimate of the power consumption proportions for particular RE (for example, wind, hydro, solar) based on a thorough and comprehensive consideration of the demand side and supply side. Therefore, using this multi-objective programming model, provincial governments may create an effective yearly power plan for the RPS policy that will encourage RE use and a smooth transition of the energy system.

11.3.6 Model Solving

As Guangdong's provincial government is obliged to implement the central government's RPS policy, the priority is to reasonably allocate the RE consumption proportion to the power users to reduce tariffs and guarantee the stability of the society. The government must also take into account the power generation side because the RPS policy's objective is to boost RE development. Previous research has shown that the government's values toward the supply and demand side might be subjective (Pramanik and Roy 2008), hence three fuzzy membership functions are employed to describe the government's values toward the various objectives (Ilyas et al. 2020). The whole model solution and process are shown in Figure 11.2.

A membership function is incorporated to ascertain the local provincial government's satisfaction with the acceptance and cost objectives in order to reduce the policy acceptance objective on the power demand side and the cost objective on the power supply side.

The membership function for the policy acceptance and cost goals are established in Eqs. (11.20)–(11.21), where $v_1(C_T)$ is the government's satisfaction with the acceptance objective, and $C_{T,\min}$ is the minimization of the objective. Similarly, v_2E_T is the government's satisfaction with the cost objective, and $E_{T,\min}$ is the minimization of the objective.

$$\begin{cases} C_T = \dfrac{1}{Q_1^T}\displaystyle\sum_{i=1}^{3} Q_i\left[1 - \dfrac{1}{P_i}\sum_{j=1}^{3}(x_{ij}b_j^z + y_{ij}b_j^y + z_{ij}b_j^z)\right]^2 \\ v_1C_T = \begin{cases} 1, C_T \le C_{T,\min} \\ \dfrac{C_{T,\max} - C_T}{C_{T,\max} - C_{T,\min}}, C_{T,\min} \le C_T \le C_{T,\max} \\ 0, C_T \ge C_{T,\max} \end{cases} \end{cases} \tag{11.20}$$

$$\begin{cases} E_T = \displaystyle\sum_{i=1}^{3}\sum_{j=1}^{3} Q_1^p\left[x_{ij}c_j^x + y_{ij}(c_j^y + \mathrm{FD}_j) + z_{ij}c_j^z\right] \\ v_2E_T = \begin{cases} 1, E_T \le E_{T,\min} \\ \dfrac{E_{T,\max} - E_T}{E_{T,\max} - E_{T,\min}}, E_{T,\min} \le E_T \le E_{T,\max} \\ 0, E_T \ge E_{T,\max} \end{cases} \end{cases} \tag{11.21}$$

A membership function is developed to ascertain whether the local provincial government is satisfied with the RPS proportion objective in order to optimize

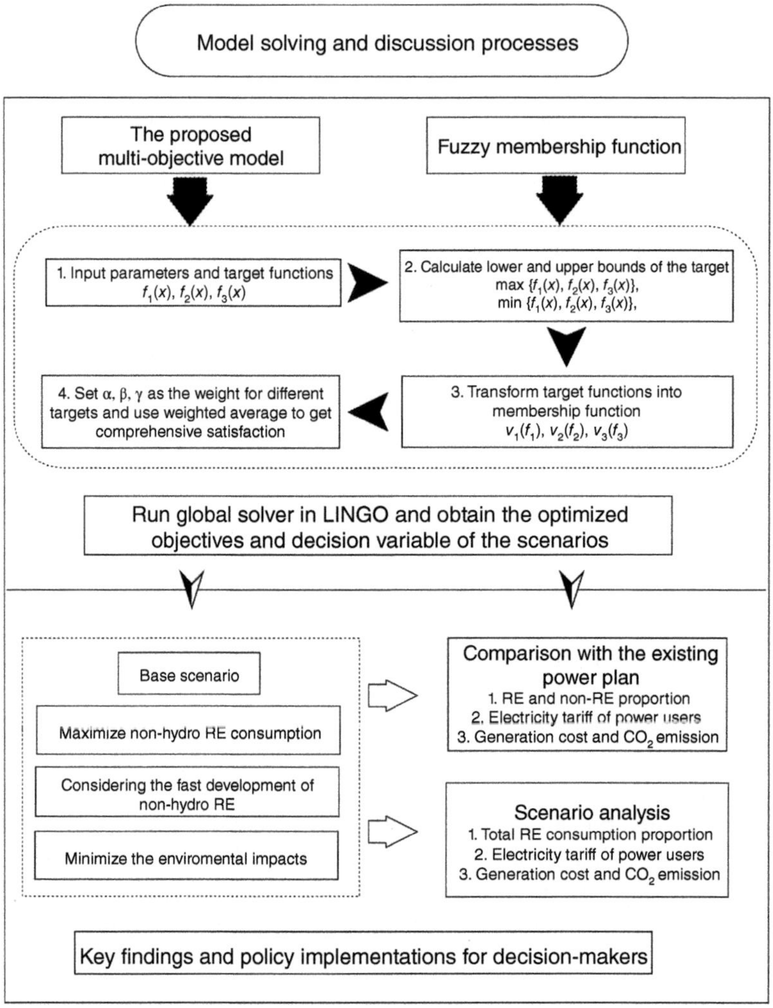

Figure 11.2 Model solving and result discussion processes.

the provincial RPS proportion target, as shown in Eq. (11.22), where $\upsilon_3 R_T$ is the government's satisfaction with the RPS proportion objective, and $f_{3,\max}$ is the maximization of the objective.

$$\begin{cases} R_T = \sum_{i=1}^{3}\sum_{j=1}^{3}(x_{ij} + z_{ij}) \\ \upsilon_3 R_T = \begin{cases} 1, R_T \geq R_{T,\max} \\ \dfrac{R_T - R_{T,\min}}{R_{T,\max} - R_{T,\min}}, R_{T,\min} \leq R_T \leq R_{T,\max} \\ 0, R_T \leq R_{T,\min} \end{cases} \end{cases} \tag{11.22}$$

In this chapter, fuzzy membership functions are utilized to handle each optimization goal and determine the maximum value for the weighted sum of the local provincial government's satisfaction with each goal in order to solve the proposed multi-objective optimization model. The objectives in Eq. (11.19) are transformed into a single objective using Eq. (11.23). The government's relative focus on the RPS policy acceptability goals is indicated by the parameter α variance, which ranges from 0 to 1, with 1 denoting the highest level of policy approval; parameters β and γ have a similar meaning.

$$\mathrm{Max}U = \alpha v_1 C_T + \beta v_2 E_T + \gamma v_3 R_T \tag{11.23}$$

Therefore, the multi-objective model can be transformed in its equivalent form as Eq. (11.24). Depending on the different decision-maker policy preferences, attitude parameters α, β, and γ can have different values. With the prepared case data and based on the existing solution algorithms for this programming, by running it in LINGO software, the solution to the social–economic–environmental equilibrium model can be determined.

$$\begin{aligned}
&\mathrm{Max}U = \alpha v_1 C_T + \beta v_2 E_T + \gamma v_3 R_T\\
&\text{s.t.}\begin{cases}
v_1 C_T = \begin{cases} 1, C_T \le C_{T,\min} \\ \dfrac{C_{T,\max} - C_T}{C_{T,\max} - C_{T,\min}}, C_{T,\min} \le C_T \le C_{T,\max} \\ 0, C_T \ge C_{T,\max} \end{cases}\\
v_2 E_T = \begin{cases} 1, E_T \le E_{T,\min} \\ \dfrac{E_{T,\max} - E_T}{E_{T,\max} - E_{T,\min}}, E_{T,\min} \le E_T \le E_{T,\max} \\ 0, E_T \ge E_{T,\max} \end{cases}\\
v_3 R_T = \begin{cases} 1, R_T \ge R_{T,\max} \\ \dfrac{R_T - R_{T,\min}}{R_{T,\max} - R_{T,\min}}, R_{T,\min} \le R_T \le R_{T,\max} \\ 0, R_T \le R_{T,\min} \end{cases}\\
C_T = \dfrac{1}{Q_1^T}\displaystyle\sum_{i=1}^{3} Q_i \left[1 - \dfrac{1}{P_i}\sum_{j=1}^{3}(x_{ij}b_j^z + y_{ij}b_j^y + z_{ij}b_j^z)\right]^2\\
E_T = \displaystyle\sum_{i=1}^{3}\sum_{j=1}^{3} Q_1^p \left[x_{ij}c_j^x + y_{ij}(c_j^y + FD_j) + z_{ij}c_j^z\right]\\
R_T = \displaystyle\sum_{i=1}^{3}\sum_{j=1}^{3}(x_{ij} + z_{ij})\\
0 \le x_{ij}, y_{ij}, z_{ij} \le 1\\
\text{Eqs. (11.9)–(11.18)}
\end{cases}
\end{aligned} \tag{11.24}$$

11.4 Case Study

To demonstrate the practical applicability and efficiency of the above-proposed multi-objective model, Guangdong Province, China, is used as a case example.

11.4.1 Case Description

Guangdong Province is located in the southernmost part of China (Figure 11.3). It is one of the most developed provinces in China, with an area of 179,800 km^2 and a population of 115.21 million. Guangdong's economy has prospered and undergone fast industrialization since the implementation of the opening-up policy and the region's favorable geographic proximity to Macao and Hong Kong (Zhou et al. 2018). However, this growth has come at the expense of severe pollution and heavy resource use, particularly in the area of power use.

Guangdong Province is the largest consumer of power in China, with social power consumption being 692.12 billion kWh in 2020, or 9.26% of the country's overall consumption (China Electric Power Press 2019). Only 70.1% of the power in Guangdong is produced locally; the rest must be imported. By the end of 2020, the Guangdong Power Grid had been linked to the central and western power grids via "eight to nine direct" high-voltage transmission lines. Guangdong has also been linked to the Yunnan power grid via five DC circuits, the Guizhou power grid via eight AC circuits, the Guangxi power grid via three DC circuits, and the State power grids via one DC circuit (Guangdong Power Exchange Center 2019).

Guangdong has played a significant role in the process of national reform as both China's largest provincial economy and its greatest consumer of electricity. In order to enable large industrial customers to directly contract with generators or buy power through power companies, Guangdong launched a wholesale power competition reform pilot in 2015 and encouraged the development of a power market system that combined medium- and long-term transactions with spot transactions (Lin et al. 2019). Guangdong develops pilot programs and market operating guidelines for the power spot market in order to facilitate this market. These guidelines cover market transaction behavior, market force prevention, financial credit risk avoidance, and market emergency preparations (National Energy Administration 2017).

11.4.2 Data Collection

The base year for this chapter was set for 2019. The data for power consumption, generation, the average feed-in tariff, and the average transmission tariff were taken from the 2019 China Electric Power Book (China Electric Power Press 2019) and the 2020 China Power Statistical Yearbook (China Electricity Council 2020), with some 2019 data being calculated based on 2018 growth rate data. The information on the power users' transmission costs was taken from related papers and consists of the costs of transmissions in the power supply province, the power demand province, and the cross-provincial transmission costs (National Development and Reform

Figure 11.3 Some basic information about Guangdong province.

Table 11.1 Parameters of different energies.

Provinces	Power sources	Average feed-in-tariff CNY\MWh	Average sailing tariff CNY\MWh	Generation quantity GWh	Generation cost CNY\MWh
Guangdong	Hydro	369.02	608.45	422,500	152
	Thermal	441.72	686.96	39,700	292.32
	Other RE	713.31	980.28	23,000	453.17
Guizhou	Hydro	287.63	752.71	139,000	162
	Thermal	348.17	818.1	76,900	271.35
	Other RE	720.84	1220.58	9800	439.3
Yunnan	Hydro	192.2	633.3	31,800	140
	Thermal	470.25	933.59	285,400	423.23
	Other RE	509.45	975.93	29,000	424

Commission 2019). The generation costs are represented by the LCOE, which was mainly based on the research of Fan et al. (2019). The line loss per unit was set at 8% and the government funds were 40 CNY per MWh. Parameters of different energies are shown in Table 11.1.

The 2020 China Power Statistical Yearbook provided information on the three provinces' combined electricity consumption in 2019. The 2019 China Electric Power Book provided information on the three provinces' combined coal consumption. The CO_2 emissions factor was 2.14 (Fan et al. 2018), and the unit emission costs for the various provinces were calculated using Guangdong's average carbon price in 2019 (Guangzhou Carbon Emission Certificate Trading Center 2020). The transmission capacities from Yunnan and Guizhou to Guangdong were based on the study by Wang et al. (2019), all of which are shown in Table 11.2.

The various industries split the power consumption among the various power consumers. The mining industry, heavy manufacturing industry, and energy generation industry were classified in this study as large-scale industries, while all other industries were classified as general industry. The power consumption was then calculated based on the proportion of the various industries in Guangdong (Guangdong Bureau of Statistics 2020). The respective power consumption for general commerce

Table 11.2 Parameters of different provinces.

Provinces	Electricity consumption GWh	Unit coal consumption kg/MWh	Unit emission cost CNY/MWh	Transmission capacity GWh
Guangdong	669,600	289	17.65	/
Guizhou	154,100	302	18.45	318,864
Yunnan	181,200	313	19.12	178,704

and industry, residents, and large-scale industry in 2019 was 205,830.98, 144,331.89, and 319,422.24 GWh, with the respective average prices being 707.24, 815.24, and 596 CNY per MWh. This study set a discount rate of 5% to get the average price in 2019.

11.4.3 Calculation Results

In this section, two calculation results for the provincial RPS policy implementation are analyzed: the power users' consumption and the Guangdong Province power sources.

11.4.3.1 Details of Power Consumption for Three Groups of Users

Here, the consumption of the various power consumers is shown so that the RE proportions in their combined energy use can be examined. Figure 11.4 displays the proportions of hydropower, non-hydro RE, and thermal power consumption that were determined using the multi-objective model. The total amount of hydropower used by households is 56,685.21 GWh (all hydro). The total amount of power used by general commercial and industrial power users in a year is 205,830.98 GWh, including 36,068.09 GWh of hydropower and 28,820.66 GWh of non-hydropower RE.

Table 11.3 shows the details of power consumption of three groups of power users, from which two main findings can be obtained. First, with 100% of their power coming from hydropower, the multi-objective approach increased the pressure on residents to consume RE compared to the typical government plan allocations. This was due to the low cost of the resident's tariff and the hydropower, which amply demonstrated the importance of residential power users in the planning of the new schedule. All non-hydro RE is used by general commercial and industrial power users, which was also due to price. Second, the large-scale industrial users are under

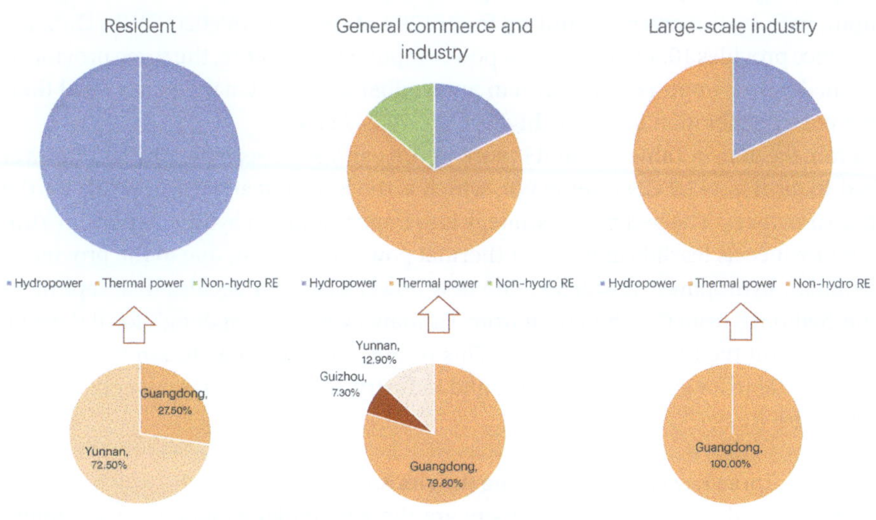

Figure 11.4 Electricity consumption of power users.

Table 11.3 Allocated hydro, thermal, and non-hydro RE power consumption of power users from different provinces (GWh).

Power sources provinces		Residents	General commerce and industry	Large-scale industry	Sum
Hydro	Guangdong	39,705	0	0	39,705.05
	Guizhou	0	36,068.09	0	36,068.09
	Yunnan	104,613	0	56,685.21	161,298.5
	Sum	144,318	36,068.09	56,685.21	237,071.64
Thermal	Guangdong	0	107491.66	262,755.4	370,247.07
	Guizhou	0	33,445.75	0	33,445.75
	Yunnan	0	0	0	0
	Sum	0	140,937.42	262,755.4	403,692.82
Non-hydro RE	Guangdong	0	23,002.92	0	23,002.92
	Guizhou	0	2095.27	0	2095.27
	Yunnan	0	3722.47	0	3722.47
	Sum	0	28,820.66	0	28,820.66

less pressure to consume RE compared to the general commercial and industrial power consumers; specifically, the proportion for general commerce and industry is 31.53%, while for the large-scale industry, it is only 17.75%.

11.4.3.2 Details of Guangdong Province's Power Schedule

The generation quantities and power imports into Guangdong province are illustrated in Figure 11.5. According to the annual power plan for Guangdong's provincial power grid corporation, Yunnan province provides 24.64% of the overall power input, with hydropower accounting for 161298 of the total 165,020 GWh. Guizhou province provides 10.69% of the total power input. Furthermore, the three provinces' thermal power imports from Yunnan are negligible, while Guangdong's local thermal power generation was at its highest, at 370,247 GWh.

From the data in Table 11.3 and Figure 11.5, it can also be seen that the hydropower feed-in tariff is 192.2 CNY per MWh, which is the lowest of all feed-in tariffs for the three provinces. The best power strategy is to import as much hydropower from Yunnan province as feasible and export thermal power to Yunnan due to the province's abundant hydropower resources. The new power plan first decides to accept more non-hydro RE from Guizhou than from Yunnan, even if Guangdong's local thermal power is still the first supply option. This is done in order to maintain the electricity tariff stable for general commerce and industry, which is the largest consumer of non-hydro RE.

11.4.3.3 Three Key Findings From the Results Analysis

First, the residential power consumers are the key implementers of Guangdong's RPS policy, according to the optimal distribution plan for Guangdong's input

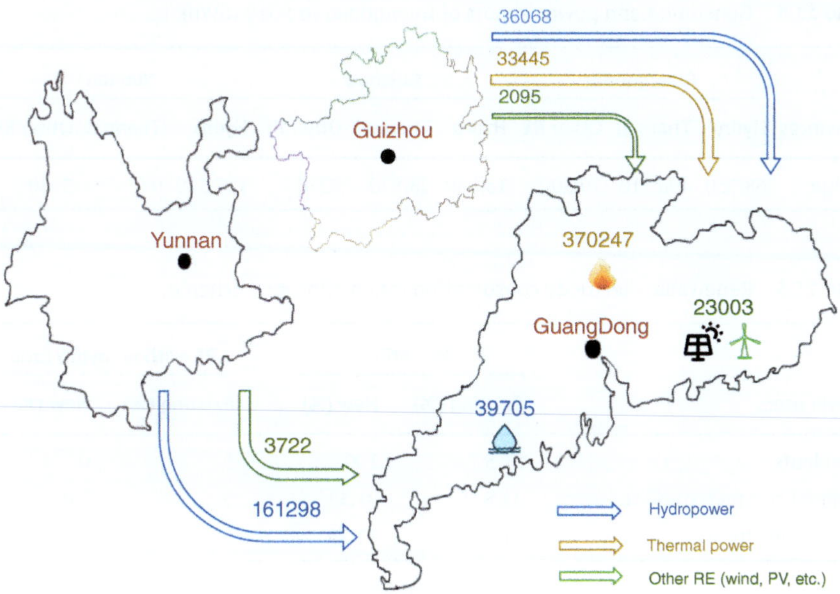

Figure 11.5 Electricity sources of Guangdong province (GWh).

hydropower from Yunnan to the inhabitants. Second, general commercial and industrial power customers are consuming non-hydro RE, which essentially raises the RPS's overall level of acceptance. Third, because it is less expensive and more reliable, Guangdong's indigenous thermal power is targeted toward large-scale industrial power customers.

11.5 Discussion

In this section, a comprehensive discussion of the optimal model calculation results, a comparative study, a scenario analysis, and policy implications are given.

11.5.1 Comparison With the Existing Schedule

This section compares the existing plan with the new plan calculated by the proposed economic–social–efficiency model. According to the monitoring and evaluation report published by China National Energy Administration (National Energy Administration 2019b) and the "2019 Southern Region Cross-Regional and Cross-Provincial Electricity Market Operation Report" from Guangzhou power trading center (Guangzhou Carbon Emission Certificate Trading Center 2020), Yunnan and Guizhou are the most important provinces that contribute to Guangdong's local generation. Guangdong's government power plan and average tariff in 2019 are shown in Table 11.4. The Implementation Plan for the Guarantee of Renewable Energy Power Consumption in Guangdong Province (Guangdong

Table 11.4 Generation and power import of Guangdong in 2019 (GWh).

	Guangdong			Guizhou			Yunnan		
Provinces	**Hydro**	**Thermal**	**Other RE**	**Hydro**	**Thermal**	**Other RE**	**Hydro**	**Thermal**	**Other RE**
Value	68,750	406,210	19,300	15,880	28,870	3370	117,570	0	5930

Table 11.5 Renewable electricity consumption ration allocation scheme.

	RE-ratio		RE-without hydro ratio	
Power users	**Existing (%)**	**New (%)**	**Existing (%)**	**New (%)**
Residents	31.5	100	5	0
General commerce and industry	31.5	31.53	5	14%
Large-scale industry	31.5	17.75	5	0

Province Energy Administration 2020), the average RE consumption undertaken for the new allocation plan, and the existing plan are shown in Table 11.5.

11.5.1.1 Comparison of Power Tariff and Policy Acceptance

In this section, the value and ratio of the various power users' tariff are plotted in a stacked area chart, as shown in Figure 11.6. It can be seen that all power users' power expenditures change after the implementation of the RPS policy. The average resident power tariff, which is the highest change among all power users, rises from 596 to 626.46 CNY per MWh, while the average rates of the other power customers decline slightly; for example, the general commercial and industrial users' average price decreases by 6.38% and the large-scaleindustry average price decreases by 4.21%.

From Figure 11.6, it can be seen that electricity users with a higher proportion of RE usage pay higher tariffs. For instance, the multi-objective model's RE consumption proportion for residential power customers is 100%, while their tariff increase ratio is 5.11%. This increase is due to the government's special concern for residents and the historically low power tariff. To meet the RPS objective set for Guangdong province under the new allocation plan, the tariff for all power consumers is generally smooth and stable, which further demonstrates the model's effectiveness.

11.5.1.2 Generation Costs and CO_2 Emissions

The supply-side RPS policy implementation effect on generation costs for the different energy types in the different provinces are shown in Figure 11.7. It is clear that Guangdong's hydro generation costs are higher in the old allocation plan – 10.45 billion CNY – than they are in the current one – 6.03 billion CNY. Compared to the 16.46 billion CNY in the previous plan, the generation expenses of hydropower from

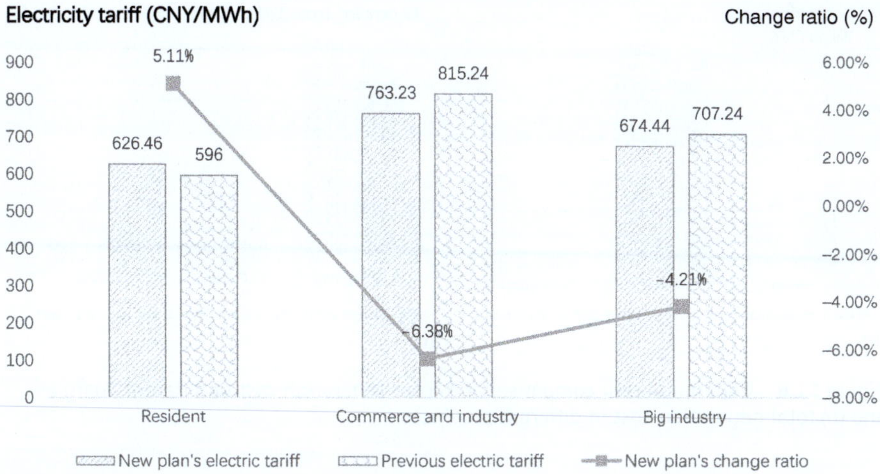

Figure 11.6 Electricity tariff of power users.

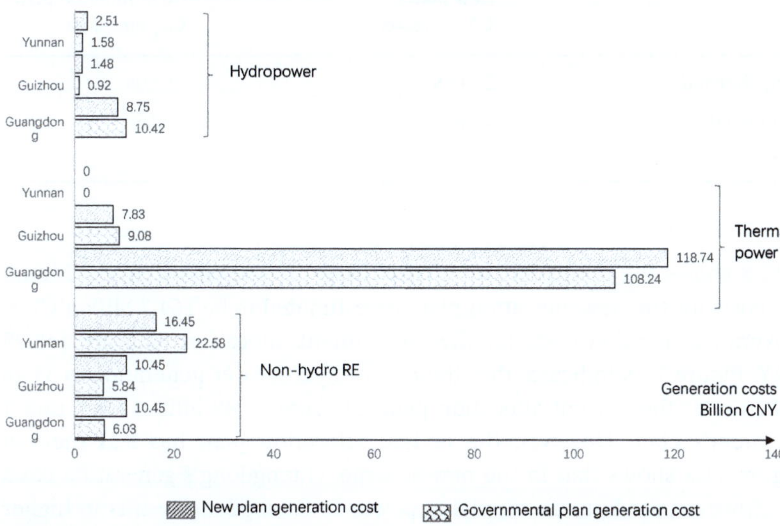

Figure 11.7 Generation costs of different energies in provinces.

Yunnan to Guangdong shared 22.58 billion CNY under the current plan. The only difference between the two plans' non-hydro RE generation costs was that Guangdong's were 10.42 billion CNY in the new allocation plan as opposed to 8.75 billion CNY in the earlier one, and that Guangdong's local thermal power plant generation costs dropped by 10.5 billion CNY.

From Figure 11.7 and Table 11.3, it can be seen that the expensive thermal power from Yunnan is not used in either plan. The government of Guangdong must purchase the remaining non-hydro RE from Yunnan and Guizhou in order to satisfy the non-hydro RE consumption targets because the allocation plan's low non-hydro RE quantity is a limitation.

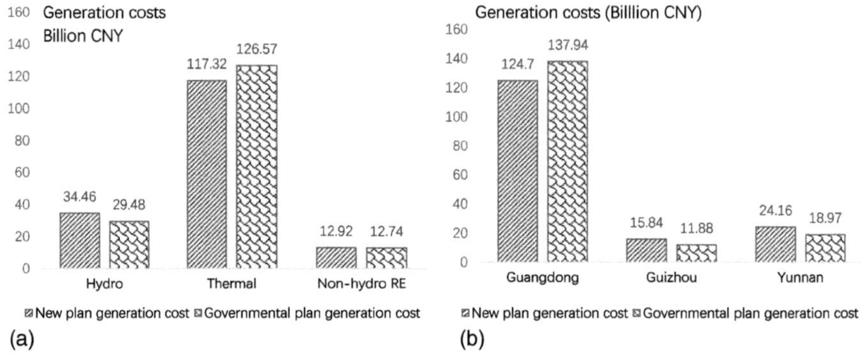

Figure 11.8 Generation cost comparison. (a) Total generation cost of different energies and (b) total generation cost in different provinces.

Table 11.6 Electricity generation related CO_2 emissions in each province.

Provinces	New plan's CO_2 emission	Governmental plan's CO_2 emission
Guangdong-thermal	239.68	262.96
Guizhou-thermal	22.62	19.53
Yunnan-thermal	0	0

Figure 11.8 shows the total generation costs from diverse perspectives. Overall generation costs for the new allocation plan are estimated to be 164.7 billion CNY, whereas overall generation costs for the government allocation plan are 168.79 billion CNY. Figure 11.8 indicates that the cost of hydropower generation is 34.46 billion CNY under the current allocation plan, up from 29.48 billion CNY under the government's plan. However, the revised allocation plan has less thermal power. Figure 11.8 shows that in the new scheme, Guangdong's generation costs are lower. However, for Yunnan supplying power to Guangdong results in higher generation cost.

Table 11.6 demonstrates that the new allocation plan emits significantly less CO_2 than the old allocation plan, particularly in Guangdong, with just 23.28 million tonnes produced under the new allocation plan. This might greatly support Guangdong's environmental preservation efforts. Overall, the new allocation plans have lower total generation costs than the current allocation method.

11.5.2 Scenario Analysis

The previous analysis had the same preferences for the three objectives of limiting tariff increases, lowering total generation costs, and increasing the proportion of RE consumption. This indicates that the acceptance, cost, and environmental considerations are equally important when developing an RPS plan, with the respective weights for [α, β, γ] being [0.333, 0.333, 0.333] (Wang et al. 2020a). Governments, however, may have varying target preferences. As a result, the power plans of the decision-maker were examined in several scenarios. Taking the same preferences for all three targets as the baseline scenario, three meaningful scenarios were established.

Scenario 1: Maximize the non-hydro RE consumption. Since Non-hydro RE is extremely important to the long-term development of RE (Johnsson et al. 2019), the National Energy Administration's RPS strategy set a single objective for non-hydro RE. In this case, a non-hydro RE consumption maximization scenario is set up to switch the third aim from maximizing RE consumption to maximizing non-hydro RE consumption in order to illustrate the government's various preferences under realistic circumstances, as follows:

$$\text{Max}RPS_{Non-hydro} = \sum_{i=1}^{3}\sum_{j=1}^{3} Z_{ij}$$

Scenario 2: Maximize RE consumption with the fast development of non-hydro RE. The Guangdong government is forced to buy excess non-hydro RE from Yunnan and Guizhou because non-hydro RE is not readily available. An abundant non-hydro RE scenario is constructed to broaden the government's options and offer guidance for future scheduling as a result of the non-hydro RE industry's quick development and encouragement from Oliver and Khanna (2018). The non-hydro RE quantities of the different provinces in different scenarios are shown in Table 11.7.

Scenario 3: Minimize the environmental impacts. In this scenario, the government has a stricter attitude toward the environment and cares more about the CO_2 emissions (Liu et al. 2017); therefore, to find the best allocation strategy, the environment index is increased from 10 to 20, with the revised objective being as follows:

$$\text{Min}E_T = \sum_{i=1}^{3}\sum_{j=1}^{3} Q_1^p \left[X_{ij}c_j^x + Y_{ij}c_j^y + Z_{ij}c_j^z + 20Y_{ij}D_j\right]$$

11.5.2.1 RE Consumption Proportion Results

Table 11.8 shows the RE and non-hydro RE allocation schemes for the different scenarios. In every scenario, the residential power consumers consume 100% RE,

Table 11.7 The quantity of non-hydro RE of different provinces in different scenarios (GWh).

Provinces	Practical situation in 2019	Abundant non-hydro RE scenario
Guangdong	23,000	46,000
Guizhou	2095	4190
Yunnan	8162	16,324
Sum	33,257	66,514

Table 11.8 Renewable energy allocation scheme.

	Renewable energy allocation				Non-hydro renewable energy allocation			
Power users	**Baseline (%)**	**Scenario 1 (%)**	**Scenario 2 (%)**	**Scenario 3 (%)**	**Baseliney (%)**	**Scenario 1 (%)**	**Scenario 2 (%)**	**Scenario 3 (%)**
Residents	100	100	100	100	0	0	0	0
General commerce and industry	31.53	33.68	34.87	31.53	14	16.16	17.35	14
Large-scale industry	17.75	16.36	17.75	17.75	0	0	0	0

while the costlier non-hydro RE is distributed to all general commercial and industrial users. In particular, scenario 1's goal of maximizing only non-hydro RE yields a 2.16% increase in the non-hydro RE allocation, but results in a fall in hydropower due to the limited supply of non-hydro RE. Scenario 2 demonstrates that the share of RE and non-hydro RE consumed by consumers in general commerce and industry is 34.87% and 17.35%, respectively, under the assumption of ample non-hydro RE.

11.5.2.2 Power Tariff Results

Figure 11.9 shows the power tariffs for different scenarios. The cost of electricity for large industrial power consumers is approximately 677.44 CNY per MWh in scenarios, while scenario 2 has the highest average power price for general commerce and industry at 771.93 CNY per MWh due to the higher percentage of non-hydro RE. Residential power users pay the same rate in all scenarios because most of Yunnan's hydropower is allocated to the residents. Scenario 3, which has a tougher environmental stance, has the lowest power tariff at 741.92 CNY per MWh because of Guangdong's affordable thermal power.

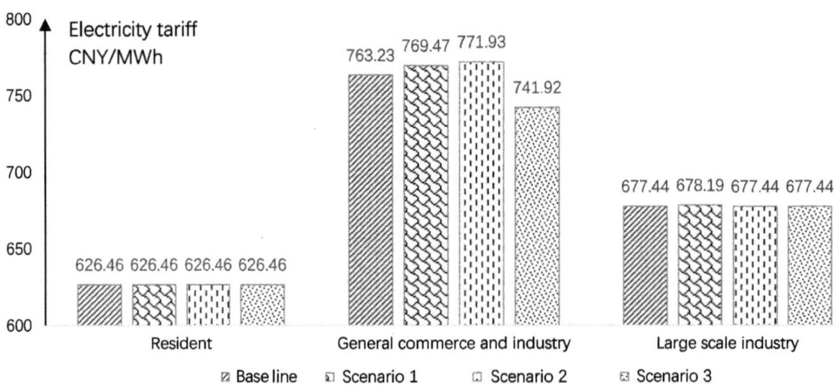

Figure 11.9 Electricity tariff in different scenarios.

11.5.2.3 Generation Cost and CO_2 Emission Results

The stacked area and line chart show the province's generation costs for supplying Guangdong's power in the different scenarios. As shown in Figure 11.10, the highest generation costs are found in scenario 1, particularly in Yunnan, in part because Guangdong imported more of Yunnan's thermal electricity. However, scenario 2's overall generation costs rise by 1.09 billion CNY in comparison to the baseline scenario, despite the procurement of more expensive non-hydro RE. The strategy in scenario 3 consumes more local thermal power in Guangdong, which raises generation costs there but lowers them overall for the three provinces.

The provincial CO_2 emissions for supplying power to Guangdong in the different scenarios are shown in Figure 11.11. Guangdong emits 233.21 and 223.87 million tonnes of CO_2 in scenarios 1 and 2, respectively. This is less than in the other scenarios since more RE is used to replace thermal power, which lowers emissions. Although Guangdong's CO_2 emissions increase in scenario 3 because the government solely selects local thermal power, they are still lower than that in the baseline scenario.

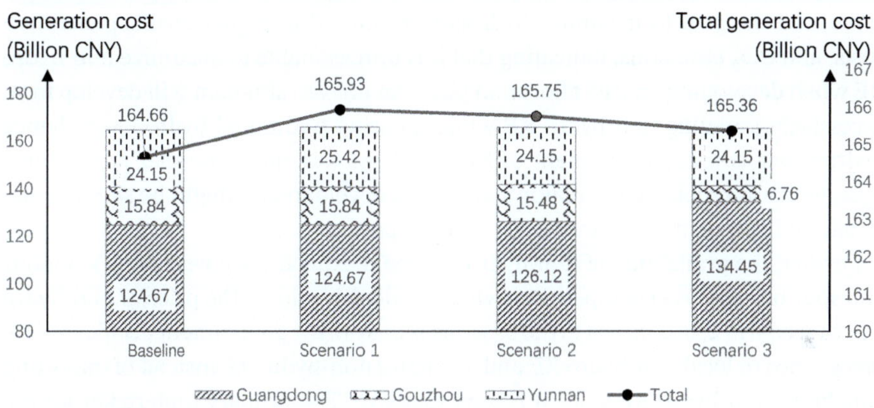

Figure 11.10 Generation costs in different scenarios.

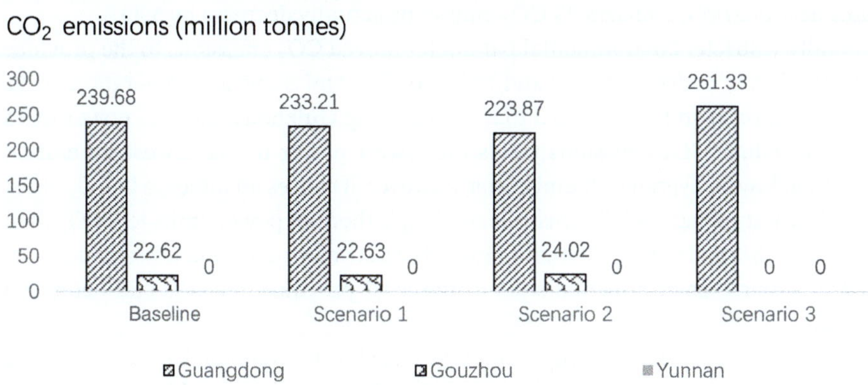

Figure 11.11 Generation CO_2 emissions in different scenarios.

11.5.3 Key Finding

Firstly, the suggested model can more successfully maintain electricity tariff stability than the existing strategy and has lower generation costs and CO_2 emissions. The suggested strategy can reduce power rates for general commercial and industrial, and large-scale power users by 6.38% and 4.21%, respectively, with only a 5.11% rise in the price of electricity for residents. In contrast to the current plan, the ideal model places the primary responsibility for hydropower consumption on residents and the primary responsibility for non-hydro RE consumption on general commerce and industry, translating to a Guangdong RPS of 35.4% and a non-hydro RPS of 4.3%. The optimal model also decreases generation costs by 4.09 billion CNY in total three provinces. The optimal model decreases CO_2 emissions by 23.28 million tonnes because more RE is consumed.

Secondly, maximizing non-hydro RE resulted in increased generation costs and tariffs but did not significantly boost non-hydro RE consumption. Due to the relative underdevelopment of non-hydro RE, the allocation plan that aimed to maximize non-hydro RE did not significantly increase non-hydro RE consumption. Instead, it reduced the goal to maximize hydropower and led to higher tariffs, generation costs, and CO_2 emissions, indicating that it is unreasonable to maximize non-hydro RE when developing an RPS allocation plan. An uneven situation will develop from excessively pursuing non-hydro RE while ignoring traditional hydropower, hence hydropower is a key component of RPS strategy. Guizhou's thermal power business, which is crucial to the region's economic development, might be significantly impacted if Guangdong abandons its thermal power.

Further, the availability of non-hydro RE could expand the government's options and spur more market competition, which could drive down the price of non-hydro RE. In scenario 2, in contrast to the baseline scenario, the government decides to give precedence to local non-hydro RE and Yunnan's non-hydro RE instead of importing Guizhou's non-hydro RE since it is too expensive. The primary undertaker for the non-hydro RE, with a tariff cut of 5.31%, is the general commercial and industrial power users. Although scenario 2's generation costs are marginally higher than the baseline scenario's, scenario 2's CO_2 emissions actually decrease by 5.49%.

Finally, a stricter environmental parameter moved CO_2 emissions to the province with the lower emission factor and reduced CO_2 emissions and generation costs. The allocation plan in scenario 3 that avoids using Guizhou's thermal power in an effort to reduce CO_2 emissions yields the lowest power tariffs, lowest generation costs, and lowest overall CO_2 emissions; however, it causes an increase in CO_2 emissions in Guangdong itself because Guangdong's thermal power emits less CO_2 than Guizhou's thermal power. This case shows that a stricter environmental attitude may lead to a worse environmental state in the local province, but it is still beneficial overall.

These findings provide an important reference for the National Development and Reform Commission of China in determining the annual renewable energy power consumption responsibility weights for each province (The General Office of the

National Development and Reform Commission, the Comprehensive Department of the National Energy Administration 2023).

References

Ang, B.W. and Su, B. (2016). Carbon emission intensity in electricity production: a global analysis. *Energy Policy* 94: 56–63.

Barbose, G., Bird, L., Heeter, J. et al. (2015). Costs and benefits of renewables portfolio standards in the United States. *Renewable and Sustainable Energy Reviews* 52: 523–533.

Barbose, G., Wiser, R., Heeter, J. et al. (2016). A retrospective analysis of benefits and impacts of US renewable portfolio standards. *Energy Policy* 96: 645–660.

Basnet, S.M.S., Aburub, H., and Jewell, W. (2015). Effect of demand response on residential energy efficiency with direct load control and dynamic price control. *2015 IEEE Eindhoven PowerTech*, 1–6. IEEE.

Bird, L., Chapman, C., Logan, J. et al. (2011). Evaluating renewable portfolio standards and carbon cap scenarios in the US electric sector. *Energy Policy* 39 (5): 2573–2585.

Bu, Y. and Zhang, X. (2020). The prospect of new provincial renewable portfolio standard in China based on structural data analysis. *Frontiers in Energy Research* 8: 59.

Bu, Y. and Zhang, X. (2021). On the way to integrate increasing shares of variable renewables in China: activating nearby accommodation potential under new provincial renewable portfolio standard. *Processes* 9 (2): 361.

Carley, S., Davies, L.L., Spence, D.B., and Zirogiannis, N. (2018). Empirical evaluation of the stringency and design of renewable portfolio standards. *Nature Energy* 3 (9): 754–763.

Chen, C., Wiser, R., Mills, A., and Bolinger, M. (2009). Weighing the costs and benefits of state renewables portfolio standards in the United States: a comparative analysis of state-level policy impact projections. *Renewable and Sustainable Energy Reviews* 13 (3): 552–566.

China Electric Power Press (2019). *China Electric Power Book*.

China Electricity Council (2020). *2020 China Power Statistical Yearbook*. China Statistics Press.

Choi, D.G., Park, S.Y., and Hong, J.C. (2015). Quantitatively exploring the future of renewable portfolio standard in the Korean electricity sector via a bottom-up energy model. *Renewable and Sustainable Energy Reviews* 50: 793–803.

Considine, T.J. and Manderson, E.J.M. (2015). The cost of solar-centric renewable portfolio standards and reducing coal power generation using Arizona as a case study. *Energy Economics* 49: 402–419.

Fan, J.L., Wang, J.X., Hu, J.W. et al. (2019). Optimization of China's provincial renewable energy installation plan for the 13th five-year plan based on renewable portfolio standards. *Applied Energy* 254 (15): 113757.1–113757.16.

Fan, J.-L., Wei, S.-J., Xu, M. et al. (2018). The LCOE of Chinese coal-fired power plants with CCS technology: a comparison with natural gas power plants. *Energy Procedia* 154: 29–35.

Guangdong Bureau of Statistics (2020). Output value of industrial sales by sector in 2020. http://www.cnemission.com/article/jydt/scyj/202012/20201200002056.shtml (accessed 20 October 2023).

Guangdong Power Exchange Center (2019). Guangdong power market annual report 2019.

Guangdong Province Energy Administration (2020). Guangdong Province 2020 annual renewable energy electricity consumption guarantee implementation plan. http://drc.gd.gov.cn/snyj/tzgg/content/post_3128089.html.

Guangzhou Carbon Emission Certificate Trading Center (2020). Guangdong carbon market 2019 annual performance trading data report. http://www.cnemission.com/article/jydt/scyj/202012/20201200002056.shtml (accessed 20 October 2023).

Hao, P., Guo, J.-P., Chen, Y., and Xie, B.C. (2020). Does a combined strategy outperform independent policies? Impact of incentive policies on renewable power generation. *Omega* 97: 102100.

Harvey, L.D.D. (2020). Clarifications of and improvements to the equations used to calculate the levelized cost of electricity (LCOE), and comments on the weighted average cost of capital (WACC). *Energy* 207: 118340.

Ilyas, M.A., Abbas, G., Alquthami, T. et al. (2020). Multi-objective optimal power flow with integration of renewable energy sources using fuzzy membership function. *IEEE Access* 8: 143185–143200.

Ji, L., Huang, G.-H., Xie, Y.-L. et al. (2017). Explicit cost-risk tradeoff for renewable portfolio standard constrained regional power system expansion: a case study of Guangdong Province, China. *Energy* 131: 125–136.

Jiao, B., Xu, K., Wu, S. et al. (2018). Optimal operation of park-based integrated energy system. *Energy Procedia* 152: 89–94.

Johnsson, F., Kjärstad, J., and Rootzén, J. (2019). The threat to climate change mitigation posed by the abundance of fossil fuels. *Climate Policy* 19 (2): 258–274.

Joshi, J. (2021). Do renewable portfolio standards increase renewable energy capacity? Evidence from the United States. *Journal of Environmental Management* 287: 112261.

Kagan, J.A. (2019). Multiple streams in hawaii: how the aloha state adopted a 100% renewable portfolio standard. *Review of Policy Research* 36 (2): 217–241.

Kim, J., Park, J., Kim, H., and Heo, E. (2012). Assessment of Korean customers' willingness to pay with RPS. *Renewable and Sustainable Energy Reviews* 16 (1): 695–703.

Kung, H.H. (2012). Impact of deployment of renewable portfolio standard on the electricity price in the State of Illinois and implications on policies. *Energy Policy* 44: 425–430.

Lee, J., Moon, H.B., and Lee, J. (2020). Consumers' heterogeneous preferences toward the renewable portfolio standard policy: an evaluation of Korea's energy transition policy. *Energy & Environment*, https://doi.org/10.1177/0958305X20949467.

Li, X., Wang, W., Wang, H. et al. (2020). Dynamic environmental economic dispatch of hybrid renewable energy systems based on tradable green certificates. *Energy* 193: 116699.

Li, C., Yao, Y., Zhao, R., and Xie, K. (2021). Optimization model for time-varying settlement of renewable energy consumption considering accommodation difficulty and supply—demand interaction. *International Journal of Electrical Power & Energy Systems* 125: 106469.

Lin, B. and Jiang, Z. (2012). Designation and influence of household increasing block electricity tariffs in China. *Energy Policy* 42: 164–173.

Lin, J., Kahrl, F., Yuan, J. et al. (2019). Economic and carbon emission impacts of electricity market transition in China: a case study of Guangdong Province. *Applied Energy* 238: 1093–1107.

Liu, F., Lyu, T., Pan, L., and Wang, F. (2017). Influencing factors of public support for modern coal-fired power plant projects: an empirical study from China. *Energy Policy* 105: 398–406.

Munoz, F.D., Pumarino, B.J., and Salas, I.A. (2017). Aiming low and achieving it: a long-term analysis of a renewable policy in Chile. *Energy Economics* 65: 304–314.

National Development and Reform Commission (2019). Transmission price of trans-provincial and trans-regional special projects. http://www.gov.cn/xinwen/2019-05/16/5392079/files/44babf47eaa34371807f2a06a8be048d.pdf (accessed 20 October 2023).

National Energy Administration (2017). The establishment of an electricity spot market pilot in Guangdong enters into a substantive stage (in Chinese).

National Energy Administration (2019a). Notice on the establishment and improvement of the renewable energy power consumption guarantee mechanism. https://www.gov.cn/zhengce/zhengceku/2019-09/25/content_5432993.htm.

National Energy Administration (2019b). 2018 National renewable energy power development monitoring and evaluation report. http://zfxxgk.nea.gov.cn/auto87/201906/t20190610_3673.htm.

Oliver, A. and Khanna, M. (2018). The spatial distribution of welfare costs of renewable portfolio standards in the United States electricity sector. *Letters in Spatial and Resource Sciences* 11 (3): 269–287.

Pramanik, S. and Roy, T.K. (2008). Multiobjective transportation model with fuzzy parameters: priority based fuzzy goal programming approach. *Journal of Transportation Systems Engineering and Information Technology* 8 (3): 40–48.

Rouhani, O.M., Niemeier, D., Gao, H.O., and Bel, G. (2016). Cost-benefit analysis of various California renewable portfolio standard targets: is a %33 RPS optimal? *Renewable and Sustainable Energy Reviews* 62: 1122–1132.

Rountree, V. (2019). Nevada's experience with the renewable portfolio standard. *Energy Policy* 129: 279–291.

Shin, J., Woo, J.R., Huh, S.-Y. et al. (2014). Analyzing public preferences and increasing acceptability for the renewable portfolio standard in Korea. *Energy Economics* 42: 17–26.

Stokes, L.C. and Warshaw, C. (2017). Renewable energy policy design and framing influence public support in the United States. *Nature Energy* 2 (8): 1–6.

Tan, Q., Ding, Y., Zheng, J. et al. (2021). The effects of carbon emissions trading and renewable portfolio standards on the integrated wind—photovoltaic—thermal power-dispatching system: real case studies in China. *Energy* 222: 119927.

The General Office of the National Development and Reform Commission, the Comprehensive Department of the National Energy Administration (2023). Notice on renewable energy power consumption in 2023 notice on the weight of responsibility and related matters. https://www.ndrc.gov.cn/xwdt/tzgg/202308/t20230804_1359103.html (accessed 20 October 2023).

Wang, D., Liu, Y., Wang, Y. et al. (2020a). Allocation of coal de-capacity quota among provinces in China: a bi-level multi-objective combinatorial optimization approach. *Energy Economics* 87: 104709.

Wang, H., Su, B., Mu, H. et al. (2020b). Optimal way to achieve renewable portfolio standard policy goals from the electricity generation, transmission, and trading perspectives in Southern China. *Energy Policy* 139: 111319.

Wang, G., Zhang, Q., Li, Y., and Mclellan, B.C. (2019). Efficient and equitable allocation of renewable portfolio standards targets among China's Provinces. *Energy Policy* 125: 170–180.

Wang, C., Zhou, K., and Yang, S. (2017). A review of residential tiered electricity pricing in China. *Renewable and Sustainable Energy Reviews* 79: 533–543.

Wendling, Z.A., Warren, D.C., Rubin, B.M. et al. (2020). A scalable energy—economy model for state-level policy analysis applied to a demand-side management program. *Economic Development Quarterly* 34 (4): 372–386.

Yao, L., He, L., Chen, X. et al. (2020). Optimal day-ahead coordination on wind-pumped-hydro system by using multiobjective multistage model. *International Journal of Energy Research* 44 (8): 6843–6859.

Yao, J., Xiao, E., Jian, X., and Shu, L. (2021). Service quality and the share of renewable energy in electricity generation. *Utilities Policy* 69: 101164.

Ying, Z., Xin-gang, Z., and Zhen, W. (2020). Demand side incentive under renewable portfolio standards: a system dynamics analysis. *Energy Policy* 144: 111652.

Zhou, C., Shi, C., Wang, S., and Zhang, G. (2018). Estimation of eco-efficiency and its influencing factors in Guangdong province based on Super-SBM and panel regression models. *Ecological Indicators* 86: 67–80.

12
Optimal RPS Implementation Strategy Considering Equity and Economy Equilibrium

A successful transition to a future net-zero emissions energy system is likely to depend on vast amounts of inexpensive, emissions-free electricity.

Steven J. Davis et al. Published on Science, 2018, 360(6396): eaas9793

To build hybrid energy systems, the policy mechanism known as the "renewable portfolio standard" has been introduced on a global scale to encourage the adoption of renewable energy sources. However, its common practice has not fully considered the interaction of the central government's equity concern and the local government's economic concern. Based on equity and economic considerations, this chapter presents a bi-level paradigm for optimizing regional renewable portfolio standard (RPS) goal allocation. The central government decides the optimal allocation targets to maximize the total renewable electricity consumption ratio while fully considering equity principles (responsibility, capacity, and equality). The local governments make decisions on their generation and trading strategy so as to satisfy the power demand and the renewable power consumption target at minimized costs. A novel interactive algorithm is developed to solve the bi-level model, and the proposed mathematical model is applied in a case study to evaluate its effectiveness. The results indicate that the current target allocation is far from an equitable allocation, and the inclusion of hydropower in the RPS targets makes equitable allocation more challenging. This chapter proposes a practical tool for the central government to promote an economical and equitable power generation restructuring.

12.1 Introduction

The renewable portfolio standard (RPS) has been adopted worldwide to encourage the use of renewables (Yu et al. 2023; Rouhani et al. 2016). It is a policy instrument that mandates a minimum market share for clean energy technologies in order to stimulate the development of those technologies (Shayegh and Sanchez 2021). Setting challenging but not overly stringent standards for each region can serve

Sustainable Hybrid Energy Systems: Carbon Neutral Approaches, Modeling, and Case Studies, First Edition. Jiuping Xu and Fengjuan Wang.

the purpose of stimulating investment and the use of renewable electricity, thus creating a virtuous cycle. On the other hand, targets that are too high or too low can discourage local governments from implementing corresponding policies because they must assess whether the time, human, material, and financial resources spent on implementing these policies match the returns they provide. Therefore, it is manifestly unscientific if the national government creates requirements for the consumption of renewable energy without taking into account the input of local governments' implementation capacity. Local governments are not simply passive acceptors and implementers of policies in this central–local policy formulation and implementation situation; they are also policy formation players with the four functions of acceptance, feedback, adjustment, and execution. Similarly, the central government does not develop consumption standards in a vacuum, but must solicit feedback and make adjustments in order to decide consumption regulations. This bi-level interaction between central and local governments is critical in decision-making, but it has been rarely included in prior studies on renewable energy consumption regulations.

Many earlier studies have examined ways to attain the RPS objectives. Some academics focus on institutional arrangements, regulatory incentives, and demand incentives as a starting point. For example, Xin-gang et al. (2014) examined the hurdles to RPS implementation in China from five perspectives and proposed relative institutional frameworks for implementation. Xin-gang et al. (2020) also suggested how the government should develop scientific policies to achieve incentive regulation for power producers under the RPS and carbon price policies. Ying et al. (2020) researched how to boost the demand side for RPS and raise the buying demand for tradable green certificates to encourage the use and advancement of renewable energy. Other studies concentrated on the application of optimization approaches to target allocation, infrastructure installation, and power consumption scheduling challenges. For example, Wang et al. (2019) provided a cost-cutting approach for allocating RPS objectives across Chinese provinces in an economically sustainable and equitable manner in order to implement this policy. A mid-to-long-term optimization model for southern China was created by Wang et al. (2020) in order to meet the aims of the RPS policy from the standpoint of electricity generation, transmission, and trade. Based on the provincial RPS targets, Fan et al. (2019) proposed an integrated model to optimize the renewable energy installation plan for provinces in China in 2020. These studies have significantly advanced the quick promotion of RPS. However, assigning an acceptable RPS target to each region is equally as critical as implementing the strategy to get the best results.

There are additional studies looking into the optimal way to allocate RPS targets. For example, Carley et al. (2018) investigated the impact of RPS stringency on several market outcomes and discovered that a one-point rise in RPS stringency leads to increases in renewable energy, solar generation, and renewable energy capacity of 0.2%, 1%, and 0.3%, respectively. Rouhani et al. (2016) conducted a cost–benefit analysis of several 2020 California RPS objectives and discovered that the mandated 33% RPS would not deliver the best cost–benefit values among the alternative targets. Siddiqui et al. (2016)demonstrated that the optimal RPS target for a perfectly competitive energy industry is greater than that for a comparable centrally managed industry, and that disregarding the relationship between RPS standards and market structure could result in suboptimal RPS objectives. Wang et al. (2018) stressed equity in terms of ability and potential, and came up with a solution for the initial distribution of RPSs from 2015 to 2030 based on local economic potential, resource endowment, infrastructure, and historical responsibility. However, none of them have considered the interaction between central and local governments.

The RPS target allocation problem (% of renewable energy power use) is actually a leader–follower game structure. The central government is the leader, as it is responsible for making the national RPS decision and allocating it to each area for implementation. The provincial government is the follower, and it must determine if it can meet the allotted objective at an acceptable cost by altering the generation scheduling of renewable and conventional resources and developing a provincial power exporting and importing plan. However, the decision-making at these two levels is not independent; rather, the decision made at the upper level affects the aim realization at the lower level, and vice versa. The bi-level optimization program was developed in 1950 to address the problem of two decision-makers (Strehlow and Stackelberg 1950), and it has since been extensively used to address issues with water resource allocation (Yao et al. 2020), vehicle pollutionrouting (Qiu et al. 2020a), airline fleet planning (Khoo2014), and the location of stone industrial parks (Gang et al. 2015). However, it has not been extensively covered in earlier RPS target allocation studies. Therefore, this chapter provides a model that incorporates these two categories of decision-makers to optimize the distribution of RPS targets, taking into account the trade-off between equity and economy.

12.2 Key Problem Statement

The bi-level relationship between the central and local governments and the trade-off between economy and equity in allocating the RPS target will be described in the following sections.

12.2.1 Bi-Level Relationship

As the upper-level decision-maker, the central government assigns RPS targets to local governments in order to optimize the amount of renewable energy use nationwide. However, this is a decision made from the standpoint of the central government, and it is still unknown whether the local government has the necessary capacity and resources to implement it. As a result, after receiving these objectives, the local government, as the lower-level decision-maker, must make judgments on its generation schedules and power importing/exporting schedules in order to

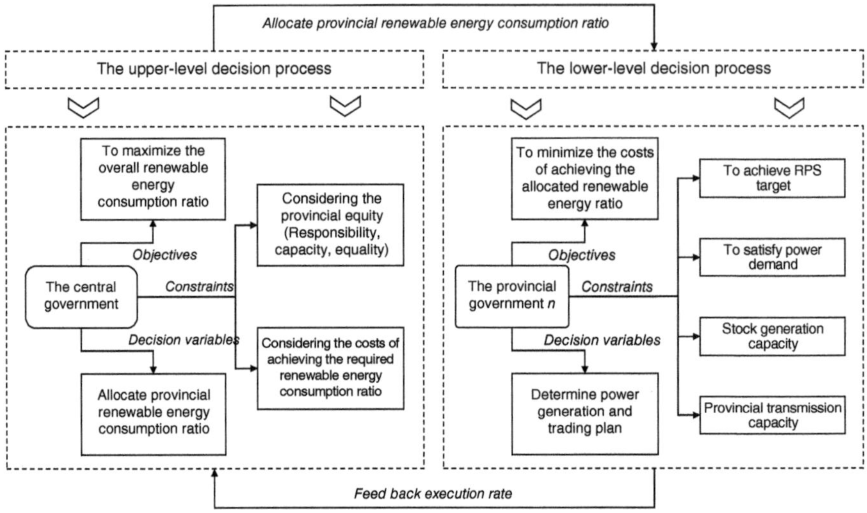

Figure 12.1 Bi-level relationship between central and local governments for RPS allocation.

provide feedback on its execution rate, taking into account the province electricity use cost. The flowchart of this decision process can be seen in Figure 12.1. Following receipt of the fed execution rate, the central government reassigns the quota to each region and completes the final cycle. Quotas that meet the goals of the national and local governments will be hammered out after multiple encounters in this manner.

12.2.2 Equity and Economy Trade-Off

The distribution of RPS targets is actually a burden distribution because renewable energy sources have the drawback of being more expensive and unstable than conventional non-renewable energy sources. In order to reach the assigned RPS targets, different local governments must incur varying expenses. For instance, in China, the average hydropower feed-in prices in Yunnan and Hainan are 198.71 and 407.91 CNY/MWh, respectively. As a result, it would cost more to assign Hainan 1MWh of consumption responsibility than Yunnan. As a result, China's RPS target distribution still heavily relies on local resource conditions; hence, regions with stronger renewable energy resources should be responsible for a larger share of consumption. Although this can lower the cost of renewable energy consumption nationally, it is clear that this is unjust because the burden of consumption will be placed on a small number of resource-rich regions, creating a serious polarization. At the same time, the resource-rich regions will also need to invest more money in order to meet a higher percentage of consumption target. While the regions with low resource circumstances only support small share of costs associated with this, they do share the growth in the ratio of renewable energy consumption brought about by the regions with rich resource conditions. This chapter makes the case that when allocating RPS targets, fairness, and economics should be taken into account in full, the gap between the RPS targets for various regions should be gradually closed rather than widened, and each region should bear equal responsibility for the consumption of renewable energy, even though this will slightly raise the cost of consumption. Therefore, it is essential to strike a balance between economy and fairness when allocating RPS targets.

12.3 Modeling

The mathematical model of the above-described optimization process and the corresponding model-solving method will be introduced in this section.

12.3.1 Notations

Indices

i	The index of non-renewable power generation technologies; for example, $i = 1, 2, 3$ represent coal, nuclear, and natural gas power generation separately in this chapter.
j	The index of renewable power generation technologies; for example, $j = 1, 2, 3, 4$ represent wind, solar, biomass, and hydro power generation separately.
m, n	Index of provinces.

Parameters

A_{in}^{N}	Unit generation cost of non-renewable source i in province n.
A_{jn}^{R}	Unit generation cost of renewable source j in province n.
A_{mn}^{L}	Unit transmission cost between provinces m and n.
A_{mn}^{T}	Power trading price between provinces m and n.
D_n	Power demand of province n.
e	Decision-maker's equity parameter.
E_{mn}	A binary variable denotes whether province m can transport electricity to province n, value 1 means it can, value 0 means it cannot.
G_{in}^{N}	Province n's total non-renewable power generation in last production cycle.
G_{n}^{R}	Province n's total renewable power generation of last production cycle.
H_{in}^{N}	The annual utilization hours of non-renewable generation technology i in province n.
H_{jn}^{R}	The annual utilization hours of renewable generation technology j in province n.
L_{mn}	Transmission line loss.
R_{mn}	The ratio of renewable power in the electricity transmitted from province m to province n.

$\overline{R_{jn}}$	The maximum renewable power utilization ratio in province n.
$\underline{R_{jn}}$	The minimum renewable power utilization ratio in province n.
β_{in}^{N}	Capacity parameter of renewable generation technology i in province n.
β_{in}^{R}	Capacity parameter of non-renewable generation technology j in province n.
ω	The provincial decision-maker's attitude parameter toward execution ratio.

Decision variables

S_n	The central government–assigned renewable power utilization target of province n (%).
C_{in}^{N}	The required non-renewable power generation capacity of province n to achieve the RPS target, MW.
C_{jn}^{R}	The required renewable power generation capacity of province n to achieve the RPS target, MW.
T_{mn}^{im}	Imported electricity from province m to province n.
T_{mn}^{ex}	Exported electricity from province m to province n.
K_n	The provincial government's execution ratio of the assigned renewable power utilization target.

12.3.2 Central Government's Equity Concern

The central government is the leader of the decision process, it seeks to maximize the overall renewable energy utilization under the constraints of equity and economy. Their objective is to maximize the assigned renewable energy ratio, as shown in Eq. (12.1).

$$\max F_1 = \frac{\sum_n S_n K_n D_n}{\sum_n D_n} \tag{12.1}$$

12.3.2.1 Equitable Allocation

The allocation should combine different fairness principles to avoid biases in favor or against certain regions. Therefore, this chapter integrated the Responsibility, Capacity, and Equality mechanisms to represent the consideration of allocation equity (Robiou du Pont et al. 2017; Höhne et al. 2014). *Responsibility* means that the regions with lower renewable electricity consumption ratio should to be assigned with higher consumption responsibility; *Capability* means regions with higher ability to pay should contribute more; and *Equality* indicates that every individual should live in an environment with high-quality air. Therefore, The historical renewable energy consumption ratio (Responsibility), the per capita gross domestic product (Capacity), and the annual average concentration of delicate particulate matter (Equality) are integrated to equally allocate the RPS targets, as shown in Eq. (12.2).

$$Z_n = \sum_k \eta_k \overline{g}_{nk} \tag{12.2}$$

where Z_n is the integrated score of province n in terms of equity principles, $\overline{g}_{nk}$ is the normalized value of each principle, and $\eta_k(0 < \eta_k < 1, \sum_k \eta_k = 1)$ is the weight of each principle. The normalized value $\overline{g}_{nk}$ can be calculated according to Eq. (12.2), where $k = 1, 2, 3$ represent Responsibility, Capacity, and Equality, respectively, and n is the index of provinces.

$$\overline{g}_{nk} = \begin{cases} \frac{g_{nk}}{\sum_n g_{nk}}, k = 1 \\ \frac{(1-g_{nk})}{\sum_n (1-g_{nk})}, k = 2, 3 \end{cases} \tag{12.3}$$

To achieve allocation equity, the relative relationship of the RPS targets assigned to the two provinces (S_m, S_n) should be consistent with the relative relationship of the integrated scores (Z_m, Z_n), as shown in Eq. (12.4).

$$e\frac{Z_n}{Z_{m,m\neq n}} \leq \frac{S_n}{S_{m,m\neq n}} \leq \frac{1}{e}\frac{Z_n}{Z_{m,m\neq n}} \tag{12.4}$$

The parameter e in Eq. (12.4) is the central government's attitude toward allocation equity Xu et al. (2019), the value of e is between 0 and 1, $e = 0$ means no equity is considered, and $e = 1$ means the central government pursues absolutely fair.

12.3.2.2 Renewable Energy Consumption Ratio

For the central government's RPS target allocation, it need to be higher than the last year's renewable energy consumption ratio $\underline{R_{jn}}$, and cannot exceed the province's maximum renewable energy providing ability $\overline{R_{jn}}$, as shown in Eq. (12.5).

$$\underline{R_{jn}} \leq S_n K_n \leq \overline{R_{jn}} \tag{12.5}$$

where S_n is the central government's decision variable, namely, the central government–allocated renewable energy consumption ratio of province n, and K_n is the decision variable of the provincial government, which is the execution rate of the province n's renewable energy utilization responsibility. The assigned rate and the executed rate should be limited within 0 and 1, as shown in Eqs. (12.6)–(12.7).

$$0 \leq S_n \leq 1 \tag{12.6}$$

$$0 \leq S_n K_n \leq 1 \tag{12.7}$$

12.3.3 The Provincial Government's Economic Concern

Local government as the follower, seeks to minimize provincial electricity utilization cost by adjusting its generation scheduling and electricity importing/exporting strategy, as shown in Eq. (12.8).

$$\min F_n = C_n^G + C_n^T \tag{12.8}$$

The generation costs (C_n^G) can be represented as Eq. (12.9), where A_{jn}^R, A_{in}^N are the unit cost of renewable and non-renewable generation technologies in province n;

C^R_{jn}, C^N_{in} are the renewable and non-renewable generation capacity in province n; and H^R_{jn}, H^N_{in} are the utilization hours of renewable and non-renewable generation technologies in province n.

$$C^G_n = \sum_j A^R_{jn} C^R_{jn} H^R_{jn} + \sum_i A^N_{in} C^N_{in} H^N_{in} \tag{12.9}$$

The revenue of power exporting and importing (C^T_n) can be written as Eq. (12.10), where E_{mn} is whether province n can transmit electricity to province m (1 means it can, and 0 means it cannot); T^{ex}_{mn} and T^{im}_{mn} are the electricity that province n export to or import from province m; A^T_{mn} denotes the electricity trading price: and A^L_{mn} is the unit transmission cost.

$$C^T_n = \sum_{n,m\neq n} E_{mn} T^{im}_{mn} \left(1 - L_{mn}\right) (A^T_{mn} + A^L_{mn}) - \sum_{n,m\neq n} E_{mn} T^{ex}_{mn} A^T_{mn} \tag{12.10}$$

12.3.3.1 The Balance of Renewable Electricity Generation and Trading

Been assigned the RPS target S_n, local governments can decide their execution rate $K_n(\underline{\omega} \leq K_n \leq \overline{\omega})$. The balance of renewable electricity generation and trading is shown in Eq. (12.11), where R_{mn} represents the ratio of renewable power in the electricity imported from province m, and R_{nm} is the ratio of renewable electricity exported from province n

$$S_n K_n D_n = \sum_j C^R_{jn} H^R_{jn} + \sum_{n,m\neq n} E_{mn} T^{im}_{mn} R_{mn} \left(1 - L_{mn}\right) - \sum_{m,m\neq n} E_{mn} T^{ex}_{mn} R_{nm} \tag{12.11}$$

12.3.3.2 The Balance of Power Supply and Demand

The local power demand is satisfied by local generated renewable electricity $\sum_j C^R_{jn} H^R_{jn}$, non-renewable electricity $\sum_i C^N_{in} H^N_{in}$, and the imported electricity $\sum_{n,m\neq n} E_{mn} T^{im}_{mn} \left(1 - L_{mn}\right)$, and the electricity exported $\sum_{m,m\neq n} E_{mn} T^{ex}_{mn} \geq D_n$ also need to be deducted, as shown in Eq. (12.12).

$$\sum_j C^R_{jn} H^R_{jn} + \sum_i C^N_{in} H^N_{in} + \sum_{n,m\neq n} E_{mn} T^{im}_{mn} \left(1 - L_{mn}\right) - \sum_{m,m\neq n} E_{mn} T^{ex}_{mn} \geq D_n \tag{12.12}$$

12.3.3.3 Limitation of Generation Capacity

In each province, there exists upper bounds of renewable power output and non-renewable power output, as shown in Eqs. (12.13)–(12.14), where G^R_{jn} and G^N_{in} represent the power output of the last production cycle, and $\underline{\beta}_j$, $\overline{\beta}_j$ are the lower and upper amount parameter, C^R_{jn} and H^R_{jn} are generation capacity and available hours of renewable power in province n.

$$\underline{\beta}^R_{jn} G^R_{jn} \leq C^R_{jn} H^R_{jn} \leq \overline{\beta}^R_{jn} G^R_{jn} \tag{12.13}$$

$$\underline{\beta}^N_{in} G^N_{in} \leq C^N_{in} H^N_{in} \leq \overline{\beta}^g_j G^N_{in} \tag{12.14}$$

12.3.3.4 Limitation of Transmission Capacity

The transmission limitation of electricity importing and exporting are shown in Eqs. (12.15)–(12.16), where $T_{nm}^{\min}$ and $T_{nm}^{\max}$ are the minimum and maximum transmission capacity from province n to province m, and $T_{mn}^{\min}$ and $T_{mn}^{\max}$ are the minimum and maximum transmission capacity from province m to province n.

$$T_{nm}^{\min} \leq T_{mn}^{ex} \leq T_{nm}^{\max} \tag{12.15}$$

$$T_{mn}^{\min} \leq T_{mn}^{im} \leq T_{mn}^{\max} \tag{12.16}$$

12.3.4 Global Model

After combining the foregoing equations, a bi-level renewable energy utilization rate allocation model between central and local governments that takes equality and economics into account can be constructed. The allocation problem is bi-level in nature, which means that both parties must engage several times in order to get a satisfying conclusion. Additionally, we provide a mathematical model of such a process in order to direct the actual analysis and application. In this paradigm, the regional renewable usage rate is decided by the central government, and the corresponding execution rate is decided by the local governments. The RPS aim and economics were the two parties' separate goals, with consideration given to how equally each region would be developed, as depicted in Eq. (12.18).

$$\max F_1 = \frac{\sum_n S_n K_n D_n}{\sum_n D_n} \tag{12.17}$$

$$\text{s.t.}\begin{cases} e\dfrac{Z_n}{Z_{m,m\neq n}} \leq \dfrac{S_n}{S_{m,m\neq n}} \leq \dfrac{1}{e}\dfrac{Z_n}{Z_{m,m\neq n}} \\ \underline{R_{jn}} \leq S_n K_n \leq \overline{R_{jn}} \\ 0 \leq S_n \leq 1 \\ 0 \leq S_n K_n \leq 1 \\ \min F_n = C_n^G + C_n^T \\ \text{s.t.}\begin{cases} C_n^G = \sum_j A_{jn}^R C_{jn}^R H_{jn}^R + \sum_i A_{in}^N C_{in}^N H_{in}^N \\ C_n^T = \sum\limits_{n,m\neq n} E_{mn} T_{mn}^{im} \left(1 - L_{mn}\right)\left(A_{mn}^T + A_{mn}^L\right) \\ \quad - \sum\limits_{n,m\neq n} E_{mn} T_{mn}^{ex} A_{mn}^T \\ S_n K_n D_n = \sum_j C_{jn}^R H_{jn}^R + \sum\limits_{n,m\neq n} E_{mn} T_{mn}^{im} R_{mn} \left(1 - L_{mn}\right) \\ \quad - \sum\limits_{m,m\neq n} E_{mn} T_{mn}^{ex} R_{mn} \\ \sum_j C_{jn}^R H_{jn}^R + \sum_i C_{in}^N H_{in}^N + \sum\limits_{n,m\neq n} E_{mn} T_{mn}^{im} \left(1 - L_{mn}\right) \\ \quad - \sum\limits_{m,m\neq n} E_{mn} T_{mn}^{ex} \geq D_n \\ \underline{\beta}_{jn}^R G_{jn}^R \leq C_{jn}^R H_{jn}^R \leq \overline{\beta}_{jn}^R G_{jn}^R \\ \underline{\beta}_{in}^N G_{in}^N \leq C_{in}^N H_{in}^N \leq \overline{\beta}_j^N G_{in}^N \\ T_{nm}^{\min} \leq T_{mn}^{ex} \leq T_{nm}^{\max} \\ T_{mn}^{\min} \leq T_{mn}^{im} \leq T_{mn}^{\max} \\ \underline{\omega} \leq K_n \leq \overline{\omega} \\ j = 1, 2, 3, 4; i = 1, 2, 3 \end{cases} \end{cases} \tag{12.18}$$

12.3.5 Model-Solving Approach

The proposed bi-level optimization mode, with interactions of electricity transmission and trading, is very difficult to find an equilibrium solution (Qiu et al. 2020b). This chapter proposes an interactive algorithm to solve this model. A detailed flowchart is shown in Figure 12.2.

Since there are transactions of electricity among local governments, their decisions are not independent. Therefore, in the solution method proposed in this chapter, it is assumed that local governments have a decision priority, which is determined by resource endowment (Zuo et al. 2021), and the electricity seller has the resource advantage, so it makes the production and selling decision first. This upper limit of electricity selling is used as a parameter and input to the buyer, so as to achieve the interaction among lower-level decision-makers in turn, as shown in the composition of gray rectangular box (Algorithm 1).

The central government, as the decision-maker, needs to consider the realistic constraints such as fairness and achievability of the target allocation; the decision-maker will maximize the RPS most target, but needs the implementation ratio of the lower

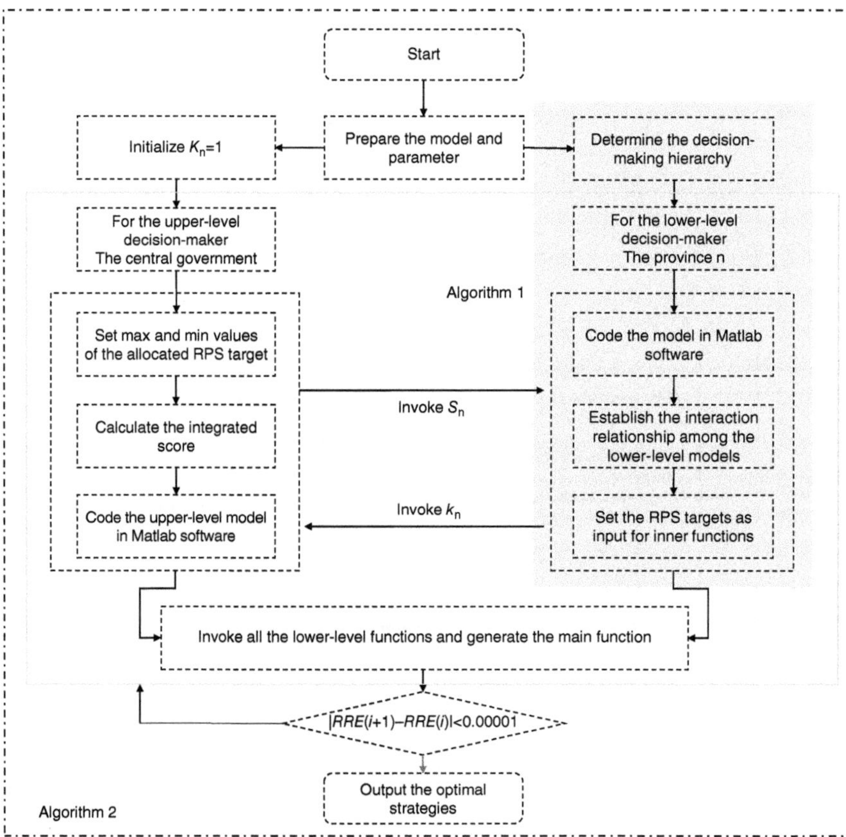

Figure 12.2 Flowchart of the interactive algorithm.

level as an important decision parameter. After setting the decision order in the first step, the optimization functions of the lower levels can be solved by calling the respective RPS objectives as input parameters. Thus, a process is generated to achieve the interaction between the upper and lower levels.

Set the initial executed ratio (K_n) to 1, compute the optimal RPS goals (S_n) taking equity into account, and then invoke the S_n into the lower level to calculate the K_n and generation and trading plans for each region. Then, invoke the new K_n into the upper level, and stop the cycle until the difference of the calculated upper-level objective value is within 0.0001. The interaction between the lower and upper levels is represented by the black dashed rectangle box (Algorithm 2).

12.4 Case Study

In this section, the basic information about the South Grid of China is given.

12.4.1 Case Description

Data from China Energy Administration shows that for Guangdong, Guangxi, Yunnan, Guizhou, and Hainan provinces, the renewable energy usage ratios in 2019 were as high as 34.40%, 43.10%, 82.90%, 35.60%, and 14.50%, respectively (Government of China 2019). However, their assigned national usage rates for 2020 were 29.50–32.50%, 50.00–55.00%, 80.0–88.00%, 31.5–34.70%, and 11.5–12.60%. When these numbers are compared, it is clear that the provinces of Guangdong, Guizhou, and Hainan can reach higher renewable energy usage ratios, whereas Guangxi may not be able to satisfy the threshold (National Energy Administration 2019a). As a result, there is a need to improve the assignment method's design.

There are eight AC paths and ten DC paths among the 18,500 kV and higher-power transmission paths in China Southern Power Grid (China South Grid 2019). The channels' combined transmission capacity is more than 50 million kilowatts. Though these lines, Yunnan province can transmit electricity to Guangdong, Guangxi and Hainan provinces, Guizhou can transmit electricity to Guangdong and Guangxi province, and Guangxi can transmit electricity to Guangdong (as shown in Figure 12.3). Here, we use the south of China as a test case for our proposed bi-level methodology. The South Grid, as the upper-level decision-maker, assigns a renewable utilization ratio for each of the five regions. The five provinces choose their own generation and trading plan as well as execution rates with the aim of reducing costs, and then provide feedback to the South Grid.

12.4.2 Data Collection

Data for Southern Grid (the upper-level decision-maker) is shown in Table 12.1. The power demand data (D_n) were collected from the government website (National Bureau of Statistics 2019) and the historical data of each province (g_{n1}, hydro included and non-hydro) were collected from the government website and

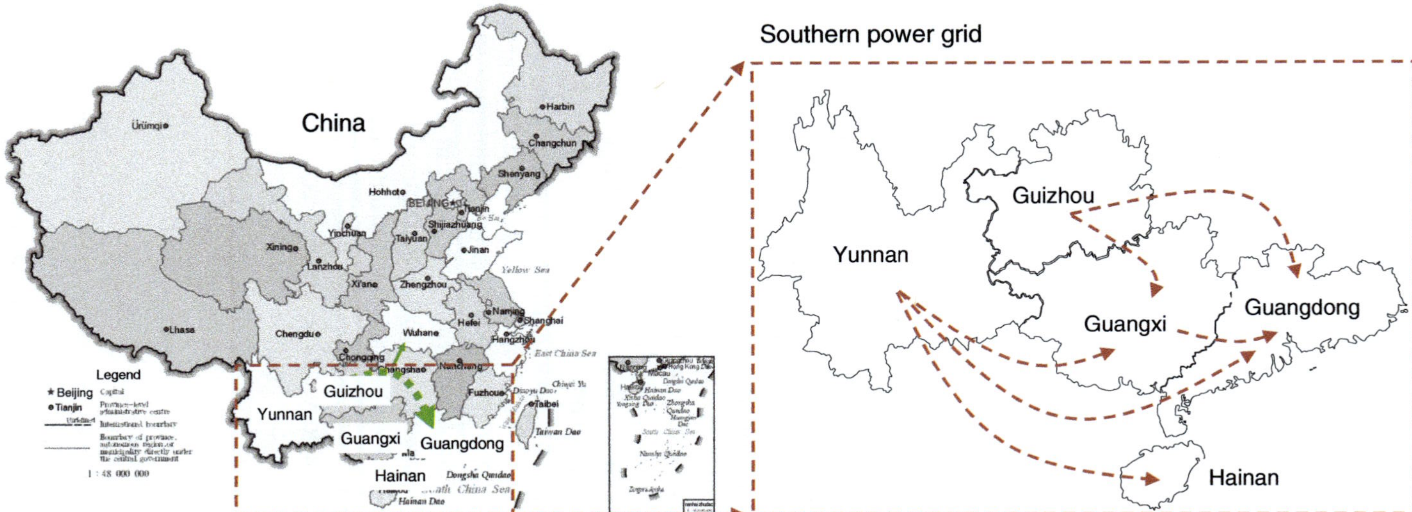

Figure 12.3 Location and electricity transmission direction for provinces in South Grid.

Table 12.1 Data for upper level.

Provinces	RE consumption ratio (2019) Hydro-included g_{n1} (%)	RE consumption ratio (2019) Non-hydro g_{n1} (%)	Min RE consumption ratio Hydro-included $\underline{R1}_{jn}$ (%)	Min RE consumption ratio Non-hydro $\underline{R2}_{jn}$ (%)	Max RE consumption ratio Hydro-included $\overline{R1}_{jn}$ (%)	Max RE consumption ratio Non-hydro $\overline{R2}_{jn}$ (%)
Guangdong	34.4	4.20	25	4.5	60	20
Guangxi	43.1	6.50	35	7	60	40
Yunnan	82.9	16.30	70	15	100	40
Guizhou	35.6	5.20	25	6	60	40
Hainan	14.5	6.80	8	6.5	50	40

Provinces	Electricity demand (2020) D_n (TWh)	Per Capita GDP g_{n2} (10^4 CNY)	Delicate particulate matter g_{n3} (g/m^3)
Guangdong	629.61	8.77	23
Guangxi	202.9	4.41	26
Yunnan	202.6	5.19	24
Guizhou	158.6	4.62	23
Hainan	36.3	5.47	14

published papers (National Energy Administration 2019b; Biomass Energy Industry Promotion Association 2019; Liu et al. 2019). The per capita GDP (g_{n2}) and delicate particulate matter (g_{n3}) were collected from China Statistical Yearbook, (Government of China 2019). Minimum and maximum renewable electricity consumption ratios ($\underline{R_{jn}}$, $\overline{R_{jn}}$) were set by the authors, considering province's renewable energy installed capacity and the announced RPS target by 2030.

The provincial power generation related data are shown in Tables 12.2–12.5. The unit provincial generation cost data (A_{jn}^{R}, A_{in}^{N}) were collected from (National

Table 12.2 Unit provincial generation cost data.

Provinces	Renewable generation (A_{jn}^{R}, CNY/MWh) Wind	Solar	Biomass	Hydro	Non-renewable generation (A_{in}^{N}, CNY/MWh) Coal	Gas	Nuclear
Guangdong	602.64	978.60	745.66	279.91	441.05	634.48	414.07
Guangxi	607.19	859.79	677.99	257.90	399.42	584.10	378.69
Yunnan	455.88	734.32	608.73	198.71	412.52	None	None
Guizhou	534.15	948.00	750.00	291.64	332.37	None	None
Hainan	607.45	442.65	677.99	407.91	438.49	658.27	428.50

Table 12.3 Last production cycle's electricity generation data.

	Renewable generation (G^R_{jn}, TWh)				Non-renewable generation (G^N_{in}, TWh)		
Provinces	**Wind**	**Solar**	**Biomass**	**Hydro**	**Coal**	**Gas**	**Nuclear**
Guangdong	7.20	3.10	5.92	39.10	290.34	63.10	96.34
Guangxi	6.13	1.36	1.50	59.34	83.81	14.78	17.15
Yunnan	24.35	4.73	0.60	285.431	31.44	0.00	0.00
Guizhou	7.80	1.41	0.56	76.94	138.96	0.00	0.00
Hainan	0.50	0.60	0.52	1.73	17.20	4.30	9.72

Table 12.4 Provincial generation technology utilization hour data.

	Renewable generation (H^R_{jn}, h/yr)				Non-renewable generation (H^N_{in} h/yr)		
Provinces	**Wind**	**Solar**	**Biomass**	**Hydro**	**Coal**	**Gas**	**Nuclear**
Guangdong	1612	1025	5181	3827	4601	2688	7548
Guangxi	2385	974	5181	3827	4601	2688	7548
Yunnan	2808	1300	5181	3827	4601	2688	7548
Guizhou	1861	837	5181	3827	4601	2688	7548
Hainan	1645	1223	5181	3827	4601	2688	7548

Table 12.5 Provincial generation capacity parameters.

	Renewable generation ($\overline{\beta}^R_{jn}$)				Non-renewable generation ($\overline{\beta}^g_j$)		
Provinces	**Wind**	**Solar**	**Biomass**	**Hydro**	**Coal**	**Gas**	**Nuclear**
Guangdong	1.113	1.349	1.212	1.276	1.000	1.000	1.235
Guangxi	1.401	1.332	1.055	1.048	1.000	1.220	1.066
Yunnan	1.091	1.256	1.220	1.103	1.200	0.000	0.000
Guizhou	1.132	1.211	1.220	1.109	1.000	0.000	0.000
Hainan	1.040	1.034	1.529	1.000	1.000	1.000	1.260

Energy Administration 2019c) and shown in Table 12.2. The last production cycle's electricity generation data (G^R_{jn}, G^N_{in}) and the provincial generation technology utilization hour data (H^R_{jn}, H^N_{in}) were collected from (Biomass Energy Industry Promotion Association 2019; Liu et al. 2019; China Nuclear Energy Association 2019) and are shown in Tables 12.3 and 12.4. The provincial generation capacity parameters (β^N_{in}, β^R_{in}) were collected from China Power Knowledge (China Power Knowledge 2019) and shown in Table 12.5.

Provincial transmission-related data are shown in Table 12.6. The electricity transmission loss data (L_{mn}) and unit transmission cost data (c_{ji}) were collected from (National Energy Administration 2019d, 2019e). The transmission capacity data (T_{mn}) were calculated using related AC and DC transmission path's capacity. The renewable electricity ratio (R_{mn}) is collected from (Biomass Energy Industry Promotion Association 2019; Liu et al. 2019) and represented by each province's electricity contribution ratio.

12.4.3 Calculation Results

In this section, the generation and trading results of each province, the minimum and maximum RPS that can be achieved by each province, and the allocation results considering equity will be presented in detail.

12.4.3.1 Generation and Trading Results for Individual Provinces

This section presents the decision-making results for each individual provinces without the involvement of the upper-level decision-maker. The results were obtained using Algorithm 1. The generation strategy with the minimized costs is first given, and then the min and max RPS that can be achieved in each region are calculated. The optimal share of renewable energy for the provinces of Guangdong, Guangxi, Yunnan, Guizhou, and Hainan is 31.2%, 39.6%, 88.4%, 47.4%, and 13.4%, respectively, according to Table 12.7. The allocation quota for Guizhou is understated while Hainan is overstated when compared to the national assigned RPS lower target and higher target (28.5% to 32.0%, 39.5% to 43.9%, 80.0% to 89.0%, 30.0% to 33.3%, and 13.5% to 14.9%). It can be seen that the accomplished percentages are consistently lower when compared with the non-hydro RPS objective in Table 12.1, demonstrating that the development of non-hydropower renewable energy sources tends to be given a lower priority due to cost considerations.

The intra-province and inter-province power generation and trade plan for both renewable and non-renewable energy sources are shown in Figure 12.4. It can be seen that hydropower has the highest priority for production, and that each province generates it at its highest level for the cheapest price. While the cost of producing wind, solar, and biomass is comparatively higher than that of producing hydroelectricity and using non-renewable resources, Guangdong, Guangxi, Yunnan, and Guizhou do so at the bare minimum. This suggests a tension between maximizing RPS and minimizing costs.

Table 12.6 Provincial transmission-related data.

	Electricity transmission loss (L_{mn},%)						Unit transmission cost (A^L_{mn}, CNY/MWh)				
Provinces	**Guangdong**	**Guangxi**	**Yunnan**	**Guizhou**	**Hainan**	**Provinces**	**Guangdong**	**Guangxi**	**Yunnan**	**Guizhou**	**Hainan**
Guangdong	0.00	0.00	0.00	0.00	0.00	Guangdong	0.0	0.0	0.0	0.0	0.0
Guangxi	3.93	0.00	0.00	0.00	0.00	Guangxi	47.1	0.0	0.0	0.0	0.0
Yunnan	6.57	2.98	0.00	0.00	7.32	Yunnan	75.5	53.8	0.0	0.0	87.1
Guizhou	7.05	3.47	0.00	0.00	0.00	Guizhou	75.5	53.8	0.0	0.0	0.0
Hainan	0.00	0.00	0.00	0.00	0.00	Hainan	0.0	0.0	0.0	0.0	0.0

	Transmission capacity (T_{mn}, TWh)						Renewable electricity ratio (R_{mn}, %)				
Provinces	**Guangdong**	**Guangxi**	**Yunnan**	**Guizhou**	**Hainan**	**Provinces**	**Guangdong**	**Guangxi**	**Yunnan**	**Guizhou**	**Hainan**
Guangdong	0	0	0	0	0	Guangdong	0.00	0.00	0.00	0.00	0.00
Guangxi	31.536	0	0	0	0	Guangxi	46.87	0.00	0.00	0.00	0.00
Yunnan	187.464	38.544	0	0	2.0	Yunnan	82.47	82.47	0.00	0.00	82.47
Guizhou	78.84	21.024	0	0	0	Guizhou	49.56	49.56	0.00	0.00	0.00
Hainan	0	0	0	0	0	Hainan	0.00	0.00	0.00	0.00	0.00

Table 12.7 Operation cost and generation capacity of each province.

Calculated results		Guangdong	Guangxi	Yunnan	Guizhou	Hainan
Total costs	Billion CNY	297.25	75.42	36.28	51.64	16.58
Required renewable generation capacity	Wind (MW)	3573	2057	6937	3353	243
	Solar (MW)	2420	1113	2913	1343	392
	Biomass (MW)	914	231	93	86	80
	Hydro (MW)	13,037	16,250	82266	22,295	451
Required non-renewable generation capacity	Coal (MW)	60,464	20,433	5467	24,162	3422
	Gas (MW)	18,780	4397			1280
	Nuclear (MW)	15,763	2423			1623
Implemented renewable energy consumption (RPS)	Hydro-included (%)	31.2	39.6	88.4	47.7	13.4
	Non-hydro (%)	2.2	3.5	11.7	3.7	3.6

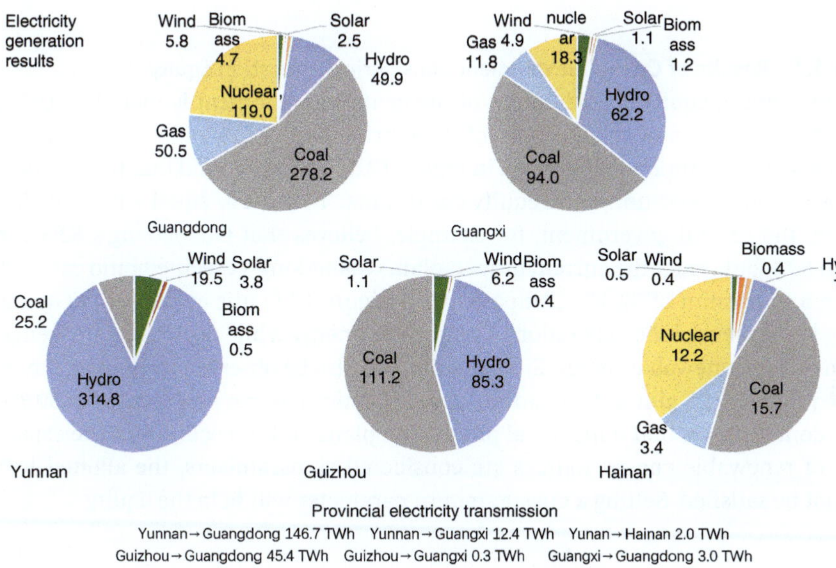

Figure 12.4 Generation and trading plans of each provinces.

12.4.3.2 The Minimum and Maximum RPS that can be Achieved for Individual Provinces

The best scenario that each region may reach under the specified renewable growth plan is estimated to investigate the RPS allocation scheme in terms of fairness. Using Algorithm 1 again, but this time modifying the objective function to maximize and minimize the implemented RPS, the solution results are displayed

Table 12.8 Minimum and maximum RPS for each region.

Calculated RPS results for each province		Guangdong (%)	Guangxi (%)	Yunnan (%)	Guizhou (%)	Hainan (%)
Min RPS	Hydro-included	28.81	31.16	82.28	31.69	9.90
	Non-hydro	2.46	2.97	11.72	2.93	3.57
Max RPS	Hydro-included	32.09	41.95	100.00	50.58	12.61
	Non-hydro	3.06	5.84	16.41	5.92	5.33

in Table 12.8. The results indicate the maximum improvement space for RPS targets of each region. The maximum implemented RPS in Guangdong and Hainan under the specified growth plan can only reach 32.09% and 12.61%, respectively. The maximum non-hydro implemented RPSs of Guangdong, Guangxi, and Hainan are lower than the 2019 practical RPSs (see Table 12.1), indicating that the overall growth rate of power consumption is greater than that of non-hydro renewable energy.

12.4.3.3 Results of Central Government Considering Allocation Equity

In this section, equity consideration of the central government is included. Using a baseline situation of equal weight of three principles of 1/3 and $e = 0.4$, the outcomes of Algorithm 2 are illustrated in Figure 12.5. It can be found that the executed ratios are quite poor under the equity constraints. To achieve equal RPS goal allocation, the central government, for example, believes that Guangdong's RPS can reach 60%, but under the current growth plan, Guangdong's executed ratio can only reach a maximum of 52.3%. Compared with Figure 12.5, it is evident that Guangdong has increased the generation of renewable energy while decreasing the generation of non-renewable energy. Similar shifts can also be observed in Hainan, where the equitable RPS aim is 50% and the greatest ratio that can be executed is 30.3%. This demonstrates that if historical production plans and projections for the expansion of renewable energy sources are considered as parameters, the allotted RPS cannot be satisfied. Setting a bigger amount parameter will help the equity.

12.5 Discussions

In this section, the trade-offs between equity and economy of RPS allocation and implementation will be discussed. The related key findings will also be presented.

12.5.1 Trade-offs Between Equity and Economy

The statistics above show that Southern Power Grid's allocation equity still requires urgent improvement. Therefore, it is essential to expand the use of renewable

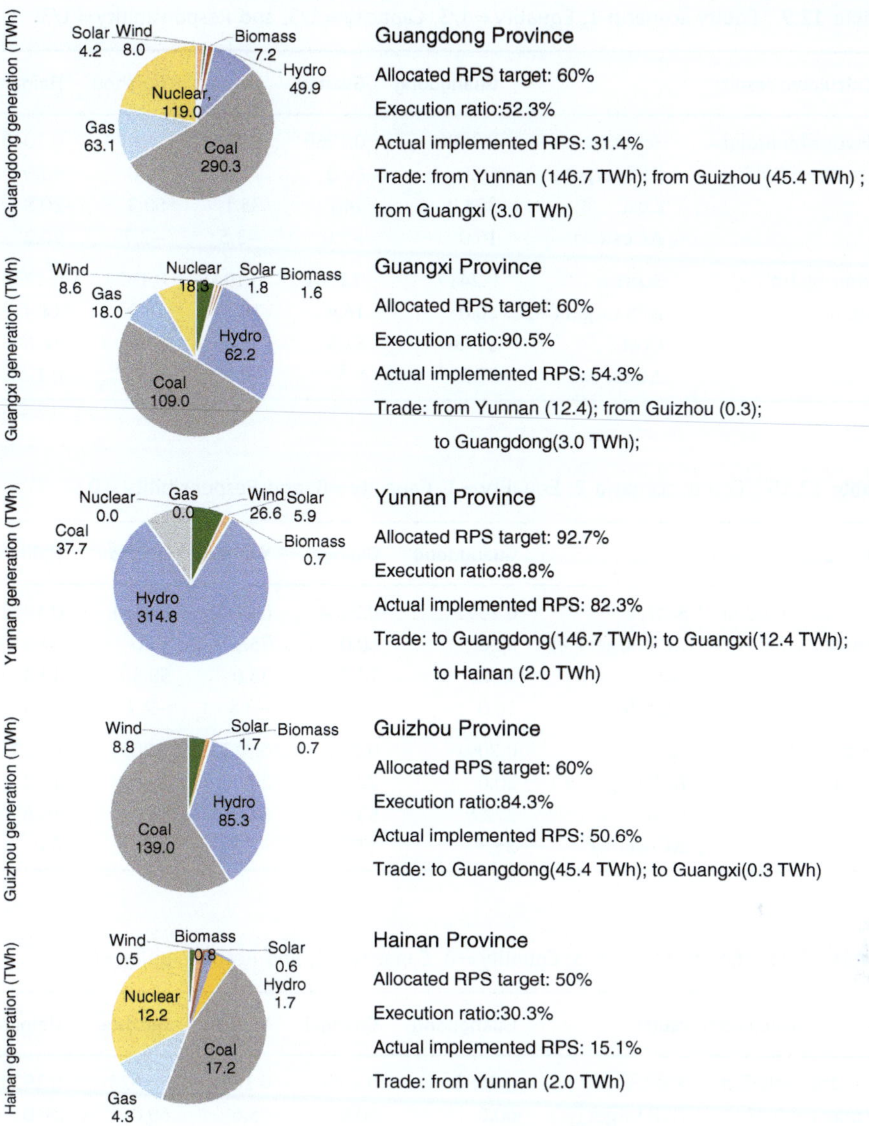

Figure 12.5 Equitable RPS target allocation and executed plans.

energy in order to examine the trade-off between equity and economy. The amount parameter of renewable energy generation technologies (hydro not included) is used as an example of 10 times the original amount parameters in the following contents. Each province's execution rate is 100%. Four scenarios are proposed based on the weights assigned to the principles: in Scenario 1, each principle is given an equal weight of 1/3; in scenarios 2, 3, and 4, only one of the Responsibility, Capacity, or Equality principles is considered, respectively. Tables 12.9–12.12 show the related

Table 12.9 Equity scenario 1: Equality = 1/3, Capacity = 1/3, and Responsibility = 1/3.

Calculated results		Guangdong	Guangxi	Yunnan	Guizhou	Hainan
Hydro-included	Scores	0.2480	0.1960	0.1532	0.1979	0.2049
(max $e = 0.49$)	RPS target (%)	60.0	60.0	75.6	60.0	50.0
	Cost	354.0	74.7	35.1	50.3	20.8
	ΔCost(%)	16.0	−1.0	−3.5	−2.7	20.2
Non-hydro	Scores	0.2417	0.1981	0.1941	0.1923	0.1738
(max $e = 1$)	RPS target (%)	20.0	16.4	16.1	15.9	14.4
	Cost	372.6	81.8	31.7	55.0	16.6
	ΔCost (%)	25.3	8.5	−12.7	6.5	0.1

Table 12.10 Equity scenario 2: Equality = 1, Capacity = 0, and Responsibility = 0.

Calculated results		Guangdong	Guangxi	Yunnan	Guizhou	Hainan
Hydro-included	Scores	0.2091	0.2364	0.2182	0.2091	0.1273
(max $e = 0.73$)	RPS target (%)	60.0	60.0	75.9	60.0	44.3
	Cost	354.0	74.7	35.0	50.3	19.4
	ΔCost(%)	16.0	−1.0	−3.8	−2.7	14.5
Non-hydro	Scores	0.2091	0.2364	0.2182	0.2091	0.1273
(max $e = 1$)	RPS target (%)	20.0	22.6	20.9	20.0	12.2
	Cost	372.6	88.8	34.2	56.6	16.6
	ΔCost (%)	25.3	17.7	−5.8	9.6	0.0

Table 12.11 Equity scenario 3: Equality = 0, Capacity = 1, and Responsibility = 0.

Calculated results		Guangdong	Guangxi	Yunnan	Guizhou	Hainan
Hydro-included	Scores	0.3082	0.1551	0.1824	0.1623	0.1920
(max $e = 0.43$)	RPS target (%)	60.0	60.0	75.6	60.0	50.0
	Cost	354.0	74.7	35.1	50.3	20.8
	ΔCost(%)	16.0	−1.0	−3.4	−2.7	20.2
Non-hydro	Scores	0.3082	0.1551	0.1824	0.1623	0.1920
(max $e = 0.78$)	RPS target (%)	20.0	12.9	15.2	13.5	16.0
	Cost	372.6	79.3	31.2	54.1	16.6
	ΔCost (%)	25.3	5.2	−14.0	4.7	0.1

Table 12.12 Equity scenario 4: Equality = 0, Capacity = 0, and Responsibility = 1.

Calculated results		Guangdong	Guangxi	Yunnan	Guizhou	Hainan
Hydro-included	Equity score	0.2266	0.1965	0.0591	0.2225	0.2953
(max $e = 0.13$)	RPS target (%)	60.0	60.0	77.0	60.0	50.0
	Cost	354.0	74.7	34.5	50.3	20.8
	ΔCost(%)	16.0	−1.0	−5.2	−2.7	20.2
Non-hydro	Scores	0.2078	0.2028	0.1816	0.2056	0.2022
(max $e = 1$)	RPS target (%)	20.0	19.5	17.5	19.8	19.5
	Cost	372.6	84.9	32.4	56.5	16.6
	ΔCost (%)	25.3	12.6	−10.7	9.4	0.3

results in each scenario. Detailed generation results for each scenario is shown in Figures 12.6 and 12.7.

12.5.1.1 Comparison of Integrated Scores

Each region has a greater variance in Scenarios 2, 3, and 4 than in scenario 1 according to the results of integrated scores for each scenario. In scenario 2, Guangxi shows the highest integrated score for this region's air quality while solely considering Equality, highlighting the pressing need to improve air quality. Guangdong receives the highest combined score in scenario 3, demonstrating its financial capability to implement renewable energy. In case 4, Hainan and Guangdong both have integrated scores of 0.2953 and 0.2266, respectively, suggesting that their current RPS levels still require improvement. While Yunnan's score is only 0.059, this shows that Yunnan has already accepted adequate responsibility for its renewable energy use.

12.5.1.2 Comparison of Maximum Equity Parameter

In cases with hydro included, e of four scenarios is 0.49, 0.73, 0.47, and 0.13; in cases without hydro, it is 1, 1, 0.78, and 1. In general, it is discovered that non-hydro cases are simpler to produce totally fair (i.e. $e = 1$) than hydro-included ones while investigating the maximum e. This shows that it is currently challenging to achieve absolute equity in the RPS allocation for the Southern Power Grid that includes hydro. By establishing non-hydro RPS, one can encourage not just the use of wind and solar power but also the equitable distribution of regional RPS.

12.5.1.3 Comparison of the Cost-Change Rate

It is clear that the hydro-included ΔCost for Guangdong and Hainan is positive, whereas this number is negative for Guangxi, Yunnan, and Guizhou. This suggests that in order to achieve a fair distribution of RPS, resource-importing regions must increase their financial investment, while resource-exporting regions, which enjoy cost advantages, will lower their electricity costs. It is important to note that all resource-exporting nations export power at full capacity, which boosts their revenue. By investing in wind, solar, and biomass power generation, all regions

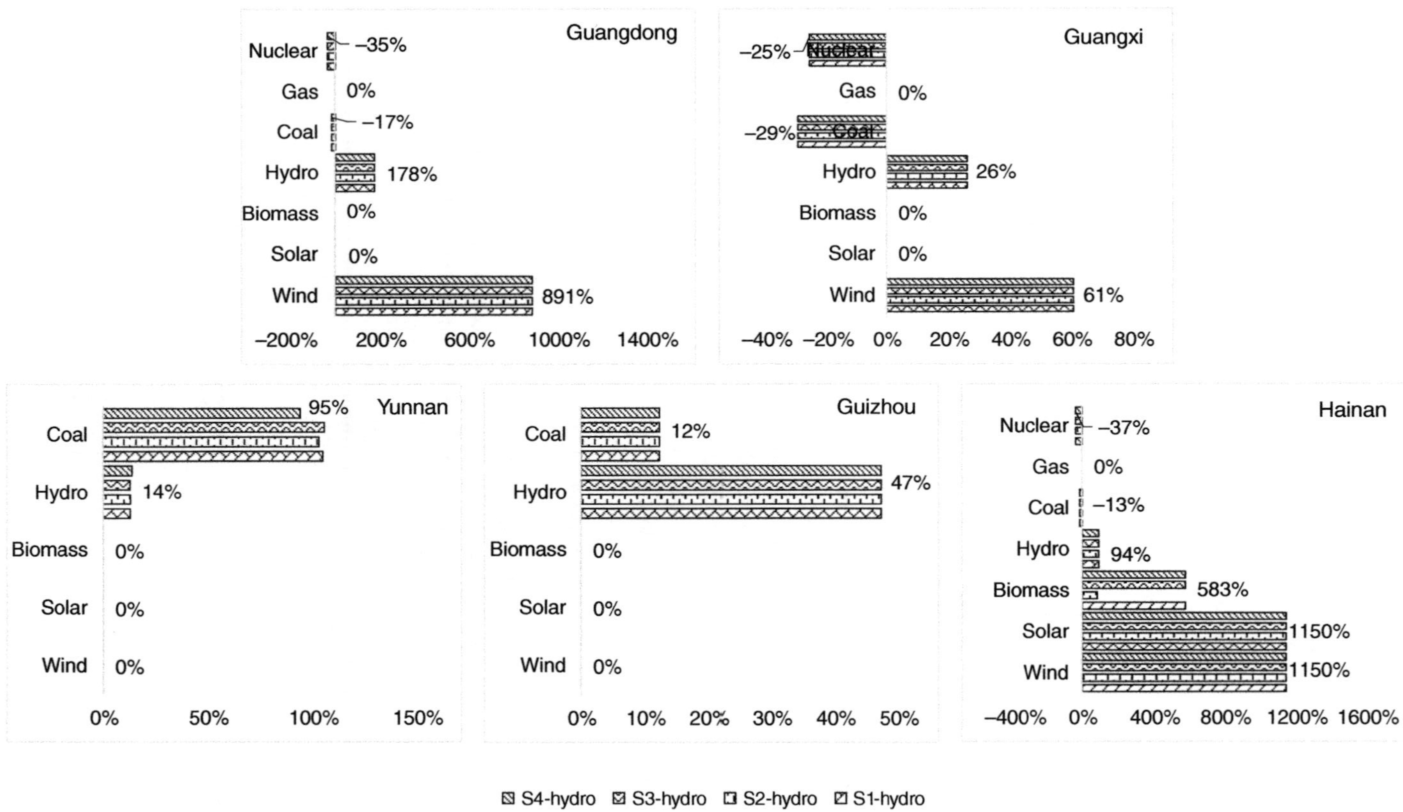

Figure 12.6 Generation plans under different equity scenarios (hydro included).

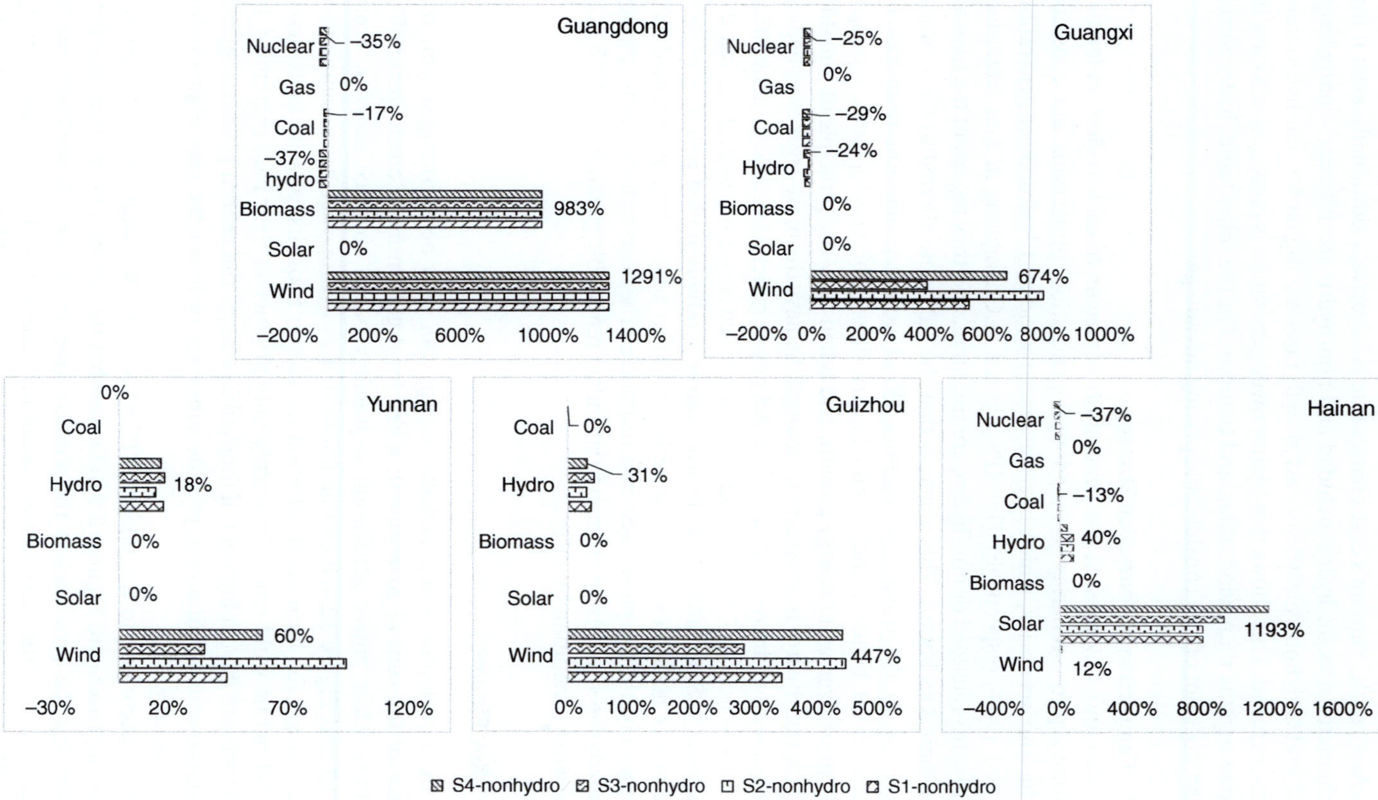

Figure 12.7 Generation plans under different equity scenarios (non-hydro).

should cost more to satisfy the non-hydro RPS in terms of non-hydro ΔCost. Yunnan can also reap the most benefits because of the volume of electricity it exports. As a result, Yunnan's cost-change rate is negative. Additionally, even if the equity characteristics for hydro-included and non-hydro are different, Guangdong, Guangxi, and Guizhou's electricity cost growth rates are higher for non-hydro than for hydro-included. This shows that hydropower generation is more cost-effective in these three regions than wind, solar, and biomass. On the other hand, hydropower in Hainan does not offer a considerable pricing advantage.

12.5.1.4 Comparison of Generation Strategy

Figures 12.6 and 12.7 present the generation strategies of each region under the above four equity scenarios. Guangdong's generation projections are constant across the four scenarios, which have increased wind and hydropower generation by 89% and 178%, respectively. The RPS allotted to Guangdong in four scenarios, whether hydro-included or non-hydro, are the same, and they represent the highest limits of their RPS bound. This suggests that the RPS now allotted to Guangdong is inequitable to the other four regions. Wind power generation resources in Guangdong have benefits and should be prioritized for development. Guangxi similarly prioritizes wind energy growth, and when only the principle of Equality (scenario 2, non-hydro) is considered, Guangxi displays the most urgency to deploy renewables, with an increase of 802% in wind power generation. Energy generation by coal is increased in Yunnan under hydro-included scenarios, indicating that the implemented RPS of Yunnan's historical energy production plan is sufficient, and thermal power generation needs to be increased to adjust the implemented RPS and further ensure the power system's resilience. The responsibility principle for non-hydro is the strongest in Hainan to stimulate renewable generating, which may increase wind generation by 1193%.

12.5.2 Key Findings

Firstly, the development of renewable energy can only be encouraged with the assistance of the central government; when local governments create generation plans, they select power generation technology based on price benefits. From Table 12.7 and Figure 12.4, it can be found that as hydropower has a competitive price advantage, the local administration decides to boost its capacity to its maximum level, while wind and solar energy only produce at their bare minimum. In order to increase the deployment of renewable energy, the central government can design improvement initiatives, provide subsidies, or lower the cost of producing renewable energy.

Secondly, according to the present RPS allocation, the equity performance has space for improvement. Figure 12.5 indicates that the optimal hydro-included RPS allocations for the provinces in the studied case are 60.0% (Guangdong), 60.0% (Guangxi), 92.7% (Yunnan), 60.0% (Guizhou), and 50.0% (Hainan), respectively. However, their maximum executed ratios are 52.3%, 90.5%, 88.8%, 84.3%, and 30.3%, resulting in an actual implemented renewable electricity consumption ratio of

31.4%, 54.3%, 82.3%, 50.6%, and 15.1%. The simulation findings revealed that the equity target forces the central government to set higher RPS allocation, despite the fact that each region's present generating capability is far from meeting the equitable distribution.

Further, renewable investment must be adjusted to ensure an equal hydro-included RPS distribution. From the hydro-included results of Tables 12.9–12.12, it can be found that in all scenarios, Guangdong and Hainan need to increase their provincial renewable generation investment, while Guangxi, Yunnan, and Guizhou should reduce the financial input of renewable energy. This also means that under the current scheme, these three regions have taken on more obligatory RPS quotas than the first two regions, regardless of the principles of Equity, Capacity, and Responsibility they are based on.

Finally, it is more difficult to accomplish equal allocation of hydro-included RPS targets than non-hydro RPS targets. In Tables 12.9–12.12, when hydro is incorporated, the maximum equity parameter are 0.49, 0.73, 0.47, and 0.13; in non-hydro scenarios, they are 1, 1, 0.78, and 1. As was already mentioned, the stricter the equity that may be reached, the higher the *e* value. Because the maximum and lower limits of the assigned RPS are established in Table 12.1, higher *e* does not result in a workable solution. To accomplish equitable allocation, the non-hydro RPS aim is more logical.

These findings provide an important reference and effective support for the National Energy Administration of China to adjust the renewable energy power consumption responsibility targets between 2022 and 2030 set for each province in China (Website of Beijixing 2021).

References

Biomass Energy Industry Promotion Association (2019). China biomass power generation industry ranking report. https://www.beipa.org.cn/productinfo/298975.html.

Carley, S., Davies, L.L., Spence, D.B., and Zirogiannis, N. (2018). Empirical evaluation of the stringency and design of renewable portfolio standards. *Nature Energy* 3 (9): 754–763.

China Nuclear Energy Association (2019). National nuclear power operations from January to December 2019. http://www.china-nea.cn/site/content/36862.html (accessed 20 October 2023).

China Power Knowledge (2019). Inventory of national electric power installation, generation, and consumption data in 2019. https://data.pkthinker.com/ (accessed 20 October 2023).

China South Grid (2019). Southern regional inter-regional and inter-provincial power market operation report. http://www.lianmenhu.com/blockchain-18586-1 (accessed 20 October 2023).

Fan, J.L., Wang, J.X., Hu, J.W. et al. (2019). Optimization of China's provincial renewable energy installation plan for the 13th five-year plan based on renewable portfolio standards. *Applied Energy* 254: 113757.

Gang, J., Tu, Y., Lev, B. et al. (2015). A multi-objective bi-level location planning problem for stone industrial parks. *Computers and Operations Research* 56: 8–21.

Government of China (2019). Notice of national energy administration on monitoring and evaluation of renewable energy power development in China in 2019.

Höhne, N., Den Elzen, M., and Escalante, D. (2014). Regional GHG reduction targets based on effort sharing: a comparison of studies. *Climate Policy* 14 (1): 122–147.

Liu, Z., Yu, C., Wang, M. et al. (2019). *2018 China Electric Power Yearbook*. China Electric Power Press.

National Bureau of Statistics (2019). China Statistical Yearbook 2019. https://www.stats.gov.cn/sj/ndsj/2019/indexch.htm.

National Energy Administration (2019a). Notice on establishing and improving the guarantee mechanism of renewable energy power consumption. https://www.gov.cn/zhengce/zhengceku/2019-09/25/content_5432993.htm.

National Energy Administration (2019b). Operation of solar power in 2019. http://www.nea.gov.cn/2020-02/28/c_138827910.htm.

National Energy Administration (2019c). 2018 National electricity price regulatory bulletin. http://www.nea.gov.cn/2019-11/05/c_138530255.htm.

National Energy Administration (2019d). Notice of the National Development and Reform Commission on reducing general industrial and commercial electricity prices. http://www.gov.cn/xinwen/2019-05/16/content_5392079.htm.

National Energy Administration (2019e). 2018 National electricity price regulatory notice. http://www.nea.gov.cn/2019-11/05/c_138530255.htm.

Qiu, R., Xu, J., Ke, R. et al. (2020a). Carbon pricing initiatives-based bi-level pollution routing problem. *European Journal of Operational Research* 286 (1): 203–217.

Qiu, R., Xu, J., Ke, R. et al. (2020b). Carbon pricing initiatives-based bi-level pollution routing problem. *European Journal of Operational Research* 286 (1): 203–217.

Robiou du Pont, Y., Jeffery, M.L., Gütschow, J. et al. (2017). Equitable mitigation to achieve the Paris agreement goals. *Nature Climate Change* 7 (1): 38–43.

Rouhani, O.M., Niemeier, D., Gao, H.O., and Bel, G. (2016). Cost-benefit analysis of various California renewable portfolio standard targets: is a 33% RPS optimal? *Renewable and Sustainable Energy Reviews* 62: 1122–1132.

Shayegh, S. and Sanchez, D.L. (2021). Impact of market design on cost-effectiveness of renewable portfolio standards. *Renewable and Sustainable Energy Reviews* 136: 110397.

Siddiqui, A.S., Tanaka, M., and Chen, Y. (2016). Are targets for renewable portfolio standards too low? The impact of market structure on energy policy. *European Journal of Operational Research* 250: 328–341.

Strehlow, H. and Stackelberg, M. (1950). Zeitschrift für Elektrochemie und angewandte physikalische Chemie. *Zur Theorie der polarographischen Kurve* 54 (1): 51–62.

Wang, H., Su, B., Mu, H. et al. (2020). Optimal way to achieve renewable portfolio standard policy goals from the electricity generation, transmission, and trading perspectives in southern China. *Energy Policy* 139: 111319.

Wang, B., Wei, Y.M., and Yuan, X.C. (2018). Possible design with equity and responsibility in China's renewable portfolio standards. *Applied Energy* 232: 685–694.

Wang, G., Zhang, Q., Li, Y., and Mclellan, B.C. (2019). Efficient and equitable allocation of renewable portfolio standards targets among China's provinces. *Energy Policy* 125: 170–180.

Website of Beijixing (2021). National Energy Administration renewable energy consumption target for comments, 2030 national unified consumption responsibility weighting 40%. https://news.bjx.com.cn/html/20210210/1135946.shtml (accessed 20 October 2023).

Xin-gang, Z., Ling, W., and Ying, Z. (2020). How to achieve incentive regulation under renewable portfolio standards and carbon tax policy? A China's power market perspective. *Energy Policy* 143 (2): 111576.

Xin-gang, Z., Tian-tian, F., Lu, C., and Xia, F. (2014). The barriers and institutional arrangements of the implementation of renewable portfolio standard: a perspective of China. *Renewable and Sustainable Energy Reviews* 30: 371–380.

Xu, Z., Yao, L., Liu, Q., and Long, Y. (2019). Policy implications for achieving the carbon emission reduction target by 2030 in Japan-Analysis based on a bilevel equilibrium model. *Energy Policy* 134: 110939.

Yao, L., Shuai, Y., and Chen, X. (2020). Regional water system vulnerability evaluation: a bi-level DEA with multi-followers approach. *Journal of Hydrology* 589: 125160.

Ying, Z., Xin-gang, Z., and Zhen, W. (2020). Demand side incentive under renewable portfolio standards: a system dynamics analysis. *Energy Policy* 144 (2): 111652.

Yu, S., Zhou, S., Dai, Y., and Zhang, J. (2023). Impacts of RPS and fit on inter-regional power transmission line layout in china: considerations of high renewable energy penetration. *Energy Policy* 178: 113615.

Zuo, Y., Zhao, X.G., and Zhang, Y. (2021). Bargaining strategies in bilateral electricity trading based on fuzzy Bayesian learning. *International Journal of Electrical Power & Energy Systems* 129: 106856.

13

Optimal RPS Implementation Strategy Considering Emission Trade and Green Certificate Trade

Streamlining the planning and permitting process for new transmission and coordinating decarbonization at the national level could enable a more efficient and rapid transition to a zero-carbon electricity system.

Patrick R. Brown and Audun Botterud
Published on Joule 2021, 5(1): 115–134

When switching to a low-carbon hybrid energy structure, market incentives are less costly. Of which, tradable green certificate (TGC) and carbon emission trading (CET)are the two most representative incentives. However, these two incentives have not been coordinated, resulting in economic and environmental problems. Therefore, this chapter aims to optimize the implementation of renewable portfolio standards (RPS) while considering the coordination of TGC and CET policies. A multi-objective equilibrium model is proposed for for a provincial hybrid power system's generation and trading plan that minimizes carbon emissions and total costs. A case study is provided to demonstrate the applicability of the proposed model, which revealed that an econo-environmental equilibrium plan could decrease carbon emissions by 13.2% while increasing total expenses by only 2.3%. Five-scenario analyses with varied decision-makers' preferences were carried out to investigate the policy consequences. It was discovered that the mixed policy reduced coal power generation while conditionally promoting solar power development and trading, depending on government preferences.

13.1 Introduction

The intensification of climate change concerns has meant that building hybrid energy system are becoming a more common solution to promote low-carbon energy transitions and reduce greenhouse gas (GHG) emissions (International Energy Agency 2022a). Developing countries like China and India, which are large carbon dioxide emitters, and have announced to become carbon neutral by

Sustainable Hybrid Energy Systems: Carbon Neutral Approaches, Modeling, and Case Studies,
First Edition. Jiuping Xu and Fengjuan Wang.

mid-century (Liu et al. 2022). Most of the GHG reductions come from the power sector (Bistline et al. 2022), and it is expected that electricity will become the main terminal energy use in the future (Liu et al. 2021b). Therefore, to stimulate hybrid energy system development in the power generation and use sector is becoming the key focus for carbon emission reduction targets and low-carbon energy transition.

Previous environmental power sector regulations has stimulated the switching to a low-carbon hybrid energy structure. Hu et al. (2020) found that high-level marketization encouraged companies to be involved in carbon dioxide emission trading and decrease emissions of hybrid energy systems. Market incentives, such as the CET policy and the TGC policy, have been found to better encourage energy efficiency and technological innovation (Pan et al. 2019), and have been implemented to drive low-carbon technological investment after the subsidies are removed (Ministry of Ecology and Environment of the People's Republic of China 2020a; National Development and Reform Commission and Ministry of Industry and Information Technology 2022; Wang et al. 2021). Meng and Yu (2023) claimed that these two policies were the most effective representative market policy incentives for China to build low-carbon national hybrid power system. However, market-based transitions are frequently influenced by reversals and price changes, and they also necessitate that industry sectors be able to meet the essential regulatory criteria (Meckling et al. 2022). It has been observed that the CET and TGC markets have unpredictable interactions that could increase social costs (Fischer et al. 2017; Fischer and Preonas 2010). Therefore, to accelerate the decarbonization of national hybrid energy system, a well-coordinated mixed policy that considers the interactions of CET and TGC policies is required (International Energy Agency 2022b).

The design, interaction, and effect of market-based policies in power industry to stimulate hybrid energy system development have been widely explored by previous studies. Many scholars focus on policy design, especially quota settings. For example, Zhao et al. (2022) developed a coevolutionary model and discovered that high quotas and price collars near to TGC values increased the cost-effectiveness of RPS. Qin et al. (2020) created a graph-restricted cooperative game model to assign responsibility for power generator to reduce carbon emissions. Chang et al. (2020) proposed a two-stage quota allocation model. Meanwhile, the interactions of different policies have also attracted some studies. For example, according to Feng et al. (2018), there is a conflict between the cohabitation of TGC and CET implementation. Feng et al. (2021) modeled the combined effects of TGC and CET on the electricity market from 2020 to 2026 and also discovered that TGC and CET have policy redundancy. Li and Wang (2022) discovered that, while CET could help reduce carbon intensity and long-term CO_2 emissions, it would be severely hampered in the long run with structural changes of hybrid energy systems. Yu et al. (2021) explored the mixed effects of carbon caps and RPS, and Yi et al. (2019) studied the synergies associated with the RPS target setting. It can be found that most previous studies tend to focus on one part or another and there is a need for, more comprehensive research to comprehensive optimize the integration of both policies.

Some progress has been made in the policy integration to optimize the hybrid power generation system. Yan et al. (2022) created a market structure for joint trading carbon emission rights and green certificates to encourage increased involvement in cross-regional trading, as well as a conservative, efficient operating scheme. Hille et al. (2020) incorporated CET into a stochastic dynamic economic dispatch model centered on wind generation in order to reduce total energy prices while meeting particular carbon emission reduction targets. Similarly, the economic goal was also included in (Li et al. 2022; Wang et al. 2019c)'s studies. The majority of the preceding studies concentrated on a particular economic or environmental standpoint. However, planning the low-carbon power industry transformation path of hybrid energy system and developing the related low-carbon regulations necessitate taking into account both carbon emission reductions and economic gains (Liang et al. 2023). As a result, effective low-carbon hybrid system transition plans must include economic and environmental trade considerations. Multi-objective optimization has been widely used to overcome econo-environmental issues in power systems. For example, in order to solve a combined economic–environmental power dispatch challenge, Sakthivel et al. (2021) suggested a new multi-objective squirrel search method; He et al. (2021) developed multi-objective capacity optimization models to simultaneously minimize the levelized cost of energy (LCOE) and loss of power supply probability (reliability), and Li et al. (2020) proposed a dynamic hybrid energy economic–environmental dispatch model using multi-objective optimization.

Based on the above discussion, this chapter considers the coordination of TGC and CET policies, and proposes a multi-objective equilibrium model for a provincial hybrid power system generation and trading plan that minimizes carbon emissions and total costs. Detailed works are therefore highlighted as follows: first, the interaction between TGC and CET policies in the power sector will be investigated. Second, to coordinate the TGC and CET policies, an economic–environmental equilibrium provincial hybrid power system generation and trading plan will be developed, which will then be applied to a case study to illustrate the feasibility of the proposed model. Third, policy-related quotas are defined, and five scenarios with varying decision-making preferences are designed to investigate the policy consequences.

13.2 Key Problem Statement

To develop the econo-environmental equilibrium power generation and trading plan for the regulation of the TGC and CET policies, the interaction mechanisms between the three markets and the power generation and trading plan are given.

13.2.1 Integration of the TGC and CET Policies

The national interprovincial power market must economically and efficiently coordinate the TGC and CET markets to achieve a low-carbon transition. The relationships between these three markets are shown in Figure 13.1. Provincial governments must use efficient market incentives to meet local power needs. On the supply side, the CET policy governs local power generation, with coal power generation linked to CET purchase (Ministry of Ecology and Environment of the People's Republic of China 2020a). Furthermore, the implementation of the

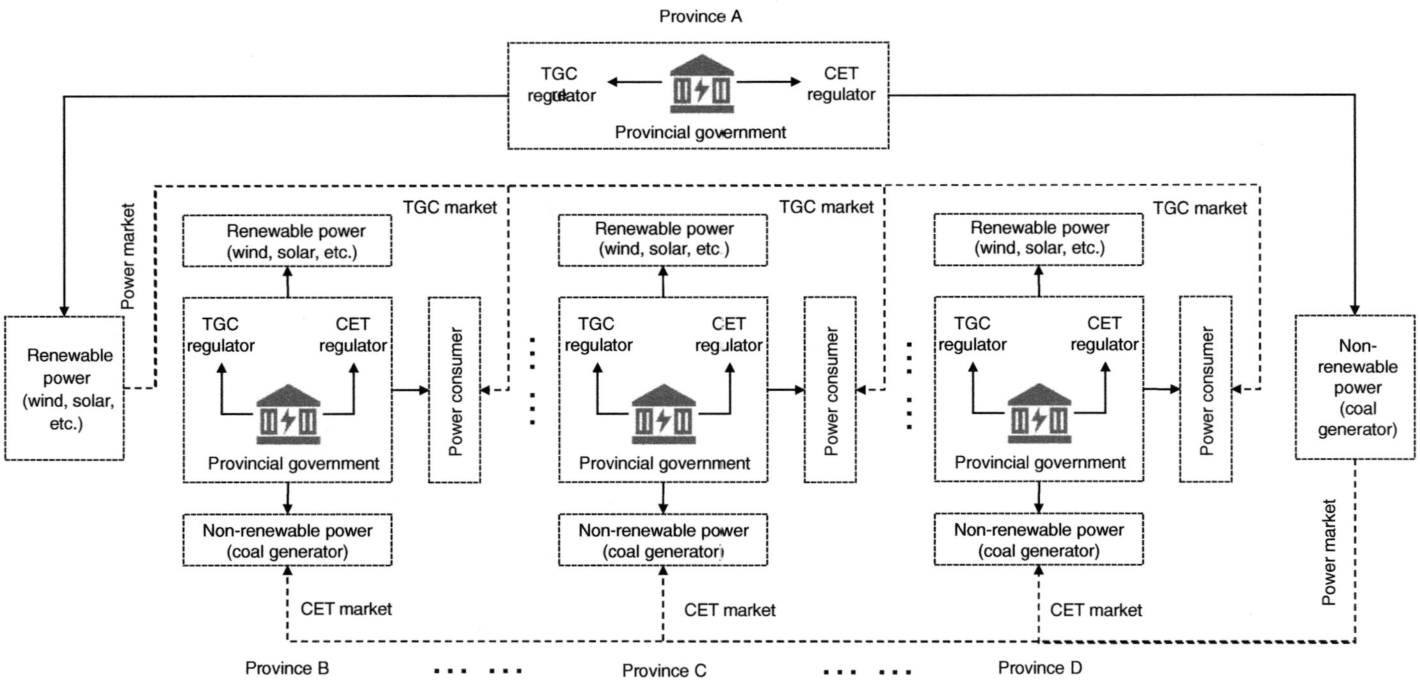

Figure 13.1 Interaction mechanism of power market coordinates with TGC and CET trading markets from the provincial government perspective.

TGC policy necessitates greater flexibility in wind and solar power supplies. The inter-provincial power market implies that other provinces can effect local power supplies; the RPS standards mean that local power demand must be supplied with a particular percentage of renewable power or TGC; and the CET limits overall power consumption carbon emissions (National Development and Reform Commission and National Energy Administration 2019). As a result of the various price-to-power technology options in the inter-provincial power market, provincial governments must meet their power generation demands by trading across all three markets.

However, the low-carbon transition necessitates the maintenance of an economic–environmental power generation, trade, and demand balance. As a result, provincial governments strive to reduce the mid- to long-term LCOE, which is also frequently used to assess techno-economic competitiveness (Li et al. 2018). Coal is less expensive than renewables but produces more pollutants. Because interprovincial power market generation technologies have diverse economic and environmental repercussions, various provinces may profit from different resource endowments (Barbose et al. 2015). For example, a province with significant coal resources may be able to provide cheaper electricity but may also fail to meet carbon emission reduction requirements, and if a province prioritizes cost reduction over environmental protection, it may import coal power after meeting its RPS targets. As a result, because pursuing economic expansion alone can exacerbate environmental concerns, an econo-environmental equilibrium provincial power plan is required (Rahman et al. 2021).

13.2.2 Interaction of Power Generation and Trading

A provincial government power generation and trading strategy to meet demand was designed to synchronize the TGC and CET power market strategies, as shown in Figure 13.2. Provincial governments must evaluate power generation and trading costs to guarantee that the RPS and CET quota requirements are met. Provincial governments plan their local coal power generation based on the area's natural resources, and the free CET quota adds extra benefits by transforming the area's wind and solar energy resources into TGC. Therefore, the original provincial TGC and CET are tied to provincial power generation, which must satisfy local demand and adhere to policy rules. If there are power shortages, power from other provinces must be obtained in a way that satisfies the RPS and carbon emissions reduction standards. As long as all policy criteria and RPS quotas are met, provincial governments are permitted to export excess power if there is a higher supply of electricity in a province than there is a need for it. In order to ensure that the overall amount of renewable energy and TGC issued equals the local RPS quota, the final provincial power demand structure must be controlled by the carbon emission limitations of the CET policy. A multi-objective programming approach was used to balance the economic and environmental goals of the power market in order to determine the econo-environmental equilibrium.

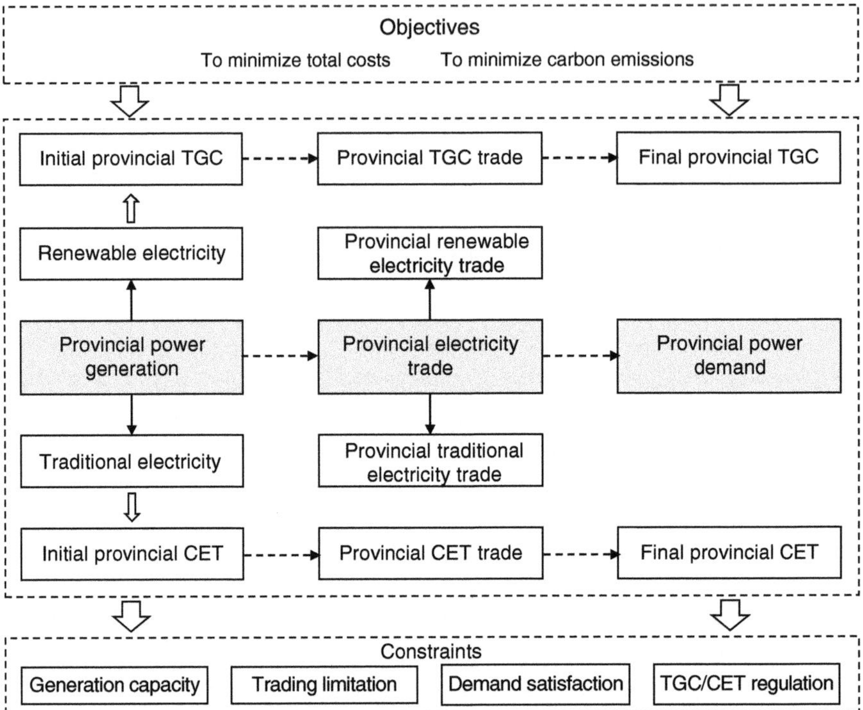

Figure 13.2 A provincial power generation and trading plan.

13.3 Modeling

In this section, an annual econo-environmental equilibrium model for a power generation and trading plan is built.

13.3.1 Assumptions

The assumptions in this chapter are as follows:

1. This is an annual power generation and trading plan for each province. Inspired by Wang et al. (2021), this chapter makes an annual plan while taking into account the coordination of TGC and CET policies in the power market at the provincial level.
2. The exact architecture of the within-province grid is not taken into account; instead, each provincial power system is treated as a single bus. Provincial power plants using the same power producing technology are treated as a single energy supplier, as in earlier policy studies (Wang et al. 2019c, 2021).
3. Power demand is assumed to be fixed and exogenous. It is believed that the decision-making process involves a trade-off between three markets: generation, trading, and demand. In China, the retail consumer power price is regulated, leading to the assumption that the power demand is constant and exogenous.

13.3.2 Notations

Indices

i Index of traditional power generation technologies, including coal ($i = 1$), nuclear ($i = 2$), and hydro ($i = 3$).

j Index of renewable power generation technologies, including wind ($j = 1$) and solar ($j = 2$).

k Index of province.

Parameters

C_i^T Unit generation cost of traditional technology i (CNY/GWh).

C_j^R Unit generation cost of renewable technology j (CNY/GWh).

IP_{ik}^T Trading price of province k's traditional electricity i (CNY/GWh).

IP_{jk}^R Trading price of province k's renewable electricity j (CNY/GWh).

EP_{ik}^T Export tariff of traditional generation technology i (CNY/GWh).

EP_{jk}^R Export tariff of renewable generation technology j (CNY/GWh).

Q_{tra}^g Unit price of TGC (CNY).

P_{tra}^e Unit price of tradable carbon emission (CNY).

$\overline{A_j^R}$ Maximum generation capacity of renewable technology j (GWh).

$\overline{A_i^T}$ Maximum generation capacity of traditional technology i (GWh).

$\overline{A_{jk}^R}$ Maximum tradable renewable electricity j in province k (GWh).

$\overline{A_{ik}^T}$ Maximum tradable traditional electricity i in province k (GWh).

IQ_k^R Total renewable power generation in province k (GWh).

IQ_k^R Non-hydro RPS policy–regulated renewable power (GWh).

$\overline{T_i}$ Transmission capacity to province k (GWh).

D Power demand of province (GWh).

LT Annual transmission loss ratio (%).

H^e Provincial free carbon emission quotas.

Q_{ini}^g Provincial initial TGC volume.

Q_{ini}^e Provincial initial CET volume.

RPS Non-hydro renewable power consumption rate (%).

α TGC conversion ratio for renewable power.

β Standard emission ratio of coal generation (%).

LT Line loss ratio (%).

E_0 Benchmark emission of coal generation (tonnes/GWh).

Decision variables

A_i^T Electricity generated by traditional technology *i* in the province (GWh).
A_j^R Total electricity generated by renewable technology *j* in the province (GWh).
D_{ik}^R Traded electricity generated by traditional technology *i* in province *k* (GWh).
D_{jk}^R Trade electricity generated by renewable technology *j* in province *k* (GWh).
Q_{tra}^g Traded TGC volume of province.
Q_{tra}^e Traded CET volume of province.

13.3.3 Power Generation and Trading Objectives

In this section, the economic performance and environmental protection objectives of the decision maker will be presented in detail.

13.3.3.1 Economic Performance

Market-based provincial incentives were necessary as electricity reform evolved in order to lessen the fiscal burden (Liu et al. 2021a). Therefore, the common decision objective is total cost minimization (Arpanahi et al. 2020), as shown in Eq. (13.1), where the generation cost C^{gen}, power trading cost C_{tra}^p, TGC trading cost C_{tra}^g and CET trading cost C_{tra}^e, are considered at the same time.

$$MinF_1 = C^{gen} + C_{tra}^p + C_{tra}^g + C_{tra}^e \tag{13.1}$$

Power generation cost (C^{gen}) is shown in Eq. (13.2), which equals to the unit cost of traditional power generation (C_i^T) and renewable power generation (C_j^R) times their generation volume A_i^T, A_j^R (Zhuo et al. 2022).

$$C^{gen} = \sum_i A_i^T C_i^T + \sum_j A_j^R C_j^R \tag{13.2}$$

Power trading cost (C_{tra}^p) was made of two parts, the traditional power trading cost B_{tra}^T and the renewable power trading cost B_{tra}^R, as shown in Eq. (13.3).

$$C_{tra}^p = B_{tra}^T + B_{tra}^R \tag{13.3}$$

The traditional and renewable power trading costs B_{tra}^T and B_{tra}^R can be further expressed by Eqs. (13.4)–(13.5), where D_{ik}^T is the trading volume of traditional power in province *k*. The positive D_{ik}^T means province *k* import traditional power, and the cost equals to the volume times import tariff (IP_{ik}^T). A negative D_{ik}^T means export behavior was executed, with the cost equal to the volume times export tariff (EP_{ik}^T). Traditional costs of renewable power have similar situation, where D_{jk}^R is the trading volume of renewable power in province *k*, with the import tariff being IP_{jk}^R and the export tariff being (EP_{jk}^R).

$$B_{tra}^T = \begin{cases} D_{ik}^R \times IP_{ik}^T, D_{ik}^R > 0 \\ 0, D_{ik}^R = 0 \\ D_{ik}^R \times EP_{ik}^T, D_{ik}^R < 0 \end{cases} \tag{13.4}$$

$$B_{tra}^{R} = \begin{cases} D_{jk}^{R} \times IP_{jk}^{R}, D_{jk}^{R} > 0 \\ 0, D_{jk}^{R} = 0 \\ D_{jk}^{R} \times EP_{jk}^{R}, D_{jk}^{R} < 0 \end{cases} \tag{13.5}$$

Provincial government TGC trading costs (C_{tra}^{g}) are as shown in Eq. (13.6). Provincial power consumption is restricted when an RPS quota is issued as power consumers can only achieve the quota goal by purchasing renewable power or green certificates (National Development and Reform Commission and Ministry of Industry and Information Technology 2022).

$$C_{tra}^{g} = Q_{tra}^{g} \times Q_{tra}^{g} \tag{13.6}$$

CET trading cost is shown in Eq. (13.7). The carbon emission quota connected with the CET policy places restrictions on conventional energy producers that enjoy comparative cost advantages. Traditional power providers may only achieve their objectives by enhancing their generation technologies or CET trading with other thermal generators after the provincial carbon emission quota is lifted.

$$C_{tra}^{e} = Q_{tra}^{e} \times P_{tra}^{e} \tag{13.7}$$

13.3.3.2 Environmental Protection

The CET policy promotes environmental protection by restricting CO_2 emissions from fossil fuel units. Therefore, CO_2 emissions related to the provincial power generation and trading plan should be minimized, as shown in Eq. (13.8).

$$MinF_2 = EF \times \left(A_1^T + \sum_k D_{1k}^T \right) \tag{13.8}$$

13.3.4 Generation and Trading Constraints

In this section, the generation capacity constraints, the demand and supply balance, the power transmission and trading limitations etc. considered by the decision maker will be presented in detail.

13.3.4.1 Renewable Power Generation Capacity Limitation

The renewable power generation (A_j^R) generated by technology j needs to be less than the generation capacity ($\overline{A_j^R}$), as shown in Eq. (13.9).

$$0 \le A_j^R \le \overline{A_j^R} \tag{13.9}$$

13.3.4.2 Traditional Power Generation Capacity Limitation

The traditional power generation (A_i^T) needs to be less than the generation capacity ($\overline{A_i^T}$), as shown in Eq. (13.10).

$$0 \le A_i^T \le \overline{A_i^T} \tag{13.10}$$

13.3.4.3 Power Demand and Supply Balance

To guarantee provincial electricity supply, a power balance constraint is formulated in Eq. (13.11). The provincial demand is represented by D; the local power supply includes local renewable power supply (A_j^R) and local traditional power supply (A_i^T). The renewable and traditional power trading volumes are represented by (D_{jk}^R) and (D_{ik}^R). LT denotes the annual transmission loss.

$$\sum_i A_i^T + \sum_j A_j^R + \left(\sum_i \sum_k D_{ik}^R + \sum_j \sum_k D_{jk}^R \right) * (1 - LT) = D \tag{13.11}$$

13.3.4.4 Power Transmission Limitation

Power purchases are limited by the capacity of transmission lines between provinces:

$$\Sigma_i D_{ik}^R + \Sigma_j D_{jk}^R \leq \overline{T_i} \times 8760 \tag{13.12}$$

13.3.4.5 Power Trading Constraints

The volume of imports from neighboring provinces is constrained due to the restricted generation capacity and RPS quotas of power generators outside the province, as shown in Eq. (13.13), where IQ_k^R is the renewable power generation of technology m. The traded traditional and renewable power are limited by the maximum tradable traditional and renewable electricity j in province k, represented by $\overline{A_{ik}^T}$ and $\overline{A_{jk}^R}$, as shown in Eqs. (13.14)–(13.15).

$$\sum_j D_{jk}^R \leq IQ_k^R - \underline{IQ_k^R} \tag{13.13}$$

$$D_{jk}^R \leq \overline{A_{jk}^R} \tag{13.14}$$

$$D_{ik}^T \leq \overline{A_{ik}^T} \tag{13.15}$$

13.3.4.6 TGC Trading Constraints

The maximum export volume is restricted since the TGC received is proportional to the generation of renewable energy, as shown in Eq. (13.16). The total provincial TGC exports are restricted by the initial TGC supply (Q_{ini}^g), as shown in Eq. (13.17), where α is the TGC conversion ratio for renewable power.

$$Q_{tra}^g \geq -Q_{ini}^g \tag{13.16}$$

$$Q_{ini}^g = \sum_j C_j^R \times \alpha \tag{13.17}$$

13.3.4.7 CET Trading Constraints

As CET is based on a free quota for coal power generation, the available export volume is limited, as shown in Eq. (13.18), where Q_{ini}^e represents the free CET quota, which can be further expressed by Eq. (13.19). A_1^T is the total electricity generated by coal, E_0 is the benchmark value for the emission, and β is the free quota ratio.

$$Q^e_{tra} \geq -Q^e_{ini} \tag{13.18}$$

$$Q^e_{ini} = A^T_1 \times E_0 \times \beta \tag{13.19}$$

13.3.4.8 RPS-Bundled TGC Consumption

According to the recent published policy of the Chinese government, the market-oriented users were required to purchase green power or TGC to fulfill the RPS quota responsibilities, as shown in Eq. (13.20), where H^{RPS} is the RPS-bundled TGC consumption, which can be expressed by Eq. (13.21). *RPS* is the regulated non-hydro renewable power consumption rate.

$$\sum_j \left(C^R_j + \sum_k D^R_{jk} \right) + Q^g_{tra} \geq H^{RPS} \tag{13.20}$$

$$H^{RPS} = \mathrm{D} \times RPS \tag{13.21}$$

13.3.4.9 CET Quota Constraints

For the provincial government, it can purchase extra CET or adjust its local coal power generation (A^T_1) to meet the CET quota requirements, but need to follow the constraint shown in Eq. (13.22), where the free CET quota (H^e) is equal to the initial CET supply (Q^e_{ini}).

$$EF \times A^T_1 - Q^e_{tra} \leq H^e \tag{13.22}$$

13.3.5 Global Model

For the provincial government, a multi-objective optimization model based on econo-environmental equilibrium can be proposed by integrating Eqs. (13.1)–(13.22). This model's structure is more thorough and methodical than the prior model for policy coordination that was suggested. First, the model is more realistic since each provincial power system is viewed as a separate bus. Second, the multi-objective optimization method is adopted to seek a global optimal solution with economic benefits and environmental impact balance in a power generation and trading plan, with which the provincial government is able to develop a provincial schedule that optimizes its total costs and carbon emissions. Third, the policy connections between power generation, power trading, power demand, and policy regulation are explored using a trading process. Fourth, each province complies with the RPS policy, and

$$MinF_1 = \sum_i A^T_i C^T_i + \sum_j C^R_j C^R_j + B^T_{tra} + B^R_{tra} + Q^g_{tra} \times Q^g_{tra} + Q^e_{tra} \times P^e_{tra}$$

$$MinF_2 = EF \times \left(A^T_1 + \sum_k D^T_{1k} \right)$$

$$
\text{s.t.}\begin{cases}
B_{tra}^{T} = \begin{cases} D_{ik}^{T} \times IP_{ik}^{T}, D_{ik}^{R} > 0 \\ 0, D_{ik}^{R} = 0 \\ D_{ik}^{R} \times EP_{ik}^{T}, D_{ik}^{R} < 0 \end{cases} \\
B_{tra}^{R} = \begin{cases} D_{jk}^{R} \times IP_{jk}^{R}, D_{jk}^{R} > 0 \\ 0, D_{jk}^{R} = 0 \\ D_{jk}^{R} \times EP_{jk}^{R}, D_{jk}^{R} < 0 \end{cases} \\
0 \leq A_{i}^{T} \leq \overline{A_{i}^{T}} \\
0 \leq A_{j}^{R} \leq \overline{A_{j}^{R}} \\
\sum_{i} A_{i}^{T} + \sum_{j} A_{j}^{R} + \left(\sum_{i} \sum_{k} D_{ik}^{R} + \sum_{j} \sum_{k} D_{jk}^{R} \right) * (1 - LT) = D \\
\sum_{i} D_{ik}^{R} + \sum_{j} D_{jk}^{R} \leq \overline{T_{i}} \times 8760 \\
\sum_{j} D_{jk}^{R} \leq IQ_{k}^{R} - IQ_{k}^{R} \\
D_{jk}^{R} \leq \overline{A_{jk}^{R}} \\
D_{ik}^{T} \leq \overline{A_{ik}^{T}} \\
Q_{tra}^{g} \geq -Q_{ini}^{g} \\
Q_{tra}^{e} \geq -Q_{ini}^{e} \\
\sum_{j} \left(C_{j}^{R} + \sum_{k} D_{jk}^{R} \right) + Q_{tra}^{g} \geq \mathrm{D} \times RPS \\
EF \times A_{1}^{T} - Q_{tra}^{e} \leq H^{e}
\end{cases} \tag{13.23}
$$

trading limitations are issued to ensure coordinated regional development. As a result, the model creates a power generation plan that encourages local investments in renewable energy as well as a power trading plan to address any concerns with wind and solar power curtailment. The suggested study framework offers a novel method for coordinating policy and can be further examined to investigate the best combination of policies for low-carbon transitions, such as the carbon tax and carbon cap. As a result, the suggested model offers provincial governments a workable and efficient low-carbon transition strategy. The global mathematical model is shown in Eq. (13.23).

13.3.6 Model Solving

To describe actual annual provincial power generation and trading that scientifically and accurately considers both the TGC and CET policies, an econo-environmental equilibrium multi-objective plan model is established. However, because of conflicts, it is very difficult to determine an optimal solution that satisfies all objectives. To solve multi-objective models, the ε-constraint method is applied to transform

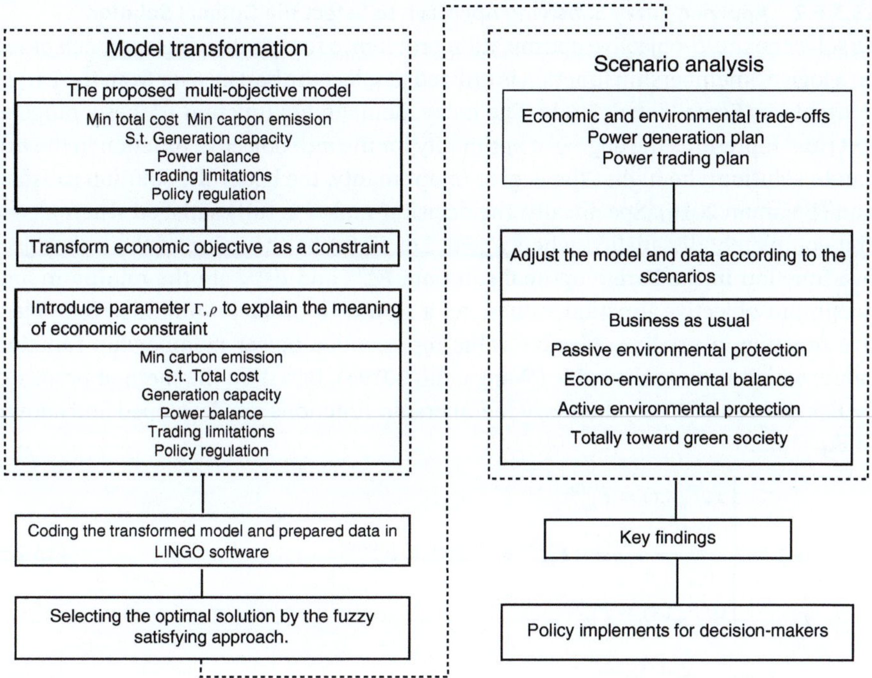

Figure 13.3 Flowchart of solution process.

the mathematical model into its equivalent single-level form. The fuzzy satisfying approach is introduced to simulate the decision-makers' attitudes and obtain the optimal solution, as shown in Figure 13.3.

13.3.6.1 Model Transformation Process

As in the previous work (Pramanik and Roy 2008), the environmental goal (f_2) is usually taken as the main objective and the economic goal (f_1) is taken as the constraint. A detailed decision-makers' survey can assist in determining a reasonable upper-bound value. Here, two parameters Γ and ρ are introduced to explain the practical meaning of the bound. With the two parameters, the model can be transformed as in Eq. (13.24).

$$MinF_2 = EF \times \left(A_1^T + \sum_k D_{1k}^T \right)$$
$$s.t. \begin{cases} \sum_i A_i^T C_i^T + \sum_j C_j^R A_j^R + B_{tra}^T + B_{tra}^R + Q_{tra}^g \times Q_{tra}^g + Q_{tra}^e \times P_{tra}^e \leq \Gamma\rho \\ \text{Eqs. (13.4)–(13.23)} \end{cases} \tag{13.24}$$

13.3.6.2 Applying Fuzzy Satisfying Approach to Select the Optimal Solution

To solve the multi-objective optimization problem, a fuzzy satisfying approach based on a logistic membership function is utilized to select the best answer from the Pareto optimal set (Esmaili et al. 2011). The fuzzy membership function μ_m^n has a range of [0,1] and represents the degree of optimality for the mth objective function in the nth Pareto solution; the higher the degree of optimality, the better the solution satisfaction (Balaman 2016). Specifically, the decision-maker is fully satisfied when μ_m^n=1, and completely dissatisfied when μ_m^n=0. Let $F_m^n(x)$ be the value of the mth objective function in the Pareto optimal solution, F_m^{Min} and F_m^{Max} are the minimum and maximum objective function values. As a result, in order to minimize the objective function, a smaller objective value represents a bigger membership function value, which is more desirable (Wang et al. 2019a). Based on the method proposed by Esmaili et al. (2009), the fuzzy membership functions are developed and shown in Eq. (13.25).

$$\mu_m^n = \begin{cases} 1; F_m(x) = F_m^{Min} \\ \dfrac{F_m^{Max} - F_m^n(x)}{F_m^{Max} - F_m^{Min}}; F_m^{Min} < F_m(x) < F_m^{Max} \\ 0; F_m(x) = F_m^{Max} \end{cases} \tag{13.25}$$

$$\mu^{\text{optimal}} = \frac{\sum_m A_m \mu_m^n}{\sum_m A_m} \tag{13.26}$$

The preferred weight vectors for the cost minimization and emission minimization objectives represent the relative importance of each objective to the decision-maker (Pramanik and Roy 2008). For each solution n, the membership degree $\mu^{optimal}$ is calculated based on its individual membership functions and the added weight factors, as shown in Eq. (13.26).

13.4 Case Study

A practical example of a provincial generation and trading plan from Guangdong, China, is provided in this section to demonstrate the applicability and efficiency of the proposed optimization method.

13.4.1 Case Description

Guangdong Province is a part of China's Southern Power Grid Region, which includes five provinces: Guangdong, Guangxi, Hainan, Guizhou, and Yunnan, see Figure 13.4. Because of transmission line limitations, Guangdong can only conduct power trades with Guangxi, Yunnan, and Guizhou. The total electricity consumption in Guangdong Province was 786.663 TWh in 2021. Guangdong Province's power generation is still dominated by coal power, accounting for 56.6% or 585.2 TWh of total power generation in 2021 (Guangdong Power Exchange Center 2022), with around 60% of its generators depending on old technology. As

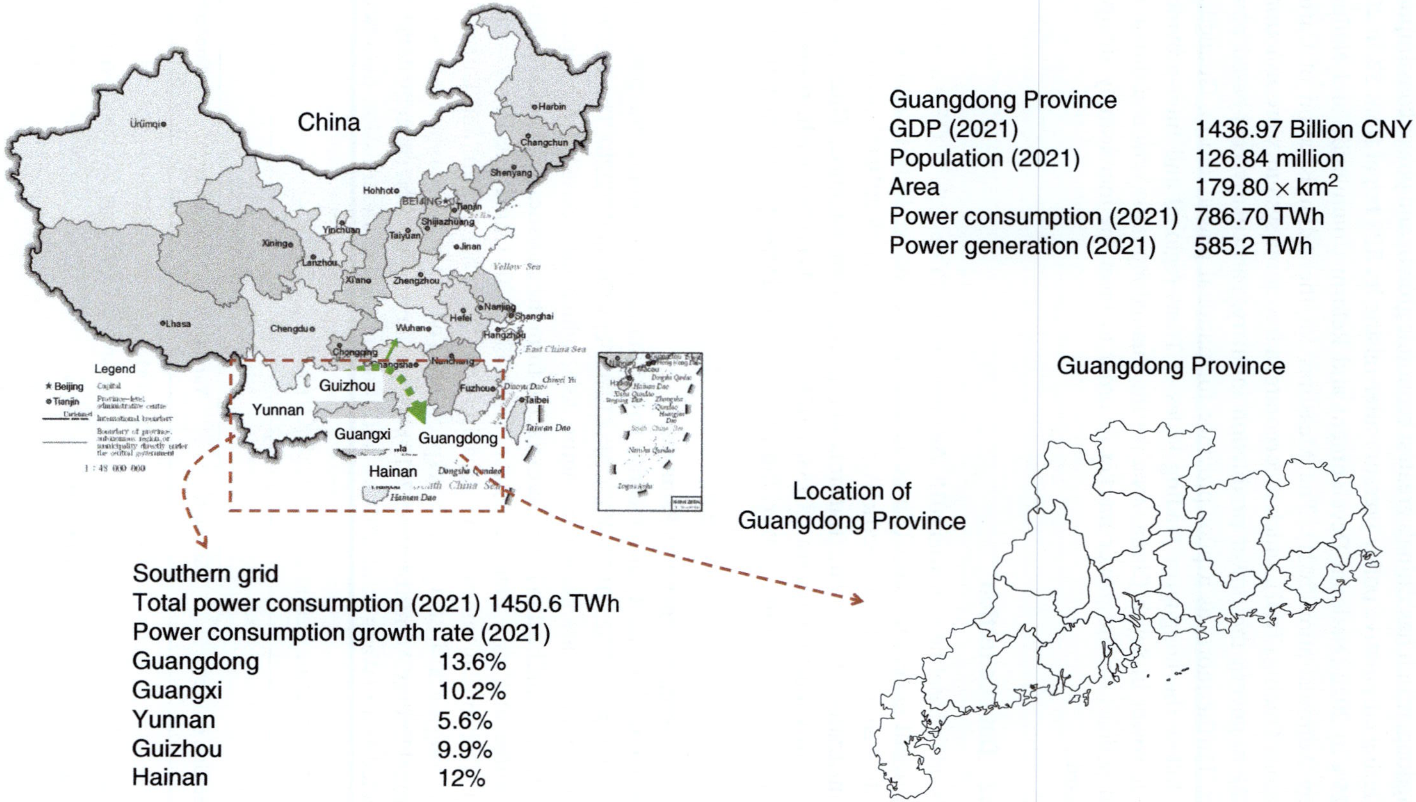

Figure 13.4 Basic description for Guangdong province.

a result, Guangdong Province requires a low-carbon transition power strategy. Guangdong, which has China's greatest economic growth and power consumption, is suffering cumulative policy pressure after raising its RPS target from 5% in 2021 to 6.25% in 2022 (National Development and Reform Commission and National Energy Administration 2021). The Measures for the Administration of Carbon Emissions Trading (Trial), which incorporates a free quota system, were announced in 2021 to govern the power production industry, particularly coal-based power plants. Furthermore, as a pilot province for national power reform, Guangdong must follow the Long-term Trading Rules of Trans-regional and Inter-provincial Electric Power in South China. As a result, Guangdong must now coordinate the mixed policies on the power market in order to meet the increasingly stringent standards.

13.4.2 Data Collection

This study's baseline year was 2018. As indicated in Table 13.1, LCOE was utilized to represent generation cost (Xu et al. 2021), and average feed-in-tariff statistics were derived from China Electric Power Statistics Yearbook 2019 (National Energy Administration 2019). The transmission prices from Guangxi, Guizhou, and Yunnan to Guangdong were 0.0755, 0.0755, and 0.0595 CNY/kWh, respectively, according to the China Power Industry Annual Development Report (China Electricity Council 2022a).

13.4.2.1 Technical Generation Parameters

According to the 2021 China Electric Power Statistical Yearbook, the power consumption is set at 6.32×10^5 GWh (China Electricity Council 2022b). The maximum power generation are shown in Figure 13.5, which are represented by installed capacity times utilization hours, as extracted from the statistics yearbooks (Editorial Committee of China Electric Power Yearbook 2020; China Electricity Council 2020). The annual transmission capacity from Guizhou, Yunan, and Guangxi to Guangdong are 20.4, 36.4, 1.8 GW, respectively. The transmission line loss was 6.02% (National Energy Administration 2019), the power transmission capacity data were extracted from (Wang et al. 2019b), and the carbon dioxide emission parameters

Table 13.1 Average feed-in-tariff in 2018 (CNY/kWh).

	Provinces			
Generation sources	**Guangxi**	**Yunan**	**Guizhou**	**Guangdong**
Coal	0.399	0.413	0.333	0.441
Hydro	0.258	0.199	0.292	0.279
Nuclear	0.379	—	—	0.414
Wind	0.607	0.456	0.534	0.603
Solar	0.56	0.734	0.948	0.978

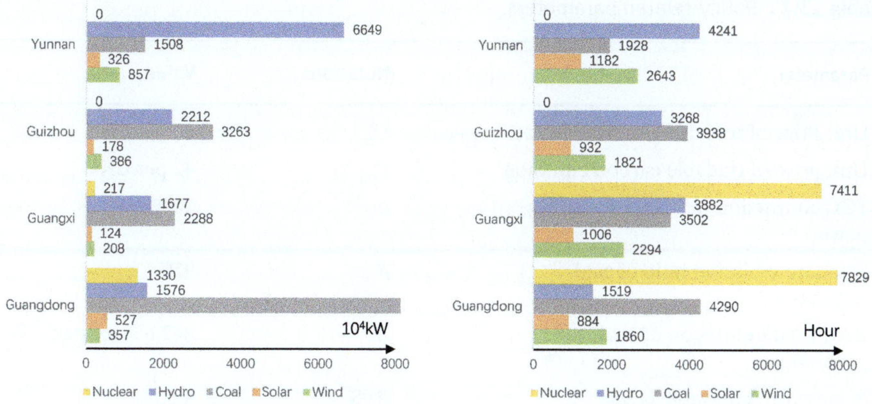

(1) Installed capacity (2) Utilization hours

Figure 13.5 Provincial installed capacity and utilization hour in 2018.

were collected from a published paper (Wang et al. 2021). Coal generation emission factor is 859.4 tonnes/GWh.

13.4.2.2 Policy-Related Parameters

The TGC and CET prices were extracted from the websites (Green Energy 2022; Tanjiaoyi.com 2022), and the conversion rates were collected from (National Development and Reform Commission 2017; Ministry of Ecology and Environment of the People's Republic of China 2020b). The benchmark emissions were calculated using the benchmark values (Zhang et al. 2023), and the coal generation unit data were extracted from the Statistical Yearbook of Electric Power (China Electricity Council 2022b). The provincial non-hydro renewable energy consumption targets were obtained from the published policy from the website (National Development and Reform Commission and National Energy Administration 2019). Table 13.2 shows the computed non-hydro RPS policy–regulated renewable power volume and the total renewable power output.

13.4.3 Calculation Results and Analysis

This section analyzes the results of Guangdong Province's generation and trading plan that includes both the TGC and CET policy considerations.

13.4.3.1 Results of Power Generation and Trading

The power generation and trading strategy was investigated from power supply to consumption in order to coordinate the interprovincial TGC and CET power markets. The multi-objective model was used to calculate the power generation and trade results for Guangdong Province, as shown in Tables 13.3 and 13.4 and Figure 13.6.

Coal power generation still leads Guangdong's local power supply with 302.98×10^3 GWh, nuclear power contributes $104.13 \times \times 10^3$ GWh, hydropower

Table 13.2 Policy-related parameters.

Parameter	Notations	Value
Unit Price of tradable green certificate	Q^{g}_{tra}	50 per CNY
Unit price of tradable carbon emission	P^{e}_{tra}	42 per CNY
TGC conversion ratio of renewable power	α	100%
Standard emission ratio of coal generation	β	95%
Benchmark emission of coal generation	E_0	842.658 tonnes/GWh
Non-hydro renewable power consumption rate	RPS	3.50%
RPS policy-regulated renewable power in Guangxi	$\underline{IQ^{R}_{1}}$	6812 GWh
RPS policy-regulated renewable power in Yunnan	$\underline{IQ^{R}_{2}}$	19,308.5 GWh
RPS policy-regulated renewable power in Guizhou	$\underline{IQ^{R}_{3}}$	6669 GWh
Total renewable generation in Guangxi	IQ^{R}_{1}	6018.96 GWh
Total renewable generation in Yunnan	IQ^{R}_{2}	26,503.83 GWh
Total renewable generation in Guizhou	IQ^{R}_{3}	8688.02 GWh

Table 13.3 Power generation plan for Guangdong Province (10^3 GWh).

Generation sources	Maximum generation capacity	Generation plan
Coal	349	302.98
Hydro	23.94	23.94
Nuclear	104.13	104.13
Wind	6.64	6.64
Solar	4.66	0

contributes 23.94×10^3 GWh, and wind contributes 6.64×10^3 GWh under the mixed policy scheme. Guangdong imports 100×10^3 GWh of hydropower and 19.02×10^3 GWh of wind power from Yunnan, 15.77×10^3 GWh of hydropower from Guangxi, and 72.29×10^3 GWh of hydropower from Guizhou to meet its power consumption and policy regulation requirements.

The local power supply is still dominated by coal-based power generation at 69%, owing to the fact that coal power production in Guangdong Province has the highest

Table 13.4 Power trading plan for Guangdong Province (10^3 GWh).

Generation sources	Guangxi	Yunnan	Guizhou	Sum
Coal	0	0	0	0
Hydro	15.77	100	72.29	188.06
Nuclear	0	0	0	0
Wind	0	19.02	0	19.02
Solar	0	0	0	0
Sum	15.77	119.02	72.29	207.08

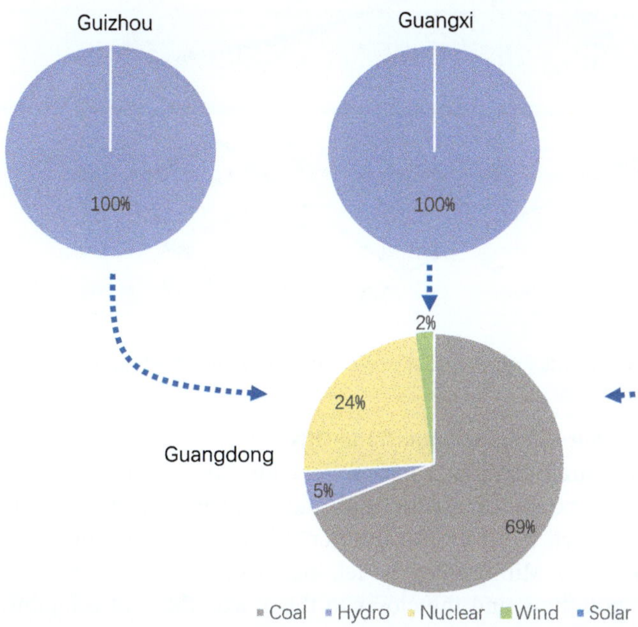

Figure 13.6 Power generation and trading plan for Guangdong province.

installed capacity and cost advantages. Second, with the exception of coal power, traditional power generation has reached its maximum generating capacity; that is, coal power output is curtailed due to environmental carbon emission regulations, while other generation technologies do not emit any emissions. Third, with imports totaling 119.02×10^3 GWh, Guangdong and Yunnan have the highest trading volumes. Yunnan Province has vast hydroelectric resources and a price advantage in wind power production. Fourth, the hydropower trading volume is the biggest among power imports, at 188.06×10^3 GWh, demonstrating that when renewable energy costs do not have a comparative advantage, hydropower is the primary alternative to meet econo-environmental demand.

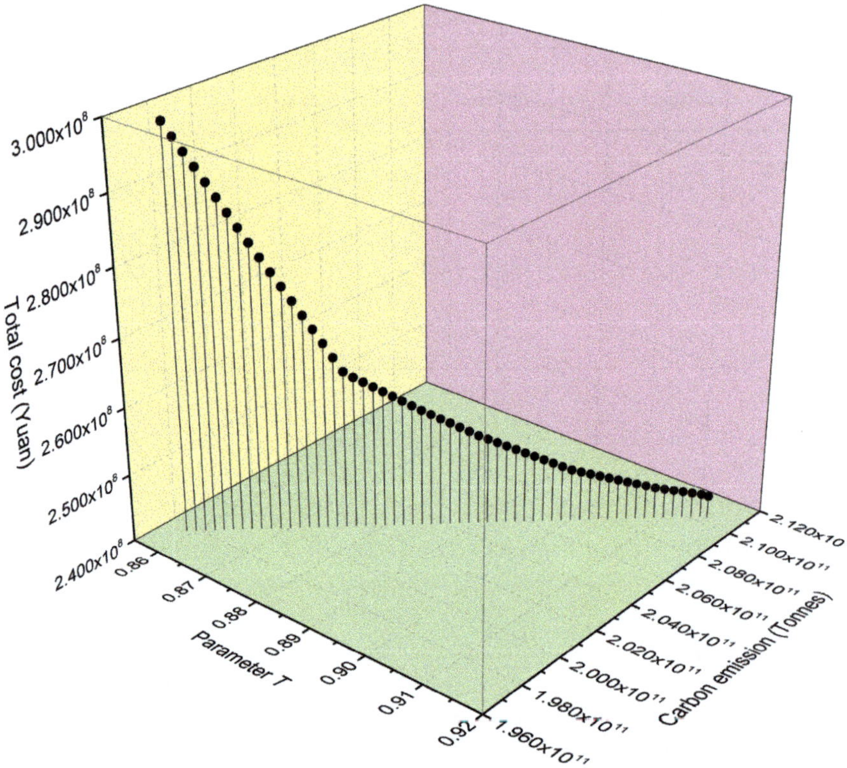

Figure 13.7 Pareto-optimal fronts for provincial power generation and trading plan, considering TGC and CET policies.

13.4.3.2 Results of Economic–Environmental Trade-Offs

Figure 13.7 depicts the Pareto-optimal solution sets for the provincial econo-environmental equilibrium power generation and trading plan, considering both the TGC and CET rules. Each point on the frontier was obtained by simulating various Γ settings. In this Min–Min frontier model, increasing the economic input reduces environmental pollution, and the closer to the *Y*-axis the unit economic input has a bigger effect. Setting the same attitude parameters for economic and environmental goals results in an econo-environmental balance, with CO_2 emissions of 260.39 billion tonnes and total expenses of 202.5 billion CNY.

13.5 Discussion

In this section, five scenarios with different policy considerations are studied and some key findings are obtained.

13.5.1 Scenario Analyses

This section examines five scenarios with varying policy considerations to assess the implications on electricity generation and trading strategies, as well as to compare cost changes and carbon emissions.

Table 13.5 Optimization models under different scenarios.

Scenarios	Details	Objective weight factor
Scenario 1 (S1)	Business as usual	min $F1$ (total costs)
Scenario 2 (S2)	Passive environment protection	min $0.75\,F1 + 0.25F2$
Scenario 3 (S3)	Econo-environmental balance	min $0.50\,F1 + 0.50F2$
Scenario 4 (S4)	Active environment protection	min $0.25\,F1 + 0.75F2$
Scenario 5 (S5)	Totally toward green society	min $F2$ (CO_2 emission)

13.5.1.1 Scenario Settings

To analyze the provincial government's economic and environmental trade-offs for their CET and TGC trading plans, the five scenarios in Table 13.5 are examined: Scenario 1 (S1) is business as usual, i.e. an optimized power generation and trading plan design centered on minimizing total annual expenses, for which the provincial government exclusively considers economic factors; Scenario 2 (S2) focuses on passive environmental conservation; Scenario 3 (S3) aims for an econo-environmental balance, with the provincial government's stance toward economic and environmental goals consistent; Scenario 4 (S4) focuses on active environmental conservation, with the provincial government preferring environmental performance over economic profit; and Scenario 5 (S5), which is entirely focused on a green society in which the government pursues environmental advantages at all costs.

13.5.1.2 Economic and Environmental Trade-Offs Under Different Scenario

The total costs and carbon emissions for the five scenarios are presented in Table 13.6. S1 has the lowest overall cost of 197.5 billion CNY but the largest carbon emissions of nearly 300 million tonnes, whereas S5 has the highest total cost of 212.5 billion CNY but the best environmental performance with only 243.3 million tonnes of CO_2. S2 costs 197.9 billion CNY and emits 296.6 million tonnes of CO_2, whereas S4 costs 209.4 billion CNY and emits 244.4 million tonnes of CO_2. Overall, environmental rules reduce carbon emissions while increasing prices.

Table 13.6 Total cost and carbon emission of different scenarios in 2018.

Scenarios	Total cost (billion CNY)	Carbon emission (million tonnes)
S1	197.5	300
S2	197.9	296.6
S3	202.5	260.3
S4	209.4	244.4
S5	212.5	243.3

The results demonstrate that S3 has the optimum econo-environmental balance, as it can reduce carbon emissions by 13.2% compared to S1 and 11.5% compared to S2, while only raising total expenses by 4.7%. S2 costs 6.8% more than S3 for comparable emissions reductions. According to the analysis, every unit cost rise in S3 can cut carbon emissions by 1.5 units, proving the efficacy of this scenario in achieving a balance between economic and environmental goals.

13.5.1.3 Power Generation Results Under Different Scenarios

The power generation plans in different scenarios are compared in Figure 13.8. In all scenarios, coal power continues to dominate power generation in Guangdong province, with solar power having the lowest generation proportion. At 422.49×10^3 GWh, S1 has the largest overall power generation while S5 has the lowest. S1 generates the most coal power (349.078×10^3 GWh) compared to S2 to S5; S2 generates 345.06×10^3 GWh; S3 generates 302.98×10^3 GWh, and S4 generates 284.39×10^3 GWh. At 4.66×10^3 GWh; S4 and S5 have the largest solar power generation. The coal proportion in the power generation structure continues to fall, from 72.2% in S1 to 67% in S5.

13.5.1.4 Power Trading Results Under Different Scenarios

The power trading plans under the different scenarios are shown in Figure 13.9. The province of Guangdong has the highest trading volume with Yunnan and the lowest with Guangxi, with hydropower trading being the highest and solar power trading being the lowest. Hydropower trading grows from S2 to S5 compared to S1, and while renewable energy is included from S3, only wind power imports from Yunnan province increase. After importing all tradable wind power, Guangdong imports solar, largely from Yunnan province. S4 imports hydropower from Guizhou province and subsequently solar electricity from Yunnan province; thus the

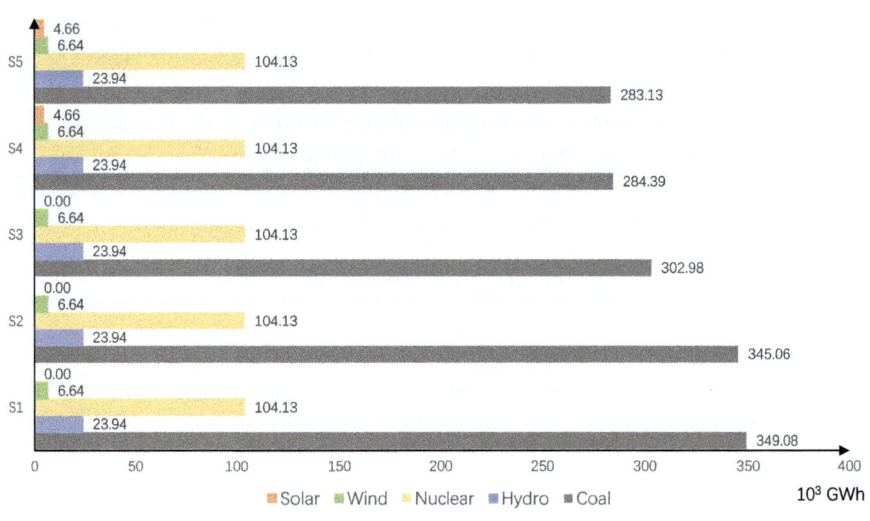

Figure 13.8 Generation plan under different scenarios.

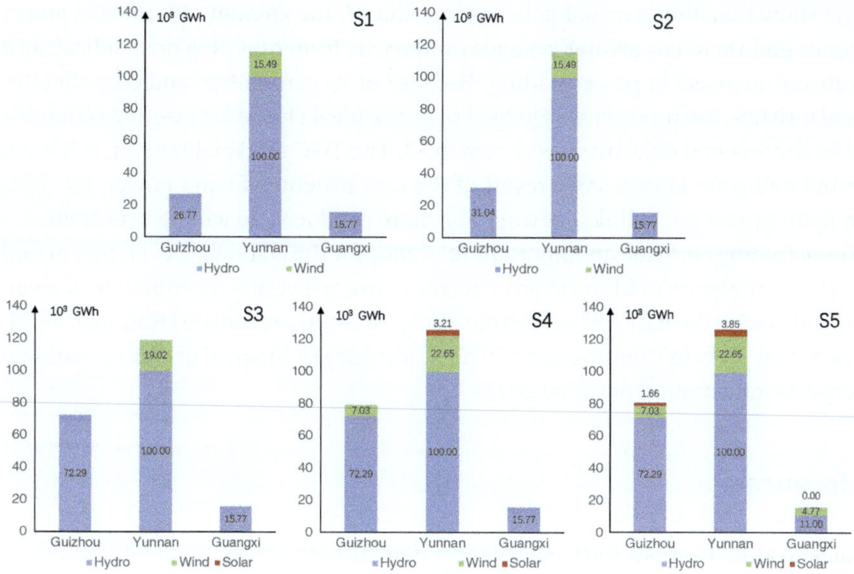

Figure 13.9 Power trading plans under different scenarios.

imported power structure from Guangxi alters to reach the greatest environmental benefits. S5 replaces Guangxi hydropower imports with nuclear (10.99×10^3 GWh) and wind power (4.77×10^3 GWh) imports. As a result, if opinions regarding the targeted shift, so do the imported power structures.

13.5.2 Key Findings

First, when mixed policies are implemented, the proposed model provides a balance between economic and environmental goals. Carbon emission reductions were accompanied by an increase in total expenditures, as demonstrated in the S1–S5 analyses. S3, for example, resulted in a 2.3% rise in overall expenses to achieve a 13.2% reduction in emissions when no explicit economic or environmental choices were present. The model was also able to accurately estimate a decline in coal power generation in Guangdong from 314.17×10^3 to 302.98×10^3 GWh in 2018.

Second, depending on the government's preferences, mixed policies can reduce the amount of coal power generation while conditionally promoting solar power generation. The mixed CET and TGC policy environment leads to a drop in local coal power output under S1 through S5. Because solar power generation is more expensive than other technologies, importing solar power from other provinces is a more practical choice. If decision-makers prioritize the environment over the economy, as shown in S4 and S5, the solar power share must increase, implying that the renewable energy supply costs would rise in order to satisfy the power system's carbon neutrality targets (Zhuo et al. 2022).

Finally, while diversified policies can boost power trading volumes, decision-makers' preferences can influence trades for various generation technologies.

S1–S5 show that after a mixed policy is introduced, the amount of local coal power declines and there is a greater reliance on imports from other regions, indicating a significant increase in power trading. Because of its carbon-free and cost-effective characteristics, hydropower was judged to be the ideal choice for meeting economic and environmental objectives, as shown in S2. The TGC market, however, is limited to wind and solar energy. As a result of the cost benefits of wind power, the TGC non-hydropower quota linkages would be more conducive to wind power trade.

These findings provide an important reference for the National Development and Reform Commission of China to promote the convergence and coordination of green certificates with the national carbon emissions trading mechanism (National Development and Reform Commission of China, Ministry of Finance of China, National Energy Administration of China 2023).

References

Arpanahi, M.K., Golshan, M.H., and Siano, P. (2020). A comprehensive and efficient decentralized framework for coordinated multiperiod economic dispatch of transmission and distribution systems. *IEEE Systems Journal* 15 (2): 2583–2594.

Balaman, Ş.Y. (2016). Investment planning and strategic management of sustainable systems for clean power generation: an ε-constraint based multi objective modelling approach. *Journal of Cleaner Production* 137: 1179–1190.

Barbose, G., Bird, L., Heeter, J. et al. (2015). Costs and benefits of renewables portfolio standards in the United States. *Renewable and Sustainable Energy Reviews* 52: 523–533.

Bistline, J., Abhyankar, N., Blanford, G. et al. (2022). Actions for reducing us emissions at least 50% by 2030. *Science* 376 (6596): 922–924.

Chang, L., Hao, X., Song, M. et al. (2020). Carbon emission performance and quota allocation in the Bohai Rim Economic Circle. *Journal of Cleaner Production* 258: 120722.

China Electricity Council (2020). *China Electric Power Statistics Yearbook 2019*. China Statistics Press.

China Electricity Council (2022a). *China Power Industry Annual Development Report 2021*. China Building Material Press.

China Electricity Council (2022b). *China Electric Power Statistics Yearbook 2021*. China Statistics Press.

Editorial Committee of China Electric Power Yearbook (2020). *2019 China Electric Power Yearbook*. China Electric Power Press.

Esmaili, M., Amjady, N., and Shayanfar, H.A. (2011). Multi-objective congestion management by modified augmented ε-constraint method. *Applied Energy* 88 (3): 755–766.

Esmaili, M., Shayanfar, H.A., and Amjady, N. (2009). Multi-objective congestion management incorporating voltage and transient stabilities. *Energy* 34 (9): 1401–1412.

Feng, T.T., Li, R., Zhang, H.M. et al. (2021). Induction mechanism and optimization of tradable green certificates and carbon emission trading acting on electricity market in China. *Resources, Conservation and Recycling* 169: 105487.

Feng, T.T., Yang, Y.S., and Yang, Y.H. (2018). What will happen to the power supply structure and CO_2 emissions reduction when TGC meets CET in the electricity market in China? *Renewable and Sustainable Energy Reviews* 92: 121–132.

Fischer, C. and Preonas, L. (2010). Combining policies for renewable energy: is the whole less than the sum of its parts? *Resource for the future discussion paper* (10–19) 2010.

Fischer, C., Preonas, L., and Newell, R.G. (2017). Environmental and technology policy options in the electricity sector: are we deploying too many? *Journal of the Association of Environmental and Resource Economists* 4 (4): 959–984.

Green Energy (2022). http://www.greenenergy.org.cn (accessed 3 March 2022).

Guangdong Power Exchange Center (2022). Annual report of Guangdong power market for 2021. https://pm.gd.csg.cn/views/page/xwzxCont-10902.html (accessed 8 February 2022).

He, Y., Guo, S., Zhou, J. et al. (2021). The quantitative techno-economic comparisons and multi-objective capacity optimization of wind-photovoltaic hybrid power system considering different energy storage technologies. *Energy Conversion and Management* 229: 113779.

Hille, E., Althammer, W., and Diederich, H. (2020). Environmental regulation and innovation in renewable energy technologies: does the policy instrument matter? *Technological Forecasting and Social Change* 153: 119921.

Hu, Y., Ren, S., Wang, Y., and Chen, X. (2020). Can carbon emission trading scheme achieve energy conservation and emission reduction? Evidence from the industrial sector in China. *Energy Economics* 85: 104590.

International Energy Agency (2022a). Greenhouse Gas Emissions from Energy. https://www.iea.org/data-and-statistics/data-product/greenhouse-gas-emissions-from-energy (accessed 20 October 2023).

International Energy Agency (2022b). Enhancing China's ETS for carbon neutrality: focus on the power sector.

Li, T., Gao, C., Chen, T. et al. (2022). Medium and long-term electricity market trading strategy considering renewable portfolio standard in the transitional period of electricity market reform in Jiangsu, China. *Energy Economics* 107: 105860.

Li, Z. and Wang, J. (2022). Spatial spillover effect of carbon emission trading on carbon emission reduction: empirical data from pilot regions in China. *Energy* 251: 123906.

Li, X., Wang, W., Wang, H. et al. (2020). Dynamic environmental economic dispatch of hybrid renewable energy systems based on tradable green certificates. *Energy* 193: 116699.

Li, Z., Zhao, Y., Wang, X. et al. (2018). Cost analysis of perovskite tandem photovoltaics. *Joule* 2 (8): 1559–1572.

Liang, H., Zeng, Y., Jiang, X., and Li, Y. (2023). Dynamic evaluation of low-carbon development in China's power industry and the impact of carbon market policies. *Heliyon* 9: e13467.

Liu, Z., Deng, Z., He, G. et al. (2022). Challenges and opportunities for carbon neutrality in China. *Nature Reviews Earth & Environment* 3 (2): 141–155.

Liu, Z.F., Li, L.L., Liu, Y.W. et al. (2021b). Dynamic economic emission dispatch considering renewable energy generation: a novel multi-objective optimization approach. *Energy* 235: 121407.

Liu, J., Yang, H., and Zhou, Y. (2021a). Peer-to-peer trading optimizations on net-zero energy communities with energy storage of hydrogen and battery vehicles. *Applied Energy* 302: 117578.

Meckling, J., Lipscy, P.Y., Finnegan, J.J., and Metz, F. (2022). Why nations lead or lag in energy transitions. *Science* 378 (6615): 31–33.

Meng, X. and Yu, Y. (2023). Can renewable energy portfolio standards and carbon tax policies promote carbon emission reduction in China's power industry? *Energy Policy* 174: 113461.

Ministry of Ecology and Environment of the People's Republic of China (2020a). Measures for the administration of carbon emission trading (for trial implementation) (in Chinese). http://www.gov.cn/zhengce/zhengceku/2021-01/06/content_5577360.htm (accessed 31 December 2020).

Ministry of Ecology and Environment of the People's Republic of China (2020b). Implementation plan for setting and allocating the total amount of national carbon emission trading quotas for 2019–2020 (power generation industry) (in Chinese). https://www.mee.gov.cn/xxgk2018/xxgk/xxgk03/202012/W020201230736907121045.pdf (accessed 30 December 2020).

National Development and Reform Commission and Ministry of Industry and Information Technology (2022). Notice on issuing the implementation plan for promoting green consumption (in Chinese). http://www.gov.cn/zhengce/zhengceku/2022-01/21/content_5669785.htm (accessed 18 January 2022).

National Development and Reform Commission and National Energy Administration (2019). Notice on establishing and perfecting the guarantee mechanism of renewable energy and electricity consumption (in Chinese). http://www.gov.cn/zhengce/zhengceku/2019-09/25/content_5432993.htm (accessed 10 May 2019).

National Development and Reform Commission and National Energy Administration (2021). Notice on the responsibility weight of renewable energy power consumption in 2021 and related matters (in Chinese). http://www.gov.cn/zhengce/zhengceku/2021-05/26/content_5612441.htm (accessed 21 May 2021).

National Development and Reform Commission, Ministry of Finance of the People's Republic of China, and National Energy Administration (2017). Notice on the trial implementation of the renewable energy green power certificate issuance and voluntary subscription trading system (in Chinese). http://www.nea.gov.cn/2017-02/06/c_136035626.htm (accessed 2 July 2017).

National Development and Reform Commission of China, Ministry of Finance of China, National Energy Administration of China (2023). Notice on doing a good job in the full coverage of renewable energy green power certificates to promote renewable energy power consumption. https://www.ndrc.gov.cn/xwdt/tzgg/202308/t20230803_1359093.html (accessed 20 October 2023).

National Energy Administration (2019). 2018 National electricity price supervision bulletin (in Chinese). http://www.nea.gov.cn/2019-11/05/c_138530255.htm (accessed 4 November 2019).

Pan, X., Ai, B., Li, C. et al. (2019). Dynamic relationship among environmental regulation, technological innovation and energy efficiency based on large scale provincial panel data in China. *Technological Forecasting and Social Change* 144: 428–435.

Pramanik, S. and Roy, T.K. (2008). Multiobjective transportation model with fuzzy parameters: priority-based fuzzy goal programming approach. *Journal of Transportation Engineering* 8 (3): 40–48.

Qin, Q., Liu, Y., and Huang, J.-P. (2020). A cooperative game analysis for the allocation of carbon emissions reduction responsibility in China's power industry. *Energy Economics* 92: 104960.

Rahman, M.M., Nepal, R., and Alam, K. (2021). Impacts of human capital, exports, economic growth and energy consumption on CO2 emissions of a cross-sectionally dependent panel: evidence from the newly industrialized countries (NICs). *Environmental Science and Policy* 121: 24–36.

Sakthivel, V.P., Suman, M., and Sathya, P.D. (2021). Combined economic and emission power dispatch problems through multi-objective squirrel search algorithm. *Applied Soft Computing* 100: 106950.

Tanjiaoyi.com (2022). http://www.tanjiaoyi.com (accessed 3 May 2022).

Wang, H., Su, B., Mu, H. et al. (2019c). Optimization of electricity generation and interprovincial trading strategies in southern China. *Energy* 174: 696–707.

Wang, D., Wan, K., Song, X., and Liu, Y. (2019a). Provincial allocation of coal de-capacity targets in China in terms of cost, efficiency, and fairness. *Energy Economics* 78: 109–128.

Wang, G., Zhang, Q., Li, Y., and Mclellan, B.C. (2019b). Efficient and equitable allocation of renewable portfolio standards targets among China's Provinces. *Energy Policy* 125: 170–180.

Wang, G., Zhang, Q., Su, B. et al. (2021). Coordination of tradable carbon emission permits market and renewable electricity certificates market in China. *Energy Economics* 93: 105038.

Xu, Y., Yang, Z., and Yuan, J. (2021). The economics of renewable energy power in China. *Clean Technology and Environmental Policy* 23 (4): 1341–1351.

Yan, S., Wang, W., Li, X., and Zhao, Y. (2022). Research on a cross-regional robust trading strategy based on multiple market mechanisms. *Energy* 261: 125253.

Yi, B.W., Xu, J.H., and Fan, Y. (2019). Coordination of policy goals between renewable portfolio standards and carbon caps: a quantitative assessment in China. *Applied Energy* 237: 25–35.

Yu, X., Dong, Z., Zhou, D. et al. (2021). Integration of tradable green certificates trading and carbon emissions trading: how will Chinese power industry do? *Journal of Cleaner Production* 279: 123485.

Zhang, H., Zhang, D., and Zhang, X. (2023). The role of output-based emission trading system in the decarbonization of China's power sector. *Renewable and Sustainable Energy Reviews* 173: 113080.

Zhao, X.G., Xu, L., and Zhou, Y. (2022). How to promote the effective implementation of China's renewable portfolio standards considering non-neutral technology? *Energy* 238: 121748.

Zhuo, Z., Du, E., Zhang, N. et al. (2022). Cost increase in the electricity supply to achieve carbon neutrality in China. *Nature Communications* 13 (1): 1–13.

14

Emerging Hybrid Energy Storage Systems

> *Across the globe, power systems are experiencing a period of unprecedented change. Low-cost renewable electricity is spreading and there is a growing urgency to boost power system resilience and enhance digitalization. This requires stockpiling renewable energy on a massive scale, notably in developing countries, which makes energy storage fundamental.*
>
> How battery energy storage can power us to Net Zero
> World Economic Forum. S ep. 2023

As renewable energy has ushered in an unprecedented development opportunity due to the rapid consumption of fossil energy and the increasingly severe impact of climate change, the third round of the energy revolution, characterized by large-scale utilization of renewable energy, is booming around the world. Therefore, converting the dependence of electricity production on traditional fossil energy to clean energy and ensuring a diversified, stable, efficient, and clean power supply are crucial to achieving the net-zero emission goal and promoting the transformation of the power structure. Due to the ease of resource availability, high technological maturity, and complete policy and market support, wind and solar power have become the most commercially potential, most dynamic, and most concerned renewable energy sources. However, unlike traditional controllable hydropower and thermal power, wind and solar generation have a strong dependence on natural resources, which makes the power output highly uncertain, resulting in waste of clean energy, loss of investment costs, and challenges to the safe and stable operation of the power grid. Therefore, it is urgent to alleviate the fluctuation of wind and solar power and solve the problem of large-scale utilization of renewable energies.

Storage technologies are accepted as one of the most important and efficient ways to facilitate stable grid operation and overcome renewable energy–related limitations. In the traditional power system, fossil fuels and water are tangible entities that are easy to store, and energy storage is mainly configured on the generation

Sustainable Hybrid Energy Systems: Carbon Neutral Approaches, Modeling, and Case Studies,
First Edition. Jiuping Xu and Fengjuan Wang.

side, such as coal yards and reservoirs. The power generation system relies on energy storage facilities to provide regulation capability for the system. However, with the increasing proportion of fluctuating wind and solar energy generation, conventional regulation capacity is decreasing, and new types of energy storage need to be introduced as a source of regulation capacity. On the power side, it assumes the role of smoothing the fluctuation of new energy processing, frequency regulation, and peak regulation; on the grid side, it provides system backup and relieves transmission and substation equipment blockage; on the user side, it can improve power quality and participate in demand response.

In recent years, a variety of storage technologies have emerged, which can be classified as electrical, electrochemical, chemical, mechanical, and thermal storage systems, as shown in Figure 14.1.

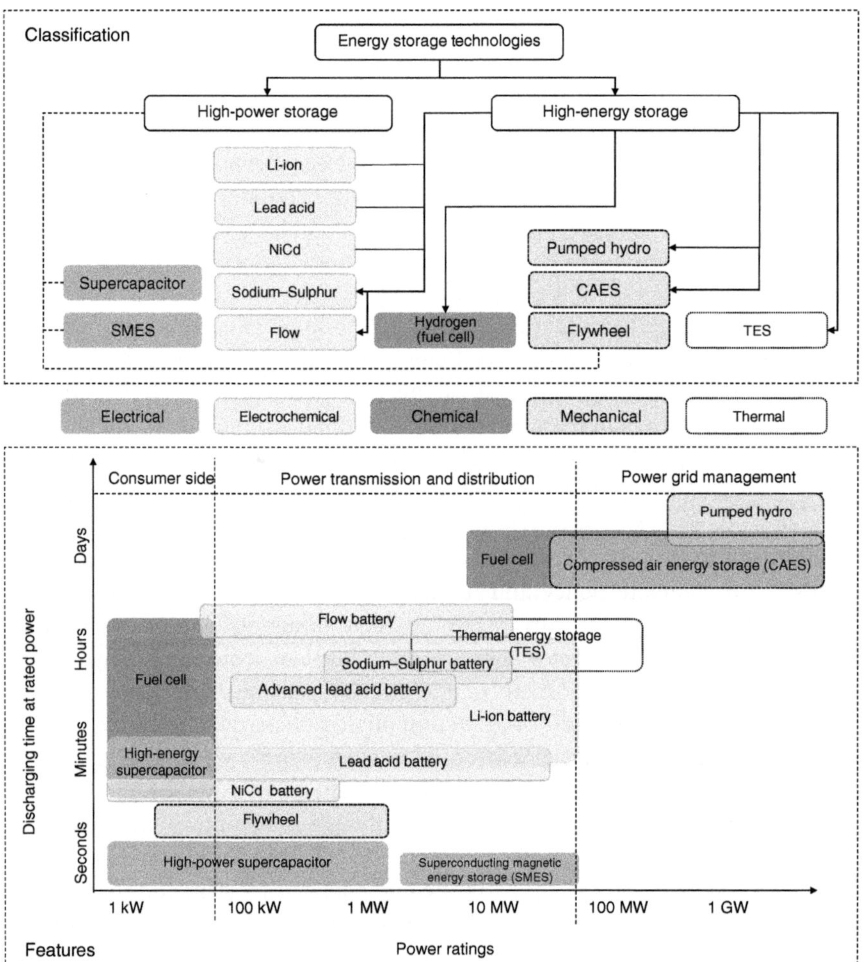

Figure 14.1 Classification of energy storage systems.

(1) Electrical storage technologies including supercapacitor (SC) and superconducting magnetic energy storage (SMES): SC is a electrical storage that has high recyclability (100,000 times) and density (7–10 kW/kg), SMES is one of the few direct electric energy storage systems that also has high density and responding speed.
(2) Electrochemical storage technologies include Li-ion battery, lead acid battery, NiCd battery, sodium–sulphur battery, and flow battery. Li-ion batteries are the technology dominating the majority of research; remarkable lifespan, nominal cell voltage modularity, and low self-discharge are their key advantages. Lead acid battery is the most mature option for stationary applications but has limited specific energy and lifespan. NiCd battery is also a mature technology but has limited cell voltage and lifespan, along with remarkable self-discharge.
(3) Chemical energy storage involves the conversion of low-energy substances into high-energy substances, which usually includes conversion into hydrogen, methane, and methanol for storage. Hydrogen storage uses electrolysis of water to produce hydrogen and store electrical energy in the form of hydrogen, which is an easy method to achieve large-scale energy storage. However, low conversion efficiency is the disadvantage of the whole process.
(4) Mechanical storage technology includes flywheel, compressed air energy storage (CAES), and pumped hydro storage (PHS). Flywheel energy storage is a fast-response mechanical system that stores electrical energy in the form of rotating kinetic energy (Alami 2019). CAES and PHS are both large-capacity mechanical energy storage technologies. CAES is based on compressing air and storing it in underground caverns, pipes, or vessels (Luo et al. 2015). PHS is based on pumping water from a lower reservoir to a higher one using off-peak power and then releasing it to power a generator (Jidai et al. 2017).
(5) Thermal energy storage uses thermal storage materials as a medium to store solar thermal, geothermal, industrial waste heat, and low-grade waste heat, or to convert electrical energy into thermal energy and release it when needed to solve the problems caused by the mismatch between thermal energy supply and demand in time, space or intensity. This process aims to maximize the energy utilization of the whole system. Molten salt is the current material of choice for high-temperature thermal storage (Global Energy Interconnetion Development and Cooperation Organization 2020).

Power density and energy density are two main characteristics of energy storages (Hajiaghasi et al. 2019). Power density is the rate of energy transfer per unit volume or mass, whereas energy density is the total amount of energy per unit volume or mass (Sutikno et al. 2022). High-power storage systems provide energy at very high rates, but typically only for a short period of time. Out of the storage technologies in Figure 14.1, SC, SMES, flywheel, and high-power batteries (such as Li-ion batteries) belong to this category. High-energy storage systems can supply energy for longer periods of time. PHS, CAES, thermal energy storage, fuel cells, and high-energy batteries belong to this category (Hemmati and Saboori 2016). The detailed energy and power densities of different storage technologies and a rough classification can be

Table 14.1 Properties of different storage technologies.

Storage technologies	Energy density	Power density	High-power storage	High-energy storage
Supercapacitor (electrical)	2.5–15 Wh/l	500–5000 W/l	✓	
SMES (electrical)	0.5–15 Wh/l	1000–4000 W/l	✓	
Flywheel (mechanical)	20–80 Wh/l	1000–2000 W/l	✓	
Pumped hydro (mechanical)	0.5–1.5 Wh/l	0.5–1.5 W/l		✓
CAES (mechanical)	3.0–6.0 Wh/l	0.5–2 W/l		✓
Fuel cell (chemical)	750 Wh/l	0.2–20 W/l		✓
Li-ion battery (electrochemical)	75–250 Wh/kg	150–315 W/kg	✓	✓
Lead acid battery (electrochemical)	30–50 Wh/kg	75–300 W/kg	✓	✓
NiCd battery (electrochemical)	50–75 Wh/kg	150–300 W/kg	✓	✓
Sodium–sulphur battery (electrochemical)	150–240 Wh/kg	90–230 W/kg	✓	✓
Flow battery (electrochemical)	15–75 Wh/kg	15–30 W/kg	✓	✓
Sensible heat (thermal)	20–30 Wh/kg	—		✓
Latent heat (thermal)	50–100 Wh/kg	—		✓
Thermochemical heat (thermal)	500–1000 Wh/kg	—		✓

seen in Table 14.1. Data for SC, SMES, flywheel, PHS, CAES, and fuel cell storages are collected from Sutikno et al. (2022). Data for electrochemical storages (battery) are collected from Pei et al. (2022). It is worth noting that because of their wide variety of properties, batteries can be used in both high-power and high-energy devices. Data for thermal storages are collected from China Energy Storage Alliance (2020).

However, none of the existing storage technologies can meet both power and energy density requirements of real applications (Hajiaghasi et al. 2019). The traditional energy storage deployment scheme composed of single energy storage technology often leads to nonoptimal solution (Leon et al. 2020). The use of single storage technology to fulfill both power and energy density requirements would lead to an oversized storage system. As an effective solution to mitigate this issue, establishing hybrid energy storage systems becomes the most feasible option. A hybrid solution is designed to take advantage of multiple energy storage technologies to achieve cost savings, efficiency, environmental performance, etc. The advantages of a hybrid energy storage system include a wide range of applications compared with a single storage system, reduced costs due to avoided oversizing (up to 15%) (Duke Energy News 2016), and an extended lifetime (up to 40%) (Vulturescu et al. 2014). Many hybrid energy storage systems have been applied worldwide. For example, Engie and Siemens developed a 137-MW pumped hydro plus 12.5-MW (13-MWh) lithium (Li)-ion battery-based hybrid energy storage system in Kraftwerksgruppe Pfreimd (Germany) for bulk management applications (Leon et al. 2020), and Duke Energy installed a supercapacitor/battery to support a 1.2-MW solar PV generation plant.

Table 14.2 The possible applications of hybrid energy storage systems in energy sectors.

Energy sector	Specific application	Hybrid energy storage systems solution
Large-generation power plants	Bulk ancillary services	Thermal + CAES Thermal + battery Pumped hydro + battery Pumped hydro + flywheel
Power transmission and distribution	Frequency regulation Load following Reserve capacity Voltage control Anti-islanding Peak shaving	Low-scale CAES + supercapacitors Battery + battery Battery + supercapacitors Battery + flywheel
Microgrids Residential Light industry Standalone systems	Deferral distribution assets Time shifting/energy arbitrage Voltage control Frequency regulation Enhanced self-consumption Grid support	Battery + hydrogen Battery + battery Battery + supercapacitors Battery + flywheel
Transportation	Electric cars, vans, and buses Catenary-free trams Full-electric ships and planes	Battery + fuel cell Battery + ultracapacitor

Therefore, a wide range of energy storage hybrids can arise as a result of the numerous types of energy storage technologies. The hybridization of different storage technologies that can be applied in different energy sectors for various applications (Leon et al. 2020) is shown in Table 14.2. It can be seen in Table 14.2 that thermal plus CAES, thermal plus battery, pumped hydro plus battery, and pumped hydro plus flywheel are commonly implemented with large-generation power plants for bulk ancillary service. Low-scale CAES plus SCs, battery plus battery, battery plus SCs, and battery plus flywheel installed for power transmission and distribution are usually applied for frequency regulation, load following, and voltage control. Battery plus hydrogen/battery/supercapacitors/flywheel applied in microgrids, residential, light industry, and standalone systems provides more functions, such as deferral distribution assets, time shifting/energy arbitrage, and grid support. Hybrid energy storage systems are also widely applied in transportation sector, such as electric cars, vans, buses, catenary-free trams, and full-electric ships and planes.

The keyword co-occurrence map of hybrid energy storage–related research is shown in Figure 14.2, The nodes indicate the keywords, with the larger the node indicating the higher frequency of the keywords, and the purple circle indicating the centrality of the node, i.e. the degree of importance. From the figure, it can

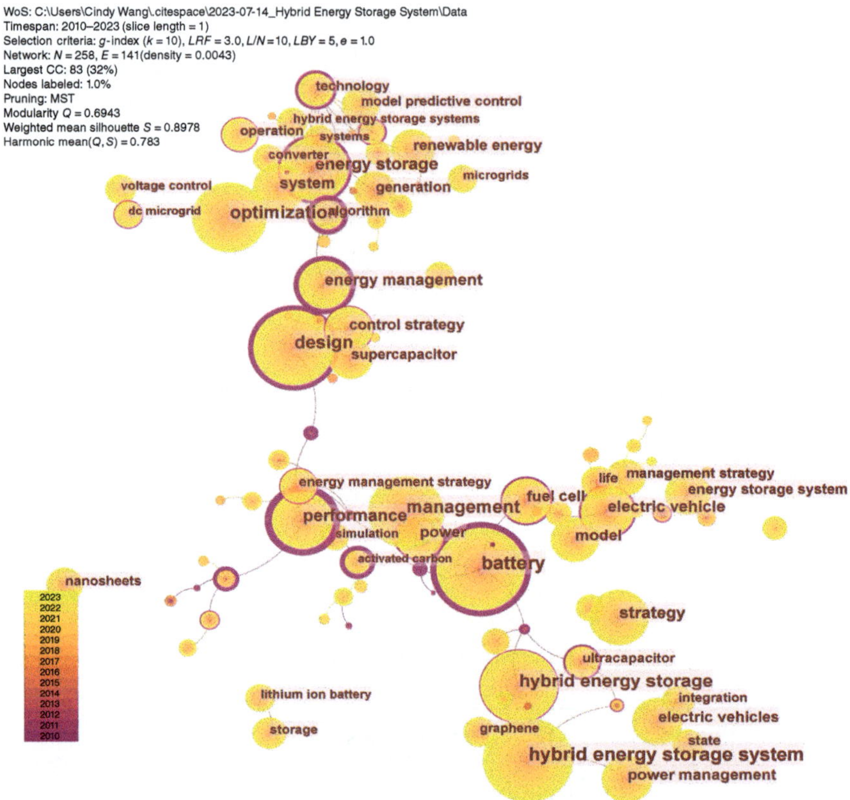

Figure 14.2 Keywords co-occurrence map of hybrid energy storage related researches.

be seen that batteries, fuel cells, and SCs are the most researched energy storage technologies, and researchers mainly study the performance, design, energy management and optimization algorithms of hybrid energy storage systems. The applications of existing hybrid energy systems are mainly microgrids and electric vehicles.

However, despite the importance of developing hybrid energy storage systems, there are still many challenges ahead (Leon et al. 2020).

(1) As the first challenge, the design of hybrid energy storage systems is quite complex because of the huge variability of the operator's objectives and the required services. The sizing of hybrid energy storage systems is normally carried out as a trade-off between costs, lifetime, efficiency, and discharging time. However, this trade-off is highly dependent on the real application. Generally, hybridization for sustainability compensation and lifespan extension has drawn wide attention and the design procedure is quite clear, but the design for peak shaving, grid support, anti-islanding, etc. has not been fully addressed. So there is a gap for multi-objective design method that sizes the hybrid energy storage system considering different factors (Hajiaghasi et al. 2019).

(2) As the second challenge, an energy management system has to be carefully designed to properly operate the hybrid energy system so that it can achieve the expected positive features. The hybridization of different energy storage technologies does not mean improved functionalities; a sophisticated system management plan need to be developed to extract the maximum advantages of different technologies (Leon et al. 2020). The control and energy management strategy of a hybrid energy system is more complicated than a single storage unit. A large number of factors from different storage technologies, including the storage technology properties, load feature, and time frames need to be taken into consideration. This includes the nonlinear and time-dependent characteristics of self-discharge, power converter efficiency, and so on.
(3) As the third challenge, government support via optimized policies and subsidies is needed to push hybrid energy storage technology development. In China, for example, more than 30 provinces and cities have issued policies requiring new energy projects to be equipped with energy storage systems in accordance with 5–30% of installed capacity; however, due to the lack of scientific proof of appropriate technology and optimal scale, the equivalent utilization factor of energy storage in new energy power plants is as low as 6.1%. In fact, with the introduction of the storage policy, the configuration of energy storage capacity is no longer independently decided by the power station, but by both government departments and power stations. However, the traditional method of energy storage allocation and storage only starts from a single decision-making subject of the energy storage project construction subsystem, which leads to a lower feasibility of the decision-making scheme. Optimized energy storage capacity installation policy that integrates the above two types of decision-makers will provide support for hybrid energy storage technology development.

The high-level integration of renewable energy sources and the reduction of non-renewable energy sources make the management of electrical energy more complex. Energy production will be more dependent on actual weather conditions and may result in critical situations where energy production and energy demand do not match (Stenclik et al. 2017). These facts inevitably lead to the use of energy storage systems as a major factor in the future development of the energy and power industry (Edmonds and Miller 2014). However, in many cases, energy storage systems of a single nature cannot fulfill all the power and energy requirements of the application. Hybrid energy storage systems, consisting of two or more energy storage technologies, are an attractive solution to meet these requirements and can even reduce the overall cost and lifetime of the system (Hajiaghasi et al. 2019). In the coming decades, the massive integration of renewable energy sources (the most important power generation systems in the grid) and the electrification of the transportation sector will lead to worldwide energy disruption (Leon et al. 2020). Hybrid energy storage systems are key to achieving this goal and deserve more attention.

References

Alami, A.H. (2019). *Flywheel Storage Systems: Mechanical Energy Storage for Renewable and Sustainable Energy Resources*, 35–49. Cham: Springer International Publishing.

China Energy Storage Alliance (2020). Current status and development trend of thermal storage technology application. https://news.bjx.com.cn/html/20200511/1070645.shtml (accessed 26 October 2023).

Duke Energy News (2016). Duke energy to put new battery and ultracapacitor system to the test in N.C. https://news.duke-energy.com/releases/duke-energy-to-put-new-battery-and-ultracapacitor-system-to-the-test-in-n-c (accessed 26 October 2023).

Edmonds, M. and Miller, T. (2014). The next 50 years: what's next for the grid? *IEEE Power and Energy Magazine* 12 (2): 92–96.

Global Energy Interconnetion Development and Cooperation Organization (2020). *Large-Scale Energy Storage Technology Development Roadmap*. China Electric Power Press.

Hajiaghasi, S., Salemnia, A., and Hamzeh, M. (2019). Hybrid energy storage system for microgrids applications: a review. *Journal of Energy Storage* 21: 543–570.

Hemmati, R. and Saboori, H. (2016). Emergence of hybrid energy storage systems in renewable energy and transport applications–a review. *Renewable and Sustainable Energy Reviews* 65: 11–23.

Jidai, W., Kunpeng, L., Lan, M. et al. (2017). Overview of compressed air energy storage and technology development. *Energies* 10 (7): 991.

Leon, J.I., Dominguez, E., Wu, L. et al. (2020). Hybrid energy storage systems: concepts, advantages, and applications. *IEEE Industrial Electronics Magazine* 15 (1): 74–88.

Luo, X., Wang, J., Dooner, M., and Clarke, J. (2015). Overview of current development in electrical energy storage technologies and the application potential in power system operation. *Applied Energy* 137: 511–536.

Pei, C., Wang, L., Wang, C., and Dai, C. (2022). Survey of application scenarios of energy storage in power system. *Power Grid Application* 41: 1–8. (in Chinese).

Stenclik, D., Denholm, P., and Chalamala, B. (2017). Maintaining balance: the increasing role of energy storage for renewable integration. *IEEE Power and Energy Magazine* 15 (6): 31–39.

Sutikno, T., Arsadiando, W., Wangsupphaphol, A. et al. (2022). A review of recent advances on hybrid energy storage system for solar photovoltaics power generation. *IEEE Access* 10: 42346–42364.

Vulturescu, B., Butterbach, S., and Forgez, C. (2014). Experimental considerations on the battery lifetime of a hybrid power source made of ultracapacitors and lead-acid batteries. *IEEE Journal of Emerging and Selected Topics in Power Electronics* 2 (3): 701–709.

Index

Sustainable Hybrid Energy Systems: Carbon Neutral Approaches, Modeling, and Case Studies,
First Edition. Jiuping Xu and Fengjuan Wang.